Library of
Davidson College

THERMODYNAMICS OF POINT DEFECTS AND THEIR RELATION WITH BULK PROPERTIES

SERIES DEFECTS IN SOLIDS

Editors:

**S. AMELINCKX
R. GEVERS
J. NIHOUL**

Studiecentrum voor kernenergie
Centre d'étude de l'énergie nucléaire
C.E.N./S.C.K., Mol, Belgium

**NORTH-HOLLAND
AMSTERDAM · OXFORD · NEW YORK · TOKYO**

THERMODYNAMICS OF POINT DEFECTS AND THEIR RELATION WITH BULK PROPERTIES

PANAYIOTIS A. VAROTSOS
KESSAR D. ALEXOPOULOS

Department of Physics
University of Athens
GREECE

1986

NORTH-HOLLAND
AMSTERDAM · OXFORD · NEW YORK · TOKYO

© Elsevier Science Publishers B.V., 1986

All rights reserved. No part of this publication may be reproduced, stored in a retrieval system, or transmitted, in any form or by any means, electronic, mechanical, photocopying, recording or otherwise, without the prior permission of the publisher, Elsevier Science Publishers B.V. (North-Holland Physics Publishing Division), P.O. Box 103, 1000 AC Amsterdam, The Netherlands.

Special regulations for readers in the USA: This publication has been registered with the Copyright Clearance Center Inc. (CCC), Salem, Massachusetts. Information can be obtained from the CCC about conditions under which photocopies of parts of this publication may be made in the USA.
All other copyright questions, including photocopying outside of the USA, should be referred to the publisher.

ISBN: 0 444 86944 1

Published by:

NORTH-HOLLAND PHYSICS PUBLISHING

a division of
Elsevier Science Publishers B.V.
P.O. Box 103
1000 AC Amsterdam
The Netherlands

Sole distributors for the U.S.A. and Canada:

Elsevier Science Publishing Company, Inc.
52 Vanderbilt Avenue
New York, N.Y. 10017
U.S.A.

Library of Congress Cataloging in Publication Data

Varotsos, Panayiotis A.
  Thermodynamics of point defects and their relation with bulk properties.

  (Defects in solids; 14)
  (Bibliography: p.
  Includes indexes.
  1. Point defects. 2. Thermodynamics. I. Alexopoulos, Kessar D. II. Title. III. Series: Defects in crystalline solids; v. 14.
  QD921.V34 1985   548'.842   85-13868
  ISBN 0-444-86944-1

Printed in The Netherlands

PREFACE

During the past few years the authors have published a number of papers relating to the study of point defects in solids. This book is to a large extent a unified compilation of these publications. Some of them dealt purely with thermodynamics; within the context many so-called "anomalous effects" have found a natural explanation. In most cases the difficulties in the interpretation of the experimental data arose from assumptions uncritically adopted in standard thermodynamic concepts. It therefore became evident that a careful separation of the thermodynamical definitions of defect parameters from any assumptions whatsoever was absolutely necessary. Furthermore thermodynamics impose some fundamental constraints on the various defect parameters, whose consideration is essential not only to experimentalists but also to theorists.

In recent publications the authors also studied the problem whether an explicit connection between defect parameters and bulk macroscopic properties really exists. Zener and others made such an attempt long ago. The question is of technological interest because one could then predict the temperature and pressure dependence of the parameters knowing only macroscopic properties as the volume and the elastic constants. The authors have suggested that such a connection really exists. By comparing the values of various defect parameters with bulk properties they noticed that certain connections repeatedly appeared. With time an empirical law was found describing correctly a large quantity of experimental data, many of them obtained through completely independent techniques. The basic relation is the formula $g = cB\Omega$, where g is the Gibbs defect energy, B is the isothermal bulk modulus and Ω the "mean" atomic volume. The coefficient c is practically temperature and pressure independent and is fixed only by the defect mechanism and the matrix material. The authors labelled the above relation the $cB\Omega$-model, and for a long time they considered it an empirical law.

The consistency with which this model applies to increasingly more cases led to the idea that we were not dealing with a coincidence, but that a deeper general law lay hidden within. A revision of the thermodynamics of defects showed that a hitherto unknown formula exists connecting defect parameters (Gibbs energy and volume) to bulk properties (B, and the pressure derivative dB/dP). Inserting usual numerical values into this formula it simplifies, to a very good approximation, to the $cB\Omega$-model, which can therefore be considered as a special case of a general thermodynamical law. The bounds of its errors can be determined and were found to permit the use of this approximation for nearly all experimental situations.

In view of the above remarks the text of this book has been divided into two relatively independent sections. In Part 1 we give a strict review of the thermodynamics of point defects and the basic philosophy for the correct analysis of the experiments, while carefully avoiding unwarranted assumptions. It starts with a brief review of general thermodynamics of solids in order to aid readers who are less familiar with the subject. Part 2 studies the connections of defect parameters with bulk properties, which is in essence the $cB\Omega$-model. In the first chapters of this part it is presented as an empirical model supported by a large quantity of experimental data. A strict theoretical basis of the model is given in the last chapter. This division enables the reader to select the point he is specially interested in without having to go through the whole book.

The $cB\Omega$-model finds important practical applications in Metallurgy, as it permits the prediction of diffusion coefficients under conditions of temperature and pressure when diffusion measurements are difficult. For such extreme conditions the measurements can be replaced by the much easier elastic and expansivity experiments. As the $cB\Omega$-model develops new theoretical connections between the defect parameters, a number of experiments is proposed on various aspects of defects.

Another interesting application is found in the field of Geophysics; certain solids emit electrical currents when the pressure reaches the value at which the relaxation time for the attainment of thermodynamic equilibrium becomes very short. This theoretical result became the impulse for experiments on the variation of the electrical field in the earth during periods of high seismicity. They led to the detection of transient electric pulses that can serve for the prediction of earthquakes.

Since the completion of the main text of this book a number of experiments has been published referring to point defects. The connection between defects and bulk properties seems to become a promising tool in the study of geophysical questions. Furthermore, pressure-induced currents observed in the earth have been exceedingly well exploited in making short-term predictions of the epicenter and the magnitude of earthquakes. Some of these newest data are described in the Appendix.

Professors David Lazarus and Larry Slifkin went through a large part of the manuscript and have made many instructive comments.

April 1985   P.A.V. and K.D.A.

To the memory of my father Antonios (P.A.V.)

CONTENTS

Preface — v

1 Introduction — 1
  1.1 Scope and organisation of the book — 1
  1.2 Notations — 6
  1.3 Symbols — 8

Part 1 – Thermodynamics — 9

2 Thermodynamic functions — 11
  2.1 Conditions for equilibrium — 11
  2.2 Temperature dependence of thermodynamic functions of solids — 13
  2.3 Pressure dependence of thermodynamic functions of solids — 16

3 Formation of vacancies — 23
  3.1 Isobaric and isochoric perfect crystals — 24
  3.2 Parameters from the comparison with the isobaric perfect crystal — 25
  3.3 Parameters from the comparison with the isochoric perfect crystal — 37
  3.4 Relation between isobaric and isochoric parameters — 44
  3.5 Statistical approach to vacancy parameters — 47
    3.5.1 Perfect lattice — 49
    3.5.2 Imperfect lattice — 50
  3.6 Types of experiments leading to the determination of formation parameters — 52
    3.6.1 Heating under constant pressure — 53
    3.6.2 Heating under constant volume — 54
    3.6.3 Isothermal pressurising — 55
    3.6.4 Comparison of the two types of heating experiments — 58
    3.6.5 Isobaric heating experiments under various pressures — 60

4	**Formation of other point defects**	63
	4.1 Schottky defects	63
	4.2 Frenkel defects in ionic solids of NaCl-structure	66
5	**Migration**	71
	5.1 Introduction	71
	5.2 Rate theory and the dynamical theory of migration	74
	5.2.1 The rate theory approximation	74
	5.2.2 The dynamical theory	80
6	**Thermodynamics of the specific heat**	85
	6.1 Introduction	85
	6.2 Contribution of vacancies to an isobaric perfect crystal	88
	6.2.1 Isobaric specific heat of an isobaric perfect crystal	89
	6.2.2 Isochoric specific heat of an isobaric perfect crystal	91
	6.3 Contribution of vacancies to an isochoric perfect crystal	94
	6.4 Specific heat of one vacancy	96
	6.4.1 Real crystal	96
	6.4.2 Isobaric perfect crystal	98
	6.4.3 Isochoric perfect crystal	101
	6.5 Connections between the specific heats of perfect crystals	102
	6.6 Specific heats of other defects	103
	6.7 Criteria of microscopic calculations from specific heats	105
7	**Analysis of experiments yielding defect parameters**	107
	7.1 X-ray parameters	107
	7.1.1 X-ray volume	108
	7.1.2 X-ray thermal expansion coefficients	108
	7.1.3 X-ray isothermal compressibility	110
	7.1.4 Determination of the formation volume from X-ray measurements	112
	7.2 Analysis of specific heat measurements	112
	7.2.1 Analysis of the C_P-curve assuming a linear extrapolation of $C_P^0 = f(T)$	113
	7.2.2 Analysis of the C_P-curve when C_V^{0*} is known from theoretical calculations	115
	7.2.3 Analysis of $C_P = f(T)$ when the X-ray expansivity and the X-ray compressibility have been measured	116

7.3	Self-diffusion	117
	7.3.1 Introduction	117
	7.3.2 Comments on the temperature dependence of the self-diffusion coefficient	121
	7.3.3 Proof of the proposal of analysing diffusion plots in the quasiharmonic approximation	125
	7.3.4 Analysis of the isothermal plot $\ln D$ versus P	126
	7.3.5 Isotope effect	127
	7.3.6 Experimental techniques for obtaining self-diffusion coefficients	128
7.4	Ionic conductivity and reorientation of dipoles	128
	7.4.1 Ionic conductivity	128
	7.4.2 Study of the reorientation of dipoles in crystals containing point defects	130
7.5	Comments on solubility	142

Part 2 – Defect parameters as a function of bulk properties: the $cB\Omega$-model   **145**

8	Connection of defect parameters to bulk properties	147
8.1	Introduction	147
8.2	Parameters by comparing with the isobaric perfect crystal	148
8.3	Parameters by comparing with the isochoric perfect crystal	158

9	The $cB\Omega$-model: Defect entropy and enthalpy. Self-diffusion	163
9.1	Introduction	163
9.2	fcc metals	168
	9.2.1 Aluminum	169
	9.2.2 Copper	176
	9.2.3 Silver	184
	9.2.4 Lead	190
9.3	bcc metals	195
	9.3.1 Sodium	195
	9.3.2 Lithium	208
	9.3.3 Potassium	214
	9.3.4 Tungsten	218
	9.3.5 Niobium	225
9.4	Hexagonal and tetragonal metals	226
	9.4.1 Zinc	226
	9.4.2 White tin	232

	9.5	Noble gas solids	242
		9.5.1 Neon	242
		9.5.2 Argon	243
		9.5.3 Krypton	244
		9.5.4 Xenon	245
	9.6	Alkali halides	246
		9.6.1 Sodium fluoride	246
		9.6.2 Potassium chloride	251
		9.6.3 Sodium chloride	253
		9.6.4 Potassium bromide	255
		9.6.5 Sodium iodide	257
		9.6.6 Potassium iodide	260
	9.7	Silver halides	260
	9.8	Summary of the progress in the study of point defects in ionic crystals	267
10	The $cB\Omega$-model: Defect volume and Gibbs energy		269
	10.1	Introduction	269
	10.2	Metals	275
		10.2.1 fcc metals	275
		10.2.2 bcc metals	275
		10.2.3 Tetragonal metals	279
	10.3	Noble gas solids	283
	10.4	Ionic crystals	284
		10.4.1 Ionic crystals with CsCl-structure	284
		10.4.2 Alkali halides with NaCl-structure	287
		10.4.3 Silver halides	295
		10.4.4 Lead fluoride	301
	10.5	Correlation between the defect volume and the enthalpy	303
11	The $cB\Omega$-model: Heterodiffusion		307
	11.1	Temperature dependence of heterodiffusion	307
	11.2	Correlation between the diffusion coefficients of atoms diffusing in a given matrix	309
		11.2.1 fcc metals	310
		11.2.2 bcc metals	314
		11.2.3 Ionic crystals	319
		11.2.4 Linearity of the log D_0-values as function of enthalpies	322

12	**Mixed alkali and silver halides**	**325**
	12.1 Variation of the mean atomic (or molecular) volume with composition	325
	12.2 Variation of the bulk modulus with composition	329
	12.3 Composition for maximum conductivity and diffusivity from the $cB\Omega$-model	335
	12.4 Thermal expansivity	341
	12.4.1 Mixed alkali halides	341
	12.4.2 Ionic solids doped with aliovalent impurities	344
13	**Explanation of various empirical laws**	**349**
	13.1 Connection of the enthalpy to the melting point	349
	13.1.1 Self-diffusion in metals	350
	13.1.2 Vacancy formation in metals	351
	13.1.3 Formation and migration of defects in alkali halides	351
	13.1.4 Organic compounds	351
	13.1.5 An alternative explanation for the proportionality of formation enthalpy and the melting temperature	352
	13.1.6 Relation between formation parameters and the Debye temperature in alkali halides	353
	13.2 Explanation of the empirical laws connecting activation entropy and enthalpy to the activation volume	354
	13.3 Connection of defect parameters to the Grüneisen constant	355
	13.3.1 Connection of the defect entropy to the Grüneisen constant; correlation between the defect enthalpy and the expansion coefficient	355
	13.3.2 Relation between the migration volume and the Grüneisen constant	357
	13.3.3 Relation between the migration volumes and the corresponding enthalpies for a given solid	358
	13.4 Compatibility with the Jost-model of the dielectric continuum	358
	13.4.1 Formation entropy	359
	13.4.2 Formation volume	360
	13.5 Correlation of activation parameters resulting from dielectric loss or ITC-experiments	361
14	**Theoretical basis of the $cB\Omega$-model**	**363**
	14.1 Introduction	363
	14.2 Preliminary aspects	365
	14.2.1 Evidence for a limiting value of the compressibility of defect volumes	365

14.2.2 The limiting values of the quantity
$-B(d^2B/dP^2)[(dB/dP)_T - 1]^{-1}$ 369
14.2.3 The pressure variation of the quantity
$(v/\Omega)[(dB/dP) - 1]^{-1}$ 370
14.3 Proof of the $cB\Omega$-model with a pressure-independent c 371
  14.3.1 Introduction 371
  14.3.2 Proof of the thermodynamical identity $[dg/d(B\Omega)]_T = (v/\Omega)[(dB/dP) - 1]^{-1}$ 372
  14.3.3 Possible error of g by considering c as pressure independent 373
  14.3.4 Proof of the equations $g = [(dB/dP) - 1]^{-1}Bv$ and $g(P, T) = c(T)B\Omega$ 376
  14.3.5 Alternative derivation of the relation $g(P, T) = c(T)B\Omega$ 378
  14.3.6 Microscopic comments related to the $cB\Omega$-model 380
  14.3.7 Justification of the $cB\Omega$-formula from the Jost-model 384
  14.3.8 Justification of the $cB\Omega$-model for the migration process in the framework of dynamical theory 387
14.4 On the temperature dependence of c 388
  14.4.1 The temperature dependence of c 389
  14.4.2 Bounds for the possible temperature variation of c 390
  14.4.3 Bounds of the formation entropy 392
  14.4.4 Alternative discussion of the possible temperature dependence of c 394
14.5 Survey of the proof of the $cB\Omega$-formula 396
  14.5.1 Summary and comments on the proof of the relation $g(V, T) = c(T)B\Omega$ 397
  14.5.2 Discussion and comments on the proof of the relation $g(V, T) = c(V)B\Omega$ 399
  14.5.3 Interconnection of the conclusions drawn in §§14.3 and 14.4 400
  14.5.4 A more direct procedure for the foundation of the $cB\Omega$-model for the formation process 401

Appendix 403
  A.1 Introduction 403
  A.2 Linear and curved Simmons–Balluffi plots 403
  A.3 Stimulated current emission in the earth 403
  A.4 Self-diffusion 406
  A.5 Discussion on the values of s/h and v/g 407

A.6	Recent developments in the detection of preseismic electric signals	410
A.7	Mixed ionic crystals	412
A.8	Calculation of defect parameters from first principles	412
A.9	Determination of the concentration of vacancies from a constant volume X-ray study at various temperatures. Thermal vacancies in solid ^{3}He	414
A.10	Self-diffusion probed by spin–lattice relaxation	414
A.11	Notes added after the completion of the main text	415
	General Thermodynamics (notes 1–3)	415
	Microscopic calculations (notes 4–7)	416
	Piezostimulated currents and related geophysical aspects (notes 8–14)	417
	Diffusion (notes 15–17)	420
	Dielectric losses and thermally stimulated currents (notes 18–22)	421
	Comments on the compressibility and expansivity of a mixed crystal (note 23)	423

References 425

List of main symbols 447

Author index 449

Subject index 463

1 | INTRODUCTION

1.1 Scope and organisation of the book

The present book starts with a part on strict definitions of the formation and migration parameters and on the laws that connect them. It is separated from the second part which is based on the "$cB\Omega$-model"; this model is simply a proposal which states that the formation g^f (or migration g^m or activation g^{act}) Gibbs energy is proportional to $B\Omega$, where B is the isothermal bulk modulus and Ω the mean volume per atom. As will be seen, the proportionality constant – labelled c^f (or c^m or c^{act}) – depends on the process but is *not* an empirical quantity; it has a definite physical meaning and can be shown to be practically temperature and pressure independent. As will be proven in ch. 14 the $cB\Omega$-model is of thermodynamical origin; in this sense the second part of the book can be considered a natural continuation of the first.

The "thermodynamical definitions" of defect parameters given in Part 1 are consistent with the general spirit of Thermodynamics and do not assume any restrictions on their temperature and pressure variation apart from those imposed by thermodynamical laws (e.g. from the third law that states that the formation entropy of a vacancy has to tend to zero for $T \to 0$ K for a pure solid etc.). These thermodynamical definitions, although being of major importance for a consistent analysis of the experimental data, do not lead by themselves to a description of the data. In an analysis one has anyhow to go one step further, i.e. to adopt some "assumptions" concerning the temperature and pressure (volume) dependence of some thermodynamical parameters. This is now the point where a dilemma emerges. In order to realise whether these assumptions are plausible or not, one usually has to go back to the microscopic picture of these parameters. This is achieved by Statistical Thermodynamics; for this reason the statistical definitions of the formation and migration parameters are also given in the first part of the book. However, there is the difficulty that the "statistical definitions" are easy only for harmonic or quasi-harmonic solids but not for real solids. This shortcoming may be serious when one is making microscopic calculations but *not* for the scope of the present book which only examines if the above assumptions are reasonable or not. The $cB\Omega$-model will be found to be able to replace them by others which are physically more acceptable. In the block diagram of fig. 1.1 an attempt is made to compare them; it is restricted to

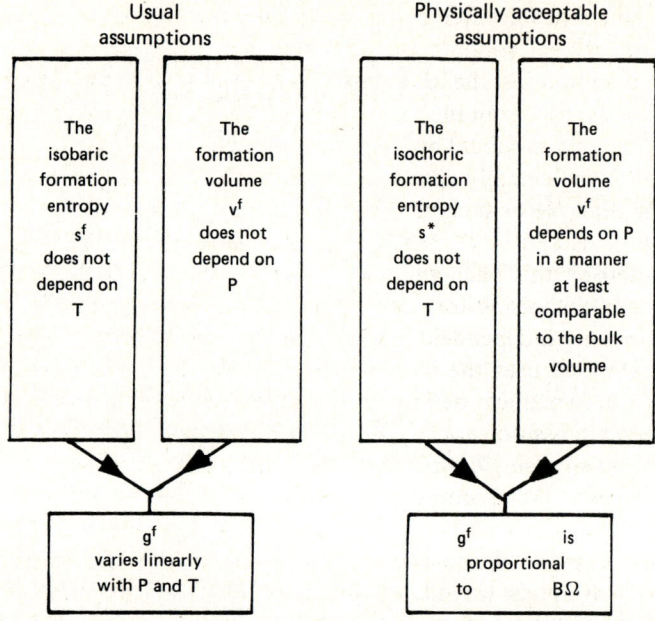

Fig. 1.1. Comparison of usual assumptions with physically acceptable assumptions.

the formation process but analogous comments are valid for the migration process. A brief discussion concerning their comparison is given below.

In the framework of the usual assumptions the formation entropy s^f and enthalpy h^f are accepted as temperature independent and the formation volume v^f is arbitrarily taken as temperature and pressure independent. This set of "current assumptions" is from a thermodynamical point of view absolutely self-consistent; attention is drawn, however, to the point that if one changes even only one of these assumptions (e.g. that s^f depends on temperature) and retains the others, the above self-consistency is immediately destroyed from a thermodynamical viewpoint. It has been repeatedly realised that the above "set of assumptions" is correct *only* for a pure harmonic solid. But we know that solids expand, that their isothermal bulk modulus B decreases with temperature, that C_P differs from C_V and so do many other properties which are characteristic of the real (i.e. anharmonic) behaviour. A direct consequence of the anharmonicity is the incontestable fact that some *isobaric* defect parameters are quite different from the *isochoric* defect parameters (see ch. 3). Therefore it is at least questionable whether harmonic assumptions can describe real (i.e. anharmonic) solids. It is reasonable to expect that they do *not*.

A strange confusion often arises concerning this point; when the experimental data of real solids cannot be satisfactorily described from these harmonic assumptions the data are considered as showing "anomalous behaviour", as for example in the case of nonlinear Arrhenius plots. However, these cases should be considered anomalous only if these harmonic assumptions were actually able to describe the behaviour of a real solid.

Another example of such an inconsiderate use of the word "anomalous" is the case of curved $\ln D$ versus P isothermal plots. In such cases many workers interpret this phenomenon as resulting from a coexistence of two cooperating mechanisms for each of which the activation volume v^{act} is assumed pressure independent. However, a definite confusion reigns in this respect. Once one uses the word "activation volume" one must recall its correct thermodynamical definition $v^{act} = (\partial g^{act}/\partial P)_T$; this definition does not preclude a pressure variation of v^{act} and hence a curved "$\ln D$ versus P plot", even when a single mechanism is operating; the "anomaly" arises at the very moment the assumption is adopted that v^{act} is pressure independent.

In order to prove the inacceptability of the current assumptions when applied to real solids we shall discuss two of them, (1) the temperature independence of the entropy s^f, (2) the pressure independence of the formation volume v^f. We will do this in the light of the quasiharmonic approximation which, as generally accepted, is undoubtedly an important progress with respect to the pure harmonic approximation.

(1) Temperature independence of s^f. The entropy s^f as mentioned is currently assumed to be temperature independent; how far is this justified? The statistical meaning of s^f is given by: $s^f = -k\Sigma_i \ln[\omega'_i(V)/\omega_i(V^0)]$, where ω'_i are the frequencies of the real volume V after the vacancy was produced and ω_i those of the "isobaric ideal lattice", i.e. at a different volume V^0. As the temperature increases while the external pressure remains constant both the volumes V and V^0 of the real and the ideal crystal (usually) increase; we are quite sure that in these two solids the frequencies ω'_i and ω_i vary upon heating predominantly due to their volume variation. It seems therefore totally unjustified or at least difficult to accept that the sum $\Sigma_i \ln[\omega'_i(V)/\omega_i(V^0)]$ does not change, e.g. when the solid is heated from $T_M/2$ up to T_M where T_M is the melting temperature. There is no physical argument to guarantee that the above complicated sum (and therefore the formation enthalpy) remains constant. In order to get a feeling how dubious this assumption is, we recall the simple Grüneisen theory: for the temperature interval from $T_M/2$ to T_M the volume of mono-atomic crystals increases by $\sim 5\%$ and hence the mean frequencies of both crystals are reduced typically by $\gamma \times 5\% \approx 15\%$ for $\gamma \approx 3$.

On the other hand one could replace the above "assumption" by another which is physically more acceptable: upon *isochoric* heating the sum $-k\Sigma_i \ln[\omega_i'(V)/\omega_i(V)]$, where the frequencies ω_i' and ω_i now refer to the same macroscopic volume V can be well considered as temperature independent. This is physically plausible because – according to the physical basis of the quasiharmonic approximation – the frequencies are explicit functions *only* of the macroscopic volume; therefore, due also to the fact that upon *isochoric* heating there is direct experimental confirmation that the frequencies exhibit only a very small explicit temperature dependence, one is led to the conclusion that the sum $-k\Sigma_i \ln[\omega_i'(V)/\omega_i(V)]$ can well be considered temperature independent. (This sum as we shall see is simply the thermal entropy s^* for an *isochoric* production of a vacancy.) Combining now this "physically acceptable assumption" with the well-known fact that $(\partial B/\partial T)_V$ is also temperature independent (which is also theoretically justified in the quasiharmonic approximation [1]) one is directly led – by thermodynamics (see ch. 14) – to the conclusion that the Gibbs formation energy g^f should be linear with respect to $B\Omega$ where the coefficient of $B\Omega$ has to be *temperature independent* (recall for $V=$ constant).

(2) *Pressure independence of v^f.* Another current assumption is that the formation volume v^f is pressure independent. The thermodynamical meaning of v^f is that it represents the variation $V - V^0$ where V^0 is the initial volume and V the real volume after the production of a vacancy under isothermal and isobaric conditions. There is no doubt that both V and V^0 have to decrease upon compression. There is no guarantee however that their difference, i.e. v^f, also remains constant. From a purely physical point of view the assumption accepting v^f as pressure independent (i.e. that the compressibility κ^f of the volume v^f is zero) is equivalent to saying that the isothermal compressibility κ of the crystal *decreases* after the introduction of vacancies under isobaric conditions. (see eq. 3.57). This is really dubious and contrary to common sense because one expects physically the compressibility to *increase* and not to decrease upon creation of vacant sites; one could assume at least that the compressibility remains almost the same which – from thermodynamics – directly leads to the conclusion that the compressibility κ^f of the formation volume is comparable to the bulk compressibility κ. When one now combines: *either* the physically plausible assumption that $\kappa^f = \kappa$ with the fact that dB/dP is pressure independent (which is exactly valid in the quasiharmonic approximation [1]), *or* the reasonable physical assumption that κ^f exceeds κ only by a few times (clearly justified from the microscopic point of view) with the well-known fact that dB/dP varies only slightly upon isothermal compression, one can directly derive from thermodynamics (see ch. 14) that g^f is linear in $B\Omega$

with a pressure (i.e. volume for $T = $ constant) independent proportionality constant.

The proportionality constant – as we have seen in the discussion of the first item – can also be considered (explicitly) temperature independent, thus arriving at the conclusion that, apart from an integration constant (see ch. 14), g^f is given by

$$g^f = c^f B \Omega,$$

with c^f being a constant which is exactly known and independent of temperature and pressure (volume); as we shall see in the last chapter of the book this relation is more general than the above "reasonable assumptions" for semi-harmonic solids would indicate. Numerous applications, presented in the second part of the book, reveal that the above relation satisfactorily describes a large body of experimental data which, in many cases, was previously called "strange" or "anomalous" or "doubly anomalous" etc. For example, the very large values for the formation volume of Schottky defects in the alkali halides, which for many years could not be accounted for, achieve a rational explanation. Similarly it offers the condition for negative values of the activation volume; such values have very recently been confirmed for AgI and a few other materials. Furthermore the temperature and pressure dependence of the formation volume can be quantitatively estimated from the bulk modulus B and the volume expansion coefficient β without having to know any other quantities. Also the temperature variation of the self-diffusion coefficient is quantitatively predicted within many orders of magnitude by using only elastic and expansivity data.

The temperature and pressure dependencies of B and β are characteristic manifestations of anharmonicity; the success of the theory therefore shows that it is of paramount importance to consider all anharmonic effects in the analysis of experimental results. This is not the first attempt to connect defect parameters with bulk properties. Zener proposed such a connection already in 1950; he related the Gibbs energy for the activation of a vacancy to the temperature dependence of the shear modulus of the solid. Later Keyes proposed a different combination of the elastic constants and Brooks presented a third combination. As will be shown in §9.2 these older attempts lead in the case of Al to results that in many cases are in direct contradiction to experimental facts. A similar disagreement is found for PbF_2 and CsCl ([1a], see also §14.1).

One further point should be stressed; by definition we have:

$$c^f = \frac{v^f}{\Omega} \left(\frac{dB}{dP} - 1 \right)^{-1} \quad \text{or} \quad c^f = -\frac{s^*}{\Omega} \left. \frac{\partial T}{\partial B} \right|_v ;$$

one sees that the relation $g^f = c^f B \Omega$ does *not* have any adjustable or empirical parameter and therefore can be subjected to a direct check by comparing with the experimental data. In essence it represents an interconnection of the macroscopic elastic and expansivity data to the thermodynamical defect parameters. Especially convincing are figs. 10.2, 10.13, 14.2, A.1, and A.2.

In brief the first part of the book (§2.1–§7.5) deals with the pure thermodynamical and statistical definitions of the defect parameters. Furthermore the various kinds of experiments are outlined (§7.1–§7.4) and comments for their appropriate analysis are made. In the second part of the book we first replace the current dubious "assumptions" concerning the defect parameters by the expression $g = cB\Omega$. In the context of this expression the various defect parameters are explicitly connected to the bulk properties (§§8.2, 8.3). In chapters 9 to 13 a great number of experiments and well-known empirical laws are explained. In the last chapter, ch. 14, a thermodynamical proof of the $cB\Omega$-model is given and the extent is estimated to which c can be considered as pressure and temperature independent. Last, we mention that especially in Part 2, a large number of new experiments are proposed.

1.2 Notations

Before entering into the contents of this review we introduce a few definitions for reasons of brevity. For instance the thermodynamic quantities pressure P, volume V and temperature T will be labelled as *natural variables*; these are the quantities that are wittingly changed when in experiments a change of state is to be studied. The remaining functions that are of interest (entropy, internal energy, etc.) will retain the usual designation of *thermodynamic functions*.* The so-called *defect parameters* are defined by comparing the thermodynamic functions of a real crystal to those of a perfect one, i.e. to a state that is free of defects. As more than one perfect crystal state can lead to a given real state a large variety of parameters emerges, a situation that has hitherto not been sufficiently realised. Of these, only two are of practical importance; these are the perfect states that have the same temperature and either the same pressure or the same volume as the state of the real solid; for brevity they will be labelled as the *isobaric* and the *isochoric perfect state*. These two cases are sufficient for the definition of the defect parameters that are needed for the main types of experiments, i.e. isobaric heating, isochoric heating and isothermal pressuris-

---

* The quantities P, V, T, S and U, H, F, G are usually labelled as *state variables* and *thermodynamic potentials*, respectively.

ing. In some cases the difference between the two families of parameters is not important (in a harmonic solid some of them coincide), but in others – as will be shown e.g. for the so-called (thermal) formation entropy – they might differ by more than a factor of two. This is the reason for the large discrepancy found between the experimental values of the formation entropy of defects in ionic solids and the theoretical values calculated from first principles. Similarly this led to wrong values (and algebraic sign) of the temperature dependence of the formation enthalpy in alkali and silver halides (§A.8).

The relations between certain thermodynamic quantities (e.g. the specific heats) and the defect parameters depend on the definition of the latter. In the literature some of these relations have been established rather loosely and have therefore led in some cases to incorrect formulae. Furthermore, many published formulae that connect defect parameters are correct but have been written intuitively by analogy to expressions that connect thermodynamic quantities without any justification. The presentation given here constitutes an internally consistent system of carefully defined parameters. The general term "energy" has been avoided; in each case the appropriate thermodynamic function (internal energy, enthalpy, etc.) is used. Finally it is emphasized that in a number of cases attention should be drawn to the fact that insufficient attention has been given to the real meaning of certain physical quantities. For example, the expression "contribution of defects" to a thermodynamic function should only be used with care. It refers to the difference between the real state and a perfect state, and therefore the contribution obviously depends on the choice of the latter.

The two important ways of defining a perfect state increase the size of the relevant thermodynamic discussions and double the number of parameters. The situation becomes confusing in the discussion of the contribution of defects to the specific heats C_P, C_V of perfect crystals, because each of them depends on the choice of the perfect state. These complications will be seen to result in an increase of the extension of the paragraphs on the thermodynamics of the specific heats and on the analysis of their measurement. At first sight such extensive discussions might appear to be only of academic interest; however, careful examination shows that the new parameters that emerge are exactly the ones that are meaningful in an analysis of specific heat experiments. Most of these parameters had not been hitherto considered, because the influence of anharmonicity on the defect parameters had not been seriously envisaged.

Some of the novel parameters developed in this paper lead to predictions that can be subjected to experimental examination and therefore might constitute incentives for further experiments.

The present review is restricted to point defects and therefore only questions relevant to this subject will be treated. It is even found that it is sufficient to make exact calculations for the formation of *vacancies*, because they can easily be extended – with slight modifications – to other types of point defects as Schottky defects, Frenkel defects, etc.

1.3 Symbols

There has been no unique way of labelling the various defect parameters. We give here a few general indications of the symbols used in the present book. Capital letters (as G, H, C) describe thermodynamic functions and can be assumed as referring to a mole of substance. A *subscript zero* or M indicates their value at absolute zero temperature or at the melting-point. A *superscript zero* indicates that it refers to a solid in an ideal state. If the ideal state has the same volume as the real crystal (*isochoric perfect crystal*) an asterisk * is added; without an asterisk the perfect state is assumed as being under the same pressure as the real crystal (*isobaric perfect crystal*). Also the symbols Δ or Δ^* usually indicate operators that give the difference between a thermodynamic function of a real crystal and an isobaric or isochoric perfect crystal.

The present review deals with formation, migration and activation processes. The relevant parameters therefore should always have a superscript f, m or act. Such superscripts however have been added only in those sections where they are necessary for the understanding of the text.

The parameters that result from the comparison of a real crystal with a perfect crystal are labelled with lower case letters as g, h, s, etc. In the literature they are often labelled as ΔG, ΔH, ΔS, etc; this is in some cases misleading, as for example the defect entropy s^f is not equal to the differences ΔS of the entropies of the two crystals. Asterisks are added to the parameters when they result from the comparison to an isochoric perfect crystal. In the chapter on specific heats c is the change of the specific heat of the solid when one defect is introduced. The symbols B of the isothermal bulk modulus and $\kappa (= 1/B)$ of the compressibility are alternatively used whenever the simplification of algebra demands it. The quantities v^f, v^m and v^{act} refer to the volumes *per defect*; they usually are given in units of Å^3. However, in chs. 9 to 14 they are also given in units of cm^3/mole; of course in the latter case the values correspond to the quantities (Nv^f), (Nv^m) and (Nv^{act}), respectively, where N denotes Avogadro's number. More information about the symbols used in this book can be found in the List of Main Symbols on p. 447 ff.

Part 1

THERMODYNAMICS

2 THERMODYNAMIC FUNCTIONS

In this chapter certain well-known thermodynamic formulae are given whereby only those are presented which are of importance for the discussion of defects.

2.1 Conditions for equilibrium

The first law of thermodynamics which expresses the conservation of energy and holds for reversible as well as for irreversible processes, is

$$\delta Q = dU + P\,dV, \tag{2.1}$$

where U is the internal energy of the system, and PdV represents the work done by the system; if this work is of a mechanical nature, P and V stand for the (external) pressure and volume respectively. When a physical system is not in thermal equilibrium (but not far from it, in a sense that thermodynamic functions can be defined) it will proceed to equilibrium in time by means of irreversible processes. The second law of thermodynamics reads

$$T\,dS \geqslant \delta Q, \tag{2.2}$$

where S denotes the entropy of the system. The equality sign holds only for reversible processes, during which, by definition, the system is constantly in equilibrium. Therefore from eq. (2.2) one may derive the conditions of equilibrium; they depend on which external constraints were kept constant during the process. We will restrict ourselves to the following three cases:

(a) *No heat-exchange with the surroundings.* A combination of $\delta Q = 0$ with eqs. (2.1) and (2.2) gives

$$T\,dS \geqslant 0, \tag{2.3}$$

which means that for a thermally isolated system the entropy has two possibilities: to increase or to remain constant; in equilibrium the entropy of such a system reaches its maximum value.

(b) *System held at constant volume and temperature.* We define $U - TS$ as the Helmholtz free energy F:

$$F \equiv U - TS. \tag{2.4}$$

A combination of $V =$ constant and $T =$ constant with eqs. (2.1) and (2.2) gives

$$d(U - TS) \leq 0, \quad \text{or} \quad dF \leq 0. \tag{2.5}$$

At equilibrium F reaches its minimum value.

(c) *System held at constant pressure and temperature.* A combination of $P =$ constant and $T =$ constant with eq. (2.2) gives

$$d(U + PV - TS) \leq 0. \tag{2.6}$$

The function $U + PV - TS$ is called the Gibbs free energy G (or the thermodynamic potential or simply the free enthalpy).

$$G \equiv U + PV - TS. \tag{2.7}$$

Its introduction into eq. (2.4) gives

$$G = F + PV. \tag{2.8}$$

In such a system, equilibrium occurs when G reaches its minimum value. The enthalpy H of the system is defined by

$$H \equiv U + PV, \tag{2.9}$$

so that eq. (2.7) can be written

$$G = H - TS. \tag{2.10}$$

In future sections we will need the following relations between the above thermodynamic functions. (Their physical meaning is explained in §A.11, note 1.) For reversible changes of a given system the following equations hold:

$$dF = -S\,dT - P\,dV, \tag{2.11}$$
$$dG = -S\,dT + V\,dP, \tag{2.12}$$
$$dH = T\,dS + V\,dP, \tag{2.13}$$
$$dU = T\,dS - P\,dV. \tag{2.14}$$

From eqs. (2.11) and (2.12) one obtains

$$S = -\left.\frac{\partial F}{\partial T}\right|_V \tag{2.15}$$

and

$$S = -\left.\frac{\partial G}{\partial T}\right|_P. \tag{2.16}$$

Similarly, by differentiating eqs. (2.10) and (2.4) and considering (2.16) and (2.15) one obtains

$$\left.\frac{\partial H}{\partial T}\right|_P = T \left.\frac{\partial S}{\partial T}\right|_P \tag{2.17}$$

and
$$\left.\frac{\partial U}{\partial T}\right|_V = T \left.\frac{\partial S}{\partial T}\right|_V. \tag{2.18}$$

Whenever the number N of the particles of the system does not remain constant, eqs. (2.11) to (2.14) become

$$dU = T\,dS - P\,dV + \mu\,dN, \tag{2.19}$$
$$dH = T\,dS + V\,dP + \mu\,dN, \tag{2.20}$$
$$dF = -S\,dT - P\,dV + \mu\,dN, \tag{2.21}$$
$$dG = -S\,dT + V\,dP + \mu\,dN. \tag{2.22}$$

The symbol μ stands for the chemical potential of the crystal and is defined by

$$\mu \equiv \left.\frac{\partial G}{\partial N}\right|_{P,T}. \tag{2.23}$$

2.2 Temperature dependence of thermodynamic functions of solids

In this section typical plots of thermodynamic functions of solids versus the temperature will be given with one of the two variables P or V being kept constant.

(a) Constant pressure. From the variety of methods of determining the temperature dependence of thermodynamic functions we choose as starting point the specific heat under constant pressure, because its measurement is a

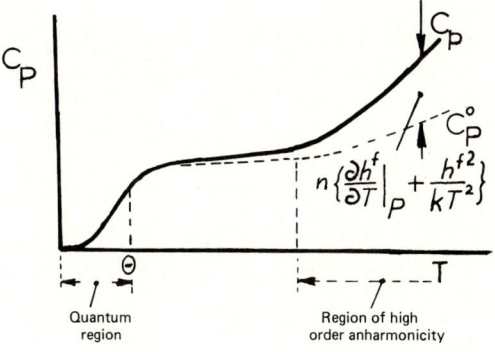

Fig. 2.1. Molar specific heat.

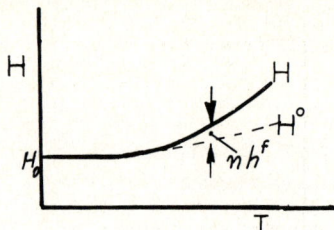

Fig. 2.2. Molar enthalpy. (Low-temperature part: small upward slope.)

simple experiment. In fig. 2.1 we give a typical result of such an experiment on a monoatomic solid. The temperature axis has been divided into regions. The low temperature region is described as the quantum region; it reaches approximately up to Θ. In the medium temperature range it increases gradually and at the high temperature end there is a region of a sudden increase of C_P [2–9] which is described as the region of "excess specific heat"; as will result from this review it could also be described as the region of high-order anharmonicity. The dotted curve in fig. 2.1 and in the following figures refers to a perfect crystal. The difference in the ordinates is due to the presence of vacancies.

The form of the enthalpy H versus T is found by integrating $C_P = (\partial H/\partial T)_P$; it is depicted in fig. 2.2. The general behaviour of the entropy S is given in fig. 2.3; it is obtained by integrating the thermodynamical relation $C_P = T(\partial S/\partial T)_P$. The form of the Gibbs free energy G is shown in fig. 2.4. It results from the two former functions according to eq. (2.10). The

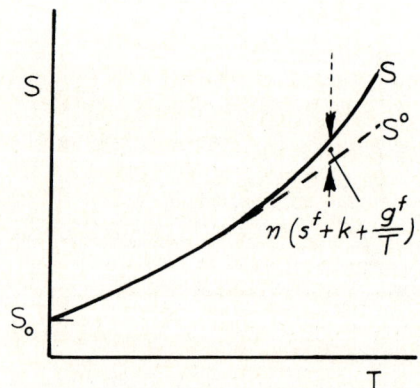

Fig. 2.3. Molar entropy.

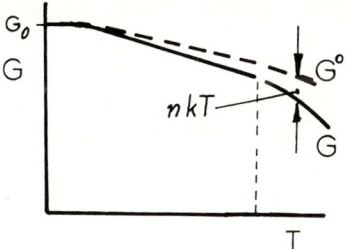

Fig. 2.4. Molar Gibbs free energy.

functions U respectively F are similar to those of H respectively G. The (macroscopic) isobaric thermal expansion coefficient β is defined by

$$\beta = \frac{1}{V}\frac{\partial V}{\partial T}\bigg|_P. \tag{2.24}$$

A typical curve of its temperature dependence is given in fig. 2.5. The isothermal * bulk modulus B and the compressibility κ are defined by

$$\kappa \equiv \frac{1}{B} = -\frac{1}{V}\frac{\partial V}{\partial P}\bigg|_T. \tag{2.25}$$

Its temperature dependence is given in fig. 2.6.

These figures are only indicative; they merely show whether the curve increases or decreases. The trend of their curvature should not be taken as depicted, because it depends on the specific solid under consideration.

(b) Constant volume. A thermodynamic function that will be needed in the discussion of specific heats is the thermal pressure P_{th}. It is defined as the pressure that develops in a solid when heated from zero temperature up

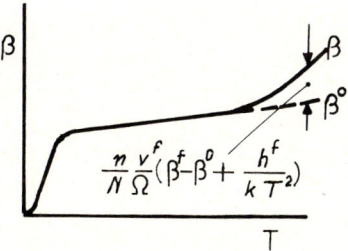

Fig. 2.5. Thermal expansion coefficient.

---

* The adiabatic bulk modulus B^S differs from B according to: $B^S - B = TC_V\gamma^2/V$ (2.25a), where $\gamma = V\beta B/C_V$ (2.25b).

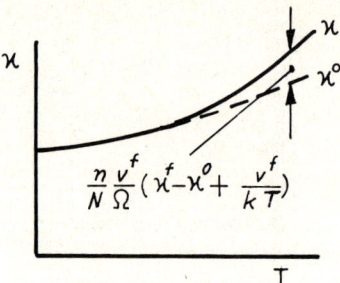

Fig. 2.6. Compressibility.

to temperature T under the constraint of constant volume:

$$P_{\text{th}} \equiv \int_0^T \left.\frac{\partial P}{\partial T}\right|_V dT. \tag{2.26}$$

By taking into consideration eqs. (2.24) and (2.25), that define β and B and the thermodynamic formula

$$-\left.\frac{\partial P}{\partial T}\right|_V = \left.\frac{\partial P}{\partial V}\right|_T \left.\frac{\partial V}{\partial T}\right|_P,$$

we get

$$\left.\frac{\partial P}{\partial T}\right|_V = \beta B. \tag{2.27}$$

The thermal pressure therefore will be

$$P_{\text{th}} = \int_0^T \beta B \, dT. \tag{2.28}$$

It is difficult to give the typical form of the plot of the isochoric specific heat of a crystal heated under constant volume V, because it has only been measured [10–14] for the rare gas solids Ne, Ar and Xe. By integrating C_V the form of the internal energy U is obtained.

2.3 Pressure dependence of thermodynamic functions of solids

In this section typical plots of the thermodynamic functions versus the pressure will be given under conditions of constant temperature. The best starting-point for such plots for solids is the Gibbs free energy G. The slope of the curve according to $(\partial G/\partial P)_T = V$ decreases with pressure P, because the volume always decreases with P (fig. 2.7). Usually (i.e. when $\beta > 0$) the entropy gives a decreasing curve (fig. 2.8), because of the thermodynamic

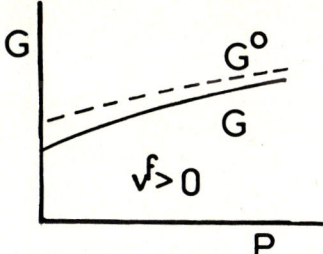

Fig. 2.7. Gibbs free energy for $v^f > 0$.

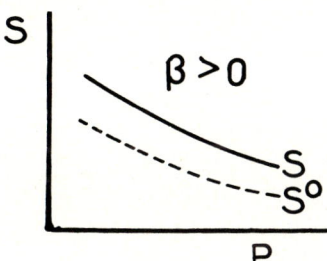

Fig. 2.8. Entropy.

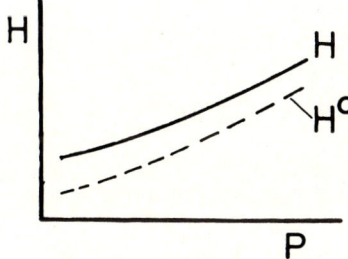

Fig. 2.9. Enthalpy.

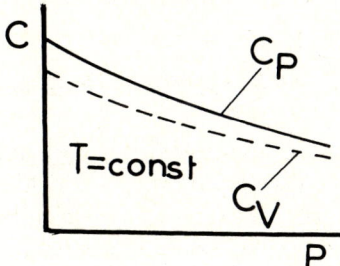

Fig. 2.10. Pressure dependence of the specific heat.

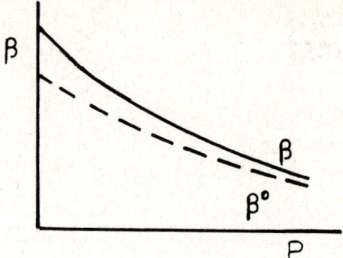

Fig. 2.11. Pressure dependence of the expansivity.

equation

$$\left.\frac{\partial S}{\partial P}\right|_T = -\left.\frac{\partial V}{\partial T}\right|_P = -\beta V.$$

As V and usually β are positive and decrease with pressure, the entropy $S = S(0) - \int_0^P \beta V \, dP$ and the absolute value of its slope decrease with pressure. The enthalpy according to $(\partial H/\partial P)_T = [\partial(G + TS)/\partial P]_T$ finally gives $(\partial H/\partial P)_T = V(1 - T\beta) \simeq V$ and therefore has a plot similar to that of G (fig. 2.9).

The specific heat C_P versus pressure is usually [15] a falling curve that tends to the value of C_V (fig. 2.10). The plot of the thermal expansivity must be a falling curve (fig. 2.11) because of the relation

$$\left.\frac{\partial \beta}{\partial P}\right|_T = -\left.\frac{\partial \kappa}{\partial T}\right|_P, \tag{2.29}$$

or

$$\left.\frac{\partial \beta}{\partial P}\right|_T = \frac{1}{B^2} \left.\frac{\partial B}{\partial T}\right|_P. \tag{2.30}$$

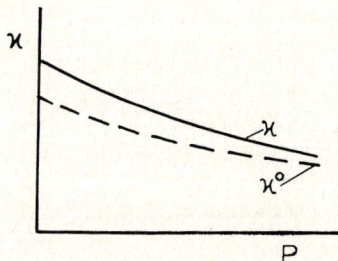

Fig. 2.12. Pressure dependence of the compressibility.

§2.3]   *Pressure dependence of thermodynamic functions of solids*   19

The variation of κ in respect to T is depicted in fig (2.6). The compressibility of almost all solids is known to fall with increasing pressure so that its plot has the form of fig. 2.12. Exceptions are: cerium [16], B4-AgI, etc. Details for these exceptions will be given in chapters 9 and 10.

The isothermal bulk modulus and its pressure derivative. From a microscopic point of view a bulk modulus can be thought of as a rough measure of the force constant between nn-atoms because it is usually calculated from an equation of the form:

$$B = \frac{(\text{number of springs}) \times (\text{force constant})}{(\text{some distance})}. \tag{2.31}$$

In a pure harmonic solid the force constant remains unchanged upon changing the pressure (volume); then $(\partial B/\partial P)_T = 0$ in a similar way as $(\partial B/\partial T)_P = 0$. In real solids *however* both derivatives are quite different from zero [1].

The quantity $(\partial B/\partial P)_T$ is extremely important in the discussion of the defect parameters. A first physical meaning of this quantity is clear by writing

$$\left.\frac{\partial B}{\partial P}\right|_T = \left.\frac{\partial B}{\partial V}\right|_T \left.\frac{\partial V}{\partial P}\right|_T = \left.\frac{\partial B}{\partial V}\right|_T \left(-\frac{V}{B}\right), \tag{2.32}$$

and hence

$$\left.\frac{dB}{dP}\right|_T = \frac{dB}{B} \bigg/ \left(-\frac{dV}{V}\right). \tag{2.33}$$

The quantity $(\partial B/\partial P)_T$ has usually values between 3.5 and 8 [17–27] which shows that upon compression the relative change of B is much larger than the relative decrease of volume; obviously this is a direct result of the anharmonicity of the solid because, as mentioned, in the harmonic solid $dB/dP = 0$.

The quantity $(dB/dP)_T$ usually slightly increases with temperature; a typical behaviour is shown in fig. 2.13 (strongly exaggerated); this is theoretically predicted [28,29] and experimentally verified [30,31]. Further the derivative $(\partial B/\partial P)_T$ varies only slightly upon compression; usually the derivative d^2B/dP^2 is negative and has a value of the order of $-B^{-1}(dB/dP)_T$. One should notice, however, that in the quasiharmonic approximation as adopted by Ludwig [1] the quantity $(\partial B/\partial P)_T$ has a concrete value not depending on temperature and pressure.

The quantity $(\partial B/\partial P)_T$ is connected to the "isobaric" and "isochoric" temperature derivative of the bulk modulus through the thermodynamical relation

$$\left.\frac{\partial B}{\partial T}\right|_P + \beta B \left.\frac{\partial B}{\partial P}\right|_T = \left.\frac{\partial B}{\partial T}\right|_V. \tag{2.34}$$

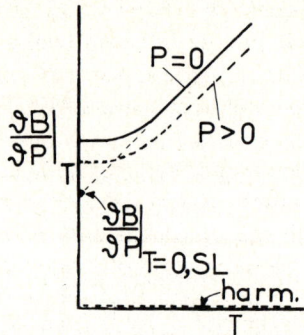

Fig. 2.13. Typical temperature dependence of $\partial B/\partial P$ in a real solid for $P \simeq 0$ (solid curve) and for $P > 0$ (dashed curve).

An alternative physical meaning of the quantity $(\partial B/\partial P)_T$. Consider a solid in which upon application of pressure the number of particles does not change. From the general thermodynamical relation (2.22) we get for an isothermal process:

$$dG = V \, dP \qquad (2.35)$$

or by introducing the definition of isothermal bulk modulus:

$$dG = -B \, dV. \qquad (2.36)$$

With the help of the relation (2.32) it gives (T = constant):

$$dG = d(BV)/[(\partial B/\partial P)_T - 1]. \qquad (2.37)$$

This general thermodynamical relation plays a major role in the derivation of the $cB\Omega$-model given in the last chapter. We turn to a brief discussion of this relation from which a direct physical meaning of $(\partial B/\partial P)_T - 1$ emerges.

It is thought that when a solid is subjected to a small uniform dilation δ, the energy density u gained from this dilation is given by [32] $u = \frac{1}{2}B\delta^2$. In this very approximate * scheme B is a measure of the density of this elastic energy and hence "BV" is a measure of the total elastic energy stored in the body. The above equation (2.37) shows that dG is *not* equal to what is usually thought as change of the "elastic" energy BV. This inequality of the absolute values of the quantities dG and d(BV) is a significant property of the anharmonic (i.e. the real) solid. Attention is drawn to the point that eq. (2.37) indicates – as physically expected – that these absolute values are equal only in the harmonic solid (set dB/dP equal to zero). Therefore the

---

* Obviously, eq. (2.36) can be written as d$G = -$d(BV) only when B is independent of pressure (volume).

physical meaning of the quantity $(\partial B/\partial P)_T$ may be given from the following rule: *In any isothermal process of a real (anharmonic) solid the ratio of the variations of the "elastic energy BV" and the Gibbs energy G is equal to $(\partial B/\partial P)_T - 1$ and not to -1, which would be valid only if the solid were harmonic.*

As a final remark it should be added: equation (2.37) indicates that in the quasiharmonic approximation (where $dB/dP =$ constant) one can always write $dG = Kd(BV)$ with the proportionality constant K being equal to $K \equiv [(\partial B/\partial P)_T - 1]^{-1}$. The integration of this equation just gives the equation of state of the solid (Murnaghan equation of state [27]).

In the last chapter we shall see that the $cB\Omega$-formula $g = cB\Omega$ is just the relation $g = Bv^f[(dB/dP) - 1]^{-1}$, with $c \equiv v^f \Omega^{-1}[(\partial B/\partial P)_T - 1]^{-1}$; it is similar to eq. (2.37) which is valid for the macroscopic body. In these relations v^f denotes the defect formation volume defined in ch. 3. It should be noted that eq. (2.34) is generally valid for any physical quantity Q which can be regarded as a function of T and P:

$$(dQ/dT)_V = (dQ/dT)_P + \beta B(dQ/dP)_T. \tag{2.34a}$$

3 | FORMATION OF VACANCIES

We consider a monoatomic solid at a temperature T which has a volume V and is under an external pressure P. Thermodynamics demands that this solid contain a number of (thermally created) vacancies. In the simplest case it contains only monovacancies.

The number of vacancies in the state of equilibrium will be labeled n_{eq} or simply n, and N the number of atoms of the solid. Usually the numbers n and N refer to the vacancies, respectively to the atoms existing in a mole of the solid (then N is Avogadro's number).

In the thermodynamics of point defects the key point is to compare the functions G, H, S etc. of the real crystal with the corresponding functions G^0, H^0, S^0 etc. of a perfect crystal (i.e. without vacancies) with the help of appropriate parameters. This introduces a difficulty because perfect crystals are not in equilibrium and therefore some thermodynamic functions cannot be defined for them. This difficulty can be overcome by suitable definitions of G^0, H^0, etc. (see §3.2). At the present point we will assume that – without loosing any generality – the three variables P^0, V^0 and T can be defined.

A given equilibrium state has a volume V corresponding to the variables P and T; in fig. 3.1 it is labelled as point A. All such real states form a surface in (P, V, T)-space, which is described by the equation of state; it has not been drawn in the figure in order not to overload it. A perfect crystal of the same pressure P and temperature T will have (usually) a smaller volume (point B). By choosing other values of P and T a surface of perfect states is produced. One sees that an infinity of (non-equilibrium) perfect states lie in the neighbourhood of point A. Of these, for reasons of simplicity, we will choose points B and C for the discussion. They both lie on the surface of the perfect state, B being a state with the same P and T, and C being a state with the same V and T as point A. The points B and C, as will be shown, have no privilege over other neighbouring points of the surface but their choice is particularly suitable for the definition of parameters, because, if the perfect state tends towards the equilibrium state along BA or CA, the paths involve processes for which P and T, or V and T remain constant; the description of such processes is convenient, because G or F will tend to minimize when evolving towards equilibrium. Depending

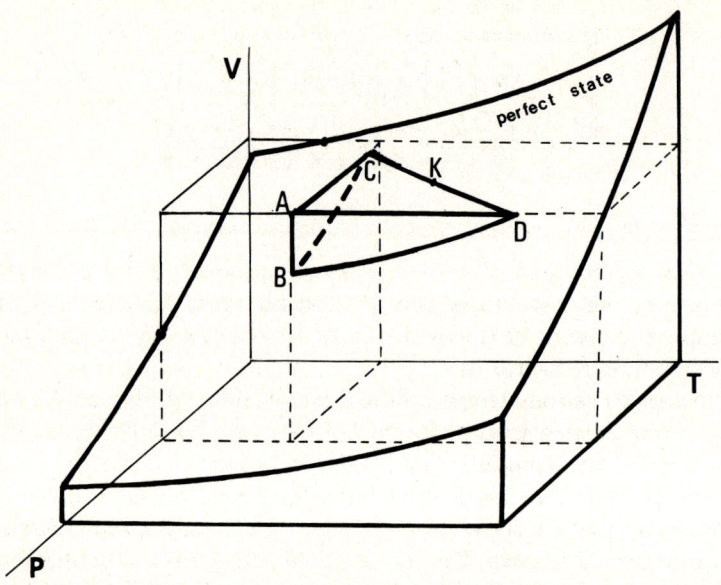

Fig. 3.1. Perfect states B, C, D in the proximity of the real state A (for point K, see §8.3).

on the choice of the neighbouring perfect state, different families of defect parameters will emerge.

3.1 Isobaric and isochoric perfect crystals

In fig. 3.2 a perfect crystal is compared with a real crystal of the same (external) pressure and temperature. Due to the existence of vacancies in the real crystal its volume V generally differs from the volume V^0 of the perfect crystal. We will call the thus defined perfect crystal the *isobaric perfect crystal*. This definition implies that the thermodynamic functions are compared by considering the following change: $P, V^0, T, n = 0 \rightarrow$

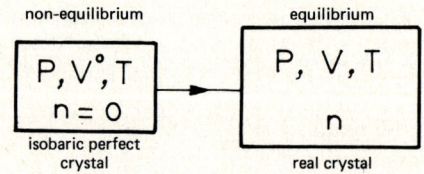

Fig. 3.2. Definition of an isobaric perfect crystal.

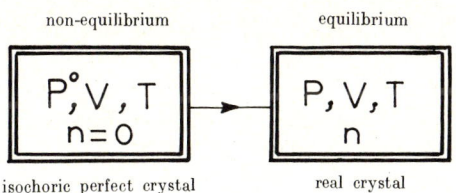

Fig. 3.3. Definition of an isochoric perfect crystal. (In §3.3 P^0 will be labelled $P^{0}*$.)

P, V, T, n, i.e. we have to consider the creation of n vacancies under conditions of constant temperature and pressure. It is obvious that the above procedure is useful in analysing experiments where the crystal is slowly heated to various temperatures under constant pressure P, because then we are interested in comparing the thermodynamic functions of the real solid in a state labelled (P, V, T, n) with the functions of a perfect solid labelled $(P, V^0, T, n = 0)$ at the same temperature and pressure.

In fig. 3.3 a perfect crystal and a real crystal of the same volume and temperature are compared. The existence of vacancies in the real crystal implies that the (external) pressure P on the walls of the real crystal generally differs from the pressure P^0 on the perfect crystal; we call the thus defined perfect crystal the *isochoric perfect crystal*. This definition implies that the comparison of the thermodynamic functions of the ideal crystal is done by considering the following change: $P^0, V, T, n = 0 \rightarrow P, V, T, n$, i.e. we have to consider the creation of n vacancies under conditions of constant temperature and volume. This procedure is usually indicated when we are dealing with experiments where the crystal is heated to various temperatures under constant volume, because we are then interested in comparing the thermodynamic functions of the solid in a state labelled (P, V, T, n) with the corresponding functions of the perfect solid (labelled $P^0, V, T, n = 0$) at the same temperature and volume.

In comparing the two perfect states, mentioned above, we emphasize that for a given real crystal (P, V, T, n) the two perfect states are *different*, and therefore the defect parameters in the two cases have different definitions and values.

3.2 Parameters from the comparison with the isobaric perfect crystal

In ch. 2 the thermodynamic functions G, H and S of a real crystal were defined. They are connected by

$$G = H - TS. \tag{3.1}$$

In order to develop the parameters of vacancies the corresponding quantities for the perfect crystal are needed. However, this is not possible because the perfect crystal is not in a state of equilibrium and therefore in principle no thermodynamic functions can be defined for it. In the present section this difficulty is surpassed to a certain degree by using quantities G^0, H^0, S^0 which are definable but are not "true" thermodynamic functions. However, a weak point still remains: it is absolutely necessary to be able to ascribe a temperature to the perfect crystal.

Consider a crystal that contains n vacancies; its Gibbs free energy can be written:

$$G = A + ng^f - TS_c. \tag{3.2}$$

TS_c is that part of G that is due to the various possible configurations of the vacancies. (We note that only vacancies contribute to a configurational entropy S_c.) The remaining part of G is $A + ng^f$ and is due to potential energies and thermal motions. With ng^f we label the contribution of the vacancies (excluding their configurational contributions). The latter is proportional to n as long as the vacancies do not interact. As the terms TS_c and ng^f become zero for $n = 0$, the term A is the value of G for $n = 0$. It is independent of n and we are justified to call it the Gibbs free energy G^0 of the perfect crystal:

$$A = G^0. \tag{3.3}$$

The same quantity G^0 has been used in the literature under the restriction that the perfect crystal is so near to an equilibrium state that thermodynamic functions can be defined.

Under the above definitions we write eq. (3.2) as

$$G = G^0 + ng^f - TS_c. \tag{3.4}$$

This equation can be considered as giving the definition of g^f.

Number of vacancies at equilibrium. The configurational entropy of n vacancies that could occupy $N + n$ sites is

$$S_c = k \ln\{(N+n)!/N!\,n!\}. \tag{3.5}$$

In ch. 2 it was stated that under constant P and T the Gibbs energy is a minimum in the equilibrium state; therefore the number of vacancies n_{eq} at equilibrium can be found from the condition

$$\left.\frac{\partial G}{\partial n}\right|_{T,P} = 0. \tag{3.6}$$

By using for S_c the Stirling approximation and differentiating eq. (3.5) for any variable y one gets

$$\frac{\partial S_c}{\partial y} = k \frac{\partial n}{\partial y} \ln \frac{N+n}{n}. \tag{3.7}$$

§3.2] *Isobaric parameters*

By introducing eq. (3.4) into eq. (3.6) and using the rule of eq. (3.7) one finally gets for $n \ll N$:

$$n_{eq} = (N + n_{eq}) \exp(-g^f/kT), \tag{3.8}$$

or

$$n_{eq} \simeq N \exp(-g^f/kT). \tag{3.9}$$

Configurational entropy at equilibrium. The configurational entropy at equilibrium $S_{c,eq}$ is found by inserting n_{eq} into eq. (3.5). One obtains

$$S_{c,eq} = n_{eq}(k + g^f/T). \tag{3.10}$$

The differentiation of $S_{c,eq}$ with regard to T will be needed in a later section; according to eq. (3.7) it gives

$$\frac{\partial S_{c,eq}}{\partial T} = \frac{g^f}{T} \frac{\partial n_{eq}}{\partial T}. \tag{3.11}$$

Gibbs energy. By introducing eq. (3.10) into eq. (3.4) we get

$$G_{eq} = G^0 - n_{eq}kT. \tag{3.12}$$

It shows that $G_{eq} < G^0$ as expected. This difference can be seen in fig. 2.4. It also shows that at equilibrium the difference between the Gibbs energy of the real and the perfect crystal is kT per vacancy. We can offer no precise physical explanation for this remarkable result. From now on the suffix eq will be dropped.

Entropy. The real crystal is in the state of equilibrium; its entropy can be found by differentiating eq. (3.2)

$$S = -\left.\frac{\partial G}{\partial T}\right|_P$$

$$= -\left.\frac{\partial G^0}{\partial T}\right|_P - n\left.\frac{\partial g^f}{\partial T}\right|_P + S_c + \left\{T\left.\frac{\partial S_c}{\partial T}\right|_P - g^f\left.\frac{\partial n}{\partial T}\right|_P\right\}. \tag{3.13}$$

The quantity in brackets is zero because of eq. (3.11). We will define a quantity S^0 by

$$S^0 \equiv -\left.\frac{\partial G^0}{\partial T}\right|_P \tag{3.14}$$

and will call it the entropy of the perfect crystal. Equation (3.13) now becomes

$$S = S^0 - n\left.\frac{\partial g^f}{\partial T}\right|_P + S_c. \tag{3.15}$$

The quantity $(-\partial g^f/\partial T)_P$ can be defined as the thermal formation entropy per defect:

$$s^f \equiv - \left.\frac{\partial g^f}{\partial T}\right|_P \tag{3.16}$$

and therefore eq. (3.15) can be written as

$$S = S^0 + ns^f + S_c. \tag{3.17}$$

With the help of eq. (3.10) the equilibrium entropy can be expressed in a form not containing S_c:

$$S = S^0 + n(s^f + k + g^f/T). \tag{3.18}$$

The difference between S and S^0 is shown in fig. 2.3. In the next section we will define a parameter h^f by $h^f \equiv g^f + Ts^f$, so that eq. (3.18) can be written

$$S = S^0 + n(k + h^f/T). \tag{3.19}$$

Equation (3.19) does not imply that $S - S^0$ is proportional to n when comparing different temperatures. It is worthwhile to note that $(S - S^0)/n$ is $k + h^f/T$ and *not* h^f/T as one would expect for a reversible isobaric process.

Enthalpy. By inserting into eq. (2.10) the values of G and S given by eqs. (3.12) and (3.18) we get:

$$H = G^0 + TS^0 + n(g^f + Ts^f). \tag{3.20}$$

We will call the quantity $G^0 + TS^0$ the enthalpy H^0 of the perfect crystal:

$$H^0 \equiv G^0 + TS^0. \tag{3.21}$$

Further we will define the quantity $g^f + Ts^f$ as the formation enthalpy h^f per defect:

$$h^f \equiv g^f + Ts^f. \tag{3.22}$$

Therefore eq. (3.20) can be written as:

$$H = H^0 + nh^f. \tag{3.23}$$

The difference between H and H^0 is depicted in fig. 2.2.

Equation (3.23) reveals the physical meaning of h^f better then eq. (3.22). In processes that are both isothermal and isobaric the variation of the enthalpy is equal to the heat that has been added. Therefore h^f (i.e. $(H - H^0)/n$) is the heat that has indispensably to be added in order for a vacancy to be produced under the constraint of constant pressure and constant temperature. A combination of eqs. (3.9) and (3.22) gives the well known result

$$n = N e^{s^f/k} e^{-h^f/kT}. \tag{3.24}$$

In later sections some derivatives will be needed:
(a) By differentiating eq. (3.22) and inserting eq. (3.16) one obtains

$$\left.\frac{\partial h^f}{\partial T}\right|_P = T \left.\frac{\partial s^f}{\partial T}\right|_P, \tag{3.25}$$

which resembles the thermodynamical eq. (2.17).

(b) By differentiating eq. (3.9) one obtains with the help of eq. (3.22)

$$\left.\frac{\partial n}{\partial T}\right|_P = \frac{nh^f}{kT^2}. \tag{3.26}$$

(c) Equation (3.24) gives by differentiation the Wigner formula:

$$\left.\frac{\partial \ln(n/N)}{\partial(1/T)}\right|_P = -\frac{h^f}{k}. \tag{3.27}$$

Volume. The volume V of the real crystal can be found by differentiating eq (3.12) with respect to pressure:

$$V \equiv \left.\frac{\partial G}{\partial P}\right|_T = \left.\frac{\partial G^0}{\partial P}\right|_T - kT \left.\frac{\partial n}{\partial P}\right|_T. \tag{3.28}$$

The derivative $(\partial G^0/\partial P)_T$ is the definition of the volume V^0 of the ideal crystal; by differentiating eq. (3.9) with respect to pressure we get

$$\left.\frac{\partial n}{\partial P}\right|_T = -\frac{n}{kT} \left.\frac{\partial g^f}{\partial P}\right|_T. \tag{3.29}$$

The volume V therefore becomes

$$V = V^0 + n \left.\frac{\partial g^f}{\partial P}\right|_T. \tag{3.30}$$

The quantity $(\partial g^f/\partial P)_T$ is called the formation volume v^f per defect:

$$v^f \equiv \left.\frac{\partial g^f}{\partial P}\right|_T \tag{3.31}$$

and hence eq. (3.30) can be written

$$V = V^0 + nv^f. \tag{3.32}$$

Note that v^f is a change of the bulk volume, if in the state of equilibrium one more vacancy were produced; it is not just the local change of volume around the vacancy as it involves contributions from all the cells. Furthermore, this is not valid for interacting vacancies.

Temperature and pressure dependence of g^f. In cases where the temperature and pressure of a real crystal change, the formation Gibbs energy g is a function of both T and P; then the following expression holds:

$$dg^f = \left.\frac{\partial g^f}{\partial T}\right|_P dT + \left.\frac{\partial g^f}{\partial P}\right|_T dP$$

or

$$dg^f = -s^f dT + v^f dP. \tag{3.33}$$

This form resembles eq. (2.12).

Internal energy. By introducing eqs. (3.23) and (3.32) into eq. (2.9) we get

$$U = H^0 - PV^0 + n(h^f - Pv^f). \tag{3.34}$$

The quantity $H^0 - PV^0$ can be called the internal energy U^0 of the perfect crystal

$$U^0 \equiv H^0 - PV^0. \tag{3.35}$$

The other quantity, $h^f - Pv^f$, is defined as the formation internal energy u^f per defect

$$u^f \equiv h^f - Pv^f \tag{3.36}$$

and hence eq. (3.34) is written

$$U = U^0 + nu^f. \tag{3.37}$$

Helmholtz free energy. The free energy F of the real crystal is $F = G - PV$; by inserting into this equation the values of G and V from eqs. (3.4) and (3.32) we get

$$F = G^0 - PV^0 + n(g^f - Pv^f) - TS_c. \tag{3.38}$$

The quantity $G^0 - PV^0$ can be defined as the free energy F^0 of the ideal crystal; the other quantity, $g^f - Pv^f$, is defined as the Helmholtz formation free energy f per defect

$$f^f \equiv g^f - Pv^f. \tag{3.39}$$

Therefore eq. (3.38) can be written as:

$$F = F^0 + nf^f - TS_c. \tag{3.40}$$

An alternative form not containing S_c can be derived by inserting eq. (3.10) into eq. (3.40) so that with the help of eq. (3.39) we get

$$F = F^0 - n(kT + Pv^f).$$

In the literature a formula has often been published of the form $F = F^0 - nkT$. In this case the quantity F^0 is not compatible with S^0, H^0, etc. of this section, as it refers to a different definition of the perfect crystal (isochoric perfect crystal); the latter case is developed in the next section.

Thermal expansion coefficients. The volume of a crystal varies with temperature due to anharmonic effects and to the thermal creation of vacancies. By differentiating eq. (3.32) with respect to temperature we get

$$\left.\frac{\partial V}{\partial T}\right|_P = \left.\frac{\partial V^0}{\partial T}\right|_P + n\left.\frac{\partial v^f}{\partial T}\right|_P + v^f\left.\frac{\partial n}{\partial T}\right|_P. \tag{3.41}$$

Beyond the macroscopic thermal expansion coefficient β of the real crystal (eq. 2.24) we define the following coefficients:

(a) The thermal expansion coefficient β^0 of the isobaric perfect crystal is defined by

$$\beta^0 \equiv \frac{1}{V^0}\left.\frac{\partial V^0}{\partial T}\right|_P. \tag{3.42}$$

(b) The thermal expansion coefficient of the formation volume is similarly defined from its temperature derivative:

$$\beta^f \equiv \frac{1}{v^f}\left.\frac{\partial v^f}{\partial T}\right|_P. \tag{3.43}$$

Due to eq. (3.26), eq. (3.41) becomes

$$\beta V = \beta^0 V^0 + n\beta^f v^f + v^f nh^f/kT^2. \tag{3.44}$$

or, due to eq. (3.32),

$$\beta = \beta^0 \frac{V^0}{V^0 + nv^f} + \frac{nv^f}{V}\left(\beta^f + \frac{h^f}{kT^2}\right). \tag{3.45}$$

We define the volume Ω^0 per atom of the perfect crystal by

$$\Omega^0 \equiv V^0/N \tag{3.46}$$

and in the same way the mean volume Ω per atom (not per unit cell) of the real crystal by

$$\Omega \equiv V/N. \tag{3.47}$$

By introducing these definitions into eq. (3.45), we obtain

$$\beta = \beta^0 \left(1 + \frac{nv^f}{\Omega^0 N}\right)^{-1} + \frac{nv^f}{\Omega N}\left(\beta^f + \frac{h^f}{kT^2}\right). \tag{3.48}$$

The formation volume v^f of single vacancies in crystals is found in experiments to be of the order of Ω or Ω^0. In this approximation and for small concentrations ($n \ll N$) we finally get (by expanding):

$$\beta \simeq \beta^0 + \frac{nv^f}{N\Omega}\left(\beta^f - \beta^0 + \frac{h^f}{kT^2}\right). \tag{3.49}$$

We note that a real crystal has a larger expansivity (see fig. 2.5) than the ideal one due to: (a) the expansivity β^f of the formation volume is larger than β^0 or β (as will be proved in ch. 8; this has also recently been verified experimentally [33]), and (b) an additional term that is due to the further creation of vacancies. The relative importance of the latter term can be estimated by comparing $\beta^f - \beta^0$ to $(1/T)(h^f/kT)$. At ambient temperatures h^f is much larger than kT and therefore in order for $\beta^f - \beta^0$ to become significant it has to be larger than $1/T$. In cases where $\beta^f - \beta^0 \ll h^f/kT^2$, eq. (3.49) simplifies to

$$\beta^0 \simeq \beta - \frac{n}{N} \frac{v^f}{\Omega} \frac{h^f}{kT^2}. \tag{3.50}$$

The term $\beta^f - \beta^0$ has often been ignored with the assumption that β^f is small (e.g. $\beta^f = 0$ or $\beta^f \simeq \beta^0 \simeq 10^{-5}$ K^{-1}); in this case it would never become larger than $1/T$. Attention is drawn to this point, because the contribution of defects to β is studied in the high temperature region, and in this region the ratio h^f/kT is much smaller, i.e. of the order of 5–10. Furthermore, as we will see in a later section, in some cases $\beta^f \gg \beta^0$, so that the term $\beta^f - \beta^0$ cannot always be neglected.

Compressibility. By differentiating eq. (3.32) with respect to pressure we get:

$$\left.\frac{\partial V}{\partial P}\right|_T = \left.\frac{\partial V^0}{\partial P}\right|_T + n \left.\frac{\partial v^f}{\partial P}\right|_T + v^f \left.\frac{\partial n}{\partial P}\right|_T. \tag{3.51}$$

Beyond the known definition (eq. 2.25) of the compressibility κ of the real crystal we make the following definitions:

$$\kappa^0 \equiv -\frac{1}{V^0} \left.\frac{\partial V^0}{\partial P}\right|_T \tag{3.52}$$

and

$$\kappa^f \equiv -\frac{1}{v^f} \left.\frac{\partial v^f}{\partial P}\right|_T \tag{3.53}$$

where the superscript 0 refers to the perfect crystal and κ^f denotes the compressibility of the formation volume. Introducing these definitions into eq. (3.51), one obtains

$$\kappa V = \kappa^0 V^0 + n\kappa^f v^f - v^f \left.\frac{\partial n}{\partial P}\right|_T. \tag{3.54}$$

Derivating eq. (3.9) and using eq. (3.31) one obtains

$$\left.\frac{\partial n}{\partial P}\right|_T = -\frac{1}{kT} n v^f. \tag{3.55}$$

Equation (3.54) therefore becomes:

$$\kappa V = \kappa^0 V^0 + n\kappa^f v^f + \frac{n(v^f)^2}{kT}.$$

By using eqs. (3.32), (3.46) and (3.47) and the considerations mentioned for β, the above equation becomes (for low concentrations)

$$\kappa \simeq \kappa^0 + \frac{nv^f}{N\Omega}\left(\kappa^f - \kappa^0 + \frac{v^f}{kT}\right). \tag{3.56}$$

We note that a real crystal has a larger compressibility (see fig. 2.6) than the ideal one for two reasons: (a) the compressibility κ^f of the formation volume is larger than κ^0 or κ (as will be shown later) and (b) the presence of an increased number of vacancies makes it more compressible. The relative importance of each term will now be discussed: A usual approximation is to take κ^f equal to zero, i.e. to take v^f as pressure independent. In the literature κ^f has also been taken equal to κ or κ^0. Due to the fact that v^f is of the order of Ω, rough calculation (under the assumption $\kappa^f = \kappa$) shows that near the melting point κ^f is a few percent of Ω/kT. As we shall show in a later section $\kappa^f > \kappa$ and therefore the contribution of the first two terms between the parentheses in some cases may not be negligible. The difference $\kappa - \kappa^0$ is indicated in fig. 2.6. The change of the compressibility for a constant number n of point defects (i.e. $(\partial n/\partial P)_T = 0$) results from eq. (3.54):

$$\kappa - \kappa^0 = \frac{n}{N}\frac{v^f}{\Omega}(\kappa^f - \kappa^0). \tag{3.57}$$

In cases where $\kappa^f - \kappa^0 \ll v^f/kT$ eq. (3.56) simplifies to

$$\kappa^0 \simeq \kappa - \frac{v^f}{\Omega}\frac{n}{N}\frac{v^f}{kT}. \tag{3.58}$$

Bulk modulus. Often calculations are made in which, instead of the compressibility, the bulk modulus B and its pressure dependence dB/dP are needed. Its connection to B^0 is best found by differentiating eq. (3.32) in respect to P.

A simple expression for B can be obtained by introducing $\kappa \equiv 1/B$ into the approximation of eq. (3.56). This gives

$$B \simeq B^0 \bigg/ \left(1 + \frac{n}{N}\frac{v^f}{\Omega}\frac{v^f}{kT}B^0\right),$$

or

$$B \simeq B^0\left(1 - \frac{n}{N}\frac{v^f}{\Omega}\frac{v^f}{kT}B^0\right). \tag{3.59}$$

After differentiation and algebraic calculations one finally gets

$$\left.\frac{\partial B}{\partial P}\right|_T \simeq \left.\frac{\partial B^0}{\partial P}\right|_T + \frac{n}{N}\frac{v^f}{\Omega}\left(\frac{B^0 v^f}{kT}\right)^2. \tag{3.60}$$

Attention is drawn to this relation; by taking $v^f \simeq \Omega$ we have, close to the melting point $B\Omega/kT \simeq 10^2$, $n/N \simeq 10^{-4}$ and hence the last term of eq. (3.60) is of the order of unity. By recalling that dB/dP is usually between 3.5 and 8 we conclude that vacancies make a *considerable* contribution to dB/dP. An experimental verification of this proposal is *highly* desirable.

Thermal pressure. The thermal pressure P_{th} of a real crystal has been defined by eq. (2.26). A similar definition can be given to the thermal pressure P_{th}^0 of the isobaric perfect crystal:

$$P_{th}^0 = \int_0^T \frac{\beta^0}{\kappa^0} dT, \tag{3.61}$$

whereas

$$P_{th} = \int_0^T \frac{\beta}{\kappa} dT. \tag{3.62}$$

It is obvious that the difference of the thermal pressures of the real and the isobaric perfect crystal is regulated by the difference $(\beta^0/\kappa^0) - (\beta/\kappa)$. By dividing eqs. (3.50) and (3.58) and disregarding the terms containing $(n/N)^2$ we get:

$$T\left(\frac{\beta^0}{\kappa^0} - \frac{\beta}{\kappa}\right) = -\frac{n}{N}\frac{v}{\Omega}\frac{1}{kT}\frac{1}{kT}\left(h^f - \frac{Tv^f\beta}{\kappa}\right). \tag{3.63}$$

This relation indicates that the algebraic sign of $(\beta^0/\kappa^0) - (\beta/\kappa)$ is regulated by the term between parentheses. In most cases and in the low temperature region, the experimental values show that $T\beta/\kappa < h^f/v^f$ and hence $\beta^0/\kappa^0 < \beta/\kappa$, or $\beta/\beta^0 > \kappa/\kappa^0$. At a certain temperature we get $T\beta/\kappa = h^f/v^f$ so that the quantities P_{th}^0 and P_{th} become practically equal. The previous considerations can be summarized by stating that in order to compare the relative contribution β/β^0 of vacancies to the expansivity or their relative contribution κ/κ^0 to the compressibility, one has to know the difference between h^f/v^f and $T\beta/\kappa$.

Chemical potential. The chemical potential μ "of an atom" has been defined by eq. (2.23). By differentiating eq. (3.12) (for low concentrations) one gets

$$\mu = \left.\frac{\partial G^0}{\partial N}\right|_{P,T} - kT\left.\frac{\partial n}{\partial N}\right|_{P,T},$$

§3.2] *Isobaric parameters* 35

or due to eq. (3.9):

$$\mu = \mu^0 - \frac{n}{N} kT, \tag{3.64}$$

where $\mu^0 \equiv (\partial G^0/\partial N)_{P,T}$ is the chemical potential "of an atom" in the ideal crystal. An exact formula for μ can be found by direct differentiation of eq. (3.4):

$$\mu = \mu^0 - kT \ln(1 + n/N). \tag{3.65}$$

It is obvious that when $n \ll N$, eq. (3.65) gives eq. (3.64). For an extensive discussion, see §7.5.

Basic differential equations for the isobaric perfect crystal. Equations (2.19) to (2.22) gave basic differential equations between the thermodynamic functions U, H, F, etc. of a real crystal. Here we will investigate analogous differential equations that connect the quantities U^0, H^0, F^0, etc. of the isobaric perfect crystal. The latter quantities are not "true" thermodynamic functions and the question arises if simple connections between their differentials exist. Due to the way in which they were defined in the present volume this becomes possible. The quantity G^0 depends on P and T and therefore we can write

$$dG^0 = \left.\frac{\partial G^0}{\partial T}\right|_P dT + \left.\frac{\partial G^0}{\partial P}\right|_T dP.$$

By introducing the definitions (3.14) and (3.28) we obtain

$$dG^0 = -S^0 dT + V^0 dP. \tag{3.66}$$

This has exactly the form of eq. (2.22). It leads to an equation that is similar to a Maxwell equation:

$$\left.\frac{\partial S^0}{\partial P}\right|_T = -\left.\frac{\partial V^0}{\partial T}\right|_P. \tag{3.67}$$

We note that according to the definition of β^0 the term on the right is equal to $-\beta^0 V^0$. By recalling the definitions of U^0, H^0 and F^0 and using eq. (3.66) one easily obtains:

$$dU^0 = T dS^0 - P dV^0, \qquad \left.\frac{\partial T}{\partial V^0}\right|_{S^0} = -\left.\frac{\partial P}{\partial S^0}\right|_{V^0}, \tag{3.68a,b}$$

$$dH^0 = T dS^0 + V^0 dP, \qquad \left.\frac{\partial T}{\partial P}\right|_{S^0} = \left.\frac{\partial V^0}{\partial S^0}\right|_P, \tag{3.69a,b}$$

$$dF^0 = -S^0 dT - P dV^0, \qquad \left.\frac{\partial S^0}{\partial V^0}\right|_T = \left.\frac{\partial P}{\partial T}\right|_{V^0} = \beta^0 V^0. \tag{3.70a,b}$$

By differentiating eq. (3.21) one easily gets

$$\left.\frac{\partial H^0}{\partial T}\right|_P = T \left.\frac{\partial S^0}{\partial T}\right|_P. \tag{3.71}$$

This equation is exact but refers to the quantities G^0, H^0, S^0, etc. that have been defined in a very specific way. In literature these quantities have been defined as thermodynamic functions in a state very near equilibrium. This latter definition would make many proofs unnecessary because then the usual thermodynamical relations would be valid. However, with such a definition and taking into account that the perfect crystal is not in equilibrium, the validity of eq. (3.71) becomes questionable [34]. This inconvenience results in an equation of the form of eq. (3.25), which would be based on the thermodynamic definitions of G^0, H^0, S^0 and might be appreciably in error.

In effect, by introducing eqs. (3.23) and (3.17) into eq. (3.1) one obtains

$$G = H^0 - TS^0 - TS_c + n(h^f - Ts^f).$$

For equilibrium $((\partial G/\partial n)_T = 0)$ this gives

$$h^f - Ts^f = T \left.\frac{\partial S_c}{\partial n}\right|_P \quad \text{or} \quad T \left.\frac{\partial S_c}{\partial T}\right|_P = (h^f - Ts^f) \left.\frac{\partial n}{\partial T}\right|_P. \tag{3.72}$$

On the other hand, introducing eqs. (3.23) and (3.17) into eq. (2.17) one gets [34]:

$$\left.\frac{\partial H^0}{\partial T}\right|_P + n \left.\frac{\partial h^f}{\partial T}\right|_P + h^f \left.\frac{\partial n}{\partial T}\right|_P$$

$$= T \left.\frac{\partial S^0}{\partial T}\right|_P + Tn \left.\frac{\partial s^f}{\partial T}\right|_P + Ts^f \left.\frac{\partial n}{\partial T}\right|_P + T \left.\frac{\partial S_c}{\partial T}\right|_P$$

or

$$\left.\frac{\partial H^0}{\partial T}\right|_P - T \left.\frac{\partial S^0}{\partial T}\right|_P + n \left(\left.\frac{\partial h^f}{\partial T}\right|_P - T \left.\frac{\partial s^f}{\partial T}\right|_P \right) = \left.\frac{\partial n}{\partial T}\right|_P (h^f - Ts^f). \tag{3.73}$$

The combination of (3.72) and (3.73) gives:

$$\left.\frac{\partial H^0}{\partial T}\right|_P - T \left.\frac{\partial S^0}{\partial T}\right|_P + n \left(\frac{\partial h^f}{\partial T} - T \frac{\partial s^f}{\partial T} \right) = 0.$$

The combination of this equation with eq. (3.71) gives eq. (3.25) which is of major importance in the analysis of experiments.

Differential equations of the vacancy parameters. It is worthwhile examining the differential equations for the vacancy parameters. From eq. (3.33)

$$dg^f = v^f dP - s^f dT \tag{3.74}$$

we get a Maxwell equation

$$\left.\frac{\partial v^f}{\partial T}\right|_P = -\left.\frac{\partial s^f}{\partial P}\right|_T. \tag{3.75}$$

According to eq. (3.43) this quantity has been defined to be equal to $\beta^f v^f$. By simple replacement of the parameters one gets:

$$df^f = -P\,dv^f - s^f\,dT, \qquad \left.\frac{\partial P}{\partial T}\right|_{v^f} = \left.\frac{\partial s^f}{\partial v^f}\right|_T, \tag{3.76a,b}$$

$$dh^f = v^f\,dP + T\,ds^f, \qquad \left.\frac{\partial v^f}{\partial s^f}\right|_P = \left.\frac{\partial T}{\partial P}\right|_{s^f}, \tag{3.77a,b}$$

$$du^f = -P\,dv^f + T\,ds^f, \qquad \left.\frac{\partial P}{\partial s^f}\right|_{v^f} = -\left.\frac{\partial T}{\partial v^f}\right|_{s^f}. \tag{3.78a,b}$$

Equation (3.78a) is not analogous to the standard first law of thermodynamics as expressed in eq. (2.14) for reversible processes. The reason is that the production of a vacancy is not a reversible process.

3.3 Parameters from the comparison with the isochoric perfect crystal

In this section the thermodynamic functions of the real state will be compared to the functions of that perfect state which has the same volume and temperature (see fig. 3.3). We start with the artificially constrained perfect crystal ($P^{0*}, V, T, n = 0$) which has a Helmholtz free energy F^{0*} and examine the free energy F of the real crystal (P, V, T, n), containing n vacancies at the same temperature and volume. * In the case of non-interacting vacancies, we can use the same procedure as in §3.2; the result is

$$F = F^{0*} + nf^* - TS_c, \tag{3.79}$$

where S_c is given by eq. (3.5) and the parameter f^* is defined by eq. (3.79). In the state of equilibrium the function F has to be a minimum and hence:

$$\left.\frac{\partial F}{\partial n}\right|_{V,T} = 0. \tag{3.80}$$

---

* The introduction of F^{0*} can also be made in the same way as the introduction of G^0 on p. 26.

Following the same considerations as in the former section we get for equilibrium:

$$n \simeq N \exp(-f^*/kT), \tag{3.81}$$

$$S_c = n(k + f^*/T), \tag{3.82}$$

$$F = F^{0*} - nkT, \tag{3.83}$$

$$S = S^{0*} - n \left.\frac{\partial f^*}{\partial T}\right|_V + S_c, \tag{3.84}$$

where S^{0*} is defined as

$$S^{0*} \equiv - \left.\frac{\partial F^{0*}}{\partial T}\right|_V, \tag{3.85}$$

i.e. the entropy of the isochoric perfect crystal. We define the thermal entropy s^* per defect produced under constant volume by

$$s^* \equiv - \left.\frac{\partial f^*}{\partial T}\right|_V \tag{3.86}$$

and hence eq. (3.84) reads:

$$S = S^{0*} + ns^* + S_c, \tag{3.87}$$

or in a form not containing S_c:

$$S = S^{0*} + n(s^* + k + f^*/T). \tag{3.88}$$

We will call the internal energy U^{0*} of the perfect crystal the quantity given by

$$U^{0*} \equiv F^{0*} + TS^{0*} \tag{3.89}$$

and label with u^* the quantity

$$u^* \equiv f^* + Ts^*. \tag{3.90}$$

The internal energy of the crystal can then be written

$$U = U^{0*} + nu^*. \tag{3.91}$$

We see that u^* is the variation of the internal energy, when one vacancy is formed under constant volume and temperature. In such processes the variation of the internal energy is exactly the heat that has indispensably to be added. The entropy S can now be written

$$S = S^{0*} + n(k + u^*/T). \tag{3.92}$$

By differentiating eq. (3.90) we get:

$$\left.\frac{\partial u^*}{\partial T}\right|_V = \left.\frac{\partial f^*}{\partial T}\right|_V + s^* + T \left.\frac{\partial s^*}{\partial T}\right|_V.$$

This simplifies because of eq. (3.86) to

$$\left.\frac{\partial u^*}{\partial T}\right|_V = T \left.\frac{\partial s^*}{\partial T}\right|_V, \tag{3.93}$$

which resembles the well known thermodynamical equation (2.18). Equation (3.90) allows an alternative expression to eq. (3.81) for n:

$$n = N e^{s^*/k} e^{-u^*/kT}. \tag{3.94}$$

By differentiating eq. (3.81) and using eqs. (3.86) and (3.90) one obtains

$$\left.\frac{\partial n}{\partial T}\right|_V = n \frac{u^*}{kT^2}. \tag{3.95}$$

In a later section the derivative of $\ln n/N$ with respect to $1/T$ will be needed. The computation follows the same method as that used for eq. (3.27) and leads to

$$\left.\frac{\partial \ln(n/N)}{\partial(1/T)}\right|_V = -\frac{u^*}{k}. \tag{3.96}$$

Pressure. By differentiating eq. (3.83) with respect to volume we get with the help of eq. (3.81)

$$\left.\frac{\partial F}{\partial V}\right|_T = \left.\frac{\partial F^{0*}}{\partial V}\right|_T + n \left.\frac{\partial f^*}{\partial V}\right|_T. \tag{3.97}$$

It is obvious that $-(\partial F/\partial V)_T$ denotes the pressure P of the real crystal, whereas

$$P^{0*} \equiv - \left.\frac{\partial F^{0*}}{\partial V}\right|_T \tag{3.98}$$

is defined as the pressure of the perfect isochoric crystal (see fig. 3.3). We define p^* by

$$p^* \equiv - \left.\frac{\partial f^*}{\partial V}\right|_T, \tag{3.99}$$

so that eq. (3.97) becomes

$$P = P^{0*} + np^*. \tag{3.100}$$

The quantity p^* is the change of pressure when one defect is formed under constant volume and temperature.

Enthalpy. The enthalpy H of the real crystal can be found with the help of eqs. (3.91) and (3.100):

$$H \equiv U + PV = U^{0*} + nu^* + (P^{0*} + np^*)V \quad \text{or} \quad H = H^{0*} + nh^*, \tag{3.101}$$

where
$$h^* \equiv u^* + p^*V \tag{3.102}$$
is defined as the increase of the enthalpy when one vacancy is formed under constant volume and temperature.

Gibbs energy. The energy G of the real crystal can be expressed with the help of eqs. (3.87) and (3.101):
$$G \equiv H - TS = H^{0*} - TS^{0*} + n(h^* - Ts^*) - TS_c$$
$$\text{or} \quad G = G^{0*} + n(h^* - Ts^*) - TS_c, \tag{3.103}$$
where
$$G^{0*} \equiv H^{0*} - TS^{0*} \tag{3.104}$$
is called the Gibbs energy of the perfect crystal at the same temperature and volume; we can define a quantity g^* by
$$g^* \equiv h^* - Ts^*. \tag{3.105}$$
Equation (3.103) can then be written:
$$G = G^{0*} + ng^* - TS_c. \tag{3.106}$$
Due to eq. (3.82) it can also be written
$$G = G^{0*} + n(g^* - kT - f^*) \tag{3.107}$$
or with the help of the equation
$$g^* - f^* = p^*V, \tag{3.108}$$
which results from eqs. (3.102), (3.90) and (3.105), we get:
$$G = G^{0*} - nkT + np^*V. \tag{3.109}$$

Compressibility. In eq. (3.56) we found a connection between the bulk compressibility κ of a real crystal and the compressibility κ^0 of its isobaric perfect crystal. The symbol κ^f was the compressibility of the formation volume. Here we determine the connection between κ and the compressibility κ^{0*} of its isochoric perfect crystal. The compressibility κ^{0*} is defined by
$$\kappa^{0*} = -\frac{1}{V}\left.\frac{\partial V}{\partial P^{0*}}\right|_T. \tag{3.110}$$
We divide eq. (2.25) by eq. (3.110) and get
$$\frac{\kappa}{\kappa^{0*}} = \left.\frac{\partial P^{0*}}{\partial P}\right|_T. \tag{3.111}$$
The value of $\partial P^{0*}/\partial P$ is obtained by differentiating eq. (3.100) with respect

to pressure. It gives

$$\left.\frac{\partial P^{0*}}{\partial P}\right|_T = 1 - p^* \left.\frac{\partial n}{\partial P}\right|_T - n \left.\frac{\partial p^*}{\partial P}\right|_T. \tag{3.112}$$

In §3.4 [see eq. (3.131)] it will be shown that

$$p^* = v^f B/V \tag{3.113}$$

so that eq. (3.100) gives:

$$P = P^{0*} + nv^f B/V. \tag{3.114}$$

By introducing this result and eqs. (3.53) and (3.55), eq. (3.111) gives:

$$\kappa = \kappa^{0*} + \frac{nv^f}{V} \left\{ \frac{\kappa^{0*}}{\kappa} \frac{v^f}{kT} + \frac{\kappa^{0*}}{\kappa} \kappa^f - \kappa^{0*} - \kappa^{0*} \left.\frac{\partial B}{\partial P}\right|_T \right\}. \tag{3.115}$$

As, to a good approximation, κ^{0*}/κ is around unity, this expression simplifies to

$$\kappa \simeq \kappa^{0*} + \frac{nv^f}{V} \left\{ \frac{v^f}{kT} + \kappa^f - \kappa^{0*} - \kappa^{0*} \left.\frac{\partial B}{\partial P}\right|_T \right\}. \tag{3.116}$$

which will be needed for the analysis of X-ray experiments under constant volume.

In order to evaluate the relative importance of the first and the fourth term in the brackets we compare the quantity $v^f/\kappa^{0*}kT$ to $(\partial B/\partial P)_T$. In the approximation $\kappa^{0*} \simeq \kappa \equiv 1/B$ we obtain for the first term in the case of Kr at 115 K the value $Bv^f/kT = 35.4$, by inserting $v \simeq \Omega$ and $B = 11.32$ kbar [14]. The experimental value for $\partial B/\partial P$ is about 7. One sees that the last term *cannot* be disregarded against the first one. In microscopic calculations one might need the pressure variation of the bulk modulus B^{0*} of the ideal isochoric lattice. Its connection to dB/dP of the corresponding real crystal is found by writing eq. (3.111) in the form of $B^{0*} = B(\partial P^{0*}/\partial P)_T$ and differentiating it with respect to pressure. One obtains

$$\left.\frac{\partial B^{0*}}{\partial P^{0*}}\right|_T = \left.\frac{\partial B}{\partial P}\right|_T + B \left.\frac{\partial^2 P^{0*}}{\partial P^2}\right|_T \bigg/ \left.\frac{\partial P^{0*}}{\partial P}\right|_T.$$

The numerator and the denominator of the last term can be obtained from eq. (3.114); one finally obtains

$$\left.\frac{\partial B}{\partial P}\right|_T = \left.\frac{\partial B^{0*}}{\partial P^{0*}}\right|_T + \frac{n}{N} \frac{v^f}{\Omega} \left\{ \frac{Bv^f}{kT} \left(\frac{Bv^f}{kT} + 3\frac{B}{B^f} - 2\frac{dB}{dP} - 2 \right) \right.$$

$$\left. + \left(\frac{B}{B^f}\right)^2 \left(\frac{dB^f}{dP} + 1\right) - 2\frac{B}{B^f}\left(\frac{dB}{dP} + 1\right) + \frac{dB}{dP} + 1 + B\frac{d^2B}{dP^2} \right\},$$

which gives the approximate relation

$$\left.\frac{\partial B}{\partial P}\right|_T \simeq \left.\frac{\partial B^{0*}}{\partial P^{0*}}\right|_T + \frac{n}{N}\frac{v^{\mathrm{f}}}{\Omega}\left(\frac{Bv^{\mathrm{f}}}{kT}\right)^2.$$

By comparing to eq. (3.60) we see that $\partial B^{0*}/\partial P^{0*} \simeq \partial B^0/\partial P$.

Thermal expansion coefficient. We will define the isobaric thermal expansion coefficient β^{0*} of the isochoric perfect crystal (P^{0*}, V, T) by

$$\beta^{0*} \equiv \frac{1}{V}\left.\frac{\partial V}{\partial T}\right|_{P^{0*}}. \tag{3.117}$$

Here we find its connection to the expansion coefficient β of the real crystal. By differentiating eq. (3.100) with respect to temperature we get:

$$\left.\frac{\partial P}{\partial T}\right|_V = \left.\frac{\partial P^{0*}}{\partial T}\right|_V + \left.\frac{\partial(np^*)}{\partial T}\right|_V.$$

The left side of this equation has been calculated in eq. (2.27). The first term on the right gives

$$\left.\frac{\partial P^{0*}}{\partial T}\right|_V = -\left.\frac{\partial P^{0*}}{\partial V}\right|_T \left.\frac{\partial V}{\partial T}\right|_{P^{0*}}$$

and therefore, according to the definitions of κ^{0*} ($\equiv 1/B^{0*}$) and β^{0*} it has the value $B^{0*}\beta^{0*}$, so that we obtain

$$\beta B = \beta^{0*}B^{0*} + p^*\left.\frac{\partial n}{\partial T}\right|_V + n\left.\frac{\partial p^*}{\partial T}\right|_V.$$

By introducing eqs. (3.95) and (3.113) and by differentiating eq. (3.113) with respect to temperature one gets

$$\beta B = \beta^{0*}B^{0*} + \frac{n}{V}\left\{v^{\mathrm{f}}\left.\frac{\partial B}{\partial T}\right|_V + B\left.\frac{\partial v^{\mathrm{f}}}{\partial T}\right|_V\right\} + \frac{nu^*}{kT^2}\frac{v^{\mathrm{f}}}{V}B.$$

The partial derivative $(\partial v^{\mathrm{f}}/\partial T)_V$ can be written as:

$$\left.\frac{\partial v^{\mathrm{f}}}{\partial T}\right|_V = \left.\frac{\partial v^{\mathrm{f}}}{\partial T}\right|_P + \left.\frac{\partial v^{\mathrm{f}}}{\partial P}\right|_T \left.\frac{\partial P}{\partial T}\right|_V.$$

By introducing eqs. (3.43), (3.53) and (2.27) we finally get

$$\beta^{0*}B^{0*} - \beta B = \frac{nv^{\mathrm{f}}}{V}\left\{\left(\frac{\kappa^{\mathrm{f}}}{\kappa} - \frac{\beta^{\mathrm{f}}}{\beta}\right)\beta B - \left.\frac{\partial B}{\partial T}\right|_V - B\frac{u^*}{kT^2}\right\}. \tag{3.118}$$

A connection between β^{0*} and β results from the introduction of the value of B^{0*} from eq. (3.115).

Basic differential equations for the isochoric perfect crystal. The quantity F^{0*} depends on T and V therefore we can write

$$dF^{0*} = \left.\frac{\partial F^{0*}}{\partial T}\right|_V dT + \left.\frac{\partial F^{0*}}{\partial V}\right|_T dV$$

and by introducing the definitions $(\partial F^{0*}/\partial T)_V = -S^{0*}$ (see eq. 3.85) and $(\partial F^{0*}/\partial V)_T = -P^{0*}$ [see eq. (3.98)] we obtain:

$$dF^{0*} = -S^{0*} dT - P^{0*} dV. \tag{3.119}$$

The corresponding Maxwell equation is

$$\left.\frac{\partial S^{0*}}{\partial V}\right|_T = \left.\frac{\partial P^{0*}}{\partial T}\right|_V. \tag{3.120}$$

The right-hand side of this equation can be written as

$$\left.\frac{\partial P^{0*}}{\partial T}\right|_V = -\left.\frac{\partial P^{0*}}{\partial V}\right|_T \left.\frac{\partial V}{\partial T}\right|_{P^{0*}} = \beta^{0*} \frac{1}{\kappa^{0*}}. \tag{3.121}$$

Other equations similar to (3.119) can be derived by introducing U^{0*}, H^{0*}, etc.

$$dU^{0*} = T dS^{0*} - P^{0*} dV, \qquad \left.\frac{\partial T}{\partial V}\right|_{S^{0*}} = -\left.\frac{\partial P^{0*}}{\partial S^{0*}}\right|_V,$$

$$dH^{0*} = T dS^{0*} + V dP^{0*}, \qquad \left.\frac{\partial T}{\partial P^{0*}}\right|_{S^{0*}} = \left.\frac{\partial V}{\partial S^{0*}}\right|_{P^{0*}},$$

$$dG^{0*} = -S^{0*} dT + V dP^{0*}, \qquad \left.\frac{\partial S^{0*}}{\partial P^{0*}}\right|_T = -\left.\frac{\partial V}{\partial T}\right|_{P^{0*}}.$$

It is remarkable that these equations as well as those for the isobaric perfect crystal have the same form as those in ch. 2, which refer to the real crystal.

The following differential equation will be needed for specific heat calculations. By differentiating eq. (3.104) with respect to T one obtains with the help of $dG^{0*} = -S^{0*} dT + V dP^{0*}$:

$$\left.\frac{\partial H^{0*}}{\partial T}\right|_{P^{0*}} = T \left.\frac{\partial S^{0*}}{\partial T}\right|_{P^{0*}}. \tag{3.122}$$

Differential equations of the vacancy parameters. As f^* depends on T and V we can write

$$df^* = \left.\frac{\partial f^*}{\partial T}\right|_V dT + \left.\frac{\partial f^*}{\partial V}\right|_T dV,$$

or, because of eqs. (3.86) and (3.99),

$$df^* = -s^* \, dT - p^* \, dV.$$

Similarly the following equations are obtained:
$$du^* = -p^* \, dV + T \, ds^*,$$
$$dh^* = V \, dp^* + T \, ds^*,$$
$$dg^* = V \, dp^* - s^* \, dT.$$

3.4 Relation between isobaric and isochoric parameters

As can be seen from figs. 3.2 and 3.3 all the parameters developed in the former sections resulted from the comparison of a given state (P, V, T) to two different perfect states. Here the relation between the two families of parameters will be developed. (See also §A.11, note 3.)

By comparing eqs. (3.81) or (3.82) to eqs. (3.9) or (3.10) one sees that:
$$f^* = g^f. \tag{3.123}$$

This equation shows that the Gibbs formation energy under constant pressure and temperature is equal to the Helmholtz formation energy under constant volume and temperature.

Relation between s^* and s^f. Equation (3.86) can be written as:
$$s^* = -\left.\frac{\partial f^*}{\partial T}\right|_V = -\left.\frac{\partial f^*}{\partial T}\right|_P - \left.\frac{\partial f^*}{\partial P}\right|_T \left.\frac{\partial P}{\partial T}\right|_V. \tag{3.124}$$

Equations (3.123) and (3.124) give:
$$s^* = -\left.\frac{\partial g^f}{\partial T}\right|_P - \left.\frac{\partial g^f}{\partial P}\right|_T \left.\frac{\partial P}{\partial T}\right|_V,$$

or by using eqs. (3.16) and (3.31) we have:
$$s^* = s^f - v^f \left.\frac{\partial P}{\partial T}\right|_V. \tag{3.125}$$

Because of the thermodynamic equation (2.27), eq. (3.125) reads:
$$s^* = s^f - v^f \beta B. \tag{3.126}$$

This equation gives the difference between the thermal entropy s^f that results from thermal experiments under constant pressure (as simultaneous X-ray and bulk-dimension studies) and s^* that results from theoretical entropy calculations, which are usually carried out for constant volume (see also §A.8). One should stress here that both the entropies s and s^* differ from an entropy defined from comparison of the real crystal (P, V, T) to an ideal one $(P, V + |\Delta V|, T)$, where ΔV is the relaxation volume $(= v^f - \Omega$

for monovacancies, and $v^f - 2\Omega$ for Schottky defects in alkali halides).

The value of s^* can be directly found by introducing the appropriate values in the right-hand side of eq. (3.126). A quick evaluation can be done as follows: The thermodynamical Grüneisen constant γ is defined as: $\gamma = \beta B V/C_V$ or $\gamma = \beta B \Omega / \tilde{c}_V$ where C_V refers to the whole solid of volume V and $\tilde{c}_V$ is the mean constant-volume specific heat per atom. In the high temperature region $\tilde{c}_V$ is close to $3k$. Therefore eq. (3.126) can be equivalently written as:

$$s^* = s^f - \gamma \frac{v^f}{\Omega} \tilde{c}_V. \tag{3.127}$$

It is easily verified from this equation that usually s^* is *negative* and that its absolute value is comparable or smaller than s. The quantities γ, v^f/Ω and s^f are roughly known for each category of solids. For alkali halides (Schottky defects) we have: $\gamma \simeq 1.7$, $v^f/\Omega \simeq 3$ and $s^f \simeq (8 \text{ to } 10)k$ and hence $s^* \simeq (-7.3 \text{ to } -5.3)k$. For the usual fcc. metals (e.g. Au, Ag,...) $\gamma \simeq 2.5$, $v^f/\Omega \simeq 0.5$ and $s^f \simeq (2 \text{ to } 3)k$ and hence $s^* \simeq (-0.8 \text{ to } -1.8)k$.

The above results are only indicative and emphasize once more the influence of anharmonicity on defect properties. In the absence of any anharmonicity ($\beta = 0$) the two entropies s and s^* coincide in contrast to the real behaviour which indicates that they differ even in their sign.

Relation between u^* and h^f. By inserting the values of f^* and s^* given by eqs. (3.123) and (3.126) into eq. (3.90), we get

$$u^* = g^f + T(s^f - \beta B v^f),$$

or due to eq. (3.22) we have

$$u^* = h^f - T\beta B v^f. \tag{3.128}$$

This formula, when considering eq. (3.134a), can be written as: $u^* = h^f + T\beta V(\partial f^*/\partial V)_T$.

The physical meaning of eq. (3.128) is easily understood: if a vacancy is produced under the constraint $P = $ constant, the heat that has to be added is h^f. During such a process the volume of the body increases by v^f. If the volume is to be returned to its original value an external pressure equal to $T\beta B$ has to be applied; for this compression the "work" $T\beta B v^f$ has to be offered to the crystal. The quantity $u^* = (U - U^{0*})/n$ therefore is the *heat* needed for an isochoric production of a vacancy; it is smaller than h^f, which is the heat needed for an isobaric production of the vacancy.

The quantity $T\beta B v^f$ increases considerably with temperature. For some solids there will therefore be some temperature for which $h^f - T\beta B v^f$ might become equal to zero. At this temperature no heat is needed for the production of a vacancy under constant volume.

It seems probable that there is a connection between the temperature for which u^* becomes zero and the melting point T_M. For instance in rare gas solids ($v^f \simeq \Omega$) one finds that $T_M \simeq h^f/\beta B v^f$ [636].

Relation between u^f and u^*. This relation is easily found as follows: $u^f - u^* = (h^f - Pv^f) - u^* = (h^f - u^*) - Pv^f$, and hence:

$$u^f - u^* = (T\beta B - P)v^f. \tag{3.129}$$

Relation between p^* and v^f. From eqs. (3.99), (3.31) and (3.123) we get

$$p^* = -\left.\frac{\partial g^f}{\partial P}\right|_T \left.\frac{\partial P}{\partial V}\right|_T = -v^f \left.\frac{\partial P}{\partial V}\right|_T. \tag{3.130}$$

Inserting the definition (2.25) of the isothermal bulk modulus into the above equation one obtains

$$p^* = \frac{v^f}{V}B. \tag{3.131}$$

This equation is a connection between the pressure developed when a vacancy is formed by heating under constant volume and the fractional increase of volume when the vacancy is produced under constant pressure. This is similar in form to the connection of the bulk modulus to changes of the macroscopic quantities P and V: $\Delta P = B\Delta V/V$. The difference $P - P^{0*}$ is the pressure that arises from the isochoric production of vacancies, as depicted in fig. 3.3; its order of magnitude can be found by introducing eq. (3.131) and $V = N\Omega$ into eq. (3.100) and making the approximation $v^f \simeq \Omega$:

$$P - P^{0*} \simeq \frac{n}{N}B. \tag{3.132}$$

Relation between h^* and h^f. By inserting into eq. (3.102) the values of u^* and p^* given by eqs. (3.128) and (3.131) we get:

$$h^* = h^f + v^f B(1 - T\beta). \tag{3.133}$$

Since $T\beta < 1$, the enthalpy h^* exceeds h^f, when v^f is positive, by a considerable amount which is of the order of ΩB. Taking into account that ΩB is usually greater than the experimental values of h^f by one order of magnitude, we conclude that the enthalpies h^f and h^* resulting from different formation processes are quite different.

Relation between g^* and g^f. By inserting into eq. (3.105) the values of h^* and s^* given by eqs. (3.133) and (3.126) we get

$$g^* = g^f + Bv^f. \tag{3.134}$$

We see that g^* is considerably larger than g^f when $v^f > 0$. The difference is of the order of $B\Omega$.

Relation between v^f and $f*$. Equation (3.31) due to eqs. (2.25) and (3.123) can also be written as:

$$v^f = -\frac{V}{B} \left.\frac{\partial f^*}{\partial V}\right|_T \tag{3.134a}$$

Comparison between the two types of perfect crystals. The two perfect crystals on the left-hand side of figs. 3.2 and 3.3 are derived from the same real crystal (P, V, T) but by means of the two different processes; they have different Gibbs free energies, G^0 and G^{0*}. Their difference can be quantitatively determined by keeping in mind that the real crystals (right-hand sides) are the same, and therefore have the same Gibbs free energy G. By setting eq. (3.12) equal to Eq. (3.109) one obtains, by using eq. (3.131):

$$G^0 - G^{0*} = nv^f B. \tag{3.135}$$

In a similar way one obtains:

$$S^0 - S^{0*} = -nv^f \beta B, \tag{3.136}$$

$$H^0 - H^{0*} = nv^f B(1 - T\beta), \tag{3.137}$$

$$F^0 - F^{0*} = nPv^f, \tag{3.138}$$

$$U^0 - U^{0*} = n(P - T\beta B)v^f. \tag{3.139}$$

Relation between κ^0 and κ^{0*}. The relation between the compressibilities of the two perfect crystals can be found by combining eqs. (3.56) and (3.115); one obtains the approximate formula $\kappa^0 \simeq \kappa^{0*}[1 - (nv^f/V) \times (\partial B/\partial P)_T]$. In a similar fashion we can find the relation between β and β^{0*}.

3.5 Statistical approach to vacancy parameters

Vacancy parameters are usually studied from a thermodynamic standpoint by considering thermodynamic functions at states near equilibrium. This approach may be sufficient for formation parameters, but it has the disadvantage [for example in the case of eq. (3.4)] to conceal the exact role of the oscillations. Here the problem is approached from a statistical point of view [35], in which the latter effect is clearly revealed. Furthermore, this method constitutes a necessary background for the study of the migration of vacancies. In certain cases it will clarify the meaning of quantities calculated theoretically from first principles and will evaluate the errors arising by disregarding oscillational quantities. In the thermodynamic approach the two families of formation parameters differ depending on the choice of the perfect crystal; in the statistical approach they will be found to depend also on the behaviour of the *oscillators* in the solid in respect to temperature and pressure. According to their behaviour two categories of solids will be distinctly differentiated: harmonic and quasi-harmonic solids. Their properties will be obtained from the study of harmonic oscillators by summing over all vibrational eigenfrequencies of the lattice in equilibrium. An extension to the study of "non-harmonic oscillations" is not yet available.

In quantum mechanics the energy levels ε of a free harmonic oscillator are given by $\varepsilon = (l + \tfrac{1}{2})\hbar\omega$, where $\omega^2 = D/m$. It is connected to its classical amplitude x_0 of the state by:

$$\varepsilon = \tfrac{1}{2}Dx_0^2 = \tfrac{1}{2}m\omega^2 x_0^2. \tag{3.140}$$

If the solid consists of a system of oscillators of frequency ω at equilibrium, the mean energy $\langle \varepsilon_\omega \rangle$ can be found by taking an average over the energy levels:

$$\langle \varepsilon_\omega \rangle = \frac{\sum_l (l + \tfrac{1}{2})\hbar\omega \exp(-l\hbar\omega/kT)}{\sum_l \exp(-l\hbar\omega/kT)} = \tfrac{1}{2}\hbar\omega \coth(\hbar\omega/2kT). \tag{3.141}$$

The suffix ω reminds us that all the oscillators vibrate with the same frequency ω. At high temperatures (i.e. $T \gg \hbar\omega/k$) the mean energy tends towards kT, as expected. The specific heat c_ω and the entropy s_ω per oscillator are found by applying the usual formulae:

$$c_\omega = \frac{\partial \langle \varepsilon_\omega \rangle}{\partial T} = k \frac{(\hbar\omega/2kT)^2}{\sinh^2(\hbar\omega/2kT)}, \tag{3.142}$$

$$s_\omega = \int_0^T \frac{c_\omega}{T}\, dT = -k\left\{ \ln(1 - e^{-\hbar\omega/kT}) + \frac{\hbar\omega}{kT}(1 - e^{\hbar\omega/kT})^{-1} \right\}. \tag{3.143}$$

This latter entropy is of purely thermal origin because, for the time being, we are not considering configurational order. At high temperatures ($kT \gg \hbar\omega$) the specific heat tends towards k, whereas entropy tends towards

$$s_\omega = k\left[1 - \ln(\hbar\omega/kT)\right]. \tag{3.144}$$

The free energy

$$f_\omega = \langle \varepsilon_\omega \rangle - Ts_\omega \tag{3.145}$$

of an oscillator is finally found to be [in analogy to eq. (2.4)]:

$$f_\omega = \tfrac{1}{2}\hbar\omega + kT \ln(1 - e^{-\hbar\omega/kT}). \tag{3.146}$$

The vacancy parameters are obtained from the comparison of a perfect

3.5.1 Perfect lattice

In the *Einstein approximation* the solid consists of $3N$ independent harmonic oscillators with the same frequency ω_E. Therefore the corresponding (molar) thermodynamical functions U, C_V, S and F can be immediately written down by multiplying the formulae (3.141) to (3.145) by $3N$ and inserting $\omega = \omega_E$. The enthalpy H and the Gibbs energy G are simply constructed from the known relations:

$$H = 3N\langle\varepsilon_\omega\rangle + PV, \tag{3.147}$$

$$G = 3N\langle\varepsilon_\omega\rangle + PV - 3NTs_\omega. \tag{3.148}$$

In the present article the expression for the entropy S will be needed only at very high temperatures, i.e. $\ln(kT/\hbar\omega) \gg 1$. Equation (3.144) leads to:

$$S = 3Nk \ln(kT/\hbar\omega_E). \tag{3.149}$$

In the *Debye approximation* we consider $3N$ harmonic oscillators with frequencies ω varying from zero to ω_D. The thermodynamic functions can be found by integrating over the frequency spectrum after considering the probability of each frequency. For the specific heat we have (harmonic approximation):

$$C_P = C_V = 3N \int_0^{\omega_D} c_\omega \frac{4\pi\omega^2}{\frac{4}{3}\pi\omega_D^3} d\omega. \tag{3.150}$$

When one now inserts the corresponding value of c_ω from eq. (3.142), the usual Debye formula for the specific heat is immediately obtained. Then the entropy can be found from the expression $S = \int_0^T C \, dT/T$. It is clear that this entropy originates only from thermal motion.

In a *lattice with a general frequency spectrum* the (thermal) internal energy and the (thermal) entropy are obtained by summing eqs. (3.141) and (3.143) over all modes. All quantities obtained until now are contributions to the thermodynamic functions arising from oscillations only. One has further to consider the non-vibrational potential energy $\mathscr{V}$ and the various contributions from the electronic gas, etc. As explained in ch. 2, after obtaining an expression for the Gibbs energy G, the other thermodynamic functions can easily be derived. We give therefore below an expression for G of a perfect crystal (after neglecting the electronic contribution) by suitably extending eq (3.146) from the Helmholtz to the Gibbs function and summing over all oscillators i:

$$G = \mathscr{V} + \tfrac{1}{2}\sum_i \hbar\omega_i + kT\sum_i \ln(1 - e^{-\hbar\omega_i/kT}) + PV. \tag{3.151}$$

The above expression still holds even if the frequencies ω_i depend on T and V, as long as the oscillations remain harmonic at each state. It should be mentioned that in the usual quasi-harmonic approximation the frequencies and the potential energy $\mathscr{V}$ are taken to be explicit functions only of V, while any explicit temperature dependencies are presumed to be zero.

3.5.2 Imperfect lattice

In order to obtain the vacancy parameters, the thermodynamic functions of a perfect lattice have to be compared with those of an imperfect lattice that has only one vacancy. We define a vector $\boldsymbol{r}^0$ that specifies the mean configuration of a perfect crystal and a vector $\boldsymbol{r}^1$ for the corresponding configuration when a vacancy is produced under conditions of the same pressure P and temperature T. According to its definition [eq. (3.4)] the quantity g^f can be found by subtracting the Gibbs energies G^0 and G^1 for the perfect and the imperfect crystal (under exclusion of the configurational part TS_c).

$$G^0 = \mathscr{V}(\boldsymbol{r}^0) + \tfrac{1}{2}\sum_i \hbar\omega_i + kT\sum_i \ln\{1 - \exp(-\hbar\omega_i/kT)\} + PV(\boldsymbol{r}^0),$$

$$G^1 = \mathscr{V}(\boldsymbol{r}^1) + \tfrac{1}{2}\sum_i \hbar\omega_i' + kT\sum_i \ln\{1 - \exp(-\hbar\omega_i'/kT)\} + PV(\boldsymbol{r}^1),$$

where $\mathscr{V}(\boldsymbol{r}^0)$ and $\mathscr{V}(\boldsymbol{r}^1)$ are the non-vibrational potential energies at the equilibrium positions of the oscillators. For high temperatures one obtains

$$g^f = G^1 - G^0 = \mathscr{V}(\boldsymbol{r}^1) - \mathscr{V}(\boldsymbol{r}^0) + kT\sum_i \ln(\omega_i'/\omega_i)$$

$$+ P\{V(\boldsymbol{r}^1) - V(\boldsymbol{r}^0)\}. \tag{3.152}$$

The frequencies ω_i' and ω_i refer to the different volumes $V(\boldsymbol{r}^1)$ and $V(\boldsymbol{r}^0)$ respectively.

Remaining always within the frame of harmonic oscillations we have to discriminate between the various types of solids depending on the influence of the volume and the temperature on the frequencies and the non-vibrational potential energy. In the *harmonic solid* the frequencies and the configurational positions $\boldsymbol{r}^0$ and $\boldsymbol{r}^1$ do not vary with temperature. The temperature derivative of eq. (3.152) gives

$$s^f = -\left.\frac{\partial g^f}{\partial T}\right|_P = -k\sum_i \ln\frac{\omega_i'}{\omega_i}. \tag{3.153}$$

This equation indicates that the formation entropy depends on the frequency shift of only those bonds which are cut around the vacancy; these shifts do not contain any contribution from the volume increase from $V(r^0)$ to $V(r^1)$ because in a harmonic solid the frequencies do not depend on the volume.

In the *quasi-harmonic solid* the frequencies and the potential energies $\mathscr{V}(r^0)$, $\mathscr{V}(r^1)$ depend by definition on volume but not explicitly on temperature and therefore a differentiation of eq. (3.152) with respect to temperature gives

$$s^f = -\left.\frac{\partial g^f}{\partial T}\right|_P = -k\sum_i \ln\frac{\omega_i'[V(r^1)]}{\omega_i[V(r^0)]}. \tag{3.154}$$

Although eqs. (3.153) and (3.154) seem to be the same, they give different results because in the latter equation the frequency shifts from $\omega_i[V(r^0)]$ to $\omega_i'[V(r^1)]$ due also to a volume variation from $V(r^0)$ to $V(r^1)$. Furthermore, this equation does not preclude a temperature dependence of s^f because at constant pressure in contrast to eq. (3.153) the volumes $V(r^0)$ and $V(r^1)$ vary with temperature.

Another important point should be noticed. Equation (3.154) resulted from the temperature derivative of the term $kT\sum_i \ln(\omega_i'/\omega_i)$ of eq. (3.152) and therefore it expresses only the phonon contribution to the formation entropy. However, the volumes $V(r^1)$ and $V(r^0)$ change with temperature so that the potential energies $\mathscr{V}(r^1)$ and $\mathscr{V}(r^0)$ vary with temperature. This alternatively means that even at "zero" pressure the above variation of the potential energies gives another contribution to the temperature derivative besides the one expressed in eq. (3.154). We therefore conclude that the usual statement in the literature that the value of s^f is described only by the variation of the vibrational spectrum due to the creation of a vacancy is not absolutely correct for a real crystal.

The isochoric Helmholtz free energy f^* [see eq. (3.79)] can be brought into a form analogous to that of eq. (3.152) for the Gibbs free energy. The present approach studies the variation of the (nonvibrational) potential energy and the frequency spectrum under constant volume. If the potential energy of a perfect crystal of volume V is labelled with $\mathscr{V}(V)$ then upon the isochoric production of one vacancy it will change to $\mathscr{V}^1(V)$. By writing the Helmholtz energies $F^0(V)$ and $F^1(V)$ for the perfect and the imperfect crystal (excluding in the latter the configurational part), one gets the quantity f^* in the high temperature approximation:

$$f^* = F^1(V) - F^0(V) = \mathscr{V}^1(V) - \mathscr{V}^0(V) + kT\sum_i \ln[\omega_i'(V)/\omega_i(V)] \tag{3.155}$$

where $\omega_i'(V)$ and $\omega_i(V)$ are the mode frequencies of the imperfect and perfect crystal for an unchanged volume. Obviously, according to eq. (3.123), the results of eq. (3.155) has to be equal to that given by eq. (3.152). We are now in a position to give a statistical expression for the entropy s^* of an isochoric production of a vacancy. It is obtained by differentiating eq. (3.155) with respect to temperature (V = constant). For the case of harmonic or quasi-harmonic approximations we get

$$s^* = - \left. \frac{\partial f^*}{\partial T} \right|_V = -k \sum_i \ln \frac{\omega_i'(V)}{\omega_i(V)}. \qquad (3.156)$$

In the *extended case of a quasi-harmonic solid* the vibrations are harmonic but contrary to the usual, restricted, Q.A. the frequencies and the potential energies depend also explicitly on T; in such cases they can change even if the volume remains constant, so that additional terms have to be included in eqs. (3.154) and (3.156). Therefore these equations are of approximative nature as long as the frequencies are explicit functions of volume. In the most general case of the *anharmonic solid*, beyond the generalisation just mentioned, the oscillations are not assumed to be harmonic. In this case analogous expressions to those mentioned in this chapter cannot be given explicitely.

3.6 Types of experiments leading to the determination of formation parameters

In the former paragraphs a number of vacancy parameters were defined by comparing a real crystal to a perfect one. In the present section we will discuss the main types of experiments which give information on these parameters. There are three simple types of experiments in which the crystal is constantly in equilibrium. In each type one of the experimental variables is kept constant during the measurement, the second variable is changed while the third one automatically adjusts itself.

The three cases are given in the following table 3.1. In the experiments of type A or B the action is a heating from T_1 to T_2 either under constant

Table 3.1
Simple types of experiments.

	type A	type B	type C
Variable held constant	P	V	T
Action	ΔT	ΔT	ΔP
Result	ΔV	ΔP	ΔV

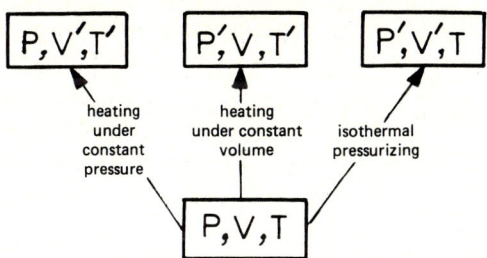

Fig. 3.4. The three main types of experiments to compare a real crystal with a perfect crystal.

pressure or constant volume. In type C the crystal is isothermally pressurised from pressure P to P'. In all three types of experiments the crystal of state (P, V, T) is subjected to some kind of action which brings it to a different state (fig. 3.4). The comparison of the three final states shows that they are different and therefore will have different numbers of vacancies.

Specific heat experiments are also an important method for determining vacancy parameters. Due to the complexity of the analysis of such experiments, their discussion is deferred to a separate section.

3.6.1 Heating under constant pressure

From a thermodynamical point of view there is no restriction as far as the temperature dependence of g^f is concerned. But once the temperature dependence of g^f is known, the parameters s^f and h^f are uniquely determined. For example, the form of the thermal entropy s^f of a vacancy as a function of temperature follows from

$$s^f = - \left.\frac{\partial g^f}{\partial T}\right|_P. \tag{3.157}$$

The entropy h^f of a vacancy can be determined from the usual relation

$$h^f = g^f + Ts^f = g^f - T\left.\frac{\partial g^f}{\partial T}\right|_P. \tag{3.158}$$

According to eq. (3.25) s^f and h^f obey the relation

$$T\left.\frac{\partial s^f}{\partial T}\right|_P = \left.\frac{\partial h^f}{\partial T}\right|_P. \tag{3.159}$$

We now examine the possible forms of the temperature dependence of these parameters: although in general there is no restriction as to the temperature dependence of g^f, a restriction emerges when one is concerned with ther-

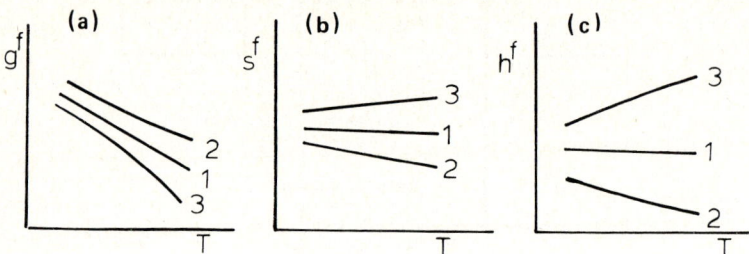

Fig. 3.5. Temperature dependence of g^f, s^f and h^f (P = constant).

mally activated processes: g^f/T has to decrease with temperature, because all such processes become stronger with temperature. By differentiating g^f/T in respect to T, one easily finds that this is equivalent to the statement $h^f > 0$. If h^f were negative, then according to eq. (3.24) the crystal would spontaneously desintegrate even at low temperature. The Gibbs formation free energy g_0^f at zero temperature also has to be positive because $g_0^f = h_0^f$. In fig. (3.5a) three forms of the curves $g^f = f(T)$ are discussed for cases of positive entropy of formation. For all three of them g falls with increasing temperature. The corresponding forms of s (fig. 3.5b) and h (fig. 3.5c) are then derived directly from eqs. (3.157) and (3.158).

(i) The curve g^f decreases linearly with temperature (curve 1). In this case the formation entropy and enthalpy are constant.

(ii) If the curve g^f versus T decreases with increasing temperature but is curved upwards (curve 2), the entropy will decrease with temperature, and so will h^f because of eq. (3.159).

(iii) If the function g^f decreases faster than linearly (curve 3) both entropy and enthalpy will *increase* with T. The question whether in fig. 3.5b, c the curves 2 and 3 bend upwards or downwards depends on the exact form of the g^f-curve. The form of the curves depicted is therefore only indicative. In fig. 3.6 ln n/N is plotted versus $1/T$. As proved by eq. (3.27) its slope always gives the enthalpy at each point. Therefore curve 3 has a form that corresponds to an enthalpy which increases with temperature. Certain published calculations of h^f based on first principles lead to a decreasing enthalpy. In some of them, however, the effect of atomic vibrations was not included (see §A.8 and §A.11, note 4).

3.6.2 Heating under constant volume

From thermodynamics, far from absolute zero, there is no restriction as to the temperature dependence of f^*. When the temperature dependence of

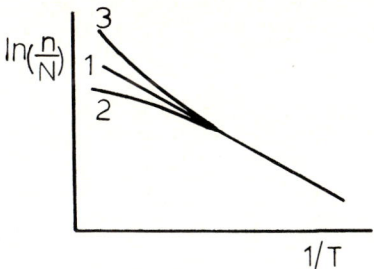

Fig. 3.6. Plots of the relative concentrations.

f^* is known, s^* can be determined from eq. (3.86) and u^* from the relation (3.90). They have to be compatible by obeying eq. (3.93). Similarly statements can be obtained for the form of the temperature dependence of h^*, g^*, $\ln n/N$, etc. It is clear that upon heating under constant volume, the pressure increases. This effect will be considered quantitatively in the discussion of the next paragraph, where also the parameters v^f, β^f, and κ^f will be treated.

3.6.3 Isothermal pressurising

Consider a crystal in equilibrium at a temperature T that has a volume V_1, when the external pressure is P_1. By isothermal compression these functions will change to P_2 and V_2, while T remains unchanged. We will calculate the effect that this change will have on the vacancy parameters.

Pressure dependence of the formation parameters. By applying a pressure P_2 the formation Gibbs energy will change from g_1 to g_2. By developing g_2 in a Taylor series around the value g_1 we have for $T = $ constant:

$$g_2^f = g_1^f + \left.\frac{\partial g^f}{\partial P}\right|_{P=P_1} \delta p + \frac{1}{2}\left.\frac{\partial^2 g^f}{\partial P^2}\right|_{P=P_1} \delta^2 P + \cdots .$$

The first order derivative $(\partial g^f/\partial p)_T$ was defined in eq. (3.31) as the formation volume v^f. The differentiation of eq. (3.22) with respect to pressure gives

$$v^f = \left.\frac{\partial h^f}{\partial P}\right|_T - T\left.\frac{\partial s^f}{\partial P}\right|_T . \tag{3.160}$$

Evidently the formation volume consists of two terms:

$$v^f = v_h^f + v_s^f , \tag{3.161}$$

where

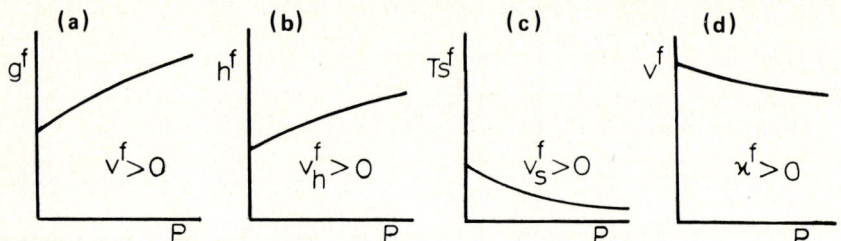

Fig. 3.7. Pressure dependence of the formation parameters.

$$v_h^f \equiv (\partial h^f/\partial P)_T \tag{3.162}$$

and

$$v_s^f \equiv -(T\partial s^f/\partial P)_T. \tag{3.163}$$

The second and third order derivatives can be expressed with the compressibility of the vacancy by

$$\left.\frac{\partial^2 g^f}{\partial P^2}\right|_T = \left.\frac{\partial v^f}{\partial P}\right|_T = -\kappa^f v^f \tag{3.164}$$

and

$$\left.\frac{\partial^3 g^f}{\partial P^3}\right|_T = \left.\frac{\partial(-\kappa^f v^f)}{\partial P}\right|_T = (\kappa^f)^2 v^f - v^f \left.\frac{\partial \kappa^f}{\partial P}\right|_T. \tag{3.165}$$

In fig. 3.7a the pressure variation of g^f has been plotted. As the formation volume has been experimentally found to be either positive or negative (for example in δ-Ce), the quantity g^f can increase or decrease with pressure, depending on the material. According to eq. (3.162) the formation enthalpy for positive formation volumes increases (fig. 3.7b) or decreases with P according to the sign of v_h^f. Correspondingly Ts^f (fig. 3.7c) varies on compression in a manner that depends on the sign of v_s^f. The compressibility κ^f of the formation volume was defined by eq. (3.53); when it is known, one can plot a curve for v^f as a function of pressure (fig. 3.7d).

When the formation volume v_1^f is known at a pressure P_1 its value at pressure P_2 can be calculated (in cases were κ^f is known as function of P) by integrating eq. (3.53):

$$v_2^f = v_1^f \exp\left(\int_{P_1}^{P_2} -\kappa^f \, dP\right). \tag{3.166}$$

Calculation of v_h^f and v_s^f. In order to calculate the connection between v^f and its two terms, we consider the expression

$$\left.\frac{\partial s^f}{\partial P}\right|_T = \frac{\partial(-\partial g^f/\partial T)}{\partial P} = -\frac{\partial(\partial g^f/\partial P)}{\partial T} = -\left.\frac{\partial v^f}{\partial T}\right|_P.$$

This expression is similar in form to the Maxwell equation $(\partial S/\partial P)_T = -(\partial V/\partial T)_P$. Equation (3.163) becomes

$$v_s^f = T \left.\frac{\partial v^f}{\partial T}\right|_P \tag{3.167}$$

and therefore

$$v_h^f = v^f - T \left.\frac{\partial v^f}{\partial T}\right|_P. \tag{3.168}$$

We see that it is sufficient to know the temperature dependence of v^f in order to determine v_s^f and v_h^f separately. They can also be connected to v^f with the help of the expansion coefficient β^f of the vacancies by considering eq. (3.43):

$$v_s^f = T v^f \beta^f \tag{3.169}$$

and

$$v_h^f = v^f (1 - \beta^f T). \tag{3.170}$$

It is easy to evaluate in a rough approximation the importance of each of them. At higher temperature (for example at 500 K, where high pressure experiments are usually carried out), by accepting roughly $\beta^f = 10^{-3}$ K^{-1}, v_s^f is about 50% of v^f. Therefore v_s^f must indispensably be considered in any high-temperature theoretical calculation. The relations (3.169) and (3.170) reveal that at absolute zero v_s^f vanishes and v_h^f coincides with v^f. At higher temperatures $v_h^f < v^f$ when β^f is positive.

The dependence of the volumes v_h^f and v_s^f on the pressure can be described with the help of isothermal compressibilities, defined as follows:

$$\kappa_h^f \equiv -\frac{1}{v_h^f} \left.\frac{\partial v_h^f}{\partial P}\right|_T, \tag{3.171}$$

$$\kappa_s^f \equiv -\frac{1}{v_s^f} \left.\frac{\partial v_s^f}{\partial P}\right|_T. \tag{3.172}$$

By integrating these equations we get:

$$v_h^f(P) = v_h^f(0) \exp\left(\int_0^P -\kappa_h^f \, dP\right), \tag{3.173}$$

$$v_s^f(P) = v_s^f(0) \exp\left(\int_0^P -\kappa_s^f \, dP\right). \tag{3.174}$$

Equations (3.173) and (3.174) permit the evaluation of v_h^f and v_s^f at any

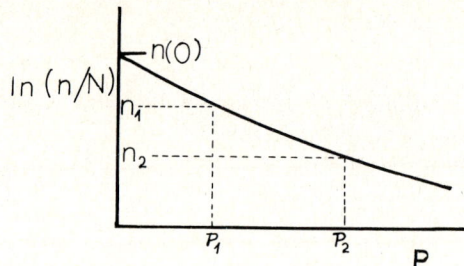

Fig. 3.8. Influence of pressure on the concentration of vacancies ($\kappa^f > 0$).

pressure P whenever their values $v_h^f(0)$ and $v_s^f(0)$ at zero pressure are known with κ_h^f and κ_s^f known functions of P.

Pressure dependence of n. From eq. (3.9) one easily obtains by differentiation in respect to pressure

$$-kT \left.\frac{\partial \ln(n/N)}{\partial P}\right|_T = v^f. \tag{3.175}$$

This means that the isothermal plot $-kT \ln(n/N)$ versus P is curved because v^f depends on pressure: its slope is equal to the formation volume at every point. A typical plot of $\ln n/N$ versus P is given in fig. 3.8; it is drawn for a positive compressibility κ^f of the formation volume.

In experiments of isothermal compression it is interesting to compare the concentrations of vacancies of a crystal that is submitted to two pressures P_1 and P_2 and that therefore has different volumes V_1 and V_2. By integrating eq. (3.175) between the boundaries P_1 and P_2 one obtains

$$\frac{n_1}{n_2} = \exp\left(-\frac{\int_{P_1}^{P_2} v^f \, dP}{kT}\right). \tag{3.176}$$

3.6.4 Comparison of the two types of heating experiments

Consider a crystal in an initial state (P, V, T) being heated up to the temperature T' in two ways, according to experiment (a) and (b) (fig. 3.9). It is often desirable to compare the numbers n_a and n_b of vacancies in the two final states. The final states (P, V', T') and (P', V, T') are different and therefore $n_a \neq n_b$. The numbers of vacancies can be calculated in each instance by inserting into eq. (3.9) the correct value of the parameter g^f for each final state. At this point one should stress the possibility of committing

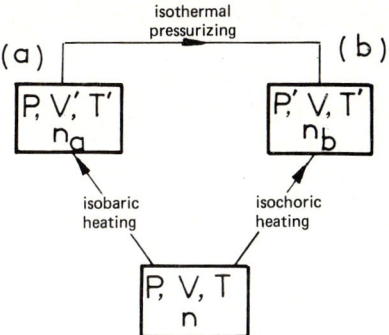

Fig. 3.9. Comparison of two heating experiments. All three processes are reversible.

an error; one is misled into applying eqs. (3.9) and (3.81) $n = N\exp(-g^f/kT')$ and $n = N\exp(-f^*/kT')$ and then using eq. (3.123): $g_a^f = f_b^*$. This is not allowed, because the equality between g and f^* holds only when the two parameters are referring to the same state, whereas the two heated states of fig. 3.9 are different.

The explicit relation between n_a and n_b can be found in a variety of ways. The simplest one is to first calculate P', which is the final pressure at the end of experiment (b) (after heating under constant volume from T to T'). It can be considered as being the sum of the initial pressure P and the thermal pressure produced by heating under constant volume from T to T'. This question has been dealt with in the previous section; if one integrates eq. (2.27) from T to T', the increase of pressure is given by

$$P' - P = \int_T^{T'} \beta B \, dT,$$

where the coefficients β and B refer to states, each of which has a different P and T. Now that the pressure P' is known, we consider an isothermal compression of the isobarically heated crystal from P to P', which changes the state (P, V', T') to the state (P', V, T') (horizontal line); the latter is exactly the state at the end of experiment (b), because the two variables P' and T' are the same. The ratio n_a/n_b can therefore be calculated from the formula (3.176) of isothermal pressurising:

$$n_a/n_b = \exp\left(-\int_P^{P'} v^f \, dP/kT'\right).$$

3.6.5 Isobaric heating experiments under various pressures

In the experiments described in the former section, one of the variables P, V, T was held constant. Here the same experiments are discussed, now with a new value of the variable that was kept constant. Figure 3.5 for example gave a plot of g^f as a function of T for an experiment in which the pressure was kept constant. Let us now consider the same experiment repeated at a different constant pressure; new values of the same parameter g^f will result. In the present section we show the interconnection of the parameters that describe such type of experiments.

Temperature dependence of the expansivities of the formation volume. By solving eq. (3.169) for β^f one obtains with the help of eq. (3.161)

$$\beta^f = \frac{1}{T} \frac{v_s^f}{v_h^f + v_s^f}.$$

As the ratio v_h^f/v_s^f changes with temperature, β^f can never be exactly proportional to $1/T$, except if $v_h^f = 0$. The temperature variation of the two terms v_s^f and v_h^f of the formation volume can be expressed with the help of expansion coefficients defined by

$$\beta_s^f \equiv \frac{1}{v_s^f} \left.\frac{\partial v_s^f}{\partial T}\right|_P \tag{3.177}$$

and

$$\beta_h^f \equiv \frac{1}{v_h^f} \left.\frac{\partial v_h^f}{\partial T}\right|_P. \tag{3.178}$$

The coefficient β_s^f can be found in terms of v^f by inserting the value of v_s^f from eq. (3.169) into eq. (3.177):

$$\beta_s^f = \left(T \left.\frac{\partial v^f}{\partial T}\right|_P\right)^{-1} \left(T \left.\frac{\partial^2 v^f}{\partial T^2}\right|_P + \left.\frac{\partial v^f}{\partial T}\right|_P\right). \tag{3.179}$$

This equation reveals the conditions under which β_s^f will be found to be equal to $1/T$; this clearly happens whenever the second temperature derivative of the formation volume is small or, more accurately, when

$$\frac{\partial^2 v^f}{\partial T^2} \ll \frac{1}{T} \frac{\partial v^f}{\partial T}.$$

An equality between β_s^f and $1/T$ has in effect been noticed. In a former publication [36] the authors stated that this law was predicted from the $cB\Omega$-model. The present calculation shows that it is a purely thermodynamic consideration. The law has been discussed (for Zn and Cd) by Gilder and Lazarus [37] (see §7.3) for the activation process.

Relation between β_s^f and β^f. By recalling the definition of β^f, eq. (3.179) can alternatively be written as

$$\beta_s^f = (T\beta^f v^f)^{-1} \left[T\beta^f (\partial v^f / \partial T)_P + Tv^f (\partial \beta^f / \partial T)_P + \beta^f v \right]$$

or

$$\beta_s^f = \beta^f + (\partial \ln \beta^f / \partial T)_P + 1/T. \qquad (3.180)$$

Equation (3.180) permits the direct evaluation of β_s^f whenever the function $\beta^f = f(T)$ (for $P = $ constant) is known.

Comparison of β_s^f to the bulk coefficient β. At low temperature experiments eq. (3.180) gives for β_s^f large values (provided that β^f and $\partial \beta^f / \partial T$ are positive) because $1/T$ lies between 10^{-3} and 10^{-1} K^{-1}; on the other hand β is in this region usually of the order of 10^{-6} K^{-1} and therefore β_s^f may exceed β by several orders of magnitude. This conclusion needs experimental confirmation.

In a later section we shall also show that the term β^f is positive for almost all materials and furthermore increases with temperature. Therefore eq. (3.180) indicates that the lower bound of β_s^f is equal to $1/T$. As in most materials the bulk expansivity β is much smaller than $1/T$ we conclude that for all temperatures β_s^f is always larger than β.

Relation between β_h^f and β^f. By differentiating eq. (3.161) with respect to temperature we get

$$\partial v^f / \partial T = \partial v_h^f / \partial T + \partial v_s^f / \partial T \quad \text{or} \quad \beta^f v^f = \beta_h^f v_h^f + \beta_s^f v_s^f$$

and therefore

$$\beta_h^f = \beta^f v^f / v_h^f - \beta_s^f v_s^f / v_h^f. \qquad (3.181)$$

With the help of eqs. (3.169) and (3.170), equation (3.181) becomes

$$\beta_h^f = \beta^f (1 - T\beta_s^f)(1 - T\beta^f)^{-1} \qquad (3.182)$$

or

$$\frac{\beta_h^f}{1 - T\beta_s^f} = \frac{\beta^f}{1 - T\beta^f}. \qquad (3.183)$$

Equation (3.183) explicitly connects the three expansion coefficients β^f, β_s^f and β_h^f. By inserting eq. (3.180) into eq. (3.182) we get an explicit relation between β_h^f and β^f:

$$\beta_h^f = -\frac{T\beta^f}{1 - T\beta^f} \left(\beta^f + \frac{\partial \ln \beta^f}{\partial T} \bigg|_P \right). \qquad (3.184)$$

In most cases β^f is positive and increases with temperature; in these cases whenever $\beta^f < 1/T$, eq. (3.184) demands that the expansion coefficient β_h^f is *negative* in contrast to β_s^f. In ref. [36] this result was found from the $cB\Omega$-model; here it is shown that it results directly from thermodynamic considerations. In order to obtain the temperature dependence of v_h^f we write this equation as

$$\frac{1}{v_h^f}\frac{\partial v_h^f}{\partial T}\bigg|_P = -\frac{T\beta^f v^f}{v^f(1-T\beta^f)}\left(\beta^f + \frac{\partial \ln \beta^f}{\partial T}\bigg|_P\right)$$

or, due to the fact that $v_h^f = v^f(1 - T\beta^f)$:

$$\frac{\partial v_h^f}{\partial T}\bigg|_P = -Tv^f\left[(\beta^f)^2 + \frac{\partial \beta^f}{\partial T}\bigg|_P\right]. \tag{3.185}$$

In view of the fact that, as mentioned, the quantities v^f and $\partial \beta^f/\partial T$ are usually positive, eq. (3.185) reveals that v_h^f decreases with temperature.

Calculation of v_{u*}^f and v_{s*}^f. In analogy to eq. (3.161) the formation volume can be alternatively considered as a sum of the following two contributions

$$v^f = v_{u*}^f + v_{s*}^f \tag{3.186}$$

where

$$v_{u*}^f \equiv \frac{\partial u^*}{\partial P}\bigg|_T, \quad \text{or } v_{u*}^f \equiv -\frac{V}{B}\frac{\partial u^*}{\partial V}\bigg|_T, \tag{3.187a, b}$$

$$v_{s*}^f \equiv -T\frac{\partial s^*}{\partial P}\bigg|_T, \quad \text{or } v_{s*}^f \equiv \frac{TV}{B}\frac{\partial s^*}{\partial V}\bigg|_T. \tag{3.188a, b}$$

By inserting eq. (3.128) into eq. (3.187a) and, respectively, eq. (3.126) into eq. (3.188a) we finally get:

$$v_{u*}^f = v^f - Tv^f\beta^f\left\{1 + \frac{\beta}{\beta^f}\left(\frac{\partial B}{\partial P}\bigg|_T - \delta - \frac{\kappa^f}{\kappa}\right)\right\} \tag{3.189}$$

$$v_{s*}^f = Tv^f\beta^f\left\{1 + \frac{\beta}{\beta^f}\left(\frac{\partial B}{\partial P}\bigg|_T - \delta - \frac{\kappa^f}{\kappa}\right)\right\} \tag{3.190}$$

where $\delta \equiv -(1/\beta B)(\partial B/\partial T)_P$.

4 | FORMATION OF OTHER POINT DEFECTS

The former sections referred only to the formation of vacancies. Solids, however, do not only form vacancies but other types of defects as well. The present discussion will restrict itself to the formation of Schottky and Frenkel defects in crystals A^+B^- with NaCl structure.

The formulae connecting the thermodynamic functions of the real crystal to the defect parameters will depend – as in the case of vacancies – on the choice of the perfect crystal.

4.1 Schottky defects

In the case of an A^+B^- solid a Schottky defect consists of one cation vacancy and one anion vacancy that do not lie in neighbouring sites. A quantity of *pure* material consisting of N "molecules" contains N_{cat} cations and N_{an} anions for which we have $N_{cat} = N_{an} = N$. The number n_{cat} of cation vacancies is distributed on these $N + n_{cat}$ sites, and similarly the number n_{an} of anion vacancies on $N + n_{an}$ sites.

In order to find the configurational entropy S_c of such a crystal we have to take the sum of the configurational entropies of the cation and anion vacancies. We denote with W_{cat} and W_{an} the number of ways of arranging n_{cat} and n_{an} vacancies. These are

$$W_{cat} = \frac{(N + n_{cat})!}{N!\, n_{cat}!} \quad \text{and} \quad W_{an} = \frac{(N + n_{an})!}{N!\, n_{an}!}$$

and hence the corresponding configurational entropy S_c is

$$S_c = k \ln \frac{(N + n_{cat})!}{N!\, n_{cat}!} + k \ln \frac{(N + n_{an})!}{N!\, n_{an}!}.$$

By applying the same method as in §3.2 and in the approximation $n_{cat} \ll N$ and $n_{an} \ll N$ we get

$$\frac{S_c}{k} = (n_{cat} + n_{an}) - n_{cat} \ln \frac{n_{cat}}{N} - n_{an} \ln \frac{n_{an}}{N}. \tag{4.1}$$

If one neglects surface boundaries and dislocations the electroneutrality condition of a pure crystal of type A^+B^- requires the formation of an equal number of vacancies ($n_{cat} = n_{an} = n$) so that eq. (4.1) becomes

$$\frac{S_c}{k} = 2n - n \ln\left(\frac{n}{N}\frac{n}{N}\right). \tag{4.2}$$

Isobaric perfect crystal. We will now determine the contribution of the Schottky defects to the various thermodynamic functions when the perfect isobaric crystal is compared with the real one. Considerations analogous to those that led to eq. (3.4) for single vacancies give:

$$G = G^0 + ng^f - TS_c, \tag{4.3}$$

where g^f is the thermal Gibbs energy for the formation of an anion vacancy plus a cation vacancy. The equilibrium value of n is obtained in terms of g^f by inserting eq. (4.1) into eq. (4.3) and applying the condition $(\partial G/\partial n)_{P,T} = 0$. It gives:

$$\left(\frac{n}{N}\frac{n}{N}\right) = \exp\left(-\frac{g^f}{kT}\right), \quad \text{or} \quad n = N\exp\left(-\frac{g^f}{2kT}\right) \tag{4.4}$$

By inserting this value of n into eq. (4.2) we obtain the configurational entropy in the equilibrium state:

$$S_c = 2nk + ng^f/T. \tag{4.5}$$

From eqs. (4.3) and (4.5) we get the Gibbs energy of the real crystal:

$$G = G^0 - 2nkT. \tag{4.6}$$

By differentiating eq. (4.6) with respect to temperature and using eq. (4.4) we get an expression for the total entropy of the real solid:

$$S = S^0 + 2nk + ns^f + ng^f/T, \tag{4.7}$$

where s^f is again defined as:

$$s^f \equiv -\left.\frac{\partial g^f}{\partial T}\right|_P. \tag{4.8}$$

By combining eqs. (4.5) and (4.7) we find

$$S = S^0 + ns^f + S_c, \tag{4.9}$$

which reveals that s^f is only the "thermal" formation entropy per Schottky defect. The definition of the formation enthalpy and the formation volume per Schottky defect has the same form as for vacancies (see eqs. 3.22 and 3.31).

The contribution to the thermal expansion coefficient will be evaluated by differentiating the relation $V = V^0 + nv^f$ in respect to temperature. The method is practically the same as for vacancies with the difference that in the present case eq. (3.26) becomes

$$\left.\frac{\partial n}{\partial T}\right|_P = \frac{nh^f}{2kT^2}. \tag{4.10}$$

Furthermore, the volume is given by $V = 2N\Omega$ where Ω is the mean volume per ion. After some algebra one obtains

$$\beta = \beta^0 \left(1 + \frac{n}{N}\frac{v^f}{2\Omega^0}\right)^{-1} + \frac{n}{N}\frac{v^f}{2\Omega}\left(\beta^f + \frac{h^f}{2kT^2}\right).$$

By expanding and disregarding the term containing $(n/N)^2$ we get the approximate relation:

$$\beta - \beta^0 \simeq \frac{n}{N}\frac{v^f}{2\Omega}\left(\beta^f - \beta^0 + \frac{h^f}{2kT^2}\right).$$

As in the case of vacancies, when $\beta^f - \beta^0 \ll h^f/2kT^2$, this simplifies to

$$\beta - \beta^0 \simeq \frac{n}{N}\frac{v^f}{2\Omega}\frac{h^f}{2kT^2}. \tag{4.11}$$

The contribution of defects to the compressibility is found by differentiating eq. (3.32) with respect to pressure. Following closely the procedure given for vacancies we get

$$\kappa = \kappa^0 \left(1 + \frac{n}{N}\frac{v^f}{2\Omega^0}\right)^{-1} + \frac{n}{N}\frac{v^f}{2\Omega}\left(\kappa^f + \frac{v^f}{2kT}\right)$$

or the approximate relation

$$\kappa - \kappa^0 \simeq \frac{n}{N}\frac{v^f}{2\Omega}\left(\kappa^f - \kappa^0 + \frac{v^f}{2kT}\right). \tag{4.12}$$

Isochoric perfect crystal. The parameters resulting from the comparison with an isochoric perfect crystal are defined in a similar way as for single vacancies. Their relations are the following

$$F = F^{0*} + nf^* - TS_c, \tag{4.13}$$

$$n = N\exp(-f^*/2kT), \tag{4.14}$$

$$H = H^{0*} + nh^*, \tag{4.15}$$

$$S_c = 2nk + nf^*/T, \tag{4.16}$$

$$S = S^{0*} + ns^* + S_c, \tag{4.17}$$

$$s^* = -(\partial f^*/\partial T)_V, \tag{4.18}$$

$$F = F^{0*} - 2nkT. \tag{4.19}$$

Equations 3.119 to 3.139, 3.157 to 3.174, 3.177 to 3.190 do not change in the above case.

4.2 Frenkel defects in ionic solids of NaCl-structure

Consider the case of Frenkel disorder in an ionic crystal A^+B^- of NaCl-structure. For simplicity we assume that the disorder is confined to the cation sublattice. As shown in fig. 4.1 a cation Frenkel defect consists of one cation vacancy and one cation interstitial at a great distance from the vacancy.

For a quantity of pure substance that contains N_{cat} cation sites and N_{an} anion sites the following holds: $N_{cat} = N_{an} = N$. The number of n_{cat} of cation vacancies is distributed on N_{cat} sites. Similarly the number n_{int} of cation interstitials is distributed on $2N$ sites, because there are two interstitial sites available for a cation. It is obvious that except near surfaces etc., the number n_{cat} is equal to the number n_{int} ($n_{cat} = n_{int} = n$). In order to find the configurational entropy S_c of such a crystal we have to take the sum of the configurational entropies of the cation vacancies and the interstitials. If W_{cat} denotes the number of ways of arranging n cation vacancies to N cation sites, we have ($N - n$ is the number of cations):

$$W_{cat} = \frac{N!}{(N-n)!n!}, \quad \text{and hence} \quad S_{c,cat} = k \ln \frac{N!}{(N-n)!n!}$$

for the configurational entropy $S_{c,cat}$ of the cation vacancies. The number W_{int} of ways of arranging n interstitial cations to $2N$ interstitial sites is given by

$$W_{int} = \frac{(2N)!}{(2N-n)!\,n!}, \quad \text{and hence} \quad S_{c,int} = k \ln \frac{(2N)!}{(2N-n)!\,n!}$$

for the configurational entropy of the interstitials. The total configurational entropy is:

$$S_c = S_{c,cat} + S_{c,int} = k \ln \left\{ \frac{N!}{(N-n)!\,n!} \frac{(2N)!}{(2N-n)!\,n!} \right\}.$$

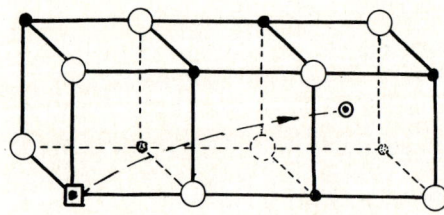

Fig. 4.1. Production of a Frenkel defect.

§4.2] Frenkel defects in NaCl-structured ionic solids

By applying Stirling's approximation and using some algebra we get

$$\frac{S_c}{k} = -N \ln\left(1 - \frac{n}{N}\right) + n \ln\left(\frac{N}{n} - 1\right) - 2N \ln\left(1 - \frac{n}{2N}\right) + n \ln\left(\frac{2N}{n} - 1\right).$$

In the approximations $n \ll N$, $\ln(1 - n/N) \simeq -n/N$, and $\ln(1 - n/2N) \simeq -n/2N$, this equation becomes:

$$\frac{S_c}{k} = 2n + n \ln\left(\frac{N}{n} - 1\right) + n \ln\left(\frac{2N}{n} - 1\right),$$

or by setting $(N/n) - 1 \simeq N/n$:

$$\frac{S_c}{k} = 2n - n \ln\left(\frac{n}{N}\frac{n}{2N}\right). \tag{4.20}$$

The definition of a defect parameter depends, as already mentioned, on the perfect crystal chosen; the latter has therefore to be accurately defined.

Isobaric perfect crystal. According to considerations analogous to those for vacancies, the expression for the Gibbs free energy of the real crystal can be written as

$$G = G^0 + ng^f - TS_c, \tag{4.21}$$

where g^f is the thermal Gibbs energy of a Frenkel defect. By differentiation we get the equilibrium value of n (when $n \ll N$):

$$n = N\sqrt{2} \exp\left(-\frac{g^f}{2kT}\right). \tag{4.22}$$

Inserting this value into eq. (4.20) we get for the configurational entropy and the Gibbs energy at equilibrium

$$S_c = 2nk + ng^f/T \tag{4.23}$$

and

$$G = G^0 - 2nkT. \tag{4.24}$$

The total entropy at equilibrium is obtained by differentiating eq. (4.24) with respect to temperature:

$$S = -\left.\frac{\partial G}{\partial T}\right|_P = -\left.\frac{\partial G^0}{\partial T}\right|_P + 2nkT + 2kT\left.\frac{\partial n}{\partial T}\right|_P. \tag{4.25}$$

The derivative $(\partial n/\partial T)_P$ can be calculated from eq. (4.22) to:

$$\left.\frac{\partial n}{\partial T}\right|_P = n\left\{-\frac{1}{2kT}\left.\frac{\partial g^f}{\partial T}\right|_P + \frac{g^f}{2kT^2}\right\}.$$

Inserting this value into eq. (4.25) and defining s^f by $s^f \equiv -(\partial g^f/\partial T)_P$ we get

$$S = S^0 + 2nk + ns^f + ng^f/T. \tag{4.26}$$

Equation (4.26) due to eq. (4.23) can be alternatively written

$$S = S^0 + ns^f + S_c.$$

It reveals that s^f is the "thermal formation entropy" per Frenkel defect.

The formation enthalpy per Frenkel defect is defined as $h^f \equiv g^f + Ts^f$. Considerations similar to those that led to eq. (3.23) give:

$$H = H^0 + nh^f. \tag{4.27}$$

The formation volume per Frenkel defect is defined as $v^f \equiv (\partial g^f/\partial P)_T$. By differentiating eq. (4.24) with respect to pressure we get

$$V \equiv \left.\frac{\partial G}{\partial P}\right|_T = \left.\frac{\partial G^0}{\partial P}\right|_T - 2kT \left.\frac{\partial n}{\partial P}\right|_T. \tag{4.28}$$

The derivative $(\partial n/\partial P)_T$ can be calculated from eq. (4.22) to $(\partial n/\partial P)_T = -nv^f/2kT$, so that eq. (4.28) becomes

$$V = V^0 + nv^f. \tag{4.29}$$

The internal energy of the crystal is found by inserting into eq. (2.9) the values of H and V given by eqs. (4.27) and (4.29); we finally get

$$U = U^0 + nu^f,$$

where U^0 denotes the internal energy of the perfect crystal, and u^f the formation internal energy defined as $u^f \equiv h^f - Pv^f$.

The Helmholtz free energy of the crystal is found by inserting into eq. (2.8) the values of G and V given by eqs. (4.24) and (4.29); we get

$$F = F^0 - 2nkT - nPv^f, \tag{4.30}$$

where $F^0 = G^0 + PV^0$ denotes the Helmholtz free energy of the perfect crystal. An alternative form for F can be derived from eq. (2.8) by using eqs. (4.21) and (4.29).

$$F = F^0 + n(g^f - Pv^f) - TS_c. \tag{4.31}$$

If now the Helmholtz formation energy per Frenkel defect is defined by $f^f \equiv g^f - Pv^f$, eq. (4.31) becomes

$$F = F^0 + nf^f - TS_c.$$

Isochoric perfect crystal. The thermodynamic functions of crystals A^+B^- containing Frenkel defects can also be described with parameters resulting from the comparison with an isochoric perfect crystal. They are

$$F = F^{0*} + nf^* - TS_c, \tag{4.32}$$

$$n = N\sqrt{2}\exp(-f^*/2kT), \tag{4.33}$$

$$S_c = 2nk + nf^*/T, \tag{4.34}$$

$$F_{eq} = F^{0*} - 2nkT, \tag{4.35}$$

$$S = S^{0*} + 2nk + ns^* + nf^*/T, \tag{4.36}$$

$$H = H^{0*} + nh^*, \tag{4.37}$$

$$U = U^{0*} + nu^*. \tag{4.38}$$

Equations 3.119 to 3.139, 3.157 to 3.174 and 3.177 to 3.190 remain unchanged.

Equilibrium fractional concentration. We define as fractional concentration x_{cat} of the cation vacancies the ratio of the number n_{cat} over the number of sites on which they can be distributed. Hence:

$$x_{cat} = n_{cat}/N_{cat} = n/N. \tag{4.39}$$

Similarly the fractional concentration x_{int} of the interstitials becomes

$$x_{int} = n/2N. \tag{4.40}$$

Equation (4.22) can be written as

$$\frac{n}{N}\frac{n}{2N} = \exp\left(-\frac{g^f}{kT}\right).$$

Inserting eqs. (4.39) and (4.40) we get

$$x_{cat}x_{int} = \exp\left(-\frac{g^f}{kT}\right). \tag{4.41}$$

Equations (4.39) and (4.40) reveal that $x_{cat} = 2x_{int}$ and therefore eq. (4.41) gives

$$x_{cat} = \sqrt{2}\exp\left(-\frac{g^f}{2kT}\right) \tag{4.42}$$

and

$$x_{int} = \frac{1}{\sqrt{2}}\exp\left(-\frac{g^f}{2kT}\right). \tag{4.43}$$

Sometimes in the literature the concentration of interstitials has been defined as $x'_{\text{int}} = n/N$ and hence $x_{\text{cat}} = x'_{\text{int}}$. In this case eqs. (4.41), (4.42) and (4.43) can be written

$$x_{\text{cat}} x'_{\text{int}} = 2 \exp\left(\frac{-g^{\text{f}}}{kT}\right), \quad x_{\text{cat}} = x'_{\text{int}} = \sqrt{2} \exp\left(\frac{-g^{\text{f}}}{2kT}\right). \quad (4.43\text{a, b})$$

Simultaneous formation of Frenkel and Schottky defects in crystals $A^+ B^-$.
If g_S^{f} denotes the Gibbs formation energy per Schottky defect we can easily show that (see §4.1):

$$x_c x_a = \exp(-g_S^{\text{f}}/kT), \quad (4.44)$$

where x_c and x_a are the fractional concentrations of cation and anion vacancies respectively.

In order to simplify the discussion we assume that: (1) the Frenkel disorder results only in the cation sublattice and (2) that the Frenkel disorder strongly predominates, i.e. the value of g_S^{f} is appreciably higher than the Gibbs formation energy g_F^{f} for (cation) Frenkel defect. If now x_i is the fractional concentration of (cation) interstitials (in the sense as used in eqs. 4.43a, b) we can again easily show that

$$x_c x_i = 2 \exp(-g_F^{\text{f}}/kT). \quad (4.45)$$

Due to the fact that the Frenkel disorder predominates we can make the approximation:

$$x_c \approx x_i. \quad (4.46)$$

It is now obvious that a combination of eqs. (4.44), (4.45) and (4.46) can lead to explicit relations for x_c, x_a and x_i in terms of g_S^{f} and g_F^{f}. By combining eqs. (4.45) and (4.46) we get:

$$x_c \approx \sqrt{2} \exp(-g_F^{\text{f}}/2kT), \quad (4.47)$$

$$x_i \approx \sqrt{2} \exp(-g_F^{\text{f}}/2kT). \quad (4.48)$$

Inserting into eq. (4.44) the value of x_c given by eq. (4.47) we have:

$$x_a \approx \frac{1}{\sqrt{2}} \exp\left(-\frac{2g_S^{\text{f}} - g_F^{\text{f}}}{2kT}\right). \quad (4.49)$$

The above equations reveal that whenever g_F^{f} and g_S^{f} decrease linearly with temperature then the plots $\ln x_c$ versus $1/T$, $\ln x_i$ versus $1/T$ and $\ln x_a$ versus $1/T$ have to be straight lines. On the other hand, when g_F^{f} decreases with temperature faster than linearly then eqs. (4.47) and (4.48) indicate that the plots $\ln x_c$ versus $1/T$ and $\ln x_i$ versus $1/T$ have to show an upward bend. In this case (4.49) reveals that the curve $\ln x_a$ versus $1/T$ shows also an upward bend only when g_S^{f} falls non-linearly with temperature to such an extent that $g_S^{\text{f}} - g_F^{\text{f}}/2$ decreases with temperature faster than linearly.

5 | MIGRATION

5.1 Introduction

For non-interacting oscillators (Einstein approach) a few well known expressions [38] will be developed: Consider the case of an interstitial vibrating in a harmonic potential as shown in fig. 5.1. It is obvious that the particle jumps to the neighbouring trough whenever its energy exceeds the barrier energy h^m separating the well from its final site (for the time being no other meaning has to be given to the symbol h^m). The energy h^m of the barrier may or may not belong to the allowed levels of the oscillator. The probability that the oscillator shall occupy an allowed state with $\varepsilon_l > h^m$ is given by:

$$P = \sum_{\varepsilon_l > h^m} \exp(-\varepsilon_l/kT) \Big/ \sum_l \exp(-\varepsilon_l/kT)$$

or approximating the sums by integrals – when $\hbar\omega$ is much larger than kT – we have:

$$P = \int_{h^m}^{\infty} \exp(-E/kT)\,dE \Big/ \int_0^{\infty} \exp(-E/kT)\,dE,$$

which gives:

$$P = \exp(-h^m/kT).$$

If the barrier is approached with an attempt frequency ν_{at}, the particle crosses the barrier at a rate w:

$$w = \nu_{at} \exp(-h^m/kT). \tag{5.1}$$

Let us assume for a moment that the argument leading to eq. (5.1) may apply to the case of atoms jumping to a vacancy. If the vacancy has z

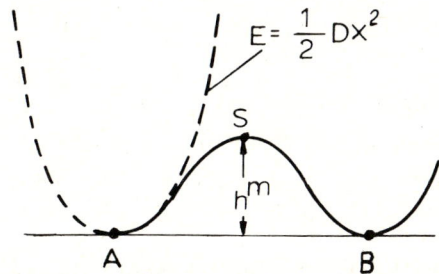

Fig. 5.1. Potential energy of a migration interstitial.

neighbours, any of which may attain the energy h^m necessary for a jump, the net jump rate is:

$$w = z\nu_{at} \exp(-h^m/kT). \tag{5.2}$$

The attempt frequency ν_{at} can be compared to the Einstein frequency ν_E by writing $\nu_{at} = \nu_E \delta$. Equation (5.2) can be written as:

$$w = z\nu_E \exp[\ln(\delta)] \exp(-h^m/kT), \tag{5.3}$$

or

$$w = z\nu_E \exp(-g^m/kT), \tag{5.4}$$

where g^m is defined as *

$$g^m = h^m - Ts^m \tag{5.5}$$

and s^m denotes the quantity:

$$s^m = k \ln(\delta). \tag{5.6}$$

Statistical mechanics of a classical system in equilibrium [38]. Consider a body containing $3N$ particles with position coordinates $x_{n,i}$ and momenta $p_{n,i}$. Obviously the "phase space" of the system is $6N$-dimensional. The probability P of finding the system in a state of energy E in the volume $dp_{11} \cdots dp_{N3} \, dx_{11} \cdots dx_{N3}$ is given by:

$$P = P_0 h^{-3N} e^{-E(p,x)/kT} dp_{11} \cdots dp_{N3} dx_{11} \cdots dx_{N3}, \tag{5.7}$$

where $P_0 h^{-3N}$ is a normalization constant. It should be noticed that eq. (5.7) is a simple statistical result if one makes the following two assumptions: (1) the probability with which a given configuration occurs depends solely on its energy and (2) the statistical weight of any volume element of the phase space is independent of its position. Now the key point is to express the energy E of the system in terms of the potential energy $\mathscr{V}$ of the system and the kinetic energy of the particles. This is easy only in the Born–Oppenheimer approximation, in which the potential energy depends only on the position of the ions

$$E = \mathscr{V}(x_{11}, \ldots, x_{N3}) + \sum_{n,i} p_{n,i}^2/2m_n. \tag{5.8}$$

The probability $P(x_{11}, \ldots, x_{N3})$ that a given region $dx_{11} \cdots dx_{N3}$ of configuration space is occupied, is obtained by inserting eq. (5.8) into eq.

---

* Here the symbols g^m, h^m and s^m do *not* correspond to the actual migration parameters Gibbs energy, enthalpy and entropy, which will be defined later.

§5.1] *Introduction* 73

(5.7) and integrating over all momentum values (i.e. between $-\infty$ and $+\infty$). This procedure finally leads to

$$P(x_{11},\ldots,x_{N3}) = P_0 h^{-3N} \prod_{n=1}^{N} (2\pi m_n kT)^{3/2} e^{-\mathscr{V}/kT} dx_{11} \cdots dx_{N3}. \tag{5.9}$$

The calculation is easy if we use mass weighted coordinates $s_{ni} = m_n^{1/2} x_{ni}$, so that:

$$dx_{11} \cdots dx_{N3} = \left\{ \prod_{n=1}^{N} m_n^{-3/2} \right\} ds_{11} \cdots ds_{N3}. \tag{5.10}$$

We now have to connect the potential energies of the system in the equilibrium configuration and the saddle point configuration. We first study the case where the volumes in the two configurations are taken as equal (*isochoric fluctuation*). If we describe with s_0 the equilibrium configuration and with s the displaced configuration the potential energy $\mathscr{V}(s)$ is connected to $\mathscr{V}(s_0)$ by the relation:

$$\mathscr{V}(s) = \mathscr{V}(s_0) + \sum_{a=1}^{3N} \tfrac{1}{2} \omega_a^2 q_a^2, \tag{5.11}$$

where q-coordinates measure the amplitudes of normal modes needed to Fourier synthesize the displacement $s_{11} \cdots s_{N3}$. By inserting eq. (5.11) and (5.10) into (5.9) we have:

$$P = P_0 h^{-3N} (2\pi kT)^{3N/2} \exp[-\mathscr{V}(s_0)/kT]$$
$$\times \int_{-\infty}^{+\infty} \cdots \int_{-\infty}^{+\infty} \exp\left(-\sum_a \omega_a^2 q_a^2 / 2kT\right) dq_{11} \cdots dq_{3N},$$

which by integration gives:

$$P = P_0 \left(\prod_{a=1}^{3N} kT/\hbar\omega_a \right) \exp[-\mathscr{V}(s_0)/kT]. \tag{5.12}$$

By expressing the product in an exponential form:

$$\prod_{a=1}^{3N} kT/\hbar\omega_a = \exp\left[-kT \sum_{a=1}^{3N} \ln(\hbar\omega_a/kT) \right],$$

eq. (5.12) can be written

$$P = P_0 \exp(-F/kT), \tag{5.13}$$

where F is defined as:

$$F \equiv \mathscr{V}(s_0) + kT \sum_{a=1}^{3N} \ln(\hbar\omega_a/kT). \tag{5.14}$$

Although eq. (5.14) is obtained from statistical mechanics it contains exactly the thermodynamic Helmholtz energy F for the case of a harmonic crystal.

We now shall study the case where the two configurations have the same pressure (*isobaric fluctuation*). The potential energy $\mathscr{V}(s_0)$ consists of two terms: a term $\mathscr{V}_0(s_0)$ related to the uncompressed crystal and a second term $pV(s_0)$ due to the pressure p; in this case by writing

$$\mathscr{V}(s_0) = \mathscr{V}_0(s_0) + pV(s_0), \tag{5.15}$$

eq. (5.13) changes to:

$$P = P_0 \, e^{-G/kT}, \tag{5.16}$$

where G is defined as:

$$G \equiv \mathscr{V}_0(s_0) + kT \sum_{a=1}^{3N} \ln(\hbar\omega_a/kT) + pV(s_0), \tag{5.17}$$

or

$$G \equiv F + pV(s_0). \tag{5.18}$$

The value of P_0 can be found from eq. (5.16) by writing $P = 1$ and hence:

$$P_0 = e^{G/kT}. \tag{5.19}$$

By inserting eqs. (5.17) and (5.19) into eq. (5.7) we find that the probability that a crystal, which is statistically fluctuating under constant temperature and pressure, can be found in a state with particular coordinates and momenta.

5.2 Rate theory and the dynamical theory of migration

5.2.1 *The rate theory approximation*

A rather complete treatment of this approximation can be found in the literature [38–40]. Here we stress only the general principles on which it is based and the corresponding results. Attention is also drawn to the point that the migration event can alternatively be considered either under *isochoric* or under *isobaric* conditions.

(i) *Isochoric parameters of migration.* Consider a lattice constrained to remain at constant volume V and temperature T (isochoric migration). For the two representative state vectors, s_A and s_B, a particle lies in the potential wells A and B; the latter are separated by a saddle-point S. When the particle occupies this point the representative configuration vector is labelled s_S. The energies $F_A \, (= F_B)$ and F_S are the F-energies as defined above for the states s_A and s_B. The jump rate w from A to B is the current

§5.2]   Rate theory / dynamical migration theory

J of representative points flowing in response to a unit occupation probability P_A of the well A [38], i.e.

$$w = J/P_A, \tag{5.20}$$

where

$$P_A = \exp[-(F_A - F)/kT]. \tag{5.21}$$

Furthermore

$$J = P_S \langle \dot{s} \rangle = P_S (kT/2\pi)^{1/2}, \tag{5.22}$$

where

$$P_S = \exp[-(F_S - F)/kT], \tag{5.23}$$

and $\langle \dot{s} \rangle = (kT/2\pi)^{1/2}$ is the mean velocity component directed from A to B.

Inserting eqs. (5.21), (5.22) and (5.23) into eq. (5.20) we finally get:

$$w = (kT/2\pi)^{1/2} \exp[-(F_S - F_A)/kT]. \tag{5.24}$$

We again stress that eq. (5.24) is an exact result of classical statistical mechanics [38]. In order to obtain an explicit relation for w we now have to calculate the F-functions at A and S. Such calculations are possible only in the quasi-harmonic approximation. In this approximation the value of F_A can be written with the help of eq. (5.14)

$$F_A = \mathscr{V}_A(V) + kT \sum_{a=1}^{3N} \ln(\hbar \omega_a(V)/kT). \tag{5.25}$$

We emphasize that eq. (5.25) contains the assumption that the frequencies ω_a and the potential minimum $\mathscr{V}_A(V)$ depend, explicitly on the volume V of the crystal only and not on the precise equilibrium position s_A.

Similarly the F-function at the saddle point can be written

$$F_S = \mathscr{V}_S(V) + kT \sum_{a=2}^{3N} \ln(\hbar \omega_a'/kT) + kT \ln[\hbar/(kT)^{1/2}], \tag{5.26}$$

where the summation in the second term does not contain the atom missing from position A and the third term comes from the integration over $\dot{s}$:

$$\int_{-\infty}^{+\infty} \exp(-\dot{s}^2/2kT)\,d\dot{s} = (2\pi kT)^{1/2}.$$

Inserting the values of F_S and F_A from eqs. (5.26) and (5.25) into eq. (5.24) we finally get:

$$w(T, V) = (kT/h) \exp(-f_0^*/kT), \tag{5.27}$$

where

$$f_0^* \equiv \mathscr{V}_S(V) + kT \sum_{a=2}^{3N} \ln[\hbar\omega_a'(V)/kT] - \mathscr{V}_A(V)$$

$$- kT \sum_{a=1}^{3N} \ln[\hbar\omega_a(V)/kT] \tag{5.28}$$

Equation (5.27) can alternatively be written as:

$$w(T, V) = \prod_{a=1}^{3N} \nu_a(V) \left\{ \prod_{a=2}^{3N} \nu_a'(V) \right\}^{-1} \exp\{-[\mathscr{V}_S(V) - \mathscr{V}_A(V)]/kT\} \tag{5.29}$$

In order to obtain an equal number of factors in the numerator and denominator we multiply both with an *arbitrary constant* frequency ν_0 and then have:

$$w(T, V) = \nu_0 \prod_{a=1}^{3N} \nu_a(V) \left\{ \prod_{a=1}^{3N} \nu_a'(V) \right\}^{-1}$$

$$\times \exp\{-[\mathscr{V}_S(V) - \mathscr{V}_A(V)]/kT\}, \tag{5.30}$$

where, by definition, ν_0 is the desired first factor $\nu_1(V)$. Equation (5.30) can now be written in the form:

$$w(T, V) = \nu_0 \exp(-f^*/kT), \tag{5.31}$$

where

$$f^* = \mathscr{V}_S(V) - \mathscr{V}_A(V) + kT \sum_{a=1}^{3N} \ln[\omega_a'(V)/\omega_a(V)]. \tag{5.32}$$

The quantity f^* is defined as the *Helmholtz free migration energy under constant volume*. We stress that the quantity f^* – as defined above – is *not* equal to the difference $F_S - F_A$, as can be seen by comparing eqs. (5.24) and (5.31); they are connected by:

$$f^* = (F_S - F_A) + kT \ln\left[\nu_0(kT/2\pi)^{-1/2}\right]. \tag{5.33}$$

We define a *migration internal energy* u^* with

$$u^*(V) \equiv \mathscr{V}_S(V) - \mathscr{V}_A(V); \tag{5.34}$$

it represents the true difference of the potential energies of the crystal when in the ground and the saddle point state.

§5.2] Rate theory / dynamical migration theory

We further define a *migration entropy under constant volume s** with

$$s^* \equiv k \sum_{a=1}^{3N} \ln[\omega_a(V)/\omega_a'(V)]. \tag{5.35}$$

We stress here that s^* – in contrast to u^* – is *not* equal to the "true" difference of the entropies of the crystal at s_A and s_S. This will be seen in eq. (5.40).

By keeping in mind that in the quasi-harmonic approximation the frequencies (and the potential energies) do not depend explicitly on temperature, eq. (5.32) gives:

$$-\left.\frac{\partial f^*}{\partial T}\right|_V = k \sum_{a=1}^{3N} \ln \frac{\omega_a(V)}{\omega_a'(V)} \tag{5.36}$$

and by further comparing to eq. (5.35) we obtain:

$$s^* = -\left.\frac{\partial f^*}{\partial T}\right|_V. \tag{5.37}$$

A combination of eqs. (5.34), (5.35) and (5.36) indicates that:

$$f^* = u^* - Ts^*. \tag{5.38}$$

By differentiating (5.38) with respect to T (V = constant) we have:

$$-\left.\frac{\partial f^*}{\partial T}\right|_V = \left.\frac{\partial u^*}{\partial T}\right|_V - T \left.\frac{\partial s^*}{\partial T}\right|_V - s^*,$$

or due to eq. (5.37):

$$\left.\frac{\partial u^*}{\partial T}\right|_V = T \left.\frac{\partial s^*}{\partial T}\right|_V. \tag{5.39}$$

The above equations for the migration parameters are similar in form in those gained in the case of formation processes. However, eq. (5.39) is restricted in the present case to a quasi-harmonic approximation: in contrast to the general eq. (3.93) of the corresponding formation parameters the definitions of u^* and s^* (see eq. (5.34) and (5.35)) imply that they do not depend explicitly on temperature and therefore eq. (5.39) is a direct consequence of the suitable definition of u^* and s^*.

An explicit relation between the entropy s^* and the "true entropy difference" $S_S - S_A = [\partial(F_S - F_A)/\partial T]_V$ of the crystal in the states s_A and s_S can now be found from a direct differentiation of eq. (5.33) with respect to T (V = constant). One finally finds:

$$s^* = (S_S - S_A) + k\left\{\partial\left[T \ln \nu_0(2\pi/kT)^{1/2}\right]/\partial T\right\}_V \tag{5.40}$$

Due to the fact that the migration event has been considered to occur under constant volume one expects the "pressures" $P = -(\partial F/\partial V)_T$ in the two configuration states to be different. This can be seen from a direct differentiation of eq. (5.33) (recall that ν_0 is arbitrarily assumed a constant)

$$-\left.\frac{\partial f^*}{\partial V}\right|_T = -\left.\frac{\partial F_S}{\partial V}\right|_T + \left.\frac{\partial F_A}{\partial V}\right|_T,$$

which gives

$$-\left.\frac{\partial f^*}{\partial V}\right|_T = P_S - P_A.$$

Therefore, if it is acceptable to interpret the quantity $-(\partial F_S/\partial V)_T = P_S$ as the "pressure" when the moving ion passes through the saddle point, the quantity $-(\partial f^*/\partial V)_T$ is truly the difference of the "pressures" of the crystal at s_S and s_A.

(ii) *Isobaric parameters of migration.* Bearing in mind the considerations leading to eq. (5.30) and following the same procedure as in the previous paragraph one can write the transition rate $w(P, T)$ for a lattice kept at constant (external) pressure and temperature

$$w = \left(\frac{kT}{2\pi}\right)^{1/2} \exp\left(-\frac{G_S - G_A}{kT}\right), \qquad (5.41)$$

where G_S and G_A denote the "Gibbs" functions of the crystal at s_A and s_S. We remember that this is a statistical result. As in the isochoric case, the functions G_A and G_S can be written in the quasi-harmonic approximation:

$$G_A = \mathscr{V}_A(V_A) + kT \sum_{a=1}^{3N} \ln[\hbar\omega_a(V_A)/kT] + pV_A, \qquad (5.42)$$

$$G_S = \mathscr{V}_S(V_S) + kT \sum_{a=2}^{3N} \ln[\hbar\omega'_a(V_S)/kT] + pV_S + kT \ln[\hbar/(kT)^{1/2}]. \qquad (5.43)$$

Inserting (5.42) and (5.43) into eq. (5.41) we get

$$w(P, T) = \frac{\prod_{a=1}^{3N} \nu_a(V_A)}{\sum_{a=2}^{3N} \nu'_a(V_S)} \exp\left[-\frac{\mathscr{V}_S(V_S) - \mathscr{V}_A(V_A) + p(V_S - V_A)}{kT}\right], \qquad (5.44)$$

where V_S and V_A denote the volume of the crystal in the two states. By defining again $\omega'_1(V_S) = 2\pi\nu_0$, eq. (5.44) takes the form

$$w(P, T) = \nu_0 \exp[-g^m(P, T)/kT], \qquad (5.45)$$

where
$$g^m \equiv \mathscr{V}_S(V_S) - \mathscr{V}_A(V_A) + kT \sum_{a=1}^{3N} \ln \frac{\omega'_a(V_S)}{\omega_a(V_A)} + p(V_S - V_A). \qquad (5.46)$$

The quantity g^m is called *Gibbs migration energy under constant pressure*. It should be emphasized that g^m is *not* equal to the difference $G_S - G_A$, as can be seen from a comparison of eq. (5.41) and (5.45). Their difference can be found from

$$g^m = G_S(V_S) - G_A(V_A) + kT \ln\left[\nu_0(2\pi/kT)^{1/2}\right]. \qquad (5.47)$$

The migration energy u^m, enthalpy h^m and entropy s^m are defined as [see eq. (5.46)]:

$$u^m \equiv \mathscr{V}_S(V_S) - \mathscr{V}_A(V_A), \qquad (5.48)$$

$$h^m \equiv \mathscr{V}_S(V_S) - \mathscr{V}_A(V_A) + p(V_S - V_A), \qquad (5.49)$$

$$s^m \equiv -k \sum_{a=1}^{3N} \ln\left[\omega'_a(V_S)/\omega_a(V_A)\right]. \qquad (5.50)$$

Equations (5.48) and (5.49) show that the quantities u^m and h^m represent the difference of the potential energy respectively enthalpy of the crystal in the two states. On the other hand eq. (5.50) indicates that the migration entropy s^m is *not* equal to the difference of the entropies of the crystal at S and A. We shall come back at this point later.

By differentiating eq. (5.47) with respect to pressure and under the restriction that ν_0 does not depend on p, the *migration volume* is defined by

$$v^m \equiv \left.\frac{\partial g^m}{\partial p}\right|_T = V_S - V_A. \qquad (5.51)$$

Equation (5.51) indicates that v^m represents the difference of the volume of the crystal at s_S and s_A.

Inserting (5.48), (5.50) and (5.51) into eq. (5.46) we get

$$g^m = u^m + pv^m - Ts^m. \qquad (5.52)$$

A combination of eqs. (5.48) and (5.49) gives

$$h^m = u^m + pv^m, \qquad (5.53)$$

and hence eq. (5.52) can also be written

$$g^m = h^m - Ts^m. \qquad (5.54)$$

By recalling that in the quasi-harmonic approximation the frequencies (and the potential energies) do not depend explicitly on temperature, the differentiation of eq. (5.46) gives:

$$-\left.\frac{\partial g^m}{\partial T}\right|_p = -k \sum_{a=1}^{3N} \ln \frac{\omega'_a(V_S)}{\omega_a(V_A)},$$

and comparing to eq. (5.50) we have:

$$s^m = -\left.\frac{\partial g^m}{\partial T}\right|_P = -k\sum_{a=1}^{3N}\ln\frac{\omega'_a(V_S)}{\omega_a(V_A)}. \tag{5.55}$$

In order to find the connection between s^m and the entropy difference $S_S - S_A$ we differentiate eq. (5.47):

$$s^m = (S_S - S_A) + \left(\partial\left\{kT\ln\left[\nu_0(2\pi/kT)^{1/2}\right]\right\}/\partial T\right)_P. \tag{5.56}$$

which – as mentioned above – emphasizes that the migration entropy s^m is *not* equal to $S_S - S_A$. This constitutes a reason for no longer retaining in the literature ΔS^m as a symbol for s^m.

By comparing eq. (5.45) and (5.31) one gets:

$$g^m = f^* \tag{5.57}$$

A combination of eqs. (5.40) and (5.56) gives:

$$s^m - s^* = S_S(V + v^m) - S_S(V), \tag{5.58}$$

where the terms in the parentheses simply indicate the volumes to which S_S corresponds. Due to

$$(\partial S/\partial V)_T = \beta B, \tag{5.59}$$

eq. (5.58) gives:

$$s^m - s^* = v^m\beta B. \tag{5.60}$$

5.2.2 The dynamical theory

This approach was initially developed by Rice [41] and Slater [42] and more recently by Glyde [43] and Flynn [38]. It treats the displacements that allow migration as a superposition of phonons in the harmonic crystal. By considering the phonon phases to be random the possibility emerges that the displacements (corresponding to various phonons) may coincide in such a way that the migration event becomes possible.

Consider the summations of N harmonically varying terms:

$$x(t) = \sum_{j=1}^{N} x_j^0 \exp(i\omega_j t).$$

If q denotes any critical value, Kac [44] has found an expression that gives the frequency w with which the function $x(t) - q$ has upzeros (i.e. passes the value zero in an increasing direction). In the case of values of q tending to zero this expression takes the approximate form (where $\omega_j = 2\pi\nu_j$)

$$w = \left\{\sum_j (\nu_j x_j^0)^2\right\}^{1/2} \left\{\sum_j (x_j^0)^2\right\}^{-1/2} \exp\left\{-q^2/\sum_j (x_j)^2\right\}. \tag{5.61}$$

As discussed by Slater [42], eq. (5.61) is asymptotically exact when the "amplitudes" x_j^0 are equal and N is large. Furthermore Feit [45] has shown that eq. (5.61) is also exact when each amplitude x_j^0 represents the projection of a classical vibrational amplitude on an arbitrary axis in the configuration space for thermal equilibrium.

Now the importance of eq. (5.61) for the present problem becomes clear: we apply it to a mobile atom and predict the jump rate w in terms of a single parameter q which measures the size of the "displacement" of the velocity or the potential necessary to allow an atomic jump. From a physical point of view the exponential in eq. (5.61) shows the probability of the system being at the critical "displacement" q due to the fluctuation and the preexponential represents a "mean attack frequency". In this discussion the term "displacement" can be interpreted either as a velocity fluctuation or a potential energy fluctuation.

(i) *Velocity fluctuations.* The velocity fluctuation q necessary for a successful jump can be specifically evaluated for the time instant when the mobile atom is at its equilibrium site [38]:

$$q = (2u/M)^{1/2}, \tag{5.62}$$

where $u = \mathscr{V}(V_S) - \mathscr{V}(V_A)$ is the height of the potential barrier under constant pressure. If the jump occurs under constant volume, u may be replaced by

$$u^* = \mathscr{V}_S(V) - \mathscr{V}_A(V). \tag{5.63}$$

In order to calculate the jump rate w we have to insert eq. (5.62) into eq. (5.61). The denominator $\Sigma_j(x_j)^2$ of the exponent is here a sum of the squares of the velocity projections.

The "displacement" u in an arbitrary direction of a phonon of frequency ω is $\boldsymbol{u} = \boldsymbol{u}^0 \, e^{-i\omega t}$; its maximum velocity is ωu^0 and its contribution to the sum is $v^0 = \omega \boldsymbol{u}^0 \cdot \hat{\boldsymbol{x}}$ where $\hat{\boldsymbol{x}}$ is the unit vector in the jump direction. The summation has to be taken over all the frequencies ω and branches λ

$$\sum_j (x_j)^2 = \sum_{\omega,\lambda} \omega^2 |\boldsymbol{u}^0_{\omega\lambda} \cdot \hat{\boldsymbol{x}}|^2.$$

In the case of an isotropic crystal we can write [38]:

$$\sum_\lambda |\boldsymbol{u}^0_{\omega\lambda} \cdot \hat{\boldsymbol{x}}|^2 = |\boldsymbol{u}^0_{\omega\lambda}|^2.$$

The amplitude u^0 of the phonon ω, λ can be calculated from its energy ε_ω by

$$\varepsilon_\omega = \tfrac{1}{2} M N \omega^2 |\boldsymbol{u}^0_{\omega\lambda}|^2.$$

We finally get:

$$\sum_j (x_j)^2 = \sum_{\omega,\lambda} \omega^2 |u^0_{\omega\lambda} \cdot \hat{x}|^2 = \sum_\omega (2\varepsilon_\omega/MN) = 2\bar{\varepsilon}/M,$$

where $\bar{\varepsilon}$ denotes the mean energy per phonon of the crystal.

The preexponential term of (5.61) can be easily calculated at high temperatures ($\bar{\varepsilon} \to kT$) by adopting the Debye approximation and therefore by replacing the sums with integrals over the Debye density of states. Such a calculation gives the value $(3/5)^{1/2} \nu_D$ where ν_D is the Debye frequency. Therefore eq. (5.61) finally gives [38]:

$$w = (3/5)^{1/2} \nu_D \exp(-u/\bar{\varepsilon}) \simeq (3/5)^{1/2} \nu_D \exp(-u/kT). \tag{5.64}$$

Now we see that eq. (5.64) has the form of the classical Arrhenius relation.

(ii) *Potential energy fluctuations.* If r_A is the equilibrium position of the diffusing atom and $r_{A,S}$ denotes the equilibrium saddle-point position with respect to r_A, then the quantities $u(r_A)$ and $u(r_{A,S})$ represent the displacements of the diffusing atom, respectively the saddle-point from their equilibrium positions. What fluctuates is the distance x:

$$x = \{u(r_A) - u(r_A + r_{A,S})\} \cdot \hat{x}, \tag{5.65}$$

i.e. the distance of the diffusing atom from the saddle-point along $\hat{x}$.

On inserting eq. (5.65) into eq. (5.61) and assuming (1) plane waves for phonons with frequency ω and wave vector k from branch λ of the spectrum, the appropriate expression for u is

$$u(r) = u^0_{\omega\lambda} \exp\{i(k \cdot r - \omega t)\},$$

and (2) a Debye approximation (i.e. $k = v/\omega$), one finally gets [38]:

$$w = (3/5)^{1/2} \bar{\nu}_D \exp\left\{ -\frac{15Mq^2}{2kT(3v_l^{-2} + v_t^{-2} + v_{t'}^{-2})} \right\}, \tag{5.66}$$

where M is $\Omega\rho$, $\bar{\nu}_D$ the square root of the average squared Debye frequency of the spectrum, t and t' denote the different transverse branches of the phonon spectrum, and l corresponds to the longitudinal branch.

In the case of the Debye spectrum we can write:

$$v^2 = c/\rho,$$

where c is the elastic constant appropriate to the branch with velocity v and crystal density ρ. Therefore it is evident that eq. (5.66) can now be written

$$w = (3/5)^{1/2} \bar{\nu}_D \exp\{-\tilde{c}\Omega\delta^2/kT\}, \tag{5.67}$$

where $\tilde{c}$ is an effective elastic modulus, $\delta \equiv q^2/s^2$ and s the saddle point displacement.

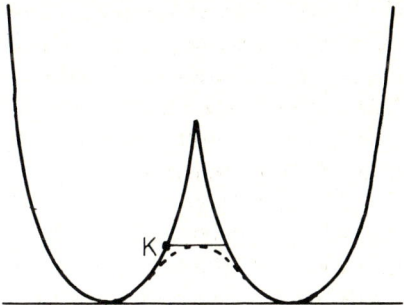

Fig. 5.2. Potential energy through the saddle point. The pure harmonic potential (solid lines) is unphysical. The dashed curve shows an approximation to the real potential [38].

The problem now is to give the definitions of the migration parameters in such a way as to make them compatible not only with the two results of the *dynamical theory* [see eqs. (5.64) and (5.67) but also with those of the *rate theory*. We have therefore to check whether the results of the two theories concerning the expressions of the jump rate w are compatible or not.

Comparison of the jump rates given by the rate theory and the dynamical theory. We remember that the dynamical approach, in contrast to the rate theory, is applicable only when the potential varies harmonically. This point is undoubtedly a serious shortcoming of this theory because, although acceptable for small displacements, it seems unphysical that $\mathscr{V}(s)$ should remain harmonic (i.e. parabolic in s) all the way from the equilibrium to the saddle point. We therefore assume that the harmonic approximation is valid only up to a certain distance. When the fluctuation reaches point K (fig. 5.2), we have a successfull jump. In this sense these two theories have a point of contact and we can therefore compare the resulting jump rates. We therefore have to compare the formulae [see eqs. (5.31) and (5.64)]:

$$w = \nu_0 \exp(-f^*/kT),$$

and

$$w = (3/5)^{1/2} \nu_D \exp(-u^*/kT).$$

Let us make for a moment the approximation that the preexponential factors ν_0 and $(3/5)^{1/2}\nu_D$ are equal and focus our attention on the comparison of the exponents. By recalling eq. (5.32) and (5.63) we see that the exponents become equal *only* when the term $-k\Sigma_a \ln[\omega'_a(V)/\omega_a(V)]$ is exactly zero. We see therefore that the dynamical theory as developed by Flynn [38] only gives the same jump rate as the rate theory when the "migration entropy under constant volume" is exactly zero. This, as will be discussed later, is not quite unreasonable because the frequency changes are expected to be quite small when the migration event occurs under constant volume. Another interesting point arising from the comparison of the two models is that the f^*-function for the isochoric migration process (or the g^m-function for the isobaric process) is equal to $c\Omega\delta^2$. This equality is significant because its differentiation with respect to the corresponding variables gives expressions for the various migration parameters. Unfortunately the latter expressions contain the unknown quantity δ, which has been empirically adjusted by Flynn [38] to be around one tenth.

Another approach to the migration process has been published by Van Vechten [46]. A common characteristic of the dynamical theory, as developed by Flynn, and the "ballistic model" of Van Vechten is that both models have to introduce some empirical constant in order to obtain numerical results for the migration parameters. The disadvantages of the dynamical theory are discussed by Van Vechten.

6 THERMODYNAMICS OF THE SPECIFIC HEAT

6.1 Introduction

Measurements of the heat necessary to increase the temperature of a given quantity of substance under conditions of constant pressure or volume give the isobaric specific heat $C_P = (\partial H/\partial T)_P$ and the isochoric specific heat $C_V = (\partial U/\partial T)_V$ (usually per mole of substance). As the measurement of the latter is difficult, it is usually obtained by the thermodynamical formula that connects these quantities:

$$C_P - C_V = TV\beta^2/\kappa.$$

In the past it was assumed that the resulting value of C_V refers to a crystal that has no vacancies. Of course both these quantities (points a and b in fig. 6.1) are thermodynamic and refer to the same given state of a real crystal; they both contain exactly the same number of vacancies.

Plots of C_P and C_V as a function of temperature when measured under constant given pressure or constant given volume will be referred to as *isobaric curves* or *isochoric curves*, respectively. In fig. 6.1 typical isobaric curves have been plotted for two different constant pressures P_1 and P_2, where $P_2 > P_1$. Furthermore, two isochoric curves have been plotted for constant volumes V_3 and V_4.

In order to study the connections between isobaric and isochoric curves, let us consider the state that corresponds to a point a (pressure P_1, temperature T_A). For simplicity we assume that the volume of this state happens to be V_4. The value of C_P which corresponds to this state is the ordinate of point a. The isochoric heat C_V of this state is given by the ordinate of point b where the abscissa T_A intersects the isochoric curve V_4. The quantities C_V and C_P belong to the same state and are connected by the thermodynamical formula given above.

We now study another state c on the same isobaric curve as before but at a higher temperature (pressure P_1, temperature T_C). The volume in this state is no longer equal to V_4, because of thermal expansivity. If now the value of C_V is calculated for state c, the result will not lie on the same isochoric

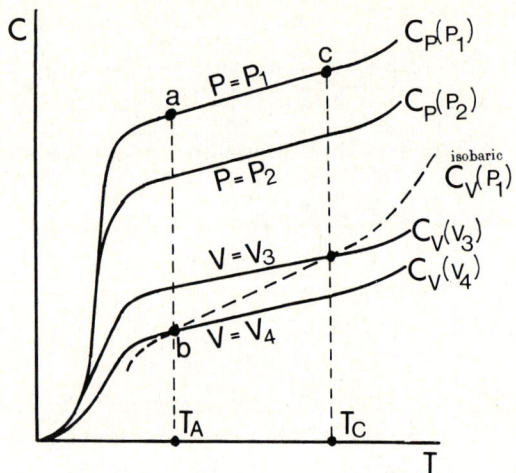

Fig. 6.1. Isobaric and isochoric curves of C_P and C_V. The dashed line represents the "isobaric curve of C_V" discussed in the text. ($P_2 > P_1$.)

curve (V_4) as before, but on an isochoric curve of larger volume (V_3). This very simple fact has not always been correctly considered in the past. If one connects the C_V-values corresponding to all the states of the isobaric curve for pressure P_1, a curve will result that we will call an "isobaric curve of C_V". Its ordinates in fig. 6.1 are labelled with $C_V(P_1)$. On the other hand the values of C_V corresponding to a given constant volume are labelled as $C_V(V)$; for example the C_V-ordinates for $V = V_3$ will be labelled $C_V(V_3)$. A curve of the latter type has been experimentally determined up to now only for some rare-gas solids; it can be called an "isochoric curve of C_V". The states to which such $C_V(V)$-values correspond have widely varying (thermal) pressures.

One defines as contribution of vacancies to the specific heat the excess of the specific heat of a real crystal over that of a perfect crystal; the latter has therefore to be definitely stated in each case. In this section, as in previous ones, two types of perfect crystals – the *isobaric* perfect crystal and the *isochoric* perfect crystal – have to be considered [637]. It will be shown that the study of only one type of perfect state leads to difficulties in finding the interconnection of the contributions of vacancies, especially when comparing the results of C_V from isochoric measurements to the corresponding values that are obtained after thermodynamic corrections of isobaric measurements. It is therefore worthwhile to recapitulate the well-known definitions of the two types of perfect crystals. On the right-hand side of fig. 6.2 a real crystal P, V, T, containing n vacancies, receives isobarically the heat

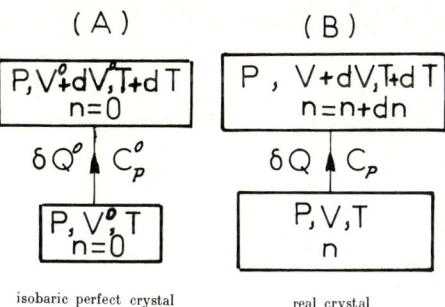

Fig. 6.2. Definition of C_P and C_P^0 by isobaric heating of a real crystal (B) and its isobaric perfect crystal (A).

δQ and therefore changes to a new equilibrium state described by P, $V + dV$, $T + dT$, that contains $n + dn$ vacancies. The heating from T to $T + dT$ of its isobaric perfect crystal under the constraint of constant pressure (fig. 6.2, left) would need the heat δQ^0. These two heats would lead to the isobaric specific heats C_P and C_P^0. A heating experiment of the same (isobaric) perfect crystal under the constraint of constant volume lead to an isochoric specific heat C_V^0.

In fig. 6.3 a real crystal P, V, T, containing n vacancies, receives isochorically the heat δQ^* and therefore changes into a new equilibrium state described by $P + dP$, V, $T + dT$, containing $n + dn$ vacancies. The heating from T to $T + dT$ of its isochoric perfect crystal under the constraint of constant volume would need the heat δQ^{0*}. These two heats lead to the isochoric specific heats C_V and C_V^{0*}. A heating experiment of the same perfect crystal under the constraint of constant pressure would lead to an isobaric specific heat C_P^{0*}.

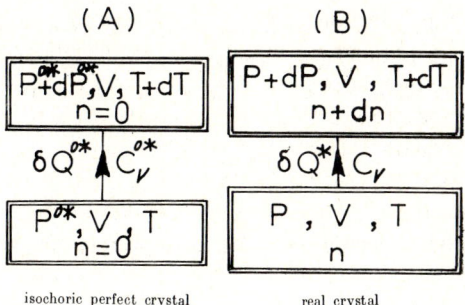

Fig. 6.3. Definition of C_V and C_V^{0*} by isochoric heating of a real crystal (B) and its isochoric perfect crystal (A). All four crystals have the same volume.

It is important to note that the perfect crystals of figs. 6.2 and 6.3 are in different states and therefore have different specific heats ($C_P^0 \neq C_P^{0*}$; $C_V^0 \neq C_V^{0*}$). We therefore have to consider six specific heats: (a) the specific heats C_P and C_V of the real crystal, (b) the specific heats C_P^0 and C_V^0 of the *isobaric* perfect crystal, and (c) the specific heats of C_P^{0*} and C_V^{0*} of the *isochoric* perfect crystal. For their differences we will use the following symbols:

$$\Delta C_P \equiv C_P - C_P^0, \qquad \Delta C_V \equiv C_V - C_V^0,$$
$$\Delta C_P^* \equiv C_P - C_P^{0*}, \qquad \Delta C_V^* \equiv C_V - C_V^{0*}.$$

Without careful separation of these terms, a consistent analysis of the variety of specific heat data is not possible.

In order to avoid any confusion arising from the profusion of symbols we recapitulate: ΔC_P and ΔC_V are the vacancy contributions that have to be added to the specific heats C_P^0 and C_V^0 of the isobaric perfect crystal in order to obtain the specific heats of the real crystal. Correspondingly ΔC_P^* and ΔC_V^* are the contributions that have to be added to the specific heats of the isochoric perfect crystal.

In the following sections the contributions of vacancies to the specific heats will be calculated in terms of the usual parameters h^f, u^*, s^f, s^*, etc. It will be found that their correct use in the analysis of specific heat experiments will give values that agree with the values resulting from the analysis of other types of experiments (such as Simmons–Balluffi experiments) to a much larger degree than formerly thought.

In order to connect the macroscopic thermodynamic functions (C_P, C_P^0, etc.) with quantities resulting from theoretical calculations on an atomic scale, the quantities c_P and c_V will be introduced. They are the specific heats necessary for the "heating of a single defect". Of special interest is the isochoric specific heat c_V^0 needed for heating a defect in the isobaric perfect crystal; it is the quantity that is the most accessible to theoretical calculations on an atomic scale. The difference $c_P - c_V$ can be expressed with experimentally accessible defect parameters from purely thermodynamic considerations (see §6.4).

6.2 Contribution of vacancies to an isobaric perfect crystal

In this section we study the excess of the specific heats of a real crystal with regard to its isobaric perfect crystal.

6.2.1 Isobaric specific heat of an isobaric perfect crystal

Calculations of C_P. The isobaric specific heat of a real crystal can be found by a direct differentiation of eq. (3.23):

$$C_P \equiv \left.\frac{\partial H}{\partial T}\right|_P = \left.\frac{\partial H^0}{\partial T}\right|_P + \left.\frac{\partial (nh^f)}{\partial T}\right|_P.$$

The quantity $(\partial H^0/\partial T)_P$ is the isobaric specific heat C_P^0 of the perfect crystal; therefore the above equation can be written as

$$C_P = C_P^0 + n\left.\frac{\partial h^f}{\partial T}\right|_P + h^f\left.\frac{\partial n}{\partial T}\right|_P. \tag{6.1}$$

Equation (6.1) shows that the contribution of vacancies to the specific heat consists of a term $n(\partial h^f/\partial T)_P$ which is due to the temperature variation of the formation enthalpy plus a term $h^f(\partial n/\partial T)_P$ which comes from the increment of the number of vacancies with increasing temperature. With the help of eq. (3.26) we obtain

$$\Delta C_P \equiv C_P - C_P^0 = n\left.\frac{\partial h^f}{\partial T}\right|_P + n\frac{(h^f)^2}{kT^2}. \tag{6.2}$$

It is worthwhile to emphasize here that the derivation of this equation is quite *general*, as it does not contain any approximation (except if the vacancies interact). In the literature it has often been stated that eq. (6.2) is valid only if $(\partial h^f/\partial T)_P$ is negligible. The general form of C_P and C_P^0 has already been depicted in fig. 2.1.

A rough estimation of C_P can be made as follows: experimental values of dh^f/dT determined from other types of experiments show that this factor is usually of minor importance in comparison to $(h^f)^2/kT^2$. In such cases eq. (6.2) can be simplified to

$$\Delta C_P \simeq \frac{n}{N}\left(\frac{h^f}{kT}\right)^2 kN.$$

As h^f exceeds kT, even at the melting point, by an order of magnitude and n/N at high temperatures is around 10^{-3} or less, one gets for ΔC_P a typical value of $0.1R$. At low temperatures ΔC_P is less, because n rapidly decreases with T.

The quantity $(\partial h^f/\partial T)_P$ is the specific heat necessary to change isobarically the "temperature of one vacancy". We will label it with

$$c_P \equiv \left.\frac{\partial h^f}{\partial T}\right|_P \tag{6.3}$$

or with its thermodynamically equivalent equation

$$c_P \equiv T \left.\frac{\partial s^f}{\partial T}\right|_P. \tag{6.4}$$

Temperature dependence of ΔC_P. By using eq. (3.24), eq. (6.2) can be written as

$$\ln(T^2 \Delta C_P) = -\frac{h^f}{kT} + \frac{s^f}{k} + \ln\left(\frac{(h^f)^2}{k} + c_P T^2\right) + \ln N. \tag{6.5}$$

We see that the plot of the left member of the equation as a function of $1/T$ is usually not a straight line, except if h^f and s^f are temperature independent. Furthermore, in the general case its intersection with the vertical axis does not give the value s^f/k.

The evaluation of the relative importance of the terms between the parentheses demands the knowledge of c_P. The value of the latter cannot be determined from general thermodynamics; however, in a later section an expression will be given in terms of other vacancy parameters (and bulk qualities) based on the $cB\Omega$-model.

By introducing experimental values of h^f and c_P eq. (6.5) can be approximated by

$$\ln(T^2 \Delta C_P) = -\frac{h^f}{kT} + \frac{s^f}{k} + \ln\frac{(h^f)^2}{k} + \ln N. \tag{6.6}$$

Pressure dependence of ΔC_P. It is obvious that the values of n, h^f and $(\partial h^f/\partial T)_P$ in eq. (6.2) depend on pressure, and therefore, in order to find the pressure dependence of ΔC_P, their pressure derivatives are needed. The value of $(\partial n/\partial P)_T$ has already been given in eq. (3.29). The derivative $(\partial h^f/\partial P)_T$ is by definition the volume v_h^f, which has been given in eq. (3.162). The derivative, $\partial(\partial h^f/\partial T)_P/\partial P$ can be calculated from

$$\frac{\partial}{\partial P}\left(\left.\frac{\partial h^f}{\partial T}\right|_P\right) = \frac{\partial}{\partial T}\left(\left.\frac{\partial h^f}{\partial P}\right|_T\right) = \left.\frac{\partial v_h^f}{\partial T}\right|_P.$$

The last derivative has been calculated in eq. (3.185). By differentiating eq. (6.2) with respect to pressure we get for monoatomic solids:

$$\left.\frac{\partial \Delta C_P}{\partial P}\right|_T = n\frac{\partial^2 h^f}{\partial T \partial P} + n\frac{2h^f}{kT^2}\left.\frac{\partial h^f}{\partial P}\right|_T + \left\{\frac{(h^f)^2}{kT^2} + \left.\frac{\partial h^f}{\partial T}\right|_P\right\}\left.\frac{\partial n}{\partial P}\right|_T,$$

or by using eqs. (3.29), (3.43) and (3.170):

$$\left.\frac{\partial \Delta C_P}{\partial P}\right|_T = -\frac{nv^f}{kT}\left\{\frac{(h^f)^2}{kT^2} + \left.\frac{\partial h^f}{\partial T}\right|_P\right\} + \frac{2nh^f v_h^f}{kT^2} - nTv^f\left\{(\beta^f)^2 + \left.\frac{\partial \beta^f}{\partial T}\right|_P\right\}.$$

With the help of eqs. (3.161) and (6.3) this equation becomes

$$\left.\frac{\partial \Delta C_P}{\partial P}\right|_T = -\frac{nv^f h^f}{kT^2}\left\{\frac{h^f}{kT} - 2\right\} - \frac{nv^f c_P}{kT} - \frac{2nv_s^f h^f}{kT^2}$$
$$- nTv^f\left\{(\beta^f)^2 + \left.\frac{\partial \beta^f}{\partial T}\right|_P\right\} \qquad (6.7)$$

In most cases the values of v^f, c_P, v_s^f, β^f and $(\partial \beta^f/\partial T)_P$ are positive; by taking also into account that $h^f \gg kT$, eq. (6.7) indicates that, as expected, the contribution of vacancies to the isobaric specific heat of the solid decreases with increasing pressure.

6.2.2 Isochoric specific heat of an isobaric perfect crystal

The isochoric specific heat C_V^0 of an isobaric perfect crystal (fig. 6.2a) is defined by

$$C_V^0 \equiv \left.\frac{\partial U^0}{\partial T}\right|_{V^0}, \qquad (6.8)$$

whereas

$$C_P^0 \equiv \left.\frac{\partial H^0}{\partial T}\right|_P = T\left.\frac{\partial S^0}{\partial T}\right|_P. \qquad (6.9)$$

Later, an expression for $C_P^0 - C_V^0$ will be needed. The following lengthy calculation will prove that it is given by an expression very similar to the well known thermodynamic formula for $C_P - C_V$.

Two of the four quantities S^0, V^0, T and P are independent; therefore we can write

$$\left.\frac{\partial S^0}{\partial T}\right|_P = \left.\frac{\partial S^0}{\partial T}\right|_{V^0} + \left.\frac{\partial S^0}{\partial V^0}\right|_T \left.\frac{\partial V^0}{\partial T}\right|_P \qquad (6.10)$$
$$\text{(a)} \qquad \text{(b)} \qquad \text{(c)}$$

According to eq. (3.42) factor (c) above gives

$$\left.\frac{\partial V^0}{\partial T}\right|_P = \beta^0 V^0. \qquad (6.11)$$

The factor (b) can be written as

$$\left.\frac{\partial S^0}{\partial V^0}\right|_T = \left.\frac{\partial S^0}{\partial P}\right|_T \left.\frac{\partial P}{\partial V^0}\right|_T. \qquad (6.12)$$

Combining eq. (3.67) and eq. (3.42) we find:

$$\left.\frac{\partial S^0}{\partial P}\right|_T = -\beta^0 V^0. \qquad (6.13)$$

Using eq. (6.13), together with eq. (3.52), we can rewrite eq. (6.12) as

$$\left.\frac{\partial S^0}{\partial V^0}\right|_T = \frac{\beta^0}{\kappa^0}. \tag{6.14}$$

According to eq. (3.68a) the value of term (a) is

$$\left.\frac{\partial S^0}{\partial T}\right|_{V^0} = \frac{1}{T} \left.\frac{\partial U^0}{\partial T}\right|_{V^0}, \tag{6.15}$$

or by introducing eq. (6.8):

$$\left.\frac{\partial S^0}{\partial T}\right|_{V^0} = \frac{C_V^0}{T}. \tag{6.16}$$

The insertion of eqs. (6.16), (6.14) and (6.11) into (6.10) gives

$$\left.\frac{\partial S^0}{\partial T}\right|_P = \frac{C_V^0}{T} + \frac{\beta^0}{\kappa^0} \beta^0 V^0. \tag{6.17}$$

The combination of this equation and eq. (6.9) finally gives

$$C_P^0 - C_V^0 = \frac{TV^0(\beta^0)^2}{\kappa^0}, \tag{6.18}$$

which has the same form as the well known thermodynamic formula

$$C_P - C_V = \frac{TV\beta^2}{\kappa}. \tag{6.19}$$

Calculation of ΔC_V. The real and perfect crystals of fig. (6.2) have the isochoric specific heats C_V and C_V^0. Their difference $\Delta C_V \equiv C_V - C_V^0$ is found by subtracting eq. (6.18) from eq. (6.19):

$$\Delta C_V = \Delta C_P - T \left\{ \frac{V\beta^2}{\kappa} - \frac{V^0(\beta^0)^2}{\kappa^0} \right\}, \tag{6.20}$$

or by considering eqs. (6.2) and (6.3):

$$\Delta C_V = n \left\{ \frac{(h^f)^2}{kT^2} + c_P \right\} - T \left\{ \frac{V\beta^2}{\kappa} - \frac{V^0(\beta^0)^2}{\kappa^0} \right\}. \tag{6.21}$$

It gives an expression of ΔC_V in terms of the parameters β^0 and κ^0 of a perfect crystal which usually cannot be measured. This difficulty can be avoided as follows: we use the differential operator Δ (as has been done until now), denoting the difference of each physical property of the real crystal in regard of the perfect one. The value of the second bracket in eq. (6.21) can now be written as

$$\Delta\left(\frac{V\beta^2}{\kappa}\right) = \frac{\beta^2}{\kappa}\Delta V + \frac{2\beta V}{\kappa}\Delta\beta - \frac{V\beta^2}{\kappa^2}\Delta\kappa. \tag{6.22}$$

ΔV has been calculated in eq. (3.32); similarly $\Delta\beta = \beta - \beta^0$ and $\Delta\kappa = \kappa - \kappa^0$ have been given by eqs. (3.49) and (3.56). Under the assumption $\beta^f - \beta^0 \ll h^f/kT^2$ and $\kappa^f - \kappa^0 \ll v^f/kT$, eqs. (3.49) and (3.56) give:

$$\Delta\beta \simeq \frac{v^f}{\Omega}\frac{n}{N}\frac{h^f}{kT^2}, \qquad \Delta\kappa \simeq \frac{v^f}{\Omega}\frac{n}{N}\frac{v^f}{kT}.$$

Taking into consideration that $V = N\Omega$, one gets

$$\Delta\left(\frac{V\beta^2}{\kappa}\right) = \frac{n}{kT^2}\left\{\frac{\beta^2 kT^2 v^f}{\kappa} + \frac{2\beta v^f h^f}{\kappa} - \frac{\beta^2 v^{f^2} T}{\kappa^2}\right\}. \tag{6.23}$$

Inserting this into eq. (6.21) one obtains

$$\Delta C_V = n\left\{\frac{1}{kT^2}\left(h^f - \frac{\beta v^f T}{\kappa}\right)^2 - \frac{\beta^2 T v^f}{\kappa} + c_P\right\}. \tag{6.24}$$

Rough calculations show that $(1/k)(h^f - \beta v^f T/\kappa)^2$ is usually larger than $\beta^2 T^3 v^f/\kappa$, but for certain materials (as gold), and for temperatures for which the contribution of vacancies to the specific heat is measured, the above expression becomes approximately zero. Under such conditions it can happen that ΔC_V will be extremely small, even if n is considerable. With the help of eqs. (3.128) and (3.95) eq. (6.24) can also be written as

$$\Delta C_V = u^* \left.\frac{\partial n}{\partial T}\right|_V + n(c_P - \beta^2 T v^f B), \tag{6.25}$$

or

$$\Delta C_V = \frac{nu^{*2}}{kT^2} + n(c_P - \beta^2 T v^f B). \tag{6.26}$$

A definite physical meaning can now be given to this new form: the first term is the heat required for the isochoric production of new vacancies. The quantity nc_P in the second term is the specific heat for "heating" n vacancies in an isobaric process and therefore involves a volume increase. In order to convert it to an isochoric process, it has to be corrected. The correction $(\beta^2 T/\kappa)nv^f$ is a result of the volume expansion nv^f. This correction is similar in form to the correction term in eq. (6.19) for a volume V.

Temperature dependence of ΔC_V. Equation (6.24) can easily be changed into

$$\ln(\Delta C_V T^2) = s^f/k - h^f/kT + \ln N$$
$$+ \ln\left\{k^{-1}(h^f - \beta T v^f B)^2 - \beta^2 T^3 v^f B + c_P T^2\right\}. \tag{6.27}$$

It is clear that the plot of the left member of eq. (6.27) versus $1/T$ will not be a straight line, even if h^f and s^f are temperature independent. Furthermore, its intercept with the vertical axis will not be s^f/k; it can be either larger or smaller, depending on the sign of the logarithm of the quantity in brackets.

Equation (6.24) reveals that ΔC_V has three contributions. The relative importance of the first two can be found from the ratio of $\beta^2 T^3 v^f B$ and $(h^f - \beta T v^f B)^2/k$. After some calculation one obtains for this ratio

$$\left(\frac{\beta T v^f B}{h^f - \beta T v^f B}\right)^2 \frac{kT}{Bv^f}.$$

The formation volume v^f is of the order of Ω and hence Bv^f is of the order of $B\Omega$. The term $B\Omega$ is roughly by one order of magnitude greater than the formation enthalpy h^f. Taking into account that h^f exceeds kT – in the high temperature region – roughly by one order of magnitude one finds that the ratio kT/Bv^f is of the order of one percent. In order to estimate the factor $[\beta T v^f B/(h^f - \beta T v^f B)]^2$ one takes into consideration that at high temperatures the terms h^f and $\beta T v^f B$ are of the same order of magnitude. This is so because the product βT is of the order of 10^{-1} whereas $v^f B$, as already mentioned, exceeds h^f roughly by one order of magnitude. The value of the denominator depends on $h^f - \beta T v^f B$. We discuss now the following cases:

(1) $h^f = \beta T v^f B$.

Then eq. (6.24) becomes:

$$\Delta C_V = n(c_P - \beta^2 T v^f B).$$

The above relation shows that at temperatures where $h^f = \beta v^f TB$, the difference $C_V - C_V^0$ is relatively small (practically zero); in this case whenever $c_P = 0$ the difference $C_V - C_V^0$ takes a small *negative* value.

(2) $h^f > \beta T v^f B$ or $h^f < \beta T v^f B$.

Whenever h^f exceeds $\beta T v^f B$ by an appreciable amount (i.e. by a factor of 3 or more), the relative importance of the second term in eq. (6.24) is of the order of 1% or even less in comparison to the first term.

6.3 Contribution of vacancies to an isochoric perfect crystal

The enthalpy and the internal energy of the isochoric perfect crystal (P^{0*}, V, T) of fig. 3.3 have been labelled with H^{0*} and U^{0*} [see eqs. (3.101) and (3.91)]. The specific heats of this crystal are defined by

$$C_P^{0*} = \left.\frac{\partial H^{0*}}{\partial T}\right|_{P^{0*}} \tag{6.28}$$

and

$$C_V^{0*} = \left.\frac{\partial U^{0*}}{\partial T}\right|_V. \tag{6.29}$$

By introducing eq. (3.122) into eq. (6.28) we get

$$C_P^{0*} = T\left.\frac{\partial S^{0*}}{\partial T}\right|_{P^{0*}}. \tag{6.30}$$

Calculation of ΔC_V^*. The connection between C_V^{0*} and C_V of the corresponding real crystal is found by differentiating eq. (3.91) with respect to temperature

$$C_V = \left.\frac{\partial U}{\partial T}\right|_V = \left.\frac{\partial U^{0*}}{\partial T}\right|_V + n\left.\frac{\partial u^*}{\partial T}\right|_V + u^*\left.\frac{\partial n}{\partial T}\right|_V. \tag{6.31}$$

By introducing eqs. (6.29) and (3.95) into eq. (6.31) we get:

$$\Delta C_V^* \equiv C_V - C_V^{0*} = n\left(\left.\frac{\partial u^*}{\partial T}\right|_V + \frac{u^{*2}}{kT^2}\right). \tag{6.32}$$

The general form of this equation is analogous to eq. (6.2). The contribution of the vacancies has two terms. The first is needed for "the heating" of n vacancies and the second for the further production of vacancies. As $\partial u^*/\partial T$ reflects the heat needed for isochorically increasing "the temperature of one vacancy" it will be labelled with:

$$c_V \equiv \left.\frac{\partial u^*}{\partial T}\right|_V, \tag{6.33}$$

and therefore eq. (6.32) becomes [637]:

$$\Delta C_V^* = n\left(c_V + \frac{u^{*2}}{kT^2}\right). \tag{6.34}$$

Calculation of $C_P^{0*} - C_V^{0*}$. From the variables S^{0*}, P^{0*}, V and T two are independent. We can therefore write:

$$\underbrace{\left.\frac{\partial S^{0*}}{\partial T}\right|_{P^{0*}}}_{(a)} = \underbrace{\left.\frac{\partial S^{0*}}{\partial T}\right|_V}_{(b)} + \underbrace{\left.\frac{\partial S^{0*}}{\partial V}\right|_T \left.\frac{\partial V}{\partial T}\right|_{P^{0*}}}_{(c)}. \tag{6.35}$$

According to eq. (6.30) the left side is equal to C_P^{0*}/T. The term (a) of the right side is found by inserting the value of U^{0*} from eq. (3.89) into eq. (6.29). One gets

$$C_V^{0*} = \left.\frac{\partial F^{0*}}{\partial T}\right|_V + T\left.\frac{\partial S^{0*}}{\partial T}\right|_V + S^{0*}.$$

The first and the third term on the right cancel because of eq. (3.85).

According to eqs. (3.120) and (3.121) the factor (b) of eq. (6.35) is equal to β^{0*}/κ^{0*}. Finally, factor (c) is $\beta^{0*}V$ by definition. Inserting all these quantities into eq. (6.35) one gets:

$$C_P^{0*} - C_V^{0*} = \frac{TV(\beta^{0*})^2}{\kappa^{0*}}. \tag{6.36}$$

A combination of eqs. (6.19), (6.32) and (6.36) can lead to the value of ΔC_P^* ($= C_P - C_P^{0*}$), which is *not the same* as that for ΔC_P.

6.4 Specific heat of one vacancy

Many theoretical calculations are aimed at calculating the change of the specific heat when the number of vacancies is increased by one. This change will be called the "specific heat of one vacancy" and will be labelled with c. Its value depends on the crystal to which the one vacancy is added and therefore three types of crystals have to be considered. They are the real crystal and the two perfect crystals on the left side of figs. 3.2 and 3.3. The resulting quantities would be c_P, c_V, c_P^0, c_V^0, c_V^{0*}, c_P^{0*}. We will see that only the following four quantities: c_P and c_V for the real crystal and c_V^0 and c_P^{0*} for the perfect crystals, are independent, because $c_P = c_P^0$ and $c_V = c_V^{0*}$.

Here we will derive valuable expressions for $c_P - c_V$, $c_P^0 - c_V^0$ and $c_V^0 - c_V^{0*}$. They are of considerable interest, because theoretical calculations on an atomic scale usually lead to expressions for c_V^0. The quantity c_P^0 could also be directly computed from c_V^0.

6.4.1 Real crystal

We consider a real crystal before and after the increase of the vacancies from n to $n+1$. As the two following types of specific heat are involved,

$$C_P = \left.\frac{\partial H}{\partial T}\right|_P \quad \text{and} \quad C_V = \left.\frac{\partial U}{\partial T}\right|_V, \tag{6.37a,b}$$

we have to study an isobaric and an isochoric introduction of the vacancy. In fig. 6.4 we have depicted the crystal before and after the introduction. The specific heats C_P and C_V will have increased to $C_P + c_p$ and $C_V + c_V$.

In the figures on the right we have noted the new value of the enthalpy and the internal energy. The new specific heats are obtained by differentiation:

$$C_P + c_p = \left.\frac{\partial(H + h^f)}{\partial T}\right|_P = C_P + \left.\frac{\partial h^f}{\partial T}\right|_P,$$

$$C_V + c_V = \left.\frac{\partial(U + u^*)}{\partial T}\right|_V = C_V + \left.\frac{\partial u^*}{\partial T}\right|_V.$$

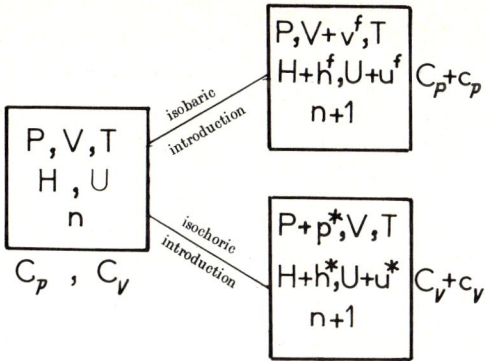

Fig. 6.4. Definition of the quantities c_P and c_V of a real crystal.

This gives:

$$c_P = \left.\frac{\partial h^f}{\partial T}\right|_P \quad \text{and} \quad c_V = \left.\frac{\partial u^*}{\partial T}\right|_V. \qquad (6.38\text{a,b})$$

Unfortunately the thermodynamic function $(\partial u^*/\partial T)_V$ is not directly accessible from measurements, so that c_V cannot be determined directly *. However, as will be shown in this section, this is possible for their difference $c_P - c_V$, so that c_V can then be evaluated if c_P is known.

Calculation of $c_P - c_V$. According to eq. (6.38b), c_V reflects the isochoric temperature dependence of u^*. It can alternatively be expressed with other vacancy parameters. By writing

$$c_V \equiv \left.\frac{\partial u^*}{\partial T}\right|_V = \left.\frac{\partial u^*}{\partial T}\right|_P + \left.\frac{\partial u^*}{\partial P}\right|_T \left.\frac{\partial P}{\partial T}\right|_V$$

and considering eqs. (2.27) and (3.128) we have

$$c_V = \left.\frac{\partial h^f}{\partial T}\right|_P - v^f \beta B - T\beta B \left.\frac{\partial v^f}{\partial T}\right|_P - Tv^f B \left.\frac{\partial \beta}{\partial T}\right|_P - Tv^f \beta \left.\frac{\partial B}{\partial T}\right|_P$$

$$+ \beta B \left\{ \left.\frac{\partial h^f}{\partial P}\right|_T - T\beta B \left.\frac{\partial v^f}{\partial P}\right|_T - Tv^f B \left.\frac{\partial \beta}{\partial P}\right|_T - Tv^f \beta \left.\frac{\partial B}{\partial P}\right|_T \right\}.$$

Taking into account the relations (see eqs. 3.162, 3.164):

$$\left.\frac{\partial h^f}{\partial T}\right|_P = c_P, \quad \left.\frac{\partial v^f}{\partial T}\right|_P = \beta^f v^f, \quad \left.\frac{\partial h^f}{\partial P}\right|_T = v_h^f, \text{ and } \left.\frac{\partial v^f}{\partial P}\right|_T = -\kappa^f v^f,$$

---

* As will be found later c_P is usually positive and c_V negative.

the previous relation becomes:

$$c_V = c_P - v^f \beta B - T\beta B \beta^f v^f - Tv^f B \left.\frac{\partial \beta}{\partial T}\right|_P - Tv^f \beta \left.\frac{\partial B}{\partial T}\right|_P$$
$$+ \beta B \left\{ v_h^f + T\beta B \kappa^f v^f - Tv^f B \left.\frac{\partial \beta}{\partial P}\right|_T - Tv^f \beta \left.\frac{\partial B}{\partial P}\right|_T \right\}.$$

By considering that $v_h^f = v^f(1 - T\beta^f)$ we finally get:

$$c_P - c_V = T\beta^2 B v^f \left(2\frac{\beta^f}{\beta} - \frac{\kappa^f}{\kappa} \right) + Tv^f \beta B \left\{ \left.\frac{\partial \ln(\beta B)}{\partial T}\right|_P + \left.\frac{\partial (\beta B)}{\partial P}\right|_T \right\}.$$
(6.39)

This purely thermodynamic formula allows the evaluation of the difference $c_P - c_V$ from experimentally accessible quantities.

The difference $c_P - c_V$ can also be set into the following form:

$$c_P - c_V = Tv^f \beta^2 B \left(2\frac{\beta^f}{\beta} - \frac{\kappa^f}{\kappa} \right)$$
$$+ Tv^f \beta^2 B \left(\frac{1}{\beta B} \left.\frac{\partial B}{\partial T}\right|_P + \frac{1}{\beta^2} \left.\frac{\partial \beta}{\partial T}\right|_P + \left.\frac{\partial B}{\partial P}\right|_T + \frac{B}{\beta} \left.\frac{\partial \beta}{\partial P}\right|_T \right),$$

so that by considering eq. (2.30) we finally get:

$$c_P - c_V = Tv^f \beta^2 B \left(2\frac{\beta^f}{\beta} - \frac{\kappa^f}{\kappa} \right)$$
$$+ Tv^f \beta^2 B \left(\frac{2}{\beta B} \left.\frac{\partial B}{\partial T}\right|_P + \left.\frac{\partial B}{\partial P}\right|_T + \frac{1}{\beta^2} \left.\frac{\partial \beta}{\partial T}\right|_P \right).$$
(6.40)

We note that for a pure harmonic crystal $c_P = c_V$ because $\beta = 0$. For a quasi-harmonic solid, see §6.7.

6.4.2 Isobaric perfect crystal

The isobaric perfect crystal has been depicted in fig. 3.2. In fig. 6.5 we show the same crystal before and after the addition of one vacancy. The right side describes the two real crystals that are derived from the perfect one by introduction of vacancies. In the center we have the result after introducing only one vacancy into the perfect crystal. In the case of an isobaric introduction the enthalpy will have increased from H^0 to $H^0 + h^f$, while in the isochoric process the internal energy will have increased from U^0 to $U^0 + u^*(V^0, T)$.

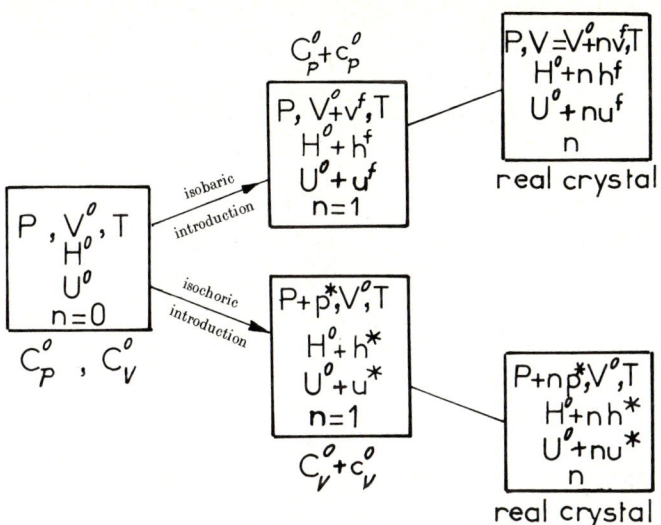

Fig. 6.5. Definition of c_P^0 and c_V^0. The real crystals are different and hence the equilibrium values n of vacancies are also different.

The changes c_P^0 and c_V^0 of the specific heat are found by differentiation:

$$c_P^0 = \left.\frac{\partial h^f}{\partial T}\right|_P, \qquad c_V^0 = \left.\frac{\partial u^*}{\partial T}\right|_{V^0}. \qquad (6.41\text{a,b})$$

We note that according to eq. (6.38a) the quantity c_P^0 of the isobaric perfect crystal is equal to c_P of the real crystal, because they refer to the same state P, V, T:

$$c_P^0 = c_P. \qquad (6.42)$$

On the other hand eq. (6.38b) shows that c_V^0 is not equal to c_V, because the former refers to a volume V^0 and not to a volume V.

Calculation of $c_P^0 - c_V^0$. The connection between C_P^0 and C_V^0 of the isobaric perfect crystal is given by eq. (6.18). The state under consideration is the one depicted on the left side of fig. 6.5. For the calculation of $c_P^0 - c_V^0$ we apply to the above equation a differential operator δ that corresponds to the variation of each physical property when a single vacancy is introduced under conditions of constant pressure and temperature (method of Gilder–Lazarus [37]):

$$\delta C_P^0 - \delta C_V^0 = \beta^{0^2} T \delta V^0 \left(\frac{2V^0 \delta \beta^0}{\beta^0 \delta V^0 \kappa^0} + \frac{1}{\kappa^0} - \frac{V^0}{(\kappa^0)^2} \frac{\delta \kappa}{\delta V^0} \right). \qquad (6.43)$$

The quantity $\delta\beta^0$ can be found by applying the operator δ to eq. (3.42)

$$\delta\beta^0 = -\frac{1}{V^0}\frac{1}{V^0}\frac{\partial V^0}{\partial T}\bigg|_P \delta V^0 + \frac{1}{V^0}\frac{\partial \delta V^0}{\partial T}\bigg|_P,$$

or

$$\delta\beta^0 = -\beta^0 \frac{\delta V^0}{V^0} + \frac{\delta V^0}{V^0}\frac{1}{\delta V^0}\frac{\partial \delta V^0}{\partial T}\bigg|_P.$$

The difference between the volume V of a real crystal and the volume V^0 of its isobaric perfect crystal is given by eq. (3.32); therefore the volume increase for the formation of one vacancy is $\delta V^0 = v^f$. Taking this into account and using the definition of β^f one gets:

$$\delta\beta^0 = -\beta^0 \frac{v^f}{V^0} + \frac{v^f}{V^0}\beta^f \quad \text{or} \quad \delta\beta^0 = (\beta^f - \beta^0)\frac{v^f}{V^0}.$$

Similarly, by applying the differential operator δ to the definition of the compressibility κ^0 of the perfect lattice one obtains

$$\delta\kappa^0 = (\kappa^f - \kappa^0)\frac{v^f}{V^0}.$$

The last two equations simply show that the expansion coefficient and the compressibility of a perfect crystal varies upon introduction of a single vacancy, if $\beta^f \neq \beta^0$ and $\kappa^f \neq \kappa^0$. By inserting the above values of $\delta\beta^0$ and $\delta\kappa^0$ into eq. (6.43) we finally get:

$$\delta C_P^0 - \delta C_V^0 = \frac{\beta^{0^2} T v^f}{\kappa^0}\left(2\frac{\beta^f}{\beta^0} - \frac{\kappa^f}{\kappa^0}\right). \tag{6.44}$$

In this formula, δC_P^0 can be replaced by c_P^0 because, according to eq. (3.23) $H - H^0$ for one vacancy is h^f and therefore $\delta C_P^0 = \partial h^f / \partial T$, which is exactly c_P^0. In the same formula δC_V^0 is c_V^0 (see fig. 6.5). It is *not* c_V because the latter refers to volume V (see fig. 6.4). Equation (6.44) therefore is

$$c_P^0 - c_V^0 = \frac{(\beta^0)^2 T v^f}{\kappa^0}\left(2\frac{\beta^f}{\beta^0} - \frac{\kappa^f}{\kappa^0}\right). \tag{6.45}$$

Because of eq. (6.42) this is also an expression for $c_p - c_V^0$. In the harmonic case ($\beta^0 = 0$) we obtain $c_P^0 = c_V^0$ as expected. For the case of a quasi-harmonic solid, see §6.7.

6.4.3 Isochoric perfect crystal

The left side of fig. 6.6 refers to the isochoric perfect crystal depicted in fig. 3.3. On the right side we describe two real crystals, which are derived from the perfect one by isothermal introduction of vacancies. In the center we have the result after introducing only one vacancy. In the isobaric introduction the enthalpy will have increased by h^f; correspondingly in the isochoric introduction the internal energy will have increased by u^*. By comparing the upper right side of the figure to the left side we note that the quantities v^f and h^f have been derived by the comparison of the perfect crystal (P^{0*}, V, T) to the real crystal ($P^{0*}, V+nv^f, T$); further the quantities p^* and u^* are derived from the comparison of the perfect crystal (P^{0*}, V, T) to the (lower) real crystal ($P^{0*}+np^*, V, T$). After the introduction of one vacancy the specific heats C_P^{0*} and C_V^{0*} have increased by

$$c_P^{0*} = \left.\frac{\partial h^f}{\partial T}\right|_{P^{0*}} \quad \text{and} \quad c_V^{0*} = \left.\frac{\partial u^*}{\partial T}\right|_V. \quad (6.46\text{a, b})$$

By comparing eq. (6.46b) to eq. (6.38b) we see that c_V^{0*} of the isochoric perfect crystal is exactly equal to c_V of the real crystal.

$$c_V^{0*} = c_V. \quad (6.47)$$

On the other hand a comparison of eq. (6.46a) to eq. (6.41a) shows that c_P^{0*} and c_P^0 are *not* equal because they refer to different pressures.

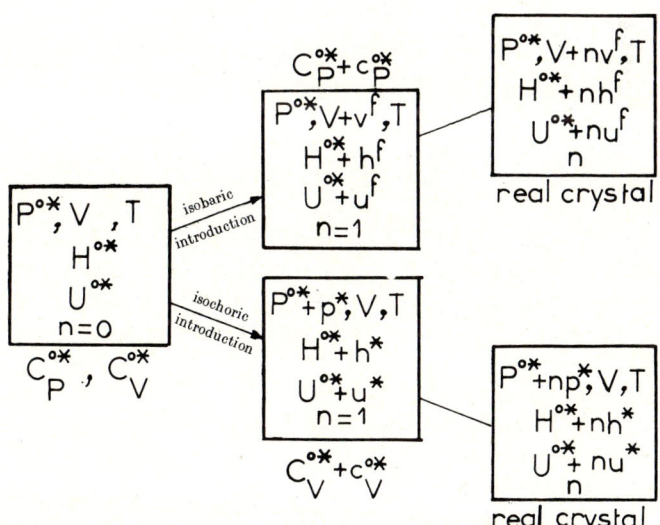

Fig. 6.6. Definition of c_P^{0*} and c_V^{0*}. Note that the two real crystals are different.

Calculation of $c_V^0 - c_V^{0*}$. Under the approximation $\kappa^f/\kappa \simeq \kappa^f/\kappa^0$ eq. (6.40) can be written as (consider also eq. 6.47)

$$c_P - c_V^{0*} \simeq Tv^f\beta^{02}B^0\left(2\frac{\beta^f}{\beta^0} - \frac{\kappa^f}{\kappa^0}\right) + Tv^f\beta^2 B\left(\frac{2}{\beta B}\left.\frac{\partial B}{\partial T}\right|_P + \left.\frac{\partial B}{\partial P}\right|_T\right.$$
$$\left. + \frac{1}{\beta^2}\left.\frac{\partial \beta}{\partial T}\right|_P\right).$$

By combining this with eq. (6.45) we get:

$$c_V^0 - c_V^{0*} \simeq Tv^f\beta^2 B\left(\frac{2}{\beta B}\left.\frac{\partial B}{\partial T}\right|_P + \left.\frac{\partial B}{\partial P}\right|_T + \frac{1}{\beta^2}\left.\frac{\partial B}{\partial T}\right|_P\right).$$

We note that in the case of an harmonic crystal the difference $c_V^0 - c_V^{0*}$ becomes zero because $\beta = 0$, whereas in an anharmonic crystal this is not correct because C_V depends on the volume.

In order to estimate the value of the difference $c_V^0 - c_V^{0*}$ we calculate it in the case of Li close to 240 K. By using the elastic and expansivity data given in Ref. [47] and taking $v^f \simeq 0.5 \, \Omega$ we get $c_V^0 - c_V^{0*} = 0.5 \, k$. This difference is very large if one recalls that Gilder and Lazarus estimated c_V^0 (at $T \gg \Theta_D$) to be of the order of $-(\Theta_D/T)^2 k$. Furthermore, if one considers that $c_V^0 - c_V^{0*}$ is usually positive and that c_V^0 has a small, negative value one sees that c_V^{0*} has also a negative value. This alternatively means that the formation energy u^* *decreases* with temperature under the constraint of constant volume.

The difference $c_V^0 - c_V^{0*}$ can alternatively be calculated as follows: the thermodynamical identity $(\partial C_V/\partial V)_T = T[\partial(\beta B)/\partial T]_V$ gives

$$\frac{c_V^0 - c_V^{0*}}{v^f} = T\left.\frac{\partial(\beta B)}{\partial T}\right|_V,$$

and hence

$$c_V^0 - c_V^{0*} = Tv^f\left.\frac{\partial(\beta B)}{\partial T}\right|_V.$$

6.5 Connections between the specific heats of perfect crystals

The two perfect crystals depicted on the left sides of figs. 6.2 and 6.3 are in different states $[(P, V^0, T)$, respectively $(P^{0*}, V, T)]$ and therefore have the different specific heats (C_P^0, C_V^0) and (C_P^{0*}, C_V^{0*}). By subtracting eq.

(6.26) from eq. (6.32) and inserting eq. (6.33) one gets

$$C_V^0 - C_V^{0*} = n(c_V - c_P) + nTv^f\beta^2/\kappa; \tag{6.48}$$

when the last term is dominant, the sign of $C_V^0 - C_V^{0*}$ depends on the sign of v^f. The direct evaluation of the difference $C_V^0 - C_V^{0*}$ is possible because the difference $c_V - c_P$ is already known from eq. (6.40). Equation (6.48) can alternatively be written as

$$C_V^0 - C_V^{0*} = n(c_V - c_P) + n\beta TBv^f\beta. \tag{6.49}$$

Taking into account that βT is at the most of the order of 10^{-1}, whereas Bv^f exceeds h^f by one order of magnitude, we get for temperatures near the melting point:

$$C_V^0 - C_V^{0*} \simeq n(c_V - c_P) + n\beta h^f. \tag{6.50}$$

In order to evaluate the importance of the selection of the perfect crystal in the analysis of experimental results we have to compare the order of magnitude of $C_V^0 - C_V^{0*}$ to the order of magnitude of $C_V - C_V^0$. The ratio of eqs. (6.48) and (6.26) gives

$$\left(c_V - c_P + \frac{Tv^f\beta^2}{\kappa}\right)\left(\frac{(u^*)^2}{kT^2} - \frac{Tv^f\beta^2}{\kappa} + c_P\right)^{-1}.$$

In restricted cases where c_V and c_P are found to be practically zero this ratio becomes:

$$\left[\frac{\kappa}{Tv^f\beta^2}\frac{1}{k}\left(\frac{h^f}{T} - v^f\beta B\right)^2 - 1\right]^{-1}.$$

We see that the above ratio strongly depends on the value of $h^f(T\beta v^f B)^{-1}$.

As an example we discuss the case of Argon at 77.7 K. By using the values $\beta = 19.4 \times 10^{-4}$ K^{-1}, $\Omega = 40.25$ Å^3, $v \simeq \Omega$, $B = 12.7$ kbar and $h^f \simeq 0.065$ eV [48], we find that the difference $C_V^0 - C_V^{0*}$ is around 10% of $C_V - C_V^0$ and is therefore *not* negligible when we are interested in an accurate analysis of experimental specific heats.

6.6 Specific heats of other defects

The formulae of crystals containing other types of defects are similar in principle but not exactly the same as those that are valid for vacancies.

(a) *Crystals of type A^+B^-, containing Schottky defects.* By following the same procedure as in the case of vacancies we obtain

$$\Delta C_P = n\left\{\left.\frac{\partial h^f}{\partial T}\right|_P + \frac{(h^f)^2}{2kT^2}\right\}. \tag{6.51}$$

Whenever $(\partial h^f/\partial T)_P \ll (h^f)^2/2kT^2$ we have the approximate relation

$$\Delta C_P \simeq N e^{s^f/2k} e^{-h^f/2kT}\left[(h^f)^2/2kT^2\right]. \tag{6.52}$$

Estimation of ΔC_P. Equation 6.51 can be written as

$$\Delta C_P = \frac{n}{N}\left(N \left.\frac{\partial h^f}{\partial T}\right|_P + \frac{1}{2}\frac{h^f}{kT}\frac{h^f}{kT}kN\right).$$

As h^f typically exceeds kT by one order of magnitude, the contribution of the second term is of the order of $10^2 R/2$. The first term usually is appreciably smaller. Near the melting point, n/N reaches values of the order of 10^{-3}. The contribution of Schottky defects therefore is of the order of $10^{-1} R$.

Temperature dependence of ΔC_P. Equation (6.52) reveals that the plot of $\ln(T^2\Delta C_P)$ versus $1/T$ is only a straight line with a gradient equal to $-h^f/2k$ if h^f (and hence s^f) is temperature independent. If, as usual, they increase with temperature, both slope and intercept increase, and therefore the plot shows an upward curvature.

(b) *Crystals of type A^+B^-, containing Frenkel defects.* By differentiating eq. (4.27) with respect to temperature one obtains:

$$\Delta C_P = n \left.\frac{\partial h^f}{\partial T}\right|_P + h^f \left.\frac{\partial n}{\partial T}\right|_P. \tag{6.53}$$

The derivative $(\partial n/\partial T)_P$ is found by differentiating eq. (4.22):

$$\left.\frac{\partial n}{\partial T}\right|_P = \frac{nh^f}{2kT^2}.$$

Therefore eq. (6.53) becomes:

$$\Delta C_P = n\left\{\left.\frac{\partial h^f}{\partial T}\right|_P + \frac{(h^f)^2}{2kT^2}\right\}. \tag{6.54}$$

The term $(\partial h^f/\partial T)_P$ is relatively small in comparison to $(h^f)^2 (2kT^2)^{-1}$, even if h^f varies (but not too strongly) with temperature. In such cases one obtains the approximate relation:

$$\Delta C_P \simeq n \frac{(h^f)^2}{2kT^2},$$

and due to eq. (4.22)

$$\Delta C_P \simeq N\sqrt{2}\, e^{s^f/2k} e^{-h^f/2kT}\left[(h^f)^2/2kT^2\right].$$

As in the case of Schottky defects, the last equation reveals that only if h^f (and hence s^f) is temperature independent, the plot of $\ln(T^2\Delta C_P)$ versus

$1/T$ would be a straight line, with a gradient equal to $-h^f(2k)^{-1}$. If, as usual, they increase with temperature, both slope and intercept increase with temperature so that the curve has to show an upward curvature.

Estimation of ΔC_P. Equation (6.54) can be written:

$$\Delta C_P = \frac{n}{N}\left\{N\left.\frac{\partial h^f}{\partial T}\right|_P + \tfrac{1}{2}\left(\frac{h^f}{kT}\right)^2 kN\right\}.$$

Due to the fact that typically h^f exceeds kT by one order of magnitude, the second contribution $\tfrac{1}{2}(h^f/kT)^2 R$ should be of the order of $10^2 R$. The other term, $N(\partial h^f/\partial T)_P$, is usually appreciably smaller. However, in some cases – as in the case of AgBr – the enthalpy h^f strongly increases with temperature so that at room temperature or higher the term $N(\partial h^f/\partial T)_P$ may be of the same order of magnitude as the second term. By also taking into account that (in these cases) the concentration n/N is very large, i.e. $\sim 10^{-2}$, we conclude that the difference ΔC_P should be of the order of a few times R. As the molar specific heat of these crystals is typically $6R$, we see that the contribution of vacancies is very important.

6.7. Criteria of microscopic calculations from specific heats

Microscopic calculations carried out within the frame of the quasi-harmonic approximation usually assume that v^f does not depend *explicitly* on temperature i.e. $(dv^f/dT)_V = 0$. In such cases the calculated values should obey the following fundamental constraints, which were hitherto not realised (compare with the comments of §A.11, note 18, valid for migration processes): Applying eq. (2.34a) for the quantity v^f and considering the definitions of β^f and κ^f we have: $(dv^f/dT)_P + \beta B(dv^f/dP)_T = 0$ and therefrom:

$$\beta^f/\beta = \kappa^f/\kappa. \tag{6.55}$$

The insertion of eq. (6.55) into eqs. (6.40) and (6.45) leads to the following relations, respectively:

$$c_P - c_V = Tv^f\beta B\beta^f + Tv^f\beta^2 B\left(\frac{2}{\beta B}\left.\frac{\partial B}{\partial T}\right|_P + \left.\frac{\partial B}{\partial P}\right|_T + \frac{1}{\beta^2}\left.\frac{\partial \beta}{\partial T}\right|_P\right), \tag{6.56}$$

$$c_P - c_V^0 \simeq T\beta\beta^f v^f B. \tag{6.57}$$

As an example eq. (6.57) states that a thermodynamically consistent microscopic calculation should lead to values of $c_P(\equiv dh^f/dT)_P$ and $(dv^f/dT)_P$ which have to be interconnected through the relation (recall that $|c_P| \gg |c_V^0|$):

$$(c_P \equiv) \left.\frac{dh^f}{dT}\right|_P \simeq T\beta B \left.\frac{dv^f}{dT}\right|_P. \tag{6.58}$$

Another example refers to those calculations which further assume that u^* does not depend explicitly on temperature; in such a case the desired relation is obtained from eq. (6.56) by setting $c_V = 0$ because $c_V = (\mathrm{d}u^*/\mathrm{d}T)_V$.

It should be mentioned that some of the recent calculations published for alkali halides, silver halides and a few fluorides, have obtained values that violate the above thermodynamical relations.

7 ANALYSIS OF EXPERIMENTS YIELDING DEFECT PARAMETERS

In principle any crystal property that responds to the presence of point defects can be used as an object for experiments on defect parameters. In this section we will briefly discuss some widely used experiments that yield parameters for the formation and activation of vacancies. Positron annihilation spectroscopy will not be discussed as some points of its analysis have not yet been generally accepted. However, extensive literature on this subject is given in the second part of the book.

7.1 X-ray parameters

In the analysis of experiments giving the concentration of vacancies, another type of volume has to be considered beyond the real volume V and the volume V_0 of the isobaric perfect crystal as defined in eq. (3.28). It will be labelled with V^X and has to do with the lattice constant a^X that results from X-ray diffraction experiments. Furthermore, two new coefficients, the X-ray thermal expansion coefficients β^X and compressibility κ^X will emerge. The volumes V and V^0 are connected by eq. (3.32). Division of this equation by N gives, after considering eqs. (3.46) and (3.47):

$$\Omega = \Omega^0 + xv^f, \tag{7.1}$$

where $x \equiv n/N$. If there were no relaxation around a vacancy, the formation volume v^f would be equal to the atomic volume Ω^0 of the perfect crystal. However, due to the occurrence of an outward or inward relaxation around the vacancy, we have $v^f \gtrless \Omega^0$ and hence can write

$$v^f = \Omega^0 + f\Omega^0, \tag{7.2}$$

i.e.,

$$f \equiv \frac{v^f}{\Omega^0} - 1, \tag{7.3}$$

where f is a measure of the mean fractional relaxation. Due to eq. (7.2) equation (7.1) can be written as

$$\Omega = \Omega^0 + x\Omega^0(1+f) \quad \text{or} \quad V = V^0 + xV^0(1+f). \tag{7.4a,b}$$

7.1.1 X-ray volume

Beyond the volumes V and V^0 an "X-ray molar volume" V^X can be deduced from X-ray diffraction experiments. These experiments deliver an average size lattice constant a^X, and hence an average atomic volume $\Omega^X = \text{const.}(a^X)^3$. The constant depends on the structure of the crystal. By multiplying with N one gets $V^X = N\Omega^X$. The volume V^X is equal to the sum of V^0 and the fractional variation due to the relaxation around the vacancies:

$$V^X = V^0 + xV^0 f \quad \text{or} \quad \Omega^X = \Omega^0 + x\Omega^0 f. \tag{7.5a,b}$$

By subtracting eq. (7.5a) from eq. (7.4b) we find

$$V - V^X = xV^0 \quad \text{or} \quad x = (V - V^X)/V^0. \tag{7.6a,b}$$

In practice the denominator V^0 is taken as the volume of the crystal at a (low) reference temperature where the concentration of vacancies is negligible and therefore $V^0 = V^{X,\text{low}}$; in this way eq. (7.6b) leads to

$$x = (V - V^X)/V^{X,\text{low}}. \tag{7.7}$$

By combining eqs. (7.4b) and (7.5a) the volume V^0 can be eliminated

$$V = V^X\left(1 + \frac{x}{1 + xf}\right). \tag{7.8}$$

For "constant volume" X-ray studies, see § A.9.

7.1.2 X-ray thermal expansion coefficients

We have already defined the expansion coefficients β and β^0 [eqs. (2.24) and (3.42)]. They are connected by eq. (3.49). It is now obvious that the temperature variation of V^X (or a^X) leads to the definition of another expansion coefficient, β^X:

$$\beta^X \equiv \frac{1}{V^X}\left.\frac{\partial V^X}{\partial T}\right|_P \quad \text{or} \quad \beta^X \equiv \frac{3}{a^X}\left.\frac{\partial a^X}{\partial T}\right|_P. \tag{7.9a,b}$$

A relation between β, β^X and β^0 can be obtained by differentiating eq. (7.6a) with respect to temperature

$$\left.\frac{\partial V}{\partial T}\right|_P = \left.\frac{\partial V^X}{\partial T}\right|_P + x\left.\frac{\partial V^0}{\partial T}\right|_P + V^0\left.\frac{\partial x}{\partial T}\right|_P,$$

which gives

$$\beta V = \beta^X V^X + xV^0\left[\beta^0 + (h^f/kT^2)\right].$$

By inserting into this equation the values of V and V^X given by eqs. (7.4b) and (7.5a) we get

$$\beta\{1 + x(1+f)\} = \beta^X(1 + xf) + x[\beta^0 + (h^f/kT^2)]. \quad (7.10)$$

This relation is an exact one and permits the evaluation of β^0 from the experimentally determinable quantities β and β^X, provided that x and f are experimentally known at every temperature.

An explicit relation between β^X and β^0 is obtained by inserting into eq. (7.9a) the value of V^X given by eq. (7.5a) and using eq. (3.26):

$$\beta^X = \frac{1}{1+xf}\left\{\beta^0(1+xf) + x\left.\frac{\partial f}{\partial T}\right|_P + fx\frac{h^f}{kT^2}\right\}. \quad (7.11)$$

In order to find the value of $\partial f/\partial T$ we differentiate eq. (7.2) with respect to temperature:

$$\beta^f v^f = (1+f)\beta^0\Omega^0 + \Omega^0\left.\frac{\partial f}{\partial T}\right|_P = \beta^0\Omega^0 + f\beta^0\Omega^0 + \Omega^0\left.\frac{\partial f}{\partial T}\right|_P$$

and therefrom by using eq. (7.3)

$$\left.\frac{\partial f}{\partial T}\right|_P = (\beta^f - \beta^0)\frac{v^f}{\Omega^0}. \quad (7.12)$$

Inserting this value into eq. (7.11) we have

$$\beta^X = \beta^0 + \frac{x}{1+xf}\left\{(\beta^f - \beta^0)\frac{v^f}{\Omega^0} + f\frac{h^f}{kT^2}\right\}. \quad (7.13)$$

Whenever $\beta^f - \beta^0$ is appreciably smaller than $(h^f/kT)(1/T)$ (which is the usual case) and taking into account that v^f/Ω^0 is of the same order of magnitude with $|f|$, eq. (7.13) (disregarding the terms containing x^2) leads to [49,50]:

$$\beta^X \simeq \beta^0 + xf\frac{h^f}{kT^2} \quad \text{or} \quad \beta^X \simeq \beta^0 + x\left(\frac{v^f}{\Omega^0} - 1\right)\frac{h^f}{kT^2}. \quad (7.14\text{a,b})$$

Whenever v^f is smaller than Ω^0 (i.e. inward relaxation around the vacancy), then f is negative; equation (7.14b) then reveals that in such cases β^X is smaller than β^0. This result was expected: the X-ray measurements give the average size of the unit cell. With increasing temperature the number of vacancies also increases and as each of them produces an inward relaxation, the total increase of volume V^X will be smaller than the increase of V^0.

The difference between β^X and β^0 can be estimated from the following considerations: it is obvious that whenever $f = 0$, the quantity β^X coincides with β^0. In the case of metals experimental data show that f is of the order of -0.5; furthermore, near the melting point h^f exceeds kT by one order of magnitude and x is 10^{-3} (or even smaller). Therefore the term $xf(h^f/kT^2)$ is of the order of $-0.05 \times 10^{-1}/T_M$, whereas β^0 is of the order of $10^{-1}/T_M$. From eq. (7.14a) we see that β^X is smaller than β^0 by a few percent. This difference, however, is significant if one takes into account the experimental accuracy in the determination of β^X.

An explicit relation between β and β^X can be found by differentiating eq. (7.8) with respect to T:

$$\beta V = \left(1 + \frac{x}{1+xf}\right)\beta^X V^X + \frac{V^X}{(1+xf)^2}\left\{(1+xf)\left.\frac{\partial x}{\partial T}\right|_P + x\left.\frac{\partial(xf)}{\partial T}\right|_P\right\},$$

or

$$\beta = \left(1 + \frac{x}{1+xf}\right)\beta^X \frac{V^X}{V}$$

$$+ \frac{V^X}{V}\frac{1}{(1+xf)^2}\left\{(1+xf)\left.\frac{\partial x}{\partial T}\right|_P + x^2\left.\frac{\partial f}{\partial T}\right|_P + xf\left.\frac{\partial x}{\partial T}\right|_P\right\}.$$

Inserting the value of V^X/V from eq. (7.8) and with the approximation $(1+xf)^2 \simeq 1 + 2xf$ this gives

$$\beta \simeq \beta^X + x\frac{h^f}{kT^2}, \tag{7.15}$$

having disregarded the x^2-terms.

This equation shows that the difference $\beta - \beta^X$ is an extra expansion due solely to the production of new vacancies [see eq. (3.26)]. It also reveals that β always exceeds β^X. By considering that near the melting point h^f exceeds kT by one order of magnitude and x is (at the most) 10^{-3}, we note that the term $x(h^f/kT^2)$ is of the order of $10^{-2}/T_M$. Taking also into account that β is of the order of $10^{-1}/T_M$, we conclude that β and β^X differ by a considerable amount (of the order of 10%), which is measurable by the current experimental accuracy.

7.1.3 X-ray isothermal compressibility

The compressibility κ^X measured by X-ray diffraction is defined by

$$\kappa^X \equiv -\frac{1}{V^X}\left.\frac{\partial V^X}{\partial P}\right|_T. \tag{7.16}$$

The current experimental techniques allow the determination of κ^X, so that it will be useful to know its connection to κ and to the compressibility κ^0 of the perfect crystal. The relation between κ^X and κ^0 is found by inserting into eq. (7.16) the value of V^X given by eq. (7.5a):

$$\kappa^X = \kappa^0 - \frac{x}{1+xf}\left.\frac{\partial f}{\partial P}\right|_T - \frac{f}{1+xf}\left.\frac{\partial x}{\partial P}\right|_T. \tag{7.17}$$

The value of $\partial f/\partial P$ is found by differentiating eq. (7.3), while the value of

$\partial x/\partial P$ is found from eqs. (3.29) and (3.31). Therefore eq. (7.17) becomes

$$\kappa^X = \kappa^0 - \frac{x}{1+xf}\left\{\frac{v^f}{\Omega^0}(\kappa^0 - \kappa^f) - \frac{fv^f}{kT}\right\}. \tag{7.18}$$

In cases where there is no relaxation ($f = 0$) this equation shows that κ^X is larger than κ^0 because, as we shall show in ch. 8, the formation volume is more compressible than the real or the perfect lattice. Due to the fact that usually the absolute value of the term $v^f(\kappa^0 - \kappa^f)/\Omega^0$ is appreciably smaller than that of fv^f/kT (even if κ^f/κ is around 4–5, as described in ch. 14), eq. (7.18), disregarding x^2 terms, gives

$$\kappa^X \simeq \kappa^0 + xfv^f(kT)^{-1}. \tag{7.19}$$

This equation reveals that whenever inward relaxation around the vacancy occurs (i.e. $f < 0$), κ^X is smaller than κ^0. In order to estimate the difference between κ^X and κ^0 we recall that near the melting point $(1/\kappa)(v^f/kT)$ is of the order of 10^2, whereas typically $f \simeq -0.5$ and x is usually at the most 10^{-3}. Equation (7.19) then reveals that κ^X is smaller than κ^0 by a few percent. At first sight this relatively small difference $\kappa^X - \kappa^0$ is unimportant; this is not so, because in real solids an increase of the compressibility by a few percent occurs when the temperature increases by a considerable amount ($\Delta T \approx T_M/5$ to $T_M/4$).

An approximate relation between κ and κ^X can be found by combining eqs. (7.19) with (3.58)

$$\kappa \simeq \kappa^X + xv^f(kT)^{-1}. \tag{7.20}$$

This equation reveals that κ always exceeds κ^X, provided v^f is positive. By considering again that near the melting point v^f/κ exceeds kT by two orders of magnitude, whereas at the most $x \simeq 10^{-3}$, we find that κ exceeds κ^X typically by 10%.

A more exact relation between κ and κ^X can be found by combining eq. (7.18) with eq. (3.56). In the approximation $x/(1+xf) \simeq x$ we get after some calculation

$$\kappa \simeq \kappa^X + x\left(\frac{v^f}{kT} + \frac{v^f}{\Omega^0}\kappa\right). \tag{7.21}$$

Due to the fact that $(1/\kappa)(v^f/kT)$ exceeds v^f/Ω^0 by two orders of magnitude, we conclude that eq. (7.20) is a very good approximation.

It is interesting to compare the derivatives $\partial \kappa/\partial P$ and $\partial \kappa^X/\partial P$ which result from different kinds of experiments. The differentiation of eq. (7.20) gives

$$\left.\frac{\partial B}{\partial P}\right|_T = \left.\frac{\partial B^X}{\partial P}\right|_T \left(\frac{B}{B^X}\right)^2 + \left(\frac{Bv^f}{kT}\right)^2 x + \frac{\kappa^f}{\kappa}\frac{Bv^f}{kT}x. \tag{7.22}$$

A rough evaluation of the second term of the right-hand side gives a value equal to unity or larger, while the third term is two orders of magnitude smaller. For example in cases where $\partial B/\partial P$ is 5, $\partial B^X/\partial P$ is 25% smaller. Such a difference can be detected experimentally. The corresponding experiment is strongly recommended.

7.1.4 Determination of the formation volume from X-ray measurements.

Korpiun and Coufal [51] have indicated a method according to which macroscopic and X-ray measurements of expansivity and compressibility can lead to the determination of the formation volume. A combination of eqs. (7.15) and (7.20) leads to:

$$\frac{\kappa - \kappa^X}{\beta - \beta^X} = \frac{v^f}{h^f} T.$$

Therefore a plot of $\kappa - \kappa^X$ versus $(\beta - \beta^X)T$ leads to v^f/h^f; as h^f is known from the Simmons–Balluffi method [51a], v^f can be directly determined. The value $v^f = (1.5 - 0.1\Omega, 1.5 + 0.3\Omega)$ mentioned for Kr in ref. [51] has been reconsidered by Schoknecht [52] $(1.4 - 0.9\Omega, 1.4 + 0.5\Omega)$ and more recently by Macrander [53] $(v^f \simeq \Omega)$. This method can be successfully applied as long as the measurements of κ, κ^X, β and β^X have been carried out in the same experimental conditions.

7.2 Analysis of specific heat measurements

As developed in ch. 6 the specific heat of a solid contains terms due to the production of vacancies and to the "heating" of the vacancies present. It is therefore desirable to use measurements of the specific heat for the determination of vacancy parameters. Most measurements refer to experiments under constant pressure, because they are more easily carried out than experiments under constant volume. A typical temperature dependence of C_P has been depicted in fig. 2.1. It is the abrupt increase in the range of high temperatures where the influence of the vacancies becomes important. Various methods have been tried for the separation of the excess specific heat of the vacancies from the contribution of the anharmonicity of the lattice or the electronic gas. Many of these have led to relatively good results concerning the formation enthalpy h^f but are usually inaccurate with regard to the formation entropy s^f. In the present section the reason for these discrepancies will be explained and also the conditions discussed under which such measurements can actually be used for the determination of h^f and s^f.

At this point it is worthwhile to note that the C_V-values, obtained from the C_P-values after the thermodynamical correction $TV\beta^2/\kappa$, belong to a real crystal. In the literature it has been claimed sometimes that these C_V-values belong to a perfect crystal with the argument that the quantities V, β^2 and κ in the correction term contain already the influence of the vacancies, and therefore the subtraction of the correction leads to a C_V of a perfect crystal. This is not correct, because all quantities involved (C_P, C_V, β, κ, etc.) are thermodynamic and therefore refer to a real crystal.

It is obvious that the extraction of the contribution $\Delta C_P = C_P - C_P^0$ of vacancies from the experimental curve $C_P = \mathrm{f}(T)$ requires the knowledge of C_P^0 which, however, is not known. In the following we shall briefly discuss various methods that have been attempted in order to overcome this difficulty.

7.2.1 Analysis of the C_P-curve assuming a linear extrapolation of $C_P^0 = f(T)$

In principle eq. (6.2) could allow the determination of h^f and s^f for each temperature from the $\Delta C_P \equiv (C_P - C_P^0)$-values. It is obvious that beyond the experimental values of C_P one needs C_P^0, which, as mentioned, is not known. In this paragraph we will discuss the implications of the usual assumptions; furthermore, within the frame of the assumption that C_P^0 is a linear function of T we indicate a method to determine its approximate values.

In fig. 7.1 the experimental C_P curve has been linearly extrapolated from a point A. It is obvious that besides the assumption that $C_P^0 = \mathrm{f}(T)$ is linear, the selection of the starting point A is also arbitrary. In the literature this arbitrarity is usually overcome by suitably selecting the starting point in such a way as to make the resulting plot of $\ln(\Delta C_P T^2)$ versus $1/T$ a straight line. Attention has to be drawn to this method because even if C_P^0 would happen to be linear, according to eq. (6.5) a straight plot would result only

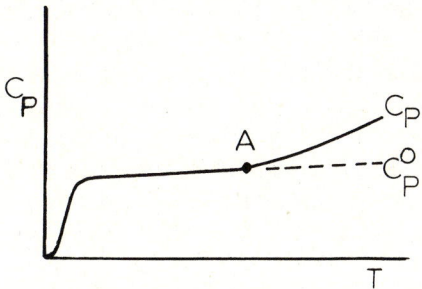

Fig. 7.1. Linear extrapolation of C_P^0 (not $\parallel$ T-axis).

if h^f and s^f were temperature independent. As this latter condition is not fulfilled the plot will be curved (fig. 7.2). The value of h^f (and approximately of s^f) can be determined as follows: From the slope of the low-temperature linear part we determine the enthalpy h_1^f and then from the intercept, according to eq. (6.6), the entropy s_1^f, which corresponds to the temperature T_1. In the same way the slope and the intercept of the high-temperature part lead to the enthalpy h_M^f and the entropy s_M^f, which correspond to a temperature close to T_M. If the selection of the starting point was correct, the following thermodynamical relation should approximately be obeyed (see eq. 3.25):

$$\frac{h_M^f - h_1^f}{T_M - T_1} \approx \langle T \rangle \frac{s_M^f - s_1^f}{T_M - T_1},$$

where $\langle T \rangle$ is some average temperature between T_1 and T_M. Obviously this procedure contains the inherent assumption that $\partial h^f / \partial T$ is temperature independent. However, this difficulty can be overcome if the above condition is checked for various intervals lying between T and T_M, so that various intermediate values (h_i, s_i) can progressively be checked. A computer fitting of the curved h_i versus T and s_i versus T in conjunction with the thermodynamical relation (3.25) leads to the functions $h = f(T)$ and $s = f(T)$. Once these functions are known, the quantity $\Delta C_P = C_P - C_P^0$ is determined by means of eq. (6.6) or approximately by:

$$\Delta C_P \approx N \frac{(h^f)^2}{kT^2} e^{s^f/k} e^{-h^f/kT}. \tag{7.23}$$

This method is acceptable only if, as mentioned above, C_P^0 is actually a linear function of T.

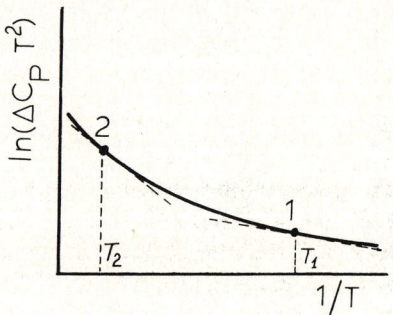

Fig. 7.2. Curved plot of $\ln(\Delta C_P T^2)$.

7.2.2 Analysis of the C_P-curve when C_V^{0*} is known from theoretical calculations

By applying the thermodynamical correction $TV\beta^2/\kappa$ to the experimental C_P-values (deduced at a pressure P_1), one can easily obtain the values of C_V as a function of temperature. It is obvious now that due to expansivity the C_V-values obtained in this way do not correspond to a discrete volume V but to a discrete pressure P_1 and therefore to various volumes; they therefore lie on an "isobaric curve" $C_V(P_1)$ of fig. 6.1. Nevertheless, as a first approximation we shall accept that all these C_V-values correspond to the volume V_1 of the lowest temperature T_1 of our experiment. Furthermore, let us assume that we know (from theoretical considerations) the isochoric specific heat C_V^{0*} of the isochoric perfect crystal * for volume V_1. According to eq. (6.34) we know that in a sufficient approximation (i.e., when $c_V \ll u^{*2}/kT^2$):

$$\left(C_V - C_V^{0*}\right)T^2 = N u^{*2} e^{s^f/k} e^{-h^f/kT}/k. \tag{7.24}$$

By considering that $u^* = h^f - Tv^f\beta B$ [see eq. (3.128)] we have

$$\left(C_V - C_V^{0*}\right)T^2 = N e^{s^f/k}\left(h^f - Tv^f\beta B\right)^2 e^{-h^f/kT}/k. \tag{7.25}$$

We now examine two cases:

(i) If the measurements extend over a small temperature range, the terms s^f, h^f and $(h^f - Tv^f\beta B)$ can roughly be considered as temperature independent. Then the plot $\ln[(C_V - C_V^{0*})T^2]$ is linear with a slope equal to $-h^f/k$ and an intercept equal to: $s^f/k + \ln[(h^f - Tv^f\beta B)^2/k]$. We therefore see that although the slopes of the curves $\ln(C_V - C_V^{0*})T^2$ versus $1/T$ and $\ln(C_P - C_P^0)T^2$ versus $1/T$ both give the formation enthalpy h^f, their intercepts are quite different. The intercept of the first curve is considerably smaller. This last remark explains why Brooks and coworkers [54–57] could not deduce a reliable value of the formation entropy from his analysis, although his h^f value was correct.

(ii) If the quantities s^f, h^f and v^f strongly depend on T – which is normally expected in the high-temperature region – and the measurements do not refer to a narrow temperature range, the plot $\ln[T^2(C_V - C_V^{0*})]$ versus $1/T$ has to show a curvature.

As temperature dependent values of s^f and h^f also produce a bend in the Arrhenius plots, this argument leads to the conclusion that materials with curved $\ln(\Delta C_P)T^2$-plots should also have curved Arrhenius plots. Inversely,

---

* The suffix V indicates that we are dealing with the isochoric specific heat $(\partial Q/\partial T)_V$, the superfix 0 indicates that it refers to a perfect crystal and the asterisk * indicates that the perfect crystal is the isochoric perfect crystal.

in materials with almost temperature independent s^f and h^f both types of plots are practically linear; as examples for curved plots we state Na, K and β-Zr.

7.2.3 Analysis of $C_P = f(T)$ when the X-ray expansivity and the X-ray compressibility have been measured

Kramer and Nölting [58] have proposed an intuitive method according to which it should be possible to determine the contribution of vacancies to the isochoric specific heat when, beyond C_P-measurements, the X-ray expansivity β^X and compressibility κ^X are available. The consideration was based on the equations

$$C_V = C_P - \frac{TV\beta^2}{\kappa} \quad \text{and} \quad C_V^X \equiv C_P - \frac{TV^X(\beta^X)^2}{\kappa^X},$$

and assumed that the difference

$$C_V - C_V^X = T\left[\frac{V^X(\beta^X)^2}{\kappa^X} - \frac{V\beta^2}{\kappa}\right] \tag{7.26}$$

represents the contribution of the defects to the isochoric specific heat. Although this procedure gave excellent enthalpy values, it is still not clear how it can lead to a reliable estimation of s^f and hence of the concentration of vacancies. The present discussion will clarify the situation: we first note that both C_V and C_V^X are quantities that refer to measurements on real solids and that therefore their difference cannot be considered as being purely a contribution of defects. If we express the values of V, β and κ from eqs. (3.32), (3.50) and (3.56) and the values of V^X, β^X and κ^X from eqs. (7.5a), (7.14a) and (7.19) after some algebra eq. (7.26) becomes

$$C_V - C_V^X = \frac{TV^0\beta^{0^2}}{\kappa^0}\left\{\frac{n}{N} - \frac{n}{N}\frac{v^f}{\kappa^0 kT} + 2\frac{n}{N}\frac{h^f}{B^0 kT^2}\right\}.$$

This formula has used the approximations $n \ll N$ and $\kappa^0 \ll v^f/kT$. The first term in the brackets can be neglected against the other two so that one obtains the final expression

$$C_V - C_V^X = \frac{TV^0\beta^{0^2}B^0}{kT^2}\{2h^f - Tv^f B^0 \beta^0\}e^{s^f/k}e^{-h^f/kT}. \tag{7.27}$$

By introducing eq. (6.18), this equation can be written

$$\ln\{(C_V - C_V^X)T^2\} = \ln\frac{C_P^0 - C_V^0}{\beta^0} + \ln(2h^f - Tv^f B^0 \beta^0) + \frac{s^f}{k} - \frac{h^f}{kT}. \tag{7.28}$$

Since both $C_P^0 - C_V^0$ and β^0 increase approximately linearly with temperature, the first term will be to a good approximation temperature independent. According to eq. (3.128) using the approximation $\beta^0 B^0 = \beta B$, the second term is $\ln(h^f + u^*)$. In a narrow temperature range both parameters h^f and u^* change very little so that the plot of $\ln[(C_V - C_V^X)T^2]$ versus $1/T$ may give a straight line from the gradient of which h^f can be determined. However, its intercept is *not* equal to s^f/k. Obviously s^f can be deduced from the intercept if the volume v^f is known, because setting $\beta^0 B^0 = \beta^X B^X$ is not a bad approximation.

7.3 Self-diffusion

7.3.1 Introduction

According to the random walk theory the x-component D_x^{SD} of the *macroscopic self-diffusion coefficient* D^{SD} is given by [59–62]:

$$D_x^{SD} = \tfrac{1}{2} \sum_{i=1}^{N} \Gamma_i \Delta x_i^2, \qquad (7.29)$$

where Γ_i denotes the number of jumps of type i made by an atom per unit time, N is the number of different types of the jumps and Δx_i is the x-projection of the corresponding jump distance. If C_i represents the atomic concentration of the defects at thermal equilibrium with the proper configuration for the occurrence of jump i, and w_i is the jump frequency involved, we can write

$$\Gamma_i = C_i w_i. \qquad (7.30)$$

Bardeen and Herring [60] have first indicated that the self-diffusion coefficient D^T obtained from tracer (T) experiments is different from D^{SD}; in *cubic* crystals they are connected (for a single mechanism) by:

$$D^T = f D^{SD}, \qquad (7.31)$$

where f is the so-called *correlation factor*. The latter is a number (between zero and unity) and depends on the kind of mechanism and the crystal system, e.g. $f = 0.781$ for monovacancy migration in a fcc lattice, $f = 0.727$ for monovacancy migration in a bcc lattice, etc. [61]. In *hexagonal* crystals one must distinguish two tensor components, the one ($f_\parallel$) parallel to the c-axis and the other ($f_\perp$) perpendicular to it [62]:

$$D_\parallel^T = f_\parallel D_\parallel^{SD}, \qquad D_\perp^T = f_\perp D_\perp^{SD}. \qquad (7.32a,b)$$

The methods of calculating correlation factors for almost all cases of practical interest have been reviewed by LeClaire [61] and by Mehrer [62]; especially for the case of hexagonal crystals we refer to Mullen [63] and to

Steiner, Mehrer and Seeger [64]. A useful review for the correlation effects in silver halides is given by Friauf [65].

The physical origin of the correlation effects is simple. As already mentioned, in the absence of external forces the defects perform a random walk; however this does not hold for the tracer atoms, since the probability that a tracer atom exchanges its site with a neighbouring vacancy is larger than the corresponding probability for the vacancy to jump to any other nearest-neighbouring site. This can be visualised as follows: For the diffusion coefficient D in an isotropic medium Einstein gives

$$D = \langle R^2 \rangle / 6t, \qquad (7.33)$$

where $\langle R^2 \rangle$ is the mean value over all possible diffusion paths of the square of the net vector displacement R of an atom in time t. If N jumps of the same length r take place in time t, according to Peterson [66] we have

$$\langle R^2 \rangle = \left\langle \left(\sum^N r \right)^2 \right\rangle = Nr^2 [1 + 2(\langle \cos \vartheta_1 \rangle + \langle \cos \vartheta_2 \rangle + \ldots)], \qquad (7.34)$$

where $\langle \cos \vartheta_i \rangle$ denotes the average value of the cosine of the angle between the first jump and the ith following jump. Notice that for a random walk – in which the direction of any jump is independent of the direction of preceding jumps – all the $\langle \cos \vartheta \rangle$-values are zero, so that it gives

$$D = \tfrac{1}{6} \frac{N}{t} r^2 \quad \text{or} \quad D = \tfrac{1}{6} \Gamma r^2, \qquad (7.35\text{a,b})$$

where Γ is the number of (successive) jumps per unit time, i.e. the jump rate and N the total number of jumps in time t. If the movements are quite random (a kind of Brownian motion) the quantity $\langle R^2 \rangle$ can be replaced by the number n of jumps multiplied with r^2, i.e. $\langle R^2 \rangle = nr^2$.

Let us now consider the case of the motion of a given tracer atom by a monovacancy mechanism [66]. When this atom has exchanged its position with a vacancy, the vacancy is immediately available to effect a second jump in the reverse direction. We see therefore that the probability is greater than random that the next jump of the tracer atom will be a reversal of the former. In other words, the term $\langle \cos \vartheta_1 \rangle$ in eq. (7.34) is negative. After the reversed jump has occurred, the vacancy remains a neighbour of the tracer atom, so that the probability that the next jump will be in the same direction as the first is greater than random; thus $\langle \cos \vartheta_2 \rangle$ is positive, but its absolute value is smaller than $\langle \cos \vartheta_1 \rangle$. The total result is that the sum of the values of $\langle \cos \vartheta \rangle$ in eq. (7.34) will have a definite negative value, in contrast to the random walk case where it is exactly zero; therefore the term included in brackets, which is by definition the "factor f", will be smaller than unity. In this case, i.e. when atomic jumps directions are correlated with one another, eq. (7.35b) has to be changed into:

$$D = \tfrac{1}{6}\Gamma r^2 f.$$

One point should be stressed: by comparing eqs. (7.29) and (7.35a) – which refer to the *macroscopic* diffusion coefficient of the uncorrelated jumps – one meets the difficulty of different numerical factors. This can be easily clarified by keeping in mind that the *microscopic* diffusion coefficient (see §A.11, note 15) for one type of isolated jumps can be always written

$$d = \lambda r^2 w, \tag{7.36}$$

(random walk theory) where the geometric factor λ represents the reciprocal of the number of possible jump directions from a given site and r the distance covered during a single jump. Obviously, when the jumps are constrained to only one dimension, λ has the value $1/2$; in a close packed cubic lattice it becomes equal to $1/6$, etc. If we are dealing with the diffusion of radiotracers the same rule holds, but then we must also include the correlation factor for the operating mechanism.

In cases where charge and mass are transported by the same mobile species, the ionic conductivity σ is related to the particle diffusion coefficient by the classical Nernst–Einstein relation:

$$\frac{\sigma}{D^{SD}} = \frac{z^2 F^2 C}{RT}, \tag{7.36a}$$

where C is the atomic concentration of the mobile species, z their charge number and F the Faraday constant. As an example, in the case of alkali halides with NaCl-structure, with a denoting the nn anion-cation distance we have $C = (2a^3)^{-1}$ (when the conductivity occurs only with cation motion, we have half a mobile ion per cube a^3), $F = Ne$, $z = 1$ and hence

$$\frac{\sigma}{D^{SD}} = \frac{e^2}{2a^3 kT}.$$

Attention is drawn to the point that in the Nernst–Einstein relation the true (i.e. uncorrelated) self-diffusion coefficient is present and not the (correlated) tracer diffusion.

The jump frequency w. In any of the three cubic Bravais lattices the jump frequency w to nearest-neighbour sites is given by (see ch. 5):

$$w = \nu \exp(-g^m/kT), \tag{7.37}$$

where g^m is the migration Gibbs energy and ν the *attempt frequency*. In the case of monatomic crystals, ν is usually arbitrarily taken equal to the Debye frequency ν_D. An alternate proposal is that of Seeger and Mehrer [67], where it is taken equal to

$$\tilde{\nu} \simeq \frac{1}{a}(h^m/M)^{1/2}, \tag{7.38}$$

where M is the atomic mass and a the lattice constant. They proposed some additional numerical factors for various structures (e.g. $(2/3)^{1/2}$) for the bcc lattice. However, in view of the fact that ν has a large uncertainty we delete in our applications these factors in order to have a common frame of reference. In the case of ionic crystals the usual approximation is $\nu \simeq \nu_D$, but it seems that $\nu \simeq \nu_{T0}$ is more justified. A plausible method for the determination of the "true" value of ν within the frame of the $cB\Omega$-model will be discussed later (see eq. 7.52 and ch. 9). By inserting the value of w into eq. (7.29) we get for the three cubic Bravais lattices and for monovacancy migration:

$$D^{SD} = xa^2\nu \exp(-g^m/kT).$$

Recalling that $x = \exp(-g^f/kT)$ this gives

$$D^{SD} = a^2\nu \exp[-(g^m + g^f)/kT], \tag{7.39}$$

whereas for the tracer diffusion coefficient:

$$D^T = fa^2\nu \exp[-(g^m + g^f)/kT], \tag{7.40a}$$

or

$$D^T = fa^2\nu \exp(-g^{act}/kT), \tag{7.40b}$$

where f is 0.781 for fcc and 0.727 [66] (or 0.723 [62]) for bcc lattices. When divacancy contribution is assumed, we refer for the corresponding expressions to Mehrer [62].

In hexagonal close-packed structures two different jump frequencies w_A and w_B must be considered; the first (w_A) corresponds to jumps within the basal plane, the other to jumps oblique to the basal plane [62]. The components of the tracer self-diffusion coefficient measured parallel to the c-axis ($D^T_{\|c}$) or perpendicular to it ($D^T_{\perp c}$) can now be written as

$$D^T_{\|c} = \tfrac{3}{4}f_\| w_B c^2 \exp(-g^f/kT) \tag{7.41}$$

and

$$D^T_{\perp c} = \tfrac{1}{2}f_\perp (3w_A + w_B)a^2 \exp(-g^f/kT), \tag{7.42}$$

where c and a denote the hexagonal lattice constants. We stress that in this case the components $f_\|$ and $f_\perp$ are functions of the ratio w_A/w_B [63]. Obviously there are different migration Gibbs energies for the various jumps; eqs. (7.41) and (7.42) hold for monovacancy mechanism only.

In alkali halides with NaCl-structure the tracer self-diffusion coefficients for the cations (D_c) and anions (D_a) are given by (we assume presence of free vacancies only):

$$D^T_c = 4a^2 f x_c w_c, \qquad w_c = \nu \exp(-g^{m,c}/kT),$$

$$D^T_a = 4a^2 f x_a w_a, \qquad w_a = \nu \exp(-g^{m,a}/kT),$$

where a is the nn-distance, x_c and x_a the concentrations of cation and anion vacancies, respectively, $f = 0.781$, and w_c and w_a the jump frequencies for the cation and anion vacancy migration. Note that we have arbitrarily assumed that the attempt frequencies for anion and cation motion into a free vacancy are equal. For the corresponding expression for a possible contribution from vacancy pairs we refer to Fredericks [68] and to Beniere, Chemla and Beniere [69]. In the case of silver halides we refer to Batra and Slifkin [70,71] and to Friauf [65].

7.3.2 Comments on the temperature dependence of the self-diffusion coefficient

Linear Arrhenius plots. In the case of linear Arrhenius plots in metals the following equation is used:

$$D^T = D_0 \exp(-h^{act}/kT), \tag{7.43}$$

which can also be written as

$$D^T = fa^2\nu \exp\left(\frac{s^f + s^m}{k}\right) \exp\left(-\frac{h^f + h^m}{kT}\right). \tag{7.44}$$

We see therefore that the following identities hold:

$$D_0 = fa^2\nu \exp(s^{act}/k) \tag{7.45}$$

and

$$h^{act} = h^f + h^m. \tag{7.46}$$

Note that in order to obtain the activation entropy from D_0 one has to assume a value of ν. In our applications this is done by setting it equal to ν_D or $\tilde{\nu}$ as discussed previously.

In the case of undoped alkali halides and for a single diffusion mechanism (vacancy process) the above relations become:

$$D_0 = 4fa^2\nu \exp(s^{act}/k) \tag{7.47}$$

and

$$h^{act} = \tfrac{1}{2}h^f + h^m, \tag{7.48}$$

where h^m and s^m refer either to cation or to anion migration, whereas h^f and s^f refer to a Schottky defect. In this case, as mentioned above, the s^{act}-value is extracted from D_0 by using ν_D or ν_{T0}.

Curved Arrhenius plots. The problem of the method of analysis for a curved self-diffusion plot is still open in the literature. Of course the most obvious interpretation is the coexistence of more than one diffusion mechanism. In the case of metals such a method of analysis has been initially forwarded by Mehrer and Seeger [72] by assuming contributions from monovacancies and divacancies. Mehrer [62] has recently refined this manner of analysis by suggesting (e.g. in the case of Na) that the divacancy contribution is described in terms of two activation enthalpies that pertain to two saddle points between the various divacancy configurations. This kind of analysis has been supported by Peterson [73] in a recent review. For the mathematical formulae of the analysis within the frame of "monovacancies + divacancies" we refer to the papers just mentioned. In alkali halides the analysis is usually carried out in terms of free (cation or anion) vacancies plus vacancy pairs. Such an analysis has been carried out for NaCl and KCl by Beniere, Chemla and Beniere [69] and has been also discussed by Fredericks [68]. In the case of silver halides we refer to Batra and Slifkin [70,71] and to Friauf [65].

A quite different method of analysis of the curved Arrhenius plots in metals has been presented by Gilder and Lazarus [37]. The basic thought in this analysis is that the temperature dependence of the activation parameters is responsible for the curvature of the Arrhenius plot. From a mathematical point of view such an analysis requires the knowledge of an explicit formula describing the temperature dependence of g^{act}. In view of the lack of such a formula, Gilder and Lazarus made the following proposition: diffusion measurements under pressure in Cd and Zn have shown that the thermal expansion coefficient β^{act} of the activation volume is approximately equal to $1/T$, i.e. much larger than β; the relation $\beta^{act} = 1/T$ gives $v^{act} \sim T$. By assuming that this relation is valid in other metals as well, Gilder and Lazarus first derive the general formula (see §A.11, note 7):

$$c_P^{act} \equiv \frac{dh^{act}}{dT}\bigg|_P = \frac{Tv^{act}\beta^{0^2}}{\kappa^0}\left\{2\frac{\beta^{act}}{\beta^0} - \frac{\kappa^{act}}{\kappa^0}\right\} + c_V^{act}. \qquad (7.49)$$

By taking $\kappa^{act}/\kappa^0 \simeq 1$ and $c_V^{act} \ll c_P^{act}$ they give the following explicit relation for the temperature dependence of h^{act} (and therefrom of s^{act}):

$$\frac{\partial h^{act}}{\partial T}\bigg|_P = 2\frac{Tv^{act}}{\kappa^0}\beta_0\frac{1}{T}.$$

This method of analysis has been applied by Gilder and Lazarus to a number of metals; in all cases – except possibly sodium – it adequately described the curvature of the Arrhenius plots, at least with the same success as the multiple mechanism model. This method of analysis seems to be more probable, although Mehrer [62] has expressed the opinion that β^{act} should be negative and not positive as adopted by Gilder and Lazarus. The weak point in the analysis, as they noticed, is the assumption that the relation $\beta^{\text{act}} = 1/T$ holds for all metals they studied. That this is not always true is strengthened by the microscopic calculations of Audit and Gilder [74] in Na, who showed that the expansion coefficient β^f of the formation volume increases with temperature (the quantity β^m of the migration volume is very difficult to calculate [638]). However, one should distinguish between two different things: the first is the physical basis of Gilder and Lazarus' proposals and second the specific model (i.e. $\beta^{\text{act}} = 1/T$) which they employed. The first (i.e. that the Arrhenius curvature is due to the thermal dependence of h^{act} and s^{act}) is quite possible to occur because thermodynamics does not preclude that $(\partial h^{\text{act}}/\partial T)_P$ [and $T(\partial s^{\text{act}}/\partial T)_P$] be appreciably different from zero. Therefore we are led to the conclusion that the physical base of Gilder and Lazarus' proposals (irrespective of the general validity of their specific relation) is undoubtedly correct.

In the following we shall add some useful comments: we shall start from general thermodynamical aspects which may help to dissolve some misunderstandings.

Within the frame of a single mechanism the slope of the diffusion plot $\ln D$ versus $1/T$ at every temperature, directly gives the activation enthalpy at this temperature:

$$\frac{d \ln D}{d(1/T)} = - \frac{h^{\text{act}}(T)}{k}. \tag{7.50}$$

This can easily be shown from a direct differentiation of eq. (7.40b) by making use of the identity $(\partial h^{\text{act}}/\partial T)_P = T(\partial s^{\text{act}}/\partial T)_P$ and the definition $s^{\text{act}} \equiv -(\partial g^{\text{act}}/\partial T)_P$. Attention is drawn to the point that the temperature dependence of the factor $fa^2\nu$ has been disregarded; this is justified in view of the fact that ν is by definition [35] assumed to be temperature independent and that a varies slightly with T. Anyhow, any possible temperature variation of $fa^2\nu$ can undoubtedly be disregarded in view of the strong upward curvature observed at high temperatures. Therefore we have arrived at the definite conclusion that – within the frame of a single mechanism – the upward curvature of the Arrhenius plot demands an *increase* of h^{act} with temperature, which means $c_P^{\text{act}} > 0$.

Let us now turn to the Gilder–Lazarus relation (7.49). There is a general agreement that c_V^{act} is appreciably smaller than c_P^{act} and that κ^{act} is positive, so that $\kappa^{\text{act}}/\kappa$ is of the order of unity. Therefore we are immediately led to the conclusion that an upward curvature of the Arrhenius plot (i.e. $c_P^{\text{act}} > 0$) is *not* compatible with a negative value of β^{act}, i.e. the activation volume has to increase and not to decrease with temperature. This is essential in view of the scarce but recent opinions expressed that β^{act} may be negative. As has become clear, a negative value of β^{act} is only compatible with a downward bending of the Arrhenius plot.

It finally remains a question of practical interest: how should an experimentalist analyse experimental data when the plot ln D versus $1/T$ turns upward. Of course a quick answer is the frame of a multiple mechanism, e.g. by a sum of two or three exponentials. However, one should always keep in mind that such an analysis – with temperature independent enthalpies and entropies – in order to be adequate, inherently assumes the *pure harmonic* description of the solid, i.e. that the solid does not expand, its elastic constants do not vary with T, etc. For example, an analysis of the upward curvature of the diffusion plot in Na [75] – within the above frame – automatically assumes that the Na-lattice does not expand with temperature, that C_P coincides with C_V, etc. In order to avoid these unrealistic consequences one has to go, at least, one step further, i.e. one has to adopt the *quasiharmonic* approximation. At once the difficulty arises how the defect parameters vary with temperature in this approximation. It is this major question to which we now turn, anticipating the contents of the second part of this book.

We recall that it is generally accepted and theoretically established that the compressibility of the formation volume of vacancies cannot exceed the bulk one by a large factor. There are good arguments (see ch. 14) that the same holds for the compressibility of the activation volume (formation plus migration process). On this point there are no differing views in the literature and furthermore it is strengthened by all the experimental data. Once this point is accepted, well-founded considerations of thermodynamics lead to the conclusion that g^{act} is proportional to $B\Omega$ (apart from an integration constant which in some cases can be proved to be zero from microscopic concepts; see the last chapter, 14)

$$g^{\text{act}} = c^{\text{act}} B\Omega. \tag{7.51}$$

Then the tracer diffusion coefficient is described by:

$$D^{\text{T}} = fa^2 \nu \exp(-c^{\text{act}} B\Omega/kT). \tag{7.52}$$

Therefore a plot of ln D^{T} versus $B\Omega/kT$ (instead of the usual ln D^{T} versus $1/T$) should give a straight line in the case of a single mechanism, even when ln D^{T} versus $1/T$ is curved. The intercept and the slope of this plot gives the quantities $fa^2\nu$ and c^{act}, respectively. The latter quantity directly gives the activation entropy, enthalpy and Gibbs energy at each temperature by the relations $s^{\text{act}} = -c^{\text{act}} \text{d}(B\Omega)/\text{d}T$, $h^{\text{act}} = c^{\text{act}}B\Omega - Tc^{\text{act}}\text{d}(B\Omega)/\text{d}T$ and eq. (7.51). The intercept also gives the possibility of obtaining the *attempt frequency* ν when the kind of mechanism (i.e. the factor f) is known; but even if the mechanism is unknown we can estimate ν with an uncertainty of a factor around 2, thanks to the fact that f varies between narrow bounds.

7.3.3 Proof of the proposal of analysing diffusion plots in the quasiharmonic approximation

It is quickly verified (see ch. 14) that the following two thermodynamical relations are valid:

$$\left.\frac{\partial g^{act}}{\partial (B\Omega)}\right|_T = \frac{v^{act}/\Omega}{(\partial B/\partial P)_T - 1}, \qquad (7.53)$$

$$\left.\frac{\partial g^{act}}{\partial (B\Omega)}\right|_V = -\frac{s^{*,act}}{\Omega(\partial B/\partial T)_V}, \qquad (7.54)$$

where $s^{*,act}$ is the "constant volume" entropy for the activation process.

Let us consider the usual assumption in the literature that κ^{act}/κ is unity; this has been adopted by Nowick and Dienes [76], by Gilder and Lazarus [37] and by others. We keep *only* this assumption and turn now to what pure thermodynamics demand when we analyse a diffusion plot – in a self-consistent way – within the frame of a quasiharmonic approximation (QA). The second member of eq. (7.53) is exactly pressure independent if one recalls that in QA the quantity dB/dP does not vary on compression [1] and that we have assumed that v^{act} varies with pressure in the same way as the bulk volume ($\kappa^{act} = \kappa$). Therefore eq. (7.53) reveals that, when the *temperature is kept constant*, the volume variation of the quantity g^{act} is exactly described by the relation: g^{act} *as a function of* $B\Omega$ *is linear*.

We turn to eq. (7.54): as in the QA the frequencies depend on the macroscopic volume one sees that $s^{*,act}$ should be temperature independent; the denominator $\Omega(\partial B/\partial T)_V$ is also temperature independent (for $V =$ constant). We are therefore led to the conclusion that, when the *volume is kept constant*, the temperature variation of the quantity g^{act} is described by the relation: g^{act} *as a function of* $B\Omega$ *is linear*.

By now combining the above conclusions from eqs. (7.53) and (7.54) we see that they are compatible with a relation of the form:

$$g^{act} = c^{act}B\Omega + K, \qquad (7.55)$$

where c^{act}, K have to be volume and temperature independent. Therefore the tracer self-diffusion coefficient in cubic crystals has the form:

$$D^T = fa^2\nu \exp\{-(c^{act}B\Omega + K)/kT\}. \qquad (7.56)$$

Although, at least in some general cases, we can prove that the integration constant K is zero, let us keep it as an unknown. Then eq. (7.56) indicates that we have to fit the diffusion data with an expression which has only three unknowns ($fa^2\nu$, c^{act} and K). It is worthwhile to stress once more that in the frame of QA the expression (7.55) is a pure result of statistical thermodynamics, apart from the point that we have accepted the current assumption $\kappa^{act} \simeq \kappa$ to be correct. As we shall see in ch. 14, even if κ^{act} exceeds κ by a factor 2–4, eq. (7.55) can again be proved to be valid.

7.3.4 Analysis of the isothermal plot $\ln D$ versus P

We shall first give in a general manner the analysis of plots of $\ln D$ versus P within the frame of a single mechanism irrespective of curvature.

The general equation (7.40b) is written as

$$\ln D^T = \ln(fa^2 v) - g^{act}/kT; \qquad (7.57)$$

by differentiating with respect to pressure (T = constant) and recalling that $v^{act} = (\partial g^{act}/\partial P)_T$ we get:

$$v^{act}(P, T) = -kT \left.\frac{\partial \ln D}{\partial P}\right|_T + kT\kappa(\gamma - \tfrac{2}{3}), \qquad (7.58)$$

where the term $\kappa(\gamma - \tfrac{2}{3})$ represents the quantity $[\partial \ln(fa^2 v)/\partial P]_T$, assuming that the volume variation of the attempt frequency v is well described by the Grüneisen constant γ. In any case the last approximation cannot introduce significant errors, because the second term is usually [37] only a few percent of the first one. Let us discuss eq. (7.58), disregarding, for reasons of convenience, the second term. It states that at any pressure the local slope of the $\ln D$ versus P plot is directly equal to $-(kT)^{-1}v^{act}(P, T)$. Only when v^{act} is pressure independent the plot is a straight line. In general it is curved because v^{act} varies on compression according to

$$v(P, T)^{act} = v(0, T)^{act} \exp\left(\int_0^P -\kappa^{act} dP\right).$$

It is therefore clear that the difficulties about the origin of the curvature arise as a direct consequence of the unjustified and *unrealistic* assumption that κ^{act} is exactly zero (i.e. v^{act} is pressure independent). In order to realise the extent of this effect, consider a typical material with $B = 50$ kbar and hence $\kappa = 2 \times 10^{-2}$ kbar^{-1}; in the simplest case, when $\kappa^{act} = \kappa$, we see that the exponential term $\exp(\int_0^P \kappa^{act} dP)$ takes a value of around 1.15 at $P = 7$ kbar, which means that the slope at 7 kbar is already 15% smaller than that corresponding to zero pressure. Obviously, for materials having high bulk moduli, as $B \simeq 1000$ kbars (i.e. $\kappa \simeq 10^{-3}$ kbar^{-1}), it is clear that in the usual pressure range up to 10 kbars the exponential term with $\kappa^{act} = \kappa$ gives a value around 1.01, thus explaining the linearity observed in these materials; recall that the usual experimental error in the determination of v^{act} is around 10%, which means that a curvature in the $\ln D$ versus P plot is detectable only when the exponential term becomes comparable – or higher – to 1.1.

The question remains in what manner an experimentalist should describe a $\ln D$ versus P plot. To a first approximation one should assume that κ^{act} is independent of pressure and expand $\exp(\int_0^P - \kappa^{act} dP)$ in the series: $1 - \kappa^{act} P + \tfrac{1}{2}(\kappa^{act} P)^2 + \dots$. By writing eq. (7.58) at zero (more accurately:

ambient) pressure we have for $\gamma \gg 2/3$:

$$v(0, T)^{act} = -kT \left.\frac{\partial \ln D(0, T)}{\partial P}\right|_T + kT\kappa(0, T)\gamma(0, T). \tag{7.59}$$

By subtracting from eq. (7.58) (in the approximation $\kappa_0\gamma_0 \simeq \kappa\gamma$):

$$\ln\{D(T, 0)/D(T, P)\} = -\left(\frac{v(0, T)^{act}}{kT} - \kappa_0\gamma_0\right)P$$

$$+ \left\{\frac{\kappa^{act}v(0, P)^{act}}{2kT}\right\}P^2. \tag{7.60}$$

Equation (7.60) can now be used to describe a $\ln D$ versus P plot; the use of this equation directly leads to the determination of $v^{act}(0, T)$ and κ^{act} for each temperature. As an example we refer the analysis of the isotherms of Mundy's [75] data for Na at $T = 288$ K and $T = 364.3$ K on the basis of eq. (7.60). Such an analysis gives [37] at 288 K:

$$v(0, T)^{act} = 11.1 \pm 0.2 \text{ cm}^3/\text{mole}, \quad \kappa^{act} = (33 \pm 5) \times 10^{-3} \text{ kbar}^{-1}$$

and at 364.3 K:

$$v(0, T)^{act} = 13.0 \pm 0.2 \text{ cm}^3/\text{mole}, \quad \kappa^{act} = (51 \pm 6) \times 10^{-3} \text{ kbar}^{-1}.$$

By considering that the bulk compressibility is around 16×10^{-3} kbar^{-1} one finds the ratio κ^{act}/κ to be 2–3, which is not unreasonable. In other words, a compressibility of the activation volume approximately twice that of the bulk can fully account, for $T = 288$ K, for all the curvature observed in the isothermal $\ln D$ versus P plot in sodium without having obligatorily to assume an arbitrary multiple mechanism.

7.3.5 Isotope effect

The most useful method for determining the correlation factor f in metals is the measurement of the isotope effect in diffusion. The diffusion coefficients D_α and D_β of two isotopes α and β with masses m_α and m_β are different. One can make the following assumptions, (a) that the Gibbs migration energies for both isotopes are equal and hence the two jump rates are connected with $w_\alpha/w_\beta = \nu_\alpha/\nu_\beta$, (b) that the attempt frequencies ν_α, ν_β obey the classical relationship $\nu_\alpha/\nu_\beta = (m_\beta/m_\alpha)^{1/2}$, and (c) that the atomic jump only involves motion of the diffusing atom, i.e. that the movement of the diffusing atom is completely decoupled from the lattice atoms. Under the above assumptions the isotope effect parameter E is given by [66] (see §A.11, note 16):

$$E \equiv \frac{(D_\alpha/D_\beta) - 1}{(m_\beta/m_\alpha)^{1/2} - 1} = f_\alpha. \tag{7.61}$$

However, if one changes assumption (c) and considers many-body effects, one gets

$$E = f_\alpha \Delta K, \qquad (7.61a)$$

where ΔK is usually interpreted as being "the fraction of the kinetic energy at the saddle point, associated with motion in the jump direction", that belongs to the diffusing atoms. Therefore ΔK is bounded between zero and unity. For an extensive discussion of the isotope effect we refer to the recent review of Peterson [66]. However, one should simultaneously consult the proposals of Gilder and Lazarus [37], who suggest that in the presence of strong anharmonicity – which represents the real behaviour of solids – the mode coupling must be greatly increased and ΔK must correspondingly be expected to decrease with increasing temperature.

Finally, we should mention that, in principle, accurate measurements of D^T and D^{SD} can give also the value of f by applying the relation $f = D^T/D^{SD}$. This method has been applied e.g. to Li and will be discussed later. The influence of the presence of divacancies in the isotope effect are extensively discussed by Mehrer [62] whereas the influence of anharmonic effects on the temperature dependence of $f \Delta K$ are discussed by Achar [77].

7.3.6 Experimental techniques for obtaining self-diffusion coefficients

A direct method for diffusivity measurements in solids is the measurement of the diffusion coefficient D^T of tracer atoms; this is obtained from a comparison of the actual tracer redistribution resulting from an anneal to the diffusion profile predicted from the second law of Fick. In the high temperature region mechanical sectioning techniques are used. In recent years *microsectioning* techniques have been developed [78–83] which permit the extension of measurements to appreciably lower temperatures.

Indirect methods for measuring self-diffusion are from NMR relaxation, Mössbauer effect and quasi-elastic neutron scattering. A brief review of the techniques is given by Mehrer [62] (see also § A.10).

7.4 Ionic conductivity and reorientation of dipoles

7.4.1 Ionic conductivity

It is far from the scope of the present review to give an extensive discussion and list of references on this subject. An extensive review up to 1972 can be found in the book of Crawford and Slifkin [84] complementary to an earlier review article by the same authors [85]. An excellent critical review of the recent developments of this field up to 1977 has been presented by Slifkin in the Catlinburg Conference [86], whereas an extensive list of references is presented by Corish, Jacobs and Radhakrishna [87] and recently by Bollmann [88]. Only a brief survey of the recent treatments for the analysis of the ionic conductivity after 1976 will be presented here.

A reliable fitting of the conductivity or of the diffusion curve even of the simplest ionic crystals has not been reported [89]. In order to simplify the discussion we shall consider in the following only the alkali halides with NaCl-structure. The various methods recently proposed for the analysis of the ionic conductivity plot can be divided into two basic categories.

(1) Besides the basic Schottky defects moving in the usual manner (i.e. free cation motion and free anion motion) additional defects are present (e.g. Frenkel defects [89]), or some additional migration mechanism is operating (e.g. the interstitialcy mechanism [90]). This proposal describes the curvature observed in conductivity plots by the coexistence of such mechanisms with the assumption of temperature independent parameters [95].

(2) The conductivity curve is described by temperature dependent defect parameters without introducing additional defects or mechanisms.

The proposals of the first category suffer from the fact that they assume temperature independent enthalpies and entropies. From a general thermodynamical viewpoint we shall show later on (see §10.4.2 and ref. [630]) that even if the thermal expansion coefficient of the formation volume of Schottky defects were equal to the bulk coefficient, there is a sufficient temperature dependence of h^f and s^f to create a detectable curvature in the conductivity plot. In view of this undoubtedly correct thermodynamical conclusion we shall not further discuss the proposals of the first category. Although they give good estimates of various defect parameters of formation and motion, they can in no case be considered as giving the exact values of these parameters, because one must then accept the harmonic approximation to be a good description of the real behaviour of alkali halides, which is undoubtedly wrong (see §A.8). We therefore turn our attention to the second category of proposals, the physical basis of which has widely been accepted during the last years.

The idea that the enthalpies (and entropies) depend on temperature has been raised in the literature many times, possibly for the first time by Mott and Gurney [91] four decades ago. However, one can cite numerous papers in which this temperature dependence has been proposed in the wrong direction. The present authors [92-94] have proposed that an upward curvature of the conductivity plot can only be explained – in the frame of a single conduction mechanism – when the formation (and migration) enthalpy and entropy increase with temperature; this is a thermodynamical necessity, because only in such a case (see § 3.6.1) the Gibbs energy decreases faster than linear, leading to an upward curvature of the conductivity (and diffusion) plot. However, there is a number of investigators who by following early suggestions, believe [96,97] that the decrease of the formation (and migration) enthalpy can explain the upward curvature (see also § A.11, note 4). In no case this can be correct because pure thermodynamics indicates that an enthalpy (and entropy) that decreases with temperature leads to a downward – and not to an upward – curvature of the conductivity (and diffusion) plot (see fig. 3.5).

By summarizing the present discussion for the analysis of the conductivity plots in alkali halides one can say the following: the idea that temperature dependent enthalpies and entropies are mainly responsible for the curvature observed in ln σT versus $1/T$ plots seems to be currently accepted. However, no detailed analysis of experimental data has appeared up to date using this concept. This is not fortuitous; such an analysis requires the knowledge of the form of the temperature dependence of the Gibbs formation and migration energy. The expressions $g^f = c^f B\Omega$ and $g^m = c^m B\Omega$ help toward this direction and we hope that such an analysis within their frame will shortly appear. In the case of silver halides the situation is more encouraging, although their conductivity curvature is appreciably stronger. This is due – as will be discussed later – to the pioneering experiments of Batra and Slifkin [98], who clearly showed that the origin of the curvature for silver halides is just the one proposed above, i.e. the temperature dependence of the defect parameters.

7.4.2 *Study of the reorientation of dipoles in crystals containing point defects*

When an external field is applied to a crystal it produces an appropriate response of the material. For instance, in all crystals stress induces an appropriate strain and this strain per unit stress is a compliance coefficient. In an analogous way an electric field applied to an insulator gives rise to a polarization and the polarization per unit electric field is an electric susceptibility coefficient. A perfect crystal responds to the external field instantaneously in the sense that it needs only the time required to establish the field. This type of polarization is connected to the so-called electronic and ionic polarizability. On the other hand, when defects are present a reversible time-dependent response may appear in addition to the "instantaneous" one mentioned above. In the case of elasticity such a response is called "anelastic relaxation" if, according to Zener, the strain tends with time to a definite value – which is proportional to the stress – and vanishes upon removal of the field. The *dielectric* and *anelastic* relaxation of crystals containing point defects has initially been reviewed by Nowick and Heller [99] and later by Nowick [100]. (See also §A.11, note 21.)

Recall that aliovalent impurities in alkali and silver halides produce vacancies, some of which are bound to the foreign atom and therefore can migrate around it, thus forming electric dipoles that can rotate. Recent studies have shown that there is a close connection between the electronic configuration of the aliovalent impurities added and the reorientation parameters of the dipoles formed [101–103].

Dielectric loss at various pressures. These data [104] can safely lead to the determination of the migration volume $v^{m,b}$ of the reorientating bound defect and of its two contributions

$$v_s^{m,b} \equiv -T \left.\frac{\partial s^{m,b}}{\partial P}\right|_T \quad \text{and} \quad v_h^{m,b} \equiv \left.\frac{\partial h^{m,b}}{\partial P}\right|_T, \qquad (7.62a,b)$$

where $h^{m,b}$ and $s^{m,b}$ denote the enthalpy and the entropy for this bound (b) defect motion. In this sense the dielectric loss technique is very useful as it has already shown that $v_s^{m,b}$ is an appreciable percentage of the total migration volume $v^{m,b}$, thus indicating the major role of anharmonicity in a migration process of ionic crystals. Such a conclusion has been drawn from an analysis [104] of the data of Fontanella and coworkers [105,106] in alkaline earth fluorides doped with monovalent or trivalent cations. For the *same* host crystal we have also shown that the quantity $s^{m,b}/h^{m,b}$ is *independent* of the dopant and has a definite value that is predetermined from bulk properties [639] as indicated from the $cB\Omega$-model (see §13.5). The *same* holds for the quantity $v^{m,b}/h^{m,b}$ (see ref. [640] and §13.3.3).

A similar experiment is suggested for alkali halides doped with divalent cations or anions; the temperature and pressure variation of the frequency at which the dielectric loss maximizes should undoubtedly lead to the values of $v_s^{m,b}$ and $v^{m,b}$, because according to ref. [104] they are given by:

$$v_s^{m,b} = kT\left\{ \left.\frac{\partial \ln \tau_0}{\partial P}\right|_T + \frac{\gamma_{T0}}{B} \right\}, \qquad (7.63a)$$

$$v^{m,b} = kT\left\{ \left.\frac{\partial \ln \tau}{\partial P}\right|_T + \frac{\gamma_{T0}}{B} \right\} \qquad (7.63b)$$

where τ_0 is the usual preexponential factor involved in the classical Arrhenius relation

$$\tau = \tau_0 \exp(E/kT). \qquad (7.64)$$

In the above relation τ denotes the relaxation time for the reorientation process and E coincides with the migration enthalpy of the process. The determination of $v_s^{m,b}$ is of high importance in the following sense: from eq.

(3.169) it must be equal to $T\beta^{m,b}v^{m,b}$, where $\beta^{m,b}$ is the volume expansion coefficient of migration volume $v^{m,b}$. If the experiment is carried out in a crystal doped with a divalent cation having a comparable ionic radius with the host cation, it is expected that the migration enthalpy h^m and volume v^m – corresponding to the free cation motion – would be comparable to those of the bound vacancy motion [101]. Therefore, if the experiment shows that $\beta^{m,b} \gg \beta$ it is very likely that this also holds for β^m (see §A.11, note 18).

Ionic thermocurrent technique. This technique (ITC) was proposed in 1964 by Bucci and Fieschi [107–109]. It is usually preferable to the dielectric loss technique mainly for two reasons: (1) it can resolve competing relaxation processes, e.g. by the usual "peak cleaning procedure" (see §A.11, note 22), and (2) it can detect a 100 times lower dipolar concentration – in alkali halides – than that detectable by the dielectric loss method. An extensive description of this method can be found in the original papers or in the article of Crawford and Slifkin [85]. Here we shall only stress the recent developments of the field and emphasize some points which need clarification.

The ITC-technique exploits the fact that the electric polarization Π reacts to a change of the electric field slowly or fast, depending on the relaxation time τ, which again depends on the temperature in an exponential way [see eq. (7.64)]. At a low temperature the relaxation time can be so long that the polarization retains its initial value, irrespective of the momentary value of the field. Upon heating the relaxation time τ decreases, so that at a certain temperature region T_C the polarization can take in a very short time the value that corresponds to equilibrium under the momentary conditions. During the change of the polarization a transient electric current density J is produced (see figs. 7.3 and 7.4).

The usual procedure in the ITC-technique is to measure the *thermally stimulated depolarization current* (TSDC). In this procedure one measures the depolarization current of previously polarized dipoles under increasing temperature after having switched off the field at a low temperature. The sequence of operations is shown in fig. 7.3: an electric field $\mathscr{E}_p$ is applied to the crystal at a relatively high temperature T_p for a time much longer than the relaxation time $\tau(T_p)$. All the dipoles have sufficient time to become oriented and thus the polarization attains the saturation value Π_0 that corresponds to $\mathscr{E}_p$ and T_p. It is given by

$$\Pi_0 = \frac{N\mu^2}{3kT_p}\mathscr{E}_p, \tag{7.65}$$

where N is the concentration of dipoles and μ the electric dipole moment; note that the above expression is valid only when the defects have cubic

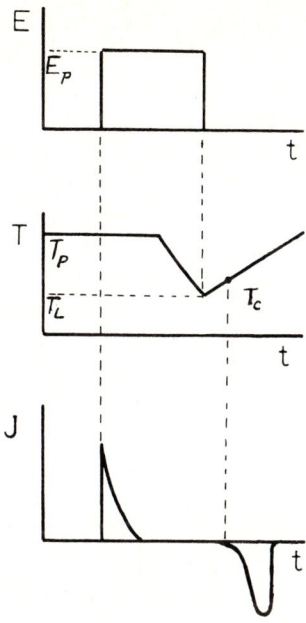

Fig. 7.3. Thermally stimulated depolarization current, $E = \mathscr{E}$, $E_P = \mathscr{E}_P$.

symmetry. When the sample is now quickly cooled down to a low temperature T_L (e.g. liquid nitrogen temperature), one would expect the "frozen in" polarization to remain equal to Π_0. However, Oliveira and Ferreira [110] indicated that it changes to Π_{fr}, depending on the cooling rate. More precisely they propose the expression

$$\Pi_{fr} = \left(N\mu^2/3kT_{eff}\right)\mathscr{E}_P,$$

where T_{eff} is an effective polarization temperature, lower than T_p, that can be explicitly calculated when the cooling programm is precisely known. This correction is small and can be disregarded in experiments in which fast cooling rates are achieved.

After the fast cooling of the sample and the interruption of the electric field, the sample is warmed at a rate dT/dt. This heating rate is usually regulated so as to be a constant b, as was initially proposed by Bucci and Fieschi. However, Müller and Teltow [111] prefer a variable rate, for which $-d(1/T)/dt$ is a constant r

$$r \equiv -d(1/T)/dt = (1/T^2)dT/dt,$$

which appreciably simplifies the expressions for the polarization Π and the current density J. In view of the advantages of the latter proposal and due to the fact that it is not extensively used, we tabulate in table 7.1 the

Table 7.1
Equations valid for ITC (depolarisation case, TSDC)

Quantity	Constant heating rate b	Variable heating rate: $r \equiv -\dfrac{d(1/T)}{dt} = $ constant [c]
Polarisation $\Pi(t)$	$\Pi_0 \exp\left\{-\tau_0^{-1} \int_0^t \exp\left[-E/(k(T_L + bt'))\right] dt'\right\}$	$\Pi_0 \exp\left\{-\dfrac{k}{\tau_0 rE} e^{-E/kT_L}(e^{rE_t/k} - 1)\right\}$
Current density $J(t)$	$\dfrac{\Pi_0}{\tau_0} \exp\left\{-\left(\dfrac{E}{kT} + \dfrac{1}{b\tau_0}\int_{T_0}^{T} e^{-E/kT'} dT'\right)\right\}$ [a]	$\dfrac{\Pi_0}{\tau_0} \exp\left\{-\left[\dfrac{E}{kT} + \dfrac{k}{\tau_0 rE}(e^{-E/kT} - e^{-E/kT_L})\right]\right\}$
Maximum current density J_m (at temperature $T = T_m$)	not existing in closed form	$\dfrac{\Pi_0 rE}{k} \exp\left(\dfrac{k}{\tau_0 rE} e^{-E/kT_L} - 1\right)$ or $\dfrac{\Pi_0 rE}{k} \exp\left\{\exp\left(\dfrac{E}{kT_m} - \dfrac{E}{kT_L}\right) - 1\right\}$
Polarisation $\Pi(T)$	$\Pi_0 - \dfrac{1}{b}\int_{T_L}^{T} J dT'' = $ not integrable in closed form [b]	$\Pi_0 \exp\left\{-\left[\dfrac{k}{\tau_0 rE}(e^{-E/kT} - e^{-E/kT_L})\right]\right\}$

[a] In the low temperature region ($T \ll T_m$) we have $J = \Pi_0 \tau_0^{-1} \exp(-E/kT)$; therefore from the slope of the straight line $\ln J$ versus $1/T$ the energy E is determined. Then the factor τ_0 can be calculated from $\tau_0 \exp(E/kT) = kT_m^2 b^{-1} E^{-1}$, which is the condition for the maximum current. The latter factor τ_0 leads to the migration entropy s^m through the relation $s^m = -k \ln(\tau_0 \lambda \nu_{T0})$, which results from a comparison of eqs. (7.67) and (7.64) by setting $E = h^m$ (see also §13.5).

[b] The value of τ at various temperatures can be obtained from $\tau(T) = \Pi(T)/J(T) = J(T)^{-1} \int_T^{\infty} J \, dT'$ (see eq. 7.68), where the latter integral is graphically estimated. Then by plotting $\ln \tau(T)$ against T^{-1} we obtain E and τ_0 (see eq. 7.64).

[c] Refs. [111,113].

resulting expressions for both cases for the sake of comparison. Attention is drawn to the point that in the derivation of these formulae two approximations have been made: (1) that $\Pi_{fr} = \Pi_0$ and (2) that the Arrhenius expression (7.64) is valid with τ_0 and E independent of temperature. Furthermore these expressions are valid as long as the dipoles do not interact. Shaw and Moghaddam [111a] found that by using heating rates that vary during the transient emission, more information can be extracted from the measurement.

Due to the fact that, as mentioned above, the first approximation does not introduce significant errors when the cooling rate is fast, we now turn our attention to the second approximation, which may well be violated, for instance in the case of materials having a very high Debye temperature and experimental temperature ranges that extend up to $\Theta_D/3$ or $\Theta_D/4$. In such cases the corresponding expressions can be derived by writing in general

$$\tau(T) = \tau_0(T) \exp[E(T)/kT] \tag{7.66}$$

or

$$\tau(T) = (\lambda \nu)^{-1} \exp(g^{m,b}/kT), \tag{7.67}$$

where λ is the number of jump paths accessible to the jumping species with an attempt frequency ν and $g^{m,b}$ the Gibbs energy for the reorientation process; obviously, when $g^{m,b}$ falls linearly with T, the above expression (7.67) reduces to the classical Arrhenius expression (7.64). Expression (7.67) becomes more complicated at very low temperatures, where quantum effects play a major role. After switching off the field the polarization decays and during the heating a current density J is produced according to the relation

$$J = -d\Pi/dt = \Pi/\tau, \tag{7.68}$$

from which one gets

$$\Pi = \Pi_0 \exp\left(-\int_0^t \frac{dt'}{\tau(t')}\right). \tag{7.69}$$

The effect vanishes when the polarization has completely decayed, thus having given a transient current, which maximizes at a temperature T_m (see fig. 7.3). An insertion of eq. (7.67) into eq. (7.69) gives the desired general expression which is not integrable (due to the unknown temperature dependence of $g^{m,b}$); however, when $g^{m,b}$ falls linearly with T, and r and ν are constants, the polarization $\Pi(t)$ can be expressed in a closed form. For the current procedure of the analysis of a TSDC-curve we refer to the footnotes of table 7.1.

An extensive list of the ionic materials studied up to 1977 can be found in the review article of Radhakrishna and Haridoss [113]. In the same article one can find an extensive discussion of ITC Data Computer Processing. A newer article by Laredo, Puma, Suarez and Figueroa [113a] refers also to computer programs. When dealing with interacting dipoles the expressions given in the table have to be corrected [114]. This effect has been extensively studied in alkaline earth fluorides doped with appreciable amounts of aliovalent impurities by Den Hartog and coworkers [115] and by Triantis [115a]. The work in fluorides was mainly initiated by Crawford and coworkers [116–119] in a series of interesting papers and continued by Jacobs and Ong [119a], by Aalbers and Den Hartog [119b] and by Den Hartog and Langevoort [119c]. Alkali halides have been studied in many papers, of which we give a brief list: NaCl [120] and KI [120a] by Cusso and Jaque; other alkali halides by the same authors [120b] and by Hernandez, Murrieta, Jaque and Rubio [120c]; NaF by Varotsos, Kostopoulos and Mourikis [120d]; NaI by Kostopoulos, Varotsos and Mourikis [120e] and by Triantis and Kostopoulos [120f]: KCl by Kostopoulos, Mourikis and Varotsos [120g]. Other ionic crystals: NH_4Cl by Kessler and Ebert [120h], $SrCl_2$ by Jacquet and Bathier [120i] and MgO by Krokoszinski and Baerner [120k]. For the semi-ionic LiH and LiD Varotsos, Kostopoulos, Mourikis and Kouremenou [120ℓ] found that the ITC-spectrum gives an unusually low value of τ_0^{-1}. TSDC-currents were studied in polymers by Zielinski, Swiderski and Kryszewski [120m]. The ITC-method can also be used in solids with inherent electric dipoles as ferroelectrics. With this method the two Curie points of Rochelle salt were separately observed by Kostopoulos, Kouremenou, Varotsos and Mourikis, [120n]. Rocks are presently being investigated by Dologlou-Revelioti [120o]. Vanderschueren and Linkens [120p] and Cusso and Jaque [120a] have compared the defect parameters obtained from ITC to those of dielectric measurements.

The ITC-technique has also been found useful by Figueroa, Laredo and Puma [120q] in detecting superstructure (Suzuki phase) in NaCl doped with $\sim 8 \times 10^{-3}$ mole-concentration of Cd^{2+}; a new ITC peak has been detected at 318 K which is attributed to the precipitates of the Suzuki phase present in the crystal. For the TSDC of $AgBr + Cd^{2+}$, see ref. [641].

Less known is the *thermally stimulated polarization current* (TSPC) technique initiated by Moran and Fields [121] and by Pfister and Ankowitz [121a] and developed by McKeever and Lilley [121b]: a constant electric field is applied on the crystal at low temperature; upon heating the randomly oriented dipoles become aligned and produce a current (fig. 7.4). An exact theoretical treatment shows that the TSPC peak should be followed by a current reversal because of the temperature dependence of the equilibrium polarization. The effect had been observed in various cases, especially by Kristianpoller and Kirsch [121c]; it was initially considered as an artifact but has been definitely recognized as a true effect [121b]. Polymers have been extensively investigated by Vanderschueren, Linkens, Gassiot, Fillard and Parot [122].

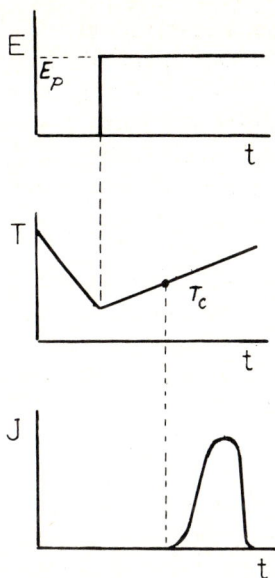

Fig. 7.4. Thermally stimulated polarization current, $E = \mathscr{E}$, $E_P = \mathscr{E}_P$.

As already mentioned an ITC-spectrum under various hydrostatic pressures has not been reported to date, although it can lead to an accurate determination of the "migration" volume for a reorientation process. Such an experiment will be useful not only for the reasons already mentioned but also in order to establish the prediction of the $cB\Omega$-model that the quantity $v^{m,b}/g^{m,b}$ for various types of dopants and dipoles in a given host material is a constant equal to $[(dB/dP) - 1]/B$. For instance, for NaCl doped with Ca^{2+}, Mg^{2+}, Sr^{+2}, etc. such an experiment seems quite easy. It is worthwhile to mention that the prediction of the $cB\Omega$-model that v/h is independent of the mechanism (see eq. 8.22.) has been verified by Andeen, Fontanella and Wintersgill [122a] for SrF_2 doped with lanthanum (see also refs. [640] and [642–644]).

The ITC-spectrum of rubidium halides doped with monovalent impurities has recently been determined [122b] under uniaxial stress, but such experiments cannot lead to the determination of $v^{m,b}$.

Piezo-stimulated Current (PSC) technique. This technique was proposed in 1975 by Bui Ai, Destruel, Gima and Loussier [123] but has not yet attracted a large interest. It is analogous to the ITC technique and consists of the following steps. An electric field is applied to a solid containing dipoles for a time appreciably longer than the relaxation time (or times) of the dipole. Let the temperature T and the pressure P_0 be the initial conditions. We label the resulting equilibrium polarization as Π_0. The pressure is then increased to a final value P_f, thus increasing the relaxation time (when $v^{m,b} > 0$) so that the dipoles are immobilized and the polarization cannot change within the duration of the experiment when the field is switched off. The pressure is then slowly decreased and when the relaxation time has been sufficiently reduced the dipoles can reorientate freely so that a depolarization current is liberated and reaches a maximum at a critical pressure P_{cr} (see also §A.11, note 8).

Let us briefly discuss this technique. As already mentioned the relaxation time of the dipoles can be written as $\tau = (\lambda \nu)^{-1} \exp(g^{m,b}/kT)$. If the migration volume $v^{m,b} \equiv (\partial g^{m,b}/\partial P)_T$ is positive, the value of τ increases upon compression because usually the preexponential factor varies only slightly, according to [104]

$$[\partial \ln(\lambda \nu)/\partial P]_T = \gamma/B, \tag{7.70}$$

where γ is the Grüneisen constant of the attempt frequency. Therefore the increase of pressure is physically similar to the cooling in the TSDC-technique; however, if the volume $v^{m,b}$ is negative one should start by applying a field at a high pressure and switching it off after the pressure has been reduced (T = constant throughout the experiment).

In a similar fashion as in the ITC-technique the decay process is given by eq. (7.68), the solution of which is

$$\Pi = \Pi_0 \exp\left\{ \frac{\lambda}{b} \int_{P_f}^{P} \nu(P) \exp\left(-\frac{g^{m,b}}{kT}\right) dP \right\}, \tag{7.71}$$

where eq. (7.65) gives the initial polarization Π_0, Π is the polarization at a pressure P and b is the rate of the decrease of pressure. The current density $J(P)$ is now directly given by:

$$J(P) = -\frac{d\Pi}{dt} = \frac{\Pi}{\tau} = \Pi_0 \lambda \nu(P)$$
$$\times \exp\left\{ -\frac{g^{m,b}}{kT} + \frac{\lambda}{b} \int_{P_f}^{P} \nu(P) \exp(-g^{m,b}/kT) dP \right\}. \tag{7.72}$$

Before proceeding to the explicit calculation of $J(P)$ we shall first stress a point which is still being confused in the literature: The energy $g^{m,b}$ can be written as

$$g^{mb} = u^{m,b} - Ts^{m,b} + Pv^{m,b}$$

(see eq. 5.52) and by differentiating it with respect to pressure:

$$\left.\frac{\partial g^{m,b}}{\partial P}\right|_T = \left.\frac{\partial (u^{m,b} - Ts^{m,b})}{\partial P}\right|_T + v^{m,b} + P\left.\frac{\partial v^{m,b}}{\partial P}\right|_T.$$

§7.4]   *Ionic conductivity and reorientation of dipoles*   139

Due to the definitions:

$$v^{m,b} \equiv \left.\frac{\partial g^{m,b}}{\partial P}\right|_T \quad \text{and} \quad \left.\frac{\partial v^{m,b}}{\partial P}\right|_T \equiv -\kappa^{m,b} v^{m,b}$$

we get:

$$\left.\frac{\partial(u^{m,b} - Ts^{m,b})}{\partial P}\right|_T = \kappa^{m,b} P v^{m,b}. \tag{7.73}$$

The last relation clearly indicates that the quantity $u^{m,b} - Ts^{m,b}$ is pressure independent only when the volume $v^{m,b}$ is arbitrarily assumed as not depending on pressure. In the latter case we have

$$g(P)^{m,b} = g(P_0)^{m,b} + v^{m,b}(P - P_0). \tag{7.74}$$

An insertion of eq. (7.74) into eq. (7.72) gives directly an explicit relation for $J(P)$ by assuming, to a first approximation, that $\nu(P)$ which appears in the integral of eq. (7.72) is a constant:

$$J(P) = \Pi_0 \lambda \nu(P) \exp\left\{-\frac{(P - P_0)v^{m,b}}{kT} - \frac{g(P_0)^{m,b}}{kT}\right.$$

$$-\frac{\lambda \nu}{b}\frac{kT}{v^m} \exp\left(-\frac{g(P_0)^{m,b}}{kT}\right)$$

$$\times \left[\exp\left(-\frac{v^{m,b}(P - P_0)}{kT}\right) - \exp\left(-\frac{v^{m,b}(P_f - P_0)}{kT}\right)\right]\right\} \tag{7.75}$$

The above expression can be thought only as a very rough approximation, because, already as mentioned, it contains the unjustified assumption that $\kappa^{m,b} = 0$. *

A more exact expression for $J(P)$ can be found by analysing $g(P)^{m,b}$ up to second order with respect to pressure:

$$g(P)^{m,b} = g(P_0)^{m,b} + \left.\frac{\partial g^{m,b}}{\partial P}\right|_{P_0}(P - P_0) + \frac{1}{2}\left.\frac{\partial^2 g^{m,b}}{\partial P^2}\right|_{P_0}(P - P_0)^2$$

or

$$g(P)^{m,b} = g(P_0)^{m,b} + v(P_0)^{m,b}(P - P_0)$$
$$- \tfrac{1}{2}\kappa^{m,b} v(P_0)^{m,b}(P - P_0)^2, \tag{7.76}$$

where the compressibility $\kappa^{m,b}$ can – to a first approximation – be considered as pressure independent. The insertion of eq. (7.76) into eq. (7.72)

---

* Equation (7.75) shows that $J(P)$ maximizes when:

$$bv^{m,b}/kT = 1/\tau(P_{cr}). \tag{7.75a}$$

immediately gives the desired expression for $J(P)$. A fitting of the experimental data of $J(P)$ versus P to this expression directly leads to the determination of $v(P_0)^{m,b}$ and $\kappa^{m,b}$.

A final comment should be added: a quick estimation of the migration volume from the experimental curve $J(P)$ versus P can be easily achieved as follows: The general expression (7.72) gives for pressures P tending towards P_f, $J(P) = \Pi_0 \lambda \nu(P) \exp(-g^{m,b}/kT)$, from which by differentiating with respect to pressure and considering eq. (7.70) we get $[\partial \ln J(P)/\partial P]_{P \to P_f} = (\gamma/B) - v(P_f)^{m,b}/kT$. This equation indicates that the slope of the curve $\ln J(P)$ versus P for pressures close to P_f immediately gives the migration volume $v(P_f)^{m,b}$, because the Grüneisen constant of the attempt frequency can be approximated in ionic crystals by γ_{TO}.

Summarizing the present discussion on the piezo-stimulated current technique we can say that it is powerful as it can easily lead to the determination of $v^{m,b}$. Such experiments in ionic crystals – doped with small amounts of aliovalent impurities – are highly desirable and we hope that they will shortly be carried out.

The piezo-stimulated polarization current has been used in geophysics as an explanation for the observed changes of the telluric current that occur several hours or days before a seismic event [123a]: Rocks usually contain piezoelectric materials, usually quartz, in a non-perfect directional distribu-

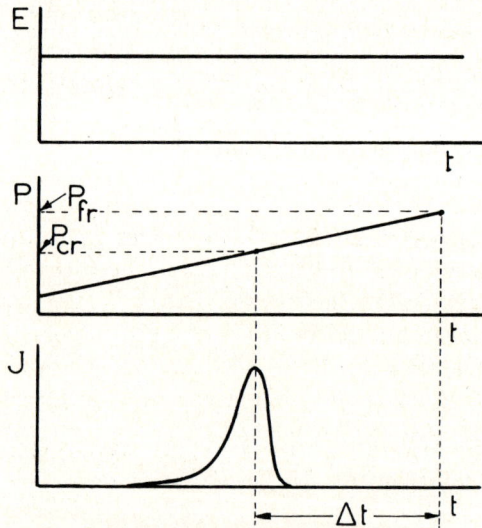

Fig. 7.5. Piezo-stimulated polarization current emitted before the fracture of rocks. $E = \mathscr{E}$. (Courtesy Physica Status Solidi; ref. [123b].)

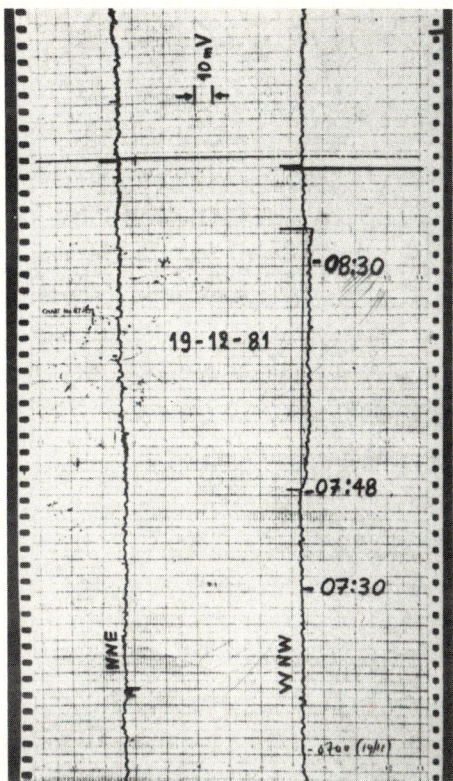

Fig. 7.6. Pressure-induced telluric current registered at Athens seven hours prior to a 7 R earthquake that occurred at a distance of 160 km. Note the strong cultural noise; the transient signal started at 07:48.

tion. In certain seismological models the stress increases gradually before an earthquake. This stress variation not only creates an electric field but also decreases the relaxation time of the dipoles in the rock (when $v^{m,b} < 0$). If the critical stress P_{cr} (fig. 7.5) lies between 0 and P_{fr}, the focal area will emit a transient electric current before the earthquake occurs. This expectation has been verified in many hundreds of seismic events in Greece. For some unknown reason the change of the telluric field occurs with a lead time between $6\frac{1}{2}$ h and one week. An example of a recording is shown in figs. 7.6 and 7.7. As changes of an electric field propagate with a velocity of the order of the velocity of light they occur almost simultaneously at any observation point. The intensity of the change decreases with distance from the epicentre. By collecting the pulses at various sites the epicenter can be determined [123c,596,597]. The magnitude is found to be proportional to

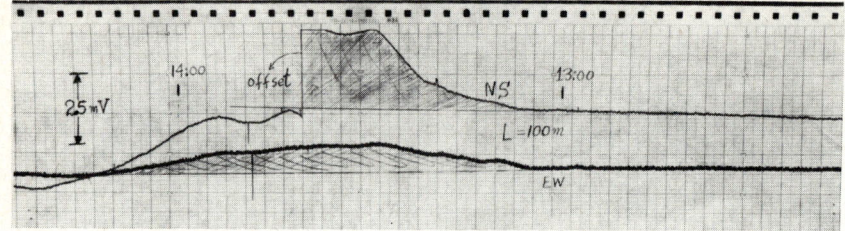

Fig. 7.7. Pressure-induced telluric current registered at Crete almost seven hours prior to a 7 R earthquake that occurred at a distance of 500 km close to Limnos Island. (Courtecy P. Ikonomeopoulos.)

the logarithm of the strength of the change. For more details on this subject see the appendix.

7.5 Comments on solubility

In any infinitesimal process in which the number of moles n_i of constituent i may be caused to change by the transfer either *to* or *from* the phase (or by the agency of a chemical reaction or both) the change of the Gibbs function is given by eq. (2.22):

$$dG = -SdT + VdP + \sum_i \mu_i dn_i. \tag{7.77}$$

We remember that μ_i is the chemical potential of the ith constituent of a phase and is defined as:

$$\mu_i \equiv \partial G/\partial n_i \tag{7.78}$$

and is a function of T, P and *all* the n's.

Consider a phase in which, at constant T and P, *all* its constituents are increased in the same proportion, i.e. $dn_1 = n_1 d\lambda, n_2 = n_2 d\lambda, \cdots, dn_c = n_c d\lambda$, where $d\lambda$ is the proportionality factor. Due to the fact that the Gibbs function G is an extensive quantity it will also be increased in the same proportion, i.e. $dG = Gd\lambda$. Therefore eq. (7.77) at $P = $ constant and $T = $ constant becomes:

$$dG = \mu_1 dn_1 + \mu_2 dn_2 + \cdots + \mu_c dn_c$$

or

$$Gd\lambda = \mu_1 n_1 d\lambda + \mu_2 n_2 d\lambda + \cdots + \mu_c n_c d\lambda.$$

We therefore conclude that

$$G = \mu_1 n_1 + \mu_2 n_2 + \cdots + \mu_c n_c$$

or
$$G = \sum_i \mu_i n_i, \quad i = 1, 2, \cdots, c. \tag{7.79}$$

This equation leads to the conclusion that the chemical potentials are intensive quantities, because if all n values are increased in the same proportion, at constant T and P, the μ values must be constant in order that G increase in the same proportion.

We now turn to the definition of μ_i; we have stated above that the chemical potential μ_i of the ith constituent is a function of T, P and *all* the n-values. But due to the fact that it is an intensive quantity, obviously the n-values must be combined in such a way that, when all of them are changed by the same factor, the value of μ_i has to remain the same. As the mole fraction of the ith constituent is given by

$$x_i \equiv n_i / \sum n_i, \tag{7.80}$$

the above requirement is fulfilled and hence we can finally say that μ_i is a function of T, P and the molar fraction x_i.

We emphasize that in the case of a phase consisting of only one constituent, for example: an *ideal lattice* having n moles, eq. (7.79) gives $G^{0,n} = \mu^0 n$, where $G^{0,n}$ denotes the Gibbs function of the whole system; therefore

$$\mu^0 = G^{0,n}/n, \tag{7.81}$$

i.e. the chemical potential in an *ideal* lattice coincides with the molar Gibbs energy $G^0 (\equiv G^{0,n}/n)$.

Definition of the ideal solution (one phase). A solution is defined as ideal when the chemical potential of each constituent i is of the form

$$\mu_i = G_i + RT \ln x_i, \tag{7.82}$$

where G_i is the Gibbs function of one mole of the ith constituent in the *pure* state.

A real crystal containing vacancies can be considered as a solution of the vacuum and of solvent atoms in the host crystal. If μ_a and μ_v denote the chemical potential of atoms (a) and of the vacancies (v), one can write in the case of the real crystal eq. (7.79) for the Gibbs energy G^S of the system as

$$G^S = n_a \mu_a + n_v \mu_v, \tag{7.83}$$

where n_a and n_v is the number of moles of atoms and vacancies, respectively. We draw attention to the fact that [see eq. (7.78)] from the definition of the chemical potential we have for the vacancy:

$$\mu_v \equiv \left. \frac{\partial G^S}{\partial n_v} \right|_{P, T, n_a},$$

which has to be zero (see ch. 3) because G^S has to minimize at the stable state. Therefore eq. (7.83) gives

$$\mu_a = \frac{G^S}{n_a} \equiv G,$$

where the ratio G^S/n_a is defined as the molar Gibbs energy G of the real crystal. As we know that $G = G^0 - nkT$ [see eq. (3.12)] the previous equation becomes $\mu_a = G^0 - nkT/N$, or considering eq. (7.81) we get $\mu_a = \mu^0 - nkT/N$, which is just eq. (3.64) of ch. 3.

On the parameters of solution. It is far from the scope of the present review to give a survey of the solubility problem. We shall stress here only some recent aspects dealing with problems which are usually not adequately treated.

In the usual solubility reports the chemical potential μ_u of a solute atom (u) is considered to be a function of temperature T, volume V and atom ratio x_u of the solute. However, it has recently been shown by Lupis [124] that this is incorrect; the quantity μ_u should be expressed as a function of the variables (T, V_S, x_u), where $V_S \equiv V/n_S$ and n_S denotes the number of moles of the solvent.

A second point to which we draw attention is that in principle – as in the case of vacancies – we could define two families of solubility parameters. The first refers to the situation where the solute atom is introduced to the solvent under constant pressure and the second family refers to the introduction of a solute atom to the solvent at constant volume. The two sets of parameters are interconnected by relations analogous to those given in ch. 3 for the case of vacancies.

For an extensive discussion of the solubility problem we refer to the recent publications [35,125–130].

Part 2

DEFECT PARAMETERS AS A FUNCTION OF BULK PROPERTIES: THE *cBΩ*-MODEL

8 CONNECTION OF DEFECT PARAMETERS TO BULK PROPERTIES

8.1 Introduction

In chapters 3 through 6 it became clear that all defect parameters can be directly determined if the Gibbs energy g is known as a function of temperature and pressure. In the present chapter we indicate a method – called the $cB\Omega$-model – which allows this determination as a function of the elastic and expansivity data of the bulk solid. This model will be given here as a postulate. Its correctness, as that of any model, will be evaluated by comparison to experimental results. In the chapters 9 to 13 a surprising agreement will be established to a wide variety of experimental facts. Many of them could hitherto receive no explanation and have often been described as "peculiar", "anomalous", etc. In the last chapter, 14, a theoretical justification of the $cB\Omega$-model is extensively treated, which actually leads to the form

$$g = cB\Omega.$$

In the present chapter we restrict to formation ("f") processes and use the form $g^f = c^f B\Omega$. In later chapters, dealing with migration ("m") and activation ("act") processes, completely analogous formulae are assumed to be valid: $g^m = c^m B\Omega$ and $g^{act} = c^{act} B\Omega$. In the elaboration of the present chapter the quantity c^f will be considered as a constant, i.e. independent of temperature and pressure. * As will be seen in ch. 14, general thermodynamics do not demand such a fact. However, there are good arguments that the possible T and P dependence of c^f is restricted to narrow bounds which are precisely known as a function of the bulk quantities β and κ. This change of c^f will be shown to be so small as not to be detectable in most

---
* Formulae that consider a temperature and pressure *dependent* c^f are given in ch. 14.

experimental situations. Under certain conditions (e.g. in a quasiharmonic approximation and when κ^f is comparable to κ) c^f is shown to be *exactly* constant.

In the present chapter, as mentioned, we give explicit formulae that are valid for the parameters (enthalpy, entropy, volume, etc.) of the *formation* process. In order to get the corresponding formulae for *migration* and *activation* processes one should simply replace the superscript "f" with "m" or "act". Attention is drawn to the point that $c^f \neq c^m \neq c^{act}$. Furthermore, we clarify that *even for the same crystal* c^f has *different* values for *different* defects (e.g. Schottky, Frenkel). The same holds for c^m (or c^{act}) when we are dealing with different migrating defects or mechanisms in the same host lattice (e.g. for CsCl the value of c^m for the cation vacancy motion is larger than that corresponding to anion vacancy motion).

8.2 Parameters by comparing with the isobaric perfect crystal

We start the $cB\Omega$-model from the form:

$$g^f = c^f B \Omega, \tag{8.1}$$

where c^f is a constant depending only on the matrix material and the type of defect.

At $T = 0$ the Gibbs formation energy becomes equal to the formation enthalpy h_0^f and therefore eq. (8.1) gives:

$$h_0^f = c^f B_0 \Omega_0 \tag{8.2}$$

or

$$c^f = h_0^f / B_0 \Omega_0. \tag{8.3}$$

In this form the constant "c^f" is expressed with the help of the defect enthalpy at zero temperature; the latter is a quantity that has been successfully calculated from first principles. By introducing this form into eq. (8.1) one obtains

$$g^f = \left(h_0^f / B_0 \Omega_0 \right) B \Omega. \tag{8.4}$$

By taking into account that

$$\Omega / \Omega_0 = b, \tag{8.5}$$

where

$$b \equiv \exp\left(\int_0^T \beta \, dT \right), \tag{8.6}$$

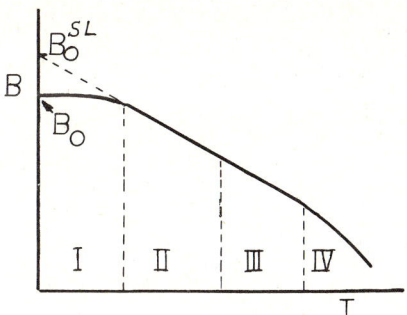

Fig. 8.1. Temperature dependence of the bulk modulus; I: quantum region; IV: region of excessive softening.

one easily gets the equivalent form

$$g^f = \frac{h_0^f}{B_0} Bb. \tag{8.7}$$

Equation (8.7) reveals that the Gibbs formation energy g^f varies in principle as the product of B and b, both of which depend on temperature and pressure.

Temperature variation of g^f. At relatively low temperature (regions I and II of fig. 8.1) the expansivity term b is always close to unity so that g^f essentially changes with temperature in the same way as the isothermal bulk modulus as shown in fig. 8.2. At very low temperatures (region I) B decreases with the temperature in a nonlinear way. Therefore g^f has to decrease nonlinearly with temperature (fig. 8.2) which means that s^f and h^f have to be temperature dependent. In region II the bulk modulus B is found to decrease linearly with T, whereas b still remains practically constant, i.e. near to unity. Therefore, in this temperature region g^f decreases almost linearly with T, and hence both s^f and h^f are practically temperature independent. In the temperature region III the temperature derivative of Bb starts deviating slightly from a constant value and hence g^f exhibits a nonlinear decrease with T. Recent experiments show that in some materials at temperatures near the melting point (region IV, the so-called region of excessive softening) B decreases with T faster than linearly, so that g^f shows a strong nonlinear temperature decrease.

Formation entropy. By inserting into eq. (3.16) the value of g^f given by eq. (8.1), we easily get

$$s^f = -c^f \Omega \left(\beta B + \frac{\partial B}{\partial T} \bigg|_P \right) \tag{8.8}$$

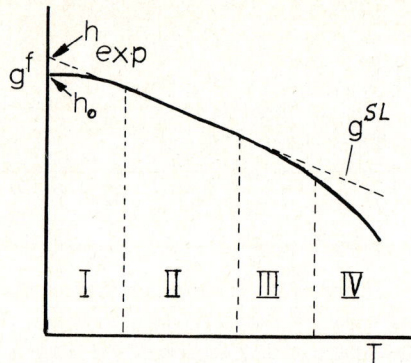

Fig. 8.2. Temperature dependence of g^f according to the $cB\Omega$-model.

or, by using the value of c^f given by eq. (8.3),

$$s^f = -\frac{h_0^f}{B_0}b\left(\beta B + \frac{\partial B}{\partial T}\bigg|_P\right). \tag{8.9}$$

The latter form is useful for numerical calculations. It is obvious from eq. (8.8) that the entropy s^f depends on temperature and pressure because of their influence on B, β and Ω. We note that the quantity s^f/h_0^f only depends on bulk properties. For different types of defects, e.g. Schottky defects, Frenkel defects, etc., occurring in a given matrix material, this quantity has the same value.

The sign of s^f can easily be estimated by using the so-called Anderson–Grüneisen parameter δ, defined as follows:

$$\delta \equiv -\frac{1}{\beta B}\frac{\partial B}{\partial T}\bigg|_P. \tag{8.10}$$

Equation (8.8) can alternatively be written as

$$s^f = c^f \beta B \Omega (\delta - 1). \tag{8.11}$$

Due to the fact that δ usually exceeds unity and β is positive, eq. (8.11) reveals s^f to be usually positive. A combination of eqs. (8.11) and (8.1) gives s^f/g^f in terms of bulk quantities only:

$$s^f/g^f = \beta(\delta - 1).$$

The typical temperature dependence of the entropy (fig. 8.3) can easily be found by recalling the considerations discussed for the temperature dependence of g^f and the definition of s^f (see §3.6.1 and fig. 3.5). In region III it increases slightly, i.e. a few percent.

The pressure variation of s^f can be described in terms of the entropic part of the formation volume, v_s^f, which will be evaluated later. We shall see

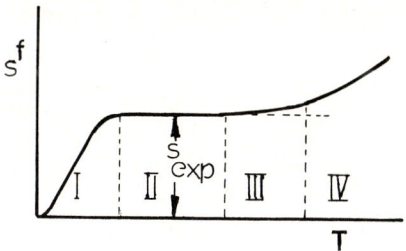

Fig. 8.3. Temperature dependence of s^f according to the $cB\Omega$-model.

that v_s^f is usually positive and very rapidly increases with temperature; a positive value of v_s^f basically means that s^f decreases with pressure, so that (see eq. 3.167) v^f increases with temperature.

Formation enthalpy. By inserting into eq. (3.22) the values of g^f and s^f, given by the relations (8.1) and (8.8), we get

$$h^f = c^f \Omega \left\{ B - T\beta B - T \left. \frac{\partial B}{\partial T} \right|_P \right\} \tag{8.12}$$

or by using the value of c^f (eq. 8.3):

$$h^f = \frac{h_0^f}{B_0} b \left\{ B - T\beta B - T \left. \frac{\partial B}{\partial T} \right|_P \right\}. \tag{8.13}$$

It is obvious from eq. (8.13) that h^f is always positive if one remembers that $T\beta$ is (at most) of the order of 10^{-1}, and that $(\partial B/\partial T)_P$ is negative. Furthermore, eq. (8.13) reveals that h^f depends on temperature and pressure.

The typical temperature dependence of h^f (fig. 8.4) can be found by considering the temperature dependence of s^f and eq. (3.25). The pressure variation of h^f can be expressed with the help of the enthalpic part of the

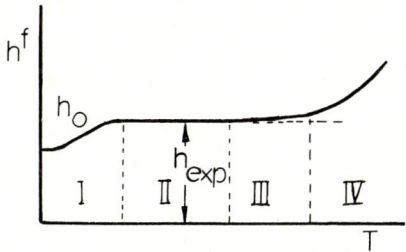

Fig. 8.4. Temperature dependence of h^f according to the $cB\Omega$-model.

formation volume $v_h^f = (\partial h^f/\partial P)_T$, (eq. 3.162), which will be evaluated later. As we shall see, v_h^f is usually positive, which means that h^f increases when the pressure increases.

Connection of s^f, h^f with the quantity h_{exp}^f. Experiments are usually carried out in the regions II and III; in region II g^f is approximately linear and s^f and h^f approximately constant. The results of the experiments in this temperature region are usually labelled with h_{exp}^f and s_{exp}^f. They have been indicated in figs. 8.3 and 8.4. As the quantity h_0^f is of large theoretical interest, we will give its connection with h_{exp}^f. In region II, B is linear, and therefore its extension in a straight line (SL) to the vertical axis gives the quantity

$$B_0^{\text{SL}} = B - T \left.\frac{\partial B}{\partial T}\right|_P, \tag{8.14}$$

as indicated in fig. 8.1. If one considers that $B_0^{\text{SL}} \gg T\beta B$ and that in region II $b \simeq 1$, eq. (8.13) gives the approximate relation:

$$h_{\text{exp}}^f \simeq \frac{B_0^{\text{SL}}}{B_0} h_0^f \tag{8.15}$$

or by using eq. (8.3)

$$h_{\text{exp}}^f \simeq c B_0^{\text{SL}} \Omega_0. \tag{8.16}$$

From eq. (8.9) and using eq. (8.15) one gets the useful connection

$$s^f \simeq -\frac{b}{B_0^{\text{SL}}} \left(\beta B + \left.\frac{\partial B}{\partial T}\right|_P\right) h_{\text{exp}}^f. \tag{8.17}$$

Due to eq. (8.15), eq. (8.7) becomes:

$$g^f \simeq h_{\text{exp}}^f Bb / B_0^{\text{SL}}. \tag{8.18}$$

Formation volume. Inserting into eq. (3.31) the value of g^f, given by eq. (8.1), we get

$$v^f = c^f \Omega \left(\left.\frac{\partial B}{\partial P}\right|_T - 1\right). \tag{8.19}$$

Due to eq. (8.2) or (8.15) this equation becomes

$$v^f = \frac{h_0^f b}{B_0} \left(\left.\frac{\partial B}{\partial P}\right|_T - 1\right) \tag{8.20}$$

or

$$v^f \simeq \frac{h_{\text{exp}}^f b}{B_0^{\text{SL}}} \left(\left.\frac{\partial B}{\partial P}\right|_T - 1\right). \tag{8.21}$$

Dividing eq. (8.1) by (8.19) we get

$$\frac{v^f}{g^f} = \frac{1}{B}\left(\left.\frac{\partial B}{\partial P}\right|_T - 1\right). \tag{8.22}$$

Equations (8.21) and (8.22) reveal that the quantities $v^f/h^f_{\exp}$ and v^f/g^f are bulk properties, independent of the kind of defect created in a given solid. It allows the prediction of v^f for materials where the latter has not been measured. As to its sign these equations reveal that, depending on the difference of $\partial B/\partial P$ from unity, v^f can be either positive, equal to zero or negative. The latter case seems to apply to some materials discussed in ch. 10.

Temperature variation of v^f. The quantities $\partial B/\partial P$ and Ω in eq. (8.19) usually increase with temperature so that the formation volume also has to increase. Furthermore, it should be noted that in these cases v^f increases faster than the bulk volume. For a large number of solids $\partial B/\partial P$ increases linearly with temperature. We will label with $(\partial B/\partial P)_{T=0,\,\text{SL}}$ the value resulting from the straight-line extension to $T=0$. In such cases (fig. 2.13)

$$\left.\frac{\partial B}{\partial P}\right|_T = \left.\frac{\partial B}{\partial P}\right|_{T=0,\,\text{SL}} + T\frac{\mathrm{d}(\partial B/\partial P)}{\mathrm{d}T}. \tag{8.23}$$

Inserting this equation into eq. (8.20) one obtains

$$v^f = \frac{h^f_0 b}{B_0}\left(\left.\frac{\partial B}{\partial P}\right|_{T=0,\,\text{SL}} + T\frac{\partial(\partial B/\partial P)}{\partial T} - 1\right). \tag{8.24}$$

At zero temperature $b=1$ and $[\partial(\partial B/\partial P)/\partial T] = 0$, so that we get:

$$v^f_0 = \frac{h^f_0}{B_0}\left(\left.\frac{\partial B}{\partial P}\right|_{T=0,\,\text{SL}} - 1\right).$$

Introducing this expression into eq. (8.24) one has

$$v^f = v^f_0 b + \frac{h^f_0 bT}{B_0}\frac{\partial(\partial B/\partial P)}{\partial T}.$$

At low temperatures b is practically equal to unity so that v must increase almost linearly with temperature and with a gradient equal to $(h^f_0/B_0)[\partial(\partial B/\partial P)/\partial T]$. In the high temperature region b has a noticeable deviation from unity so that the graph of v^f versus T shows an upward curvature. At very low temperatures the third law of thermodynamics demand that $\partial(\partial B/\partial P)/\partial T \to 0$ (as $T \to 0$) so that $\partial B/\partial P$ becomes temperature independent (see fig. 2.13). The plot can therefore be expected to become horizontal at very low temperatures. Such a deviation from linearity

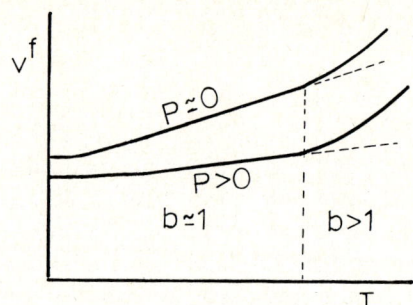

Fig. 8.5. Typical temperature dependence of v^f according to the $cB\Omega$-model (when dB/dP increases with T in a manner indicated in fig. 2.13).

has actually been observed by Raab and Peisl [131]. Figure 8.5 shows the typical form of the temperature dependence of v^f.

Pressure variation of v^f. In eq. (8.19) besides Ω also the term $\partial B/\partial P$ decreases with pressure; therefore v^f has to decrease with pressure faster than the bulk volume. Dividing the values of v^f at zero pressure and at pressure P one gets

$$v^f(P) = v^f(0) \frac{[\partial B/\partial P]_T(P) - 1}{[\partial B/\partial P]_T(0) - 1} \exp\left(\int_0^P -\kappa dP\right). \tag{8.25}$$

Figure 8.6 shows the typical variation of v^f with pressure for positive formation volumes. Details will be discussed later.

Thermal expansivity of the formation volume. Inserting into eq. (3.43) the value of v^f given by eq. (8.19) we finally get for the expansivity β^f of the formation volume

$$\beta^f = \beta + \left[\frac{d}{dT}\left(\frac{\partial B}{\partial P}\right)\right]\left[\left.\frac{\partial B}{\partial P}\right|_T - 1\right]^{-1} \tag{8.26}$$

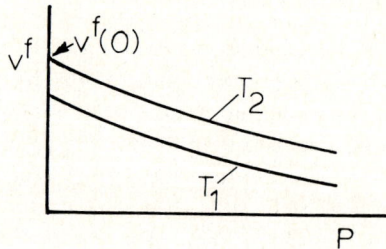

Fig. 8.6. Pressure dependence of v^f ($T_2 > T_1$).

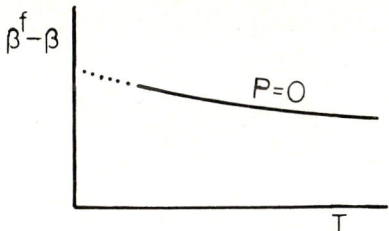

Fig. 8.7. Temperature dependence of the difference between the thermal expansion coefficients of the formation volume and the bulk volume.

or

$$\beta^f = \beta + \frac{d}{dT} \ln\left(\left.\frac{\partial B}{\partial P}\right|_T - 1\right). \tag{8.27}$$

We note that the expansivity of the defect depends only on bulk properties and not on the type of the defect. In contrast to the formation parameters s^f and v^f, it can be directly calculated from pure bulk properties. The temperature dependence of $\beta^f - \beta$ can be evaluated in the approximation of eq. (8.23) by inserting the latter equation into eq. (8.26). One gets:

$$\beta^f - \beta = \frac{\dfrac{d}{dT}\left(\left.\dfrac{\partial B}{\partial P}\right|_T\right)}{\left.\dfrac{\partial B}{\partial P}\right|_{T=0,\,SL} - 1 + T\dfrac{d}{dT}\left(\dfrac{\partial B}{\partial P}\right)}. \tag{8.28}$$

As shown in fig. 2.13 the ratio $\partial B/\partial P$ is usually found to increase to a good approximation linearly with temperature so that the numerator of eq. (8.28) is constant, while the denominator increases with temperature. Figure 8.7 gives the general form of $\beta^f - \beta$ versus T. As follows from fig. 2.5, β increases with temperature, but no definite conclusions can be drawn about an increase or decrease of β^f with temperature. However, one can easily see that the ratio β^f/β has an absolute value always larger than unity and may be very large at low temperatures.

Calculation of β_s^f and β_h^f. Inserting eq. (8.26) into eq. (3.180) we get after some calculation

$$\beta_s^f = \beta + \frac{\partial}{\partial T} \ln\left(\frac{\partial B}{\partial P} - 1\right) + \frac{\dfrac{\partial \beta}{\partial T} + \dfrac{\partial^2}{\partial T^2}\left[\ln\left(\dfrac{\partial B}{\partial P}\right) - 1\right]}{\beta + \dfrac{\partial}{\partial T}\ln\left(\dfrac{\partial B}{\partial P} - 1\right)} + \frac{1}{T}.$$

The value of β_h^f can easily be found by inserting eq. (8.27) into eq. (3.184):

$$\beta_h^f = -\frac{T\left\{\beta + \dfrac{d}{dT}\ln\left(\dfrac{\partial B}{\partial P} - 1\right)\right\}}{1 - T\left\{\beta + \dfrac{d}{dT}\ln\left(\dfrac{\partial B}{\partial P} - 1\right)\right\}}\left\{\beta + \frac{d}{dT}\left(\ln\frac{\partial^2 B}{\partial T \partial P}\right)\right\}.$$

Calculation of v_s^f and v_h^f. The value of v_s^f can be found by inserting the value of v^f given by eq. (8.19) into eq. (3.167):

$$v_s^f = Tc^f\Omega\left\{\beta\left(\left.\frac{\partial B}{\partial P}\right|_T - 1\right) + \frac{\partial^2 B}{\partial T \partial P}\right\}. \tag{8.29}$$

As already mentioned the value of $\partial B/\partial P$ is found to be greater than unity and to increase with temperature; therefore eq. (8.29) predicts that the volume v_s^f increases at a rate faster than proportional to T. The value of v_h^f is directly found from

$$v_h^f = v^f - v_s^f = c^f\Omega\left\{(1 - T\beta)\left(\left.\frac{\partial B}{\partial P}\right|_T - 1\right) - T\frac{\partial^2 B}{\partial T \partial P}\right\}. \tag{8.30}$$

The temperature dependence of v_h^f can be evaluated [in the approximation of eq. (8.23)] by:

$$v_h^f = c^f\Omega\left[\left(\left.\frac{\partial B}{\partial P}\right|_{T=0,\,\text{SL}} - 1\right) - T\beta\left(\left.\frac{\partial B}{\partial P}\right|_{T=0,\,\text{SL}} - 1\right) - T^2\beta\frac{\partial^2 B}{\partial T \partial P}\right].$$

If the last term were not present, v_h^f would be

$$c^f\Omega\left[\left(\left.\frac{\partial B}{\partial P}\right|_{T=0,\,\text{SL}} - 1\right)(1 - T\beta)\right];$$

this expression is relatively constant with increasing temperature because at the melting point the factor Ω has increased up to approximately 10% while the factor $(1 - T\beta)$ has decreased by approximately the same percentage. A temperature dependence of v_h^f therefore results mainly from the existence of the last term. According to fig. 2.13 the quantity $\partial^2 B/\partial T \partial P$ has at high T a constant positive value. By multiplying this with $T\beta = 0.1$ one finds that at high temperatures v_h^f must usually slightly decrease with T.

Compressibility of the formation volume. By inserting into eq. (3.53) the value of v^f given by eq. (8.19), we finally get

$$\kappa^f = \kappa - \frac{(\partial^2 B/\partial P^2)_T}{(\partial B/\partial P)_T - 1}. \tag{8.31}$$

Equation (8.31) permits the direct calculation of the compressibility κ^f of the formation volume from bulk properties without having to know any vacancy parameter. The experiments usually show that $\partial B/\partial P$ decreases on compression so that $\partial^2 B/\partial P^2$ is negative. Therefore κ^f must be larger than κ.

It is interesting to discuss the temperature dependence of κ^f. Experiments [30,31] show that both $\partial B/\partial P$ and κ increase with temperature but, due to large experimental errors, no direct experimental evidence exists for the temperature dependence of $\partial^2 B/\partial P^2$. By assuming the latter to be practically temperature independent, eq. (8.31) shows that κ^f increases with temperature. The temperature and pressure dependence of the ratio κ^f/κ can be found by setting eq. (8.31) into the following form:

$$\frac{\kappa^f}{\kappa} = 1 - \frac{B(\partial^2 B/\partial P^2)}{(\partial B/\partial P)_T - 1}. \tag{8.32}$$

By again assuming that $\partial^2 B/\partial P^2$ is temperature and pressure independent we can draw the following conclusions about the ratio κ^f/κ: (a) it decreases with temperature but constantly exceeds unity and (b) it increases with pressure because B increases and $(\partial B/\partial P)_T$ decreases with pressure.

Specific heat. By inserting eq. (8.12) into eq. (6.3) we get after some calculation the specific heat c_P of a vacancy:

$$c_P = -c^f \Omega T \left\{ \beta^2 B + 2\beta \left.\frac{\partial B}{\partial T}\right|_P + B \left.\frac{\partial \beta}{\partial T}\right|_P + \left.\frac{\partial^2 B}{\partial T^2}\right|_P \right\}. \tag{8.33}$$

This equation permits the direct evaluation of c_P from h_0^f or c^f (see eq. 8.3), provided that β and B are experimentally known as functions of temperature.

The sign of c_P can be determined by rearranging eq. (8.33) as follows:

$$c_P = -c^f B \Omega T \beta^2 \left\{ 1 - 2\delta + \frac{1}{\beta} \left.\frac{\partial \ln \beta}{\partial T}\right|_P + \frac{1}{\beta^2 B} \left.\frac{\partial^2 B}{\partial T^2}\right|_P \right\}, \tag{8.34}$$

where δ is the usual Anderson–Grüneisen parameter [eq. (8.10)]. The value of $(\partial^2 B/\partial T^2)/(\beta^2 B)$ in the high temperature range is usually zero or negative, whereas the value of $(\partial \ln \beta/\partial T)/\beta$ is positive. By considering that the absolute values of these two terms are comparable, we see that the value of the quantity included in the brackets of eq. (8.34) is governed by the value of $(1 - 2\delta)$, which is usually negative. Therefore c_P is usually positive.

We proceed with the calculation of c_V^0, which is the parameter that usually results from microscopic calculations from first principles. By using the approximations $\beta^f/\beta^0 \simeq \beta^f/\beta$ and $\kappa^f/\kappa^0 \simeq \kappa^f/\kappa$ eq. (6.45) gives

$$c_V^0 \simeq c_P - T v^f \beta^2 B \left[(2\beta^f/\beta) - (\kappa^f/\kappa) \right]. \tag{8.35}$$

We insert into this equation eqs. (8.27), (8.32), (8.19) and (8.34), which are the $cB\Omega$-formulae for β^f/β, κ^f/κ, v^f and c_P; one finally gets:

$$c_V^0 = c^f \Omega T \beta B \left\{ 2\beta + 2 \left.\frac{\partial \ln B}{\partial T}\right|_P + \left.\frac{\partial \ln \beta}{\partial T}\right|_P \right.$$

$$\left. + \frac{1}{\beta B} \left.\frac{\partial^2 B}{\partial T^2}\right|_P - \beta \left(\left.\frac{\partial B}{\partial P}\right|_T - B \frac{\partial^2 B}{\partial P^2} \right) + 2 \frac{d}{dT} \left.\frac{dB}{dP}\right|_T \right\}. \qquad (8.36)$$

In order to estimate c_V^0 we rearrange this equation as follows:

$$c_V^0 = c^f \gamma \tilde{c}_V T \beta \left(2 + \frac{2}{\beta} \left.\frac{\partial \ln B}{\partial T}\right|_P + \frac{1}{\beta} \left.\frac{\partial \ln \beta}{\partial T}\right|_P \right.$$

$$\left. + \frac{1}{\beta^2 B} \left.\frac{\partial^2 B}{\partial T^2}\right|_P - \left.\frac{\partial B}{\partial P}\right|_T + B \left.\frac{\partial^2 B}{\partial P^2}\right|_T + \frac{1}{\beta} \frac{\partial}{\partial T} \left.\frac{\partial B}{\partial P}\right|_T \right), \qquad (8.37)$$

where γ represents the thermodynamical Grüneisen parameter. For monoatomic solids we have $\gamma = \beta B \Omega / \tilde{c}_V$ and $\tilde{c}_V = 3k$ at high T. The sign and the value of each term can now be estimated (for monoatomic solids):

(1) Whenever B decreases linearly with T – which holds for temperatures $T > \Theta_D$ but not near the melting point – the term $(\partial^2 B/\partial T^2)/(\beta^2 B)$ seems to be of minor importance.

(2) The term $\partial \ln B/\partial T$ is negative, whereas the terms $\partial \ln \beta/\partial T$ and $\partial(\partial B/\partial P)/\partial T$ are positive. The experimental data show that the sum of these two terms are of the order of $-\partial \ln B/\partial T$.

(3) The dimensionless quantity $\partial B/\partial P$ lies usually between 3.5 and 8; the derivative $\partial^2 B/\partial P^2$ is negative so that the term $B\partial^2 B/\partial P^2$ is of the order of $-\partial B/\partial P$.

(4) At high temperatures the term βT is of the order of 10^{-1}, whereas the specific heat $\tilde{c}_V$ per volume Ω is, as already mentioned, approximately equal to $3k$.

Taking also into account that c^f usually lies between 0.1 and 0.15 and that c_P is positive, eq. (8.37) shows that (for $T > \Theta_D$) c_V^0 is negative. Its absolute value is equal to a few tenths of an entropy unit. This conclusion is in fundamental agreement with microscopic concepts [37], which show that c_V^0 is of the order of $-(\Theta_D/T)^2 k$.

8.3 Parameters by comparing with the isochoric perfect crystal

Due to eq. (3.123), $f^* = g$, eq. (8.1) gives directly

$$f^* = c^f B \Omega. \qquad (8.38)$$

Inserting the values of s^f and v^f given by eqs. (8.8) and (8.19) into the thermodynamical relation (3.126) we get for the thermal entropy

$$s^* = -c^f\Omega\left(\left.\frac{\partial B}{\partial T}\right|_P + \beta B \left.\frac{\partial B}{\partial P}\right|_T\right) \tag{8.39}$$

or

$$s^* = -c^f\Omega \left.\frac{\partial B}{\partial T}\right|_V \tag{8.40}$$

and hence

$$\frac{s^*}{g^f} = -\frac{1}{B}\left.\frac{\partial B}{\partial T}\right|_V. \tag{8.41}$$

A form more accessible to calculation is obtained by inserting eqs. (8.3) and (8.5) into eq. (8.39). The resulting equation reveals that the quantity s^*/h_0^f for each temperature and pressure is an intrinsic property of the solid irrespective of the type of defects. The sign of s^* can be found by introducing the Anderson–Grüneisen parameter δ into eq. (8.39):

$$s^* = -c^f B\Omega\beta\left(\left.\frac{\partial B}{\partial P}\right|_T - \delta\right). \tag{8.42}$$

The parameter δ is usually around 2γ (where γ is the Grüneisen parameter) so that δ is greater than unity but comparable to $\partial B/\partial P$, which for many solids has a value around four. By comparing to eq. (8.11) we find that (provided $\beta > 0$) the absolute value of s^* is smaller than s^f. The small absolute values of s^* in relation to s^f can be explained as being due to the small change of the frequency spectrum when the vacancy is produced under isochoric conditions, provided that the corresponding Grüneisen constants are positive (see §14.4).

The calculation of $s^* - s^f$ from eq. (3.126) is easy because we already know v^f from eq. (8.19). By dividing eqs. (8.42) and (8.11) we get

$$\frac{s^*}{s^f} = \frac{(\partial B/\partial P)_T - \delta}{1 - \delta}, \tag{8.43}$$

which permits the direct evaluation of the entropy s^* from the entropy s^f with the help of bulk properties only.

By inserting into eq. (3.128) the values of h^f and v^f given by eqs. (8.12) and (8.19), we get the formation internal energy

$$u^* = c^f\Omega\left\{B - T\beta B \left.\frac{\partial B}{\partial P}\right|_T - T\left.\frac{\partial B}{\partial T}\right|_P\right\}. \tag{8.44}$$

A form, more accessible to numerical calculation, is obtained by introducing eq. (8.3):

$$u^* = \frac{h_0^f b}{B_0}\left(B - T\beta B \left.\frac{\partial B}{\partial P}\right|_T - T \left.\frac{\partial B}{\partial T}\right|_P\right). \tag{8.45}$$

In order to estimate the sign of u^*, it is convenient to introduce the Anderson–Grüneisen parameter δ:

$$u^* = c^f \Omega B \left\{1 + T\beta\left(\delta - \left.\frac{\partial B}{\partial P}\right|_T\right)\right\}. \tag{8.46}$$

At zero temperature this equation gives the positive value $u^* = c^f \Omega_0 B_0 = h_0^f$. Usually $\partial B/\partial P$ is not very different from δ so that initially u^* might either increase or decrease with temperature. However, since for a given pressure $\partial B/\partial P$ increases with temperature, at a certain temperature the expression $\delta - \partial B/\partial P$ will definitely become negative. When it reaches the value $-(T\beta)^{-1}$, u^* will become zero. This unique isochoric condition for the temperature, for which $u^* = 0$, (point K of fig. 3.1) can be predicted purely from bulk properties. It indicates that in some solids under certain conditions a vacancy can be produced *without* any heat having to be added. Recent investigations [636] indicate that it occurs in some rare gas solids at temperatures close to the melting point.

It is of interest to note that by dividing eq. (8.12) by eq. (8.44) one gets

$$\frac{h^f}{u^*} = \frac{B - T\beta B - T(dB/dT)_P}{B - T\beta B(dB/dP)_T - T(dB/dT)_P}.$$

Inserting into eq. (3.133) the values of h^f and v^f given by eqs. (8.13) and (8.20) we get the formation enthalpy h^*:

$$h^* = \frac{h_0^f}{B_0} b \left\{B(1 - T\beta) \left.\frac{\partial B}{\partial P}\right|_T - T \left.\frac{\partial B}{\partial T}\right|_P\right\}. \tag{8.47}$$

Due to the fact that $\partial B/\partial P > 0$, $\partial B/\partial T < 0$ and $T\beta < 1$, this equation reveals that h^* is always positive. Subtracting eq. (8.13) from eq. (8.47) one gets

$$h^* - h = h_0^f b \frac{B}{B_0}(1 - T\beta)\left(\left.\frac{\partial B}{\partial P}\right|_T - 1\right). \tag{8.48}$$

Inserting into eq. (3.105) the values of h^* and s^* from eqs. (8.47) and (8.39) we get the connection

$$g^* = c^f B \Omega \left.\frac{\partial B}{\partial P}\right|_T. \tag{8.49}$$

§8.3] *Parameters by comparing with the isochoric perfect crystal* 161

In most cases $\partial B/\partial P$ is much greater than unity, and therefore eq. (8.49) shows g^* to be not only positive but also appreciably larger than g. Obviously g^f and g^* are directly connected by the relation

$$g^*/g^f = \partial B/\partial P. \tag{8.50}$$

Note that for materials with negative $\partial B/\partial P$ the parameter g^* has to be negative.

The pressure developed when a vacancy is formed by heating under constant volume can be calculated from the $cB\Omega$-model by introducing eq. (8.19) into eq. (3.131)

$$p^* = \frac{h_0^f b}{B_0} \left(\frac{dB}{dP} - 1 \right) \frac{B}{V}. \tag{8.51}$$

Specific heat. Equation (6.40) *is a thermodynamic expression for* $c_P - c_V$. *As* c_P *could be determined in eq.* (8.33) *with the help of the* $cB\Omega$-*model, it is now possible to obtain by subtraction the following expression:*

$$c_V = -c^f \Omega T \beta^2 B$$

$$\times \left\{ 1 + \frac{2}{\beta B} \frac{\partial B}{\partial T}\bigg|_P + \frac{1}{\beta^2} \frac{\partial \beta}{\partial T}\bigg|_P + \frac{1}{\beta^2 B} \frac{\partial^2 B}{\partial T^2}\bigg|_T - \left(\frac{\partial B}{\partial P}\bigg|_T - 1 \right) \right.$$

$$\left. \times \left(2\frac{\beta^f}{\beta} - \frac{\kappa^f}{\kappa} + \frac{2}{\beta B} \frac{\partial B}{\partial T}\bigg|_P + \frac{\partial B}{\partial P}\bigg|_T + \frac{1}{\beta^2} \frac{\partial \beta}{dT}\bigg|_P \right) \right\}. \tag{8.52}$$

This equation permits the direct calculation of c_V which is of high theoretical interest [132,37].

9 | THE $cB\Omega$-MODEL: DEFECT ENTROPY AND ENTHALPY. SELF-DIFFUSION

The $cB\Omega$-model is based on the assumption that the coefficient c in the formula $g = cB\Omega$ is pressure and temperature independent but takes different values depending on the type of defect and the solid under discussion (see §8.1). The most sensitive criterion of this constancy results from comparing experimental data of quantities that depend exponentially on c. This is the case of e.g. the vacancy concentration, the diffusion coefficients and the conductivities of ionic materials. In this chapter the quantity s/h will also be studied, because the $cB\Omega$-model predicts that it is a bulk quantity.

9.1 Introduction

According to the $cB\Omega$-model the defect entropy and enthalpy are given by:

$$s = -c \left.\frac{\partial(B\Omega)}{\partial T}\right|_P \quad \text{or} \quad s = -c\Omega\left(\beta B + \left.\frac{\partial B}{\partial T}\right|_P\right), \tag{9.1}$$

$$h = cB\Omega - Tc\left.\frac{\partial(B\Omega)}{\partial T}\right|_P \quad \text{or} \quad h = c\Omega\left(B - T\beta B - T\left.\frac{\partial B}{\partial T}\right|_P\right). \tag{9.2}$$

The above relations are assumed to be valid irrespective of the process (for a justification, see ch. 14); attention is drawn to the point that in order to get the formation parameters one has to insert the c^f value; similarly in order to get h^{act} and s^{act} one has to insert c^{act}, etc. Each of the values of c^f, c^{act}, etc. can be determined from a single experimental situation (see below).

The quantity s/h. One important point resulting from the above expressions is that in a given material the quantity

$$\frac{s}{h} = -\frac{\beta B + \left.\frac{\partial B}{\partial T}\right|_P}{B - T\beta B - T\left.\frac{\partial B}{\partial T}\right|_P} \tag{9.3}$$

must be independent of the process and equal to a bulk property. This is checked in the present chapter for a great number of materials; in order to do so, we plot s versus h for various processes and examine whether the corresponding points fall on a straight line with a slope equal to the bulk property indicated from the above relation. Obviously a strict check of eq. (9.3) requires the knowledge of the experimental values of s and h for each temperature; this is imperative in the few cases where the Arrhenius plots are curved.

In most cases the experimentalists describe an almost linear Arrhenius plot with a single value of enthalpy, which we label with $h_{\exp}$ (see fig. 8.4). In such cases the $cB\Omega$-model predicts that $h_{\exp}$ is connected to the exact value of h, for each temperature, with the following approximate relation which results from eqs. (8.13) and (8.15):

$$h \simeq h_{\exp} \frac{1}{B_0^{SL}} \left(B - T\beta B - T \frac{\partial B}{\partial T}\bigg|_P \right) \exp \int_0^T \beta \, dT.$$

Then eq. (9.3) can be simplified to:

$$\frac{s}{h_{\exp}} \simeq -\frac{\left(\beta B + \frac{\partial B}{\partial T}\bigg|_P \right) \exp \int_0^T \beta \, dT}{B_0^{SL}}. \tag{9.4}$$

When B decreases linearly with temperature the second members of eqs. (9.3) and (9.4) differ only by a few percent so that to a first approximation one can use any one of them; in practice, in the case of linear Arrhenius plots the exact value of h for each temperature differs from $h_{\exp}$ only by a few percent, which is surely covered by the experimental error.

Another point which is checked in this chapter is whether the $cB\Omega$-model gives the correct values of s and h for formation, activation and migration processes. For this we have to know the value of c for these processes (see below).

Calculation of the formation parameters. The value of c^f can be determined from the value of n/N at a single temperature. For instance by applying the relation (3.9):

$$\frac{n}{N} = \exp\left(-\frac{c^f B \Omega}{kT} \right) \tag{9.5}$$

near the melting point ($T = T_M$) one immediately gets the value of c^f; then, once c^f is known, the above equation can determine n/N for each temperature with the help of elastic and expansivity data. In other words, eqs. (9.1) and (9.2) can then determine the values of the formation parameters s^f and

Introduction

h^f. Furthermore, the formation volume can then be directly computed from the $cB\Omega$-model [see eq. (8.19)].

Calculation of the self-diffusion activation parameters from a single measurement. The tracer self-diffusion coefficient is given by (see eq. 7.40b.)

$$D = fa^2 \nu \exp\left(-\frac{g^{\text{act}}}{kT}\right) \tag{9.6}$$

or by inserting $g^{\text{act}} = c^{\text{act}} B\Omega$:

$$D = fa^2 \nu \exp\left(-\frac{c^{\text{act}} B\Omega}{kT}\right). \tag{9.7}$$

As explained in §7.3.1 the frequency factor ν can be approximated by ν_D or in monoatomic crystals by (see eq. 7.38)

$$\nu \simeq \tilde{\nu} \equiv \frac{1}{a}\left(\frac{h^m}{M}\right)^{1/2}. \tag{9.8}$$

When the D-value is known for a single temperature eq. (9.7) leads directly to a good estimation of c^{act}. Once c^{act} is known this equation can predict the D-value at any temperature and pressure [132a]; the values of h^{act} and s^{act} can then be directly computed from eqs. (9.1) and (9.2).

A quite important fact emerges from eq. (9.7): it indicates that when self-diffusion data are available for an extensive temperature range one should plot $\ln D$ versus $B\Omega/kT$ instead of the usual $\ln D$ versus $1/T$. The linearity of such a plot would indicate that a single diffusion mechanism is operating in spite of the fact that the plot $\ln D$ versus $1/T$ may show an appreciable curvature. Further advantages of plotting $\ln D$ versus $B\Omega/kT$ are the following: (a) the intercept leads to the determination of the frequency ν and (b) the slope gives c^{act}, from which one can directly compute not only the activation enthalpy h^{act} – from eq. (9.2) – but also the "true value" of the activation entropy s^{act} from eq. (9.1). As a characteristic example we give in fig. 9.1 for Na the "curved" diffusion plot $\log D$ versus $1/T$ and also $\ln D$ versus $B\Omega/kT$. One immediately sees that although the "curved" diffusion plot shows a strong upward curvature, the plot $\ln D$ versus $B\Omega/kT$ is linear within the experimental uncertainty of the various quantities used. Details will be given in §9.3.1.

The assessment of the diffusion coefficient when the bulk modulus is known can be useful in geophysics. The bulk modulus can be estimated at various depths by seismological methods which can then lead to the diffusivity in the rocks of the interior of the earth.

Throughout the book the conversion from B^S to B is done by means of the thermodynamical formula: $B^{-1} - (B^S)^{-1} = TV\beta^2/C_P$, or the equivalent: $B^S - B = TC_V\gamma^2/V$, where $\gamma = \beta VB/C_V$.

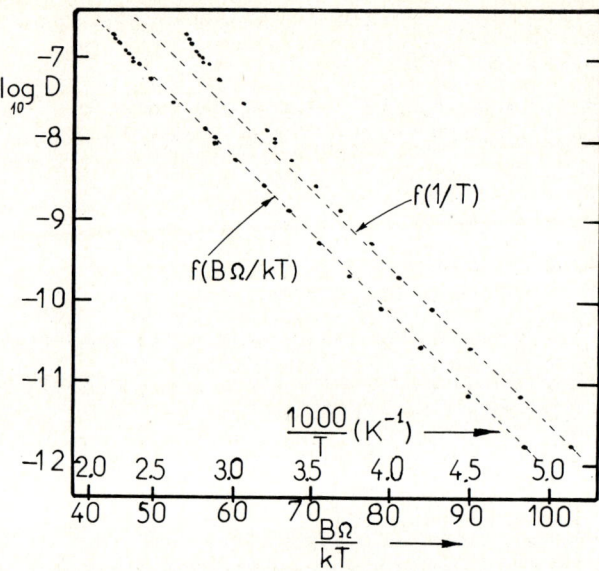

Fig. 9.1. Self-diffusion in sodium.

Features of Arrhenius plots. The Arrhenius plots for diffusion or conductivity of ionic solids are in some cases practically linear, while in others (β-Ti, β-Zr, AgBr) they have a pronounced upward curvature. According to the $cB\Omega$-model (see §8.2) the approximately linear case occurs in the temperature region where the bulk modulus decreases linearly with temperature, i.e. according to the formula $B = B_0^{SL} + T\partial B/\partial T$. We label the values of the bulk modulus and the diffusion coefficient of such linear temperature regions with B^{SL} and D^{SL} (figs. 9.2 and 9.3). When the diffusion plot bends at higher temperature, the excessive diffusivity $\delta \ln D$ can always receive an explanation from the $cB\Omega$-model by being ascribed to an excessive fall, δB, of the bulk modulus. By writing eq. (9.7) for $\ln D$, and for $\ln D^{SL}$ the value that results from a linear extrapolation of the (linear part of the) low-temperature data (see fig. 9.3) one gets:

$$\ln D = \ln fa^2\nu - \frac{cB\Omega}{kT}$$

and

$$\ln D^{SL} \simeq \ln fa^2\nu - \frac{cB^{SL}\Omega}{kT},$$

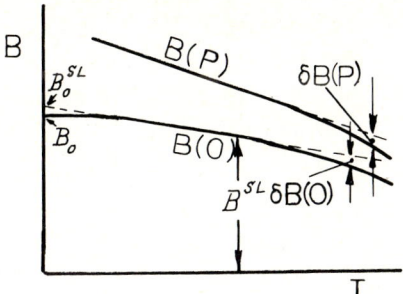

Fig. 9.2. Excessive fall δB of the bulk modulus for ambient and increased pressure (for an explanation, see §A.11, note 2).

which gives for the excessive diffusion

$$\delta \ln D \equiv \ln D - \ln D^{SL} \simeq \frac{c(\delta B)\Omega}{kT}. \qquad (9.9)$$

One sees that the excessive increase of $\ln D$ is mainly due to δB ($\equiv B^{SL} - B$).

The curvature of an Arrhenius plot can *alternatively* be described by the change of its slope. It has been shown that the slope is equal to $-h^{act}/k$, a change in slope therefore can be described by the derivative dh^{act}/dT which according to eq. (7.49) is identical to c_P^{act}. Gilder and Lazarus [37] have shown that (see eq. 7.49)

$$c_P^{act} \simeq T\beta\gamma\tilde{c}_V \frac{v^{act}}{\Omega}\left(2\frac{\beta^{act}}{\beta} - \frac{\kappa^{act}}{\kappa}\right).$$

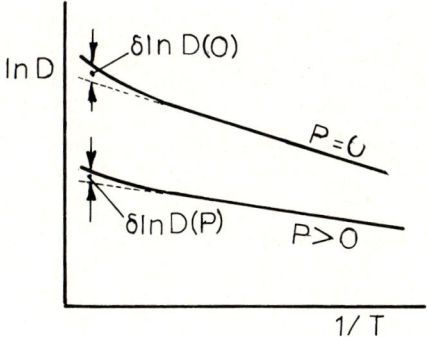

Fig. 9.3. Change of the form of the Arrhenius plot under increased pressure. The dashed line corresponds to D^{SL} (ref. [645]).

According to Nowick and Dienes [76] a curvature can be experimentally detected when c_P^{act} is larger than 0.5 k. The curvature will be important in materials for which the quantity in parentheses is large. It involves the ratios β^{act}/β and $\kappa^{\text{act}}/\kappa$, both of which can be evaluated by the $cB\Omega$-model.

We will now briefly discuss the behaviour of a few characteristic materials for which suitable data exist: the diffusion plots of In and Pb are practically linear, i.e. $\delta \ln D \approx 0$. This is in complete agreement with the present consideration, as the bulk modulus has been found to fall linearly up to the melting point. For Na, K, Ag, β-Zr and β-Ti, their self-diffusion plots exhibit a (smaller or larger) curvature; although a second operating mechanism cannot be precluded, a single monovacancy mechanism nicely explains the effect when c_P^{act} takes the values 9, 10, 15, 22 and 23 k, respectively [37]. For AgBr both diffusion and conductivity data indicate a very strong upward curvature [65]. For this material the bulk modulus shows, at high temperatures, an unusual "excessive fall" which can account for the above curvature. All the materials mentioned above (except β-Zr) are discussed in detail in this chapter. At the end of this chapter a useful summary is given for ionic crystals.

9.2 fcc metals

In order to examine the reliability of the proposed connection between defect parameters and bulk properties, we should select representative cases for which experimental data are available in a large temperature range concerning the bulk properties (elastic and expansivity data) on one hand and the defect concentration and migration on the other. Such cases are Al, Cu, Ag and Pb. The concentration of defects for the first three elements has well been studied by various experimental techniques (differential dilatometry, resistivity and positron annihilation). One can thus obtain in a self-consistent way the values of n/N in a temperature range for which the concentration varies by a few orders of magnitude and hence the correctness of the connection of g^f with $B\Omega$ can be reliably checked. Furthermore, in these materials the self-diffusion data have been extended to a wider temperature range, which gives the opportunity to examine whether the connection of g^{act} with $B\Omega$ is able to describe the temperature variation of the self-diffusion coefficient within many orders of magnitude.

Concerning Pb, in spite of the fact that both the concentration of vacancies and the self-diffusion coefficient have not been determined in the same large temperature region as the previous elements, it is included in the present discussion because solubility data are available for a large amount of impurities; it therefore provides a suitable material for the examination

9.2.1 Aluminum

This element has been studied by means of the differential dilatometry technique by three independent groups [134–136]. The formation parameters obtained by various investigators are given in table 9.1 and plotted in fig. 9.4. They were all obtained by least square fitting to a straight line of high temperature data, extending from 750 K up to 925 K, i.e. close to the melting point and therefore should be considered as the "effective formation parameters" for this temperature range.

The isothermal bulk modulus has been measured by Tallon and Wolfenden [137] from RT to just 20 K below the melting point. The values of B, reported for each temperature, are given in table 9.2. A least square fitting to a straight line of the values between 673 and 873 K gives $B_0^{SL} = 840$ kbar and $(dB/dT)_P = -0.2975$ kbar/K. For 850 K the expansion coefficient is $\sim 1.05 \times 10^{-4}$ K^{-1}, whereas $B = 617$ kbar. By inserting these data into eq. (9.4) we get for 850 K the value $s/h_{exp} = 3 \times 10^{-4}$ K^{-1} which gives the straight line drawn in fig. 9.4. By considering the experimental error one sees that the points are scattered around the line predicted from the $cB\Omega$-model.

A more strict method of checking the $cB\Omega$-model is possible. It makes use of the temperature dependence of the defect concentration n/N not only from differential dilatometry but also from electrical resistivity after quenching, so that it can be followed to relatively lower temperatures.

Table 9.1
Defect entropies and enthalpies of aluminum.

Formation:			
differential dilatometry (1)	0.81	3.12	[134] [a]
differential dilatometry (2)	0.76	2.4	[135]
differential dilatometry (3)	0.71	1.76	[136]
average of (1), (2), (3)	0.76	2.43	
$cB\Omega$-calculation	0.76	2.66	
Activation:			
tracer self-diffusion	1.476	5.25–5.82	[146] [b]
$cB\Omega$-calculation	1.47	5.65	

[a] Values from a least squares fitting to a straight line.
[b] For the extraction of the entropies from the published D_0-value, see text.

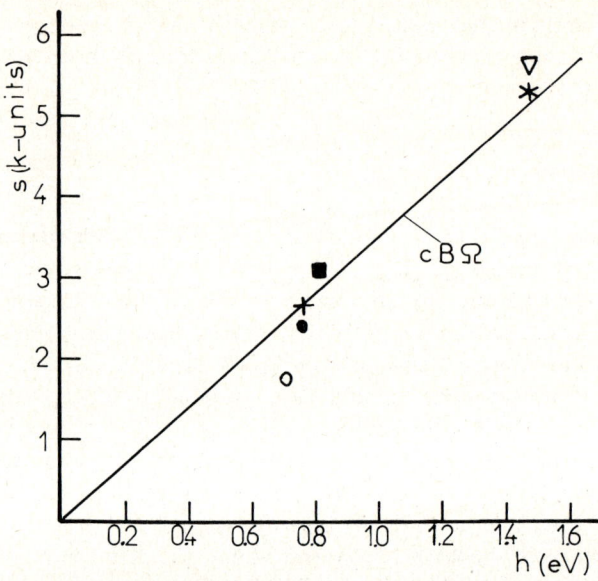

Fig. 9.4. Defect entropies and enthalpies of aluminum; formation: ■: Guerard et al. [134]; ●: Simmons–Balluffi [135]; ○: Bianchi et al. [136]; +: calculated with $cB\Omega$-model. Self-diffusion activation: *: Lundy–Murdock [146] with $\nu = \nu_D$. ▽: calculated with $cB\Omega$-model. Line: $cB\Omega$-model.

Equation (9.5) is very sensitive to the constancy of c^f and to correct B-values, because they lie in the exponent. In order to directly compare this equation to experimental data we also give in table 9.2 the $\Delta L/L_0$-values as

Table 9.2
Calculated temperature dependence of n/N for aluminum.

T (K)	B (kbar)	$\Delta L/L_0$ [a]	Ω (Å³)	n/N [b]
913	567	1.8461×10^{-2}	17.5353	8.6×10^{-4}
873	580	1.6924×10^{-2}	17.45604	$5.42^{+0.64}_{-0.58} \times 10^{-4}$
833	591	1.54778×10^{-2}	17.38165	$3.37^{+0.4}_{-0.4} \times 10^{-4}$
793	603	1.40974	17.31087	$1.96^{+0.3}_{-0.2} \times 10^{-4}$
753	614	1.27665	17.242805	$1.10^{+0.16}_{-0.14} \times 10^{-4}$
713	628	1.14743	17.17688	$5.49^{+0.86}_{-0.75} \times 10^{-5}$
673	638	1.02148	17.11280	$2.70^{+0.46}_{-0.39} \times 10^{-5}$
633	650	0.898617	17.050437	$1.17^{+0.22}_{-0.18} \times 10^{-5}$
593	661	0.778905	16.98982	$4.65^{+0.92}_{-0.79} \times 10^{-6}$
553	673	0.662593	16.93106	$1.58^{+0.34}_{-0.29} \times 10^{-6}$
513	683	0.54998	16.87430	$4.72^{+1.13}_{-0.93} \times 10^{-7}$

[a] Ref. [134].
[b] $cB\Omega$-calculation.

measured by Guerard, Peisl and Zitzmann [134] and expressed by them in a polynomial. These values lead to the mean volume per atom, Ω, also given in the table, when one uses the lattice constant 4.0493 Å reported by Cornish and Burke [138] for 21°C (this latter value practically coincides with the value 4.05109 Å reported by Guerard et al. [134]). We now determine the value of c^f at 913 K, which is the highest temperature at which B has been reported. By using the value resulting from the polynomial we obtain $n/N = 8.5776 \times 10^{-4}$; alternatively, if one uses the effective values of h^f and s^f reported by them, one gets $n/N = 7.6829 \times 10^{-4}$. The first n/N-value gives $c^f = 0.08948$ when inserted into eq. (9.5), whereas the latter leads to $c^f = 0.090877$, i.e. 1.5% larger. When inserted into eq. (9.5) the first c^f-values gives the concentrations given in table 9.2 and shown as solid dots in fig. 9.5. The lower error bar gives the resulting concentration if c^f is taken 1.5% larger. In this figure we have also inserted data from five different independent groups of experimenters. The low temperature data have been deduced from resistivity measurements [139,140] on quenched samples using $\rho = 1.0$ $\mu\Omega$ cm per atomic percent of vacancies. In the figure we have inserted only those resistivity measurements which according to Siegel [141] are the most reliable. There is an excellent agreement between the experimental n/N-values and those calculated from the $cB\Omega$-model. It is a striking fact that although we have started from the one end of the scale (i.e. 913 K), the $cB\Omega$-model can reproduce *without* any adjustable parameters the measurements of the other end of the scale, which differ by *three* orders of magnitude and have been obtained by a quite different technique.

This is a suitable point to compare the $cB\Omega$-results with those calculated from Zener's model (see ch. 14). In effect, according to table 9.2, upon cooling from 913 to 553 K the vacancy concentration decreases from 8.6×10^{-4} to 1.58×10^{-6} in good agreement with the experimental value. This result came from connecting g^f to the temperature variation of the isothermal bulk modulus from 567 to 673 kbar. On the other hand, if instead of the bulk modulus one inserts the values $c_{44} = 178$ and 240 kbar for the two temperatures just mentioned, Zener's proposal gives a decrease of the vacancy concentration from 8.6×10^{-4} to 1.48×10^{-7}; the latter value is in error by a factor of 10. Alternatively, using $(c_{11} - c_{12})/2$ their values 122 and 189 kbar give for $T = 553$ K a vacancy concentration 1.44×10^{-8}, which is 100 times too small.

We now turn to the calculation of the absolute values of the formation parameters. For instance, by considering the elastic data reported in table 9.2 for 553 and 593 K and using $c^f = 0.08948$ we get:

$$s^f = -c^f \left. \frac{d(B\Omega)}{dT} \right|_P = 2.665 \, k$$

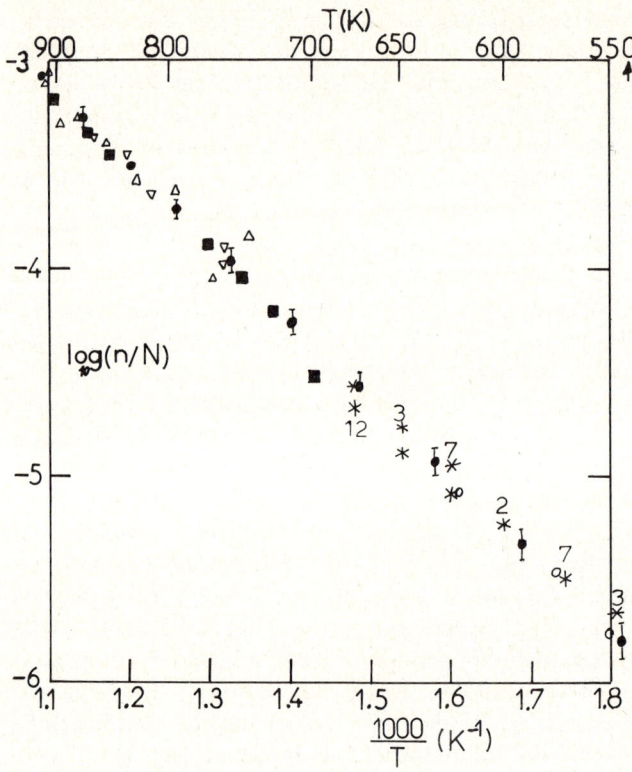

Fig. 9.5. Temperature dependence of vacancy concentration in aluminum; △: differential dilatometry, Guerard et al. [134]; ○: resistivity measurements, Bass [140]; *: resistivity measurements, Berger [139] (the numbers indicate the number of measurements); ■: Simmons–Balluffi [135]; ▽: differential dilatometry, Bianchi et al. [136]; ●: calculated with $cB\Omega$-model; the latter values are from table 9.2.

and hence at $T = 593$ K:

$$g^f = c^f B\Omega = 0.628 \text{ eV},$$

$$h^f = c^f B\Omega + Ts^f \simeq 0.76 \text{ eV}.$$

Comparing these values with the experimental ones we find a satisfactory agreement (see table 9.1).

We finally deal with the controversial question concerning the manner of analysing formation or self-diffusion data of aluminum.

(a) monovacancies + divacancies analysis: This method was applied by Seeger and Mehrer [67] and by Seeger [142] and recently analysed by Siegel

[141]. It consistently leads to the values $h^f = 0.65$–0.66 eV and $s^f = 0.6$–0.8 k. Furthermore it has recently been supported by Peterson [73].

(b) analysis in the frame of a single (monovacancy) mechanism: It leads to parameters that slightly increase with temperature. This point of view is strongly supported by Gilder and Audit's calculations [143,144]. A rough idea of the values predicted from such an analysis can best be obtained by a simultaneous fitting of the measurements carried out by the differential dilatometry technique at high temperatures. In this region such a fit was made by Hood and Schultz [145] (who did not take into account the measurements of Guerard et al. [134]) and found the values $h^f = 0.73 \pm 0.03$ eV and $s^f = 2.1 \pm 0.5$ k. Even if one accepts the lowest values of this analysis one sees that the two kinds of analyses give monovacancy entropies that differ, at least, by a factor of two and enthalpies differing by at least 6%. It was initially thought that positron annihilation spectroscopy (PAS) could help to solve the disagreement. The first measurements of this kind (see table 9.3) led to values that were, rather, in support of Seeger's proposals. However, later measurements [145] seem to be in favour of the single mechanism proposal. Hood and Schultz have stated that "there is no convincing evidence for the presence of divacancies and their results may best be considered as being substantially in accord with the single vacancy description". However, the most interesting comment of these latter workers is that "depending on the manner of analysis one can furnish 'reasonable' h^f-values anywhere in the range 0.62–0.78 eV". In view of the above remarks it is evident that PAS cannot substantially help to give an unambiguous value of h^f. Therefore we have to look for another source to dissolve the disagreement. It is this point to which we now turn. It is based

Table 9.3
Recent formation enthalpy data for aluminum.

Technique	h^f (eV)	Reference
Resistivity upon quenching	0.66 ± 0.01	[139]
	0.73	[140]
	0.70	[531]
	0.69 ± 0.03	[532]
Positron annihilation	0.66 ± 0.04	[533]
	0.67 ± 0.03	[534]
	0.66 ± 0.01	[535]
	0.66 ± 0.02	[536]
	$0.70 \pm 0.02 \sim$ $\sim 0.76 \pm 0.03$	[537]
	$0.62 \sim 0.78$	[145]

on the thermodynamical identity of eq. (3.126). The value of v^f/Ω cannot be below 0.4. This is so because self-diffusion data under pressure show that the activation volume is $\sim 0.8\ \Omega$. Seeger [142] has proposed that the best value for the monovacancy mechanism is even higher, i.e. $0.65\ \Omega$. The term βB close to 900 K is according to Tallon and Wolfenden's [137] measurements $\sim 6.3 \times 10^{-2}$ kbar/K; by considering that Ω is ~ 17.5 Å^3 we immediately find that the term $v^f\beta B$ cannot be smaller than 3.2 k. If one accepts the value $s^f \simeq 0.6$–$0.7\ k$ suggested by the analysis of "monovacancies + divacancies", one immediately has to accept from the thermodynamical identity $s^* = s^f - v^f\beta B$ that $s^* < -2.5\ k$, i.e. the absolute value of s^* *exceeds* that of s by a factor of at least around 4. This is highly improbable because the entropy s^* – which is a measure of the change of frequencies of the solid upon vacancy creation under *constant volume* – must have, at the most, an absolute value equal to the isobaric formation entropy s^f.

These purely thermodynamic arguments are extremely unlikely to be violated; therefore, we are led to the conclusion that even if divacancies do exist (this cannot be precluded), the current methods of analysis must be reconsidered.

The self diffusion of aluminum has been studied by Lundy and Murdock [146] between 729 and 916 K and later by Beyeler and Adda [147] between 673 and 883 K with the tracer technique; the self-diffusion coefficient has also been calculated from NMR experiments by Sun [148] by applying the Slichter–Ailion theory [149]. However, this theory, when used for the conversion of "T_1-measurements" below 600 K into tracer coefficients, *needs* according to Seeger [142] some refinement; therefore we shall not further discuss these values.

In order to examine whether the $cB\Omega$-model reproduces the (tracer) self-diffusion curve we start from the highest temperature (916 K) at which tracer measurements have been reported; for this temperature Lundy and Murdock published the values 1.403×10^{-8} and 1.180×10^{-8} cm^2/s and therefore we shall use their average 1.29×10^{-8} cm^2/s. Inserting this value into the eq. (9.7) we get $c^{\text{act}} = 0.1714$. We have used the values $a = 4.124$ Å, $f = 0.78$, $\nu = \nu_D = 6.8 \times 10^{12}$ s^{-1}, $\Theta_D = 328$ K [150], $B = 566$ kbar, $\Omega = 17.541$ Å^3; the latter two quantities have been calculated for 916 K from a linear extrapolation of the corresponding values given in table 9.2 for 913 and 873 K. Once c^{act} has been determined eq. (9.7) can give D at various temperatures with the help of the elastic and expansivity data given in table 9.2. In table 9.4 and fig. 9.6 we show the D-values calculated in this way. The stated uncertainty corresponds to a plausible experimental error of 1.5% in the values of $B\Omega$ used in the calculation. For many temperatures the values of B and Ω used in the calculation have been estimated by making a linear interpolation of the corresponding values given in table 9.2.

Table 9.4
Calculated tracer self-diffusion in aluminum. [a]

T (K)	D calculated from $cB\Omega$ (cm²/s)	D experimental (cm²/s)	T (K)	D calculated from $cB\Omega$ (cm²/s)	D experimental (cm²/s)
916	1.29 E−8	1.29 E−8 [b]	773	4.19±0.94 E−10	3.75 E−10 [c]
913	1.21±0.22 E−8	−	753	2.37±0.55 E−10	−
895	8.24±1.55 E−9	8.22 E−9 [b]	749	2.09±0.50 E−10	1.88 E−10 [b]
883	6.32±1.21 E−9	6.2 E−9 [b]	737	1.42±0.34 E−10	1.42 E−10 [b]
873	5.03±1 E−9	5.03 E−9 [b]	729	1.09±0.26 E−10	1.13 E−10 [b]
870	4.72±0.92 E−9	5.44 E−9 [b]	727	1.02±0.24 E−10	9.2 E−11 [c]
844	2.62±0.53 E−9	2.54 E−9 [b]	713	6.28±1.54 E−11	−
843	2.56±0.52 E−9	2.43 E−9 [c]	673	1.62±0.43 E−11	1.45 E−11 [c]
833	2.02±0.34 E−9	−	633	3.28±0.92 E−12	~4 E−12 [d]
807	1.05±0.23 E−9	9.49 E−10 [b]	593	5.55±1.65 E−13	−
803	7.3 ±1.6 E−10	8.6 E−10 [c]	553	7.01±2.24 E−14	−
793	7.21±1.6 E−10	−	513	6.95±2.38 E−15	−
776	4.55±1.0 E−10	4.2 E−10 [b]			

[a] The error bars given in the fifth column correspond to an uncertainty of 1.5% of the values of $B\Omega$ (see the text).
[b] Ref. [146]; the value at 916 K is the mean value of the measurements of these workers.
[c] Ref. [147].
[d] Extracted from fig. 7 of ref. [142].

An inspection of table 9.4 indicates that the calculated D-values agree favourably with the experimental ones if one considers the stated errors and the fact that the diffusion data have been obtained by two independent groups.

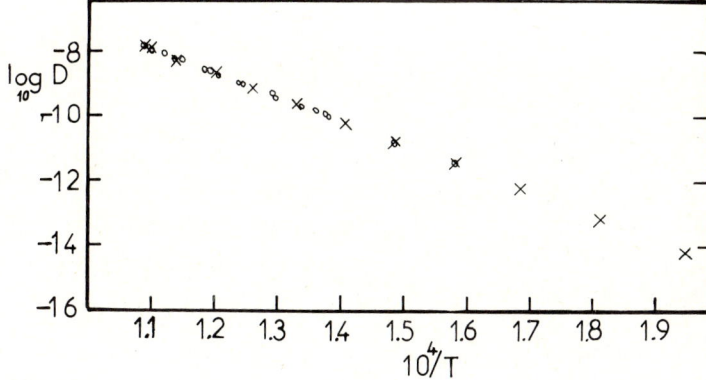

Fig. 9.6. Self-diffusion in aluminum (tracer); ○: experimental; ×: calculated with $cB\Omega$-model; the exact values are given in table 9.4.

We now proceed to the computation of the values of the activation parameters from the $cB\Omega$-formulae. By using the values of $B\Omega$ at 873 and 913 K (see table 9.2) we get (recall: $c^{act} = 0.1714$):

$$s^{act} = -c^{act}[d(B\Omega)/dT]_P \simeq 5.65\ k.$$

Similarly for $T = 873$ K we get by using $\beta = 1.13 \times 10^{-4}$ K^{-1}:

$$h^{act} = c^{act}\Omega\left(B - T\beta B - T\left.\frac{dB}{dT}\right|_P\right) \simeq 1.47\text{ eV}.$$

The calculated value of h^{act} agrees with that of 1.476 eV reported by Lundy and Murdock [146]; furthermore, these workers give $D_0 = 1.71$ cm^2/s, which – by using the value $fa^2\nu_D = 9.02 \times 10^{-3}$ cm^2/s reported above – gives $s^{act} = 5.25\ k$, in agreement with the calculated value. If one alternatively uses $\nu = \tilde{\nu} = 3.83 \times 10^{12}$ s^{-1} (with $h^m = 0.7$ eV) one finds that the experimental s^{act}-value is 5.82 k (see table 9.1). The experimental values of h^{act} and s^{act} have been inserted in fig. 9.4 and actually lie on the line predicted from the $cB\Omega$-model thus justifying the prediction that the quantity s/h – irrespective of the process (formation or activation) – is predetermined from the bulk properties.

We finally proceed to the calculation of the activation volume from the $cB\Omega$-formulae. By inserting the value $(dB/dP)_T = 5.31$ and recalling that $c^{act} = 0.1714$ the relation (8.19) gives $v^{act} = 0.74\ \Omega$, in excellent agreement with the experimental value of $v^{act} = 0.71\ \Omega \pm 18\%$ (for references for dB/dP and v^{act}, see table 10.1).

Summarizing the present discussion for aluminum we can say that the $cB\Omega$-model successfully reproduces both the concentration of vacancies and the diffusion coefficient over the whole temperature range; in essence this means that it correctly interconnects a variety of experimental data without using any further parameter, empirical or not.

9.2.2 Copper

The vacancy concentration of this element has been initially studied by Simmons and Balluffi [151] with the differential dilatometry technique. Although their data showed a large scatter they achieved to extract the values $s^f = 1.5 \pm 0.5\ k$ and $h^f = 1.17 \pm 0.11$ eV. Later Bourassa and Lengeler [152] measured the increment in the residual electrical resistivity (due to the quenched-in vacancies) in quenched low dislocation density copper single crystals from temperatures in the range 1023–1323 K. Their measurements gave a good fit to a single Arrhenius curve with slope $h^f = 1.27 \pm 0.05$ eV with no apparent curvature at the highest temperatures. These data scale well with the Simmons–Balluffi measurements when $\rho = 0.62\ \mu\Omega$ cm per

§9.2]   fcc metals   177

atomic percent of vacancies; a simultaneous linear fitting (Siegel [141]) of the differential dilatometry data and those of Bourassa and Lengeler gives $s^f = 2.35$ k and $h^f = 1.27$ eV. Later Scholz and Schuele [153] used a high temperature electrical resistivity measurement technique, but their data are based on some assumptions; due to the fact that these are still open to question (see Berger, Ockers and Siegel [154]), we shall not use them in our applications. By using a low temperature (clearly vacancy-loss) quenching technique and the same value of ρ as above they [154] determined the concentration of vacancies down to 798 K. Their data were consistent with $s^f = 2.8 \pm 0.3$ k and $h^f = 1.30 \pm 0.05$ eV. As they noticed, when their data are combined with those of Bourassa and Lengeler, a straight line fit to a single exponential function was found, exhibiting no curvature over the temperature region 923 ~ 1323 K. However, it should be mentioned that this latter conclusion is based on the above "assumed ρ-value" which has not yet been unambiguously established.

The formation parameters resulting from the data mentioned above have been collected in table 9.5. In the same table we have inserted the parameters $s^f = 2.5 \pm 0.2$ k and $h^f = 1.28 \pm 0.03$ eV which result from a simultaneous fitting of all the "formation" data obtained either from differential

Table 9.5
Formation, migration and activation parameters in copper.

h^f (eV)	s^f (k-units)	h^{act} (eV)	s^{act} (k-units)	h^m (eV)	s^m (k-units)
1.17±0.11 [a]	1.5±0.5 [a]				
1.27 [b]	2.35 [b]				
1.30±0.05 [c]	2.8±0.3 [c]			0.76±0.04 [c]	1.64 [j]
1.28±0.03 [d]	2.5±0.2 [d]	2.125 [e-i]	4.14 [e-i]	0.91±0.04 [k]	2.25 [k]
		2.19 [g]	4.75 [g]		
1.287 [ℓ]	2.64 [ℓ]	2.126 [ℓ]	4.36 [ℓ]	0.84±0.05 [ℓ]	1.72±0.1 [ℓ]

[a] Ref. [151].
[b] Ref. [152]; using $\rho = 0.62$ $\mu\Omega$ cm per atomic percent of vacancies; from a fitting of the data in [151] and [152].
[c] Ref. [154].
[d] From a fitting of the data in [151], [152] and [154].
[e] From a fitting of the data in [82], [155], [161], [538].
[f] Bartdorff, Neumann and Reimers [155].
[g] Rothman and Peterson [538].
[h] Lam, Rothman and Nowicki [161].
[i] Maier, Bassani and Schuele [82].
[j] From $s^f = 2.5$ k and $s^{act} = 4.15$ k reported in notes d and e.
[k] From the difference of the activation parameters in [538] and the formation parameters in note d.
[ℓ] Calculated from the $cB\Omega$-model.

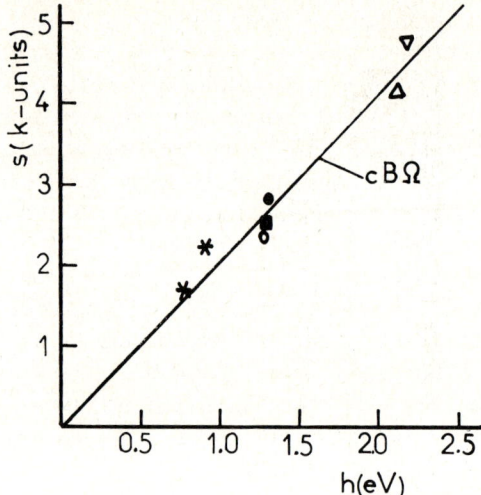

Fig. 9.7. Entropy and enthalpy of defects in copper; △: self-diffusion parameters of simultaneous analysis of all the published data; ▽: self-diffusion parameters from Rothman and Peterson [538]; ∗: migration parameters, see table 9.5 and the text; ●: formation parameters of Berger et al. [154]; ○: formation parameters from simultaneous fitting of the data of Simmons and Balluffi [151] and Bourassa and Lengeler [152]; ■: Best formation parameters (see text). Line: $cB\Omega$-model.

dilatometry or by the resistivity-technique. It seems that these parameters provide the most reliable set of "effective" parameters for all the temperature ranges studied up to date. In the same table, we have inserted the activation enthalpy resulting from a simultaneous least squares fitting to a straight line of all the recent self-diffusion data (Bartdorff, Neumann and Reimers [155]). The value of D_0 obtained in this way is 0.42 cm²/s which in conjunction to $f = 0.78$, $\nu \simeq \nu_D = 6.52 \times 10^{12}$ s^{-1} [156] and $a = 3.63542$ Å (resulting from Simmons–Balluffi data at 613.7 K) gives $s^{act} = 4.14\ k$. Furthermore, in the table 9.5 we have also inserted the migration enthalpy 0.76 ± 0.04 eV, deduced in the work of Berger et al. [154] from the "change of slope" measurements during post-quench annealing. The best value of the migration entropy results from the difference of the values $s^{act} = 4.14\ k$ and $s^f = 2.5\ k$, each one of which has been obtained, as already mentioned, from a simultaneous fitting of all the experimental data published to date. All these data are plotted in fig. 9.7.

We now turn to the comparison of the above results with those predicted from the $cB\Omega$-model. The adiabatic bulk modulus B^S has been measured by Chang and Himmel [157] only up to 800 K; in order to extend our calculations to higher temperatures we first make a least squares fitting to a

§9.2] fcc metals 179

Table 9.6
Calculated and experimental vacancy concentration and tracer self-diffusion coefficients in Cu. [g]

T (K)	$10^3 \Delta L/L$	a (Å)	Ω (Å³)	B (kbars)	$n/N\vert_{exp}$	$n/N\vert_{cB\Omega}$	$D_{cB\Omega}$ (cm²/s)	D_{exp} (cm²/2)
613.7	5.723	3.63542	12.0117	1263		$3.53 \pm 1.3 \times 10^{-10}$	1.63×10^{-18}	1.63×10^{-18}
800	9.396	3.64869	12.1437	1202	10.6×10^{-8}	10.6×10^{-8}	$2.02 \pm 0.98 \times 10^{-14}$	$2^{a} \sim 1.65^{b} \times 10^{-14}$
900	11.488	3.65626	12.2194	1169	8.33×10^{-7}	$8.53 \pm 2 \times 10^{-7}$	$6.42 \pm 2.6 \times 10^{-13}$	$7.5^{a} \sim 4.3^{c} \times 10^{-13}$
1000	13.675	3.66416	12.2988	1135	4.34×10^{-6}	$4.60 \pm 0.9 \times 10^{-6}$	$1.04 \pm 0.36 \times 10^{-11}$	$8.28^{d} \times 10^{-11}$
1100	15.972	3.67246	12.3826	1103	1.67×10^{-5}	$1.79 \pm 0.3 \times 10^{-5}$	$9.86 \pm 3 \times 10^{-11}$	$9.9^{e} \sim 7.3^{c} \times 10^{-11}$
1200	18.364	3.68111	12.4703	1071	5.15×10^{-5}	$5.55 \pm 0.9 \times 10^{-5}$	$6.43 \pm 1.8 \times 10^{-10}$	$5.15^{f} \times 10^{-10}$
1300	20.928	3.69038	12.5647	1039	1.33×10^{-4}	$1.45 \pm 0.2 \times 10^{-4}$	$3.15 \pm 0.75 \times 10^{-9}$	$2.64^{f} \times 10^{-9}$
1356	22.44	3.69584	12.6206	1021	2.14×10^{-4}	$2.32 \pm 0.3 \times 10^{-4}$	$6.91 \pm 1.6 \times 10^{-9}$	$5.93^{f} \times 10^{-9}$

[a] Ref. [82].
[b] From a simultaneous fitting to a straight line reported in refs. [161] and [82].
[c] Ref. [538].
[d] From a fitting of the data in [155], [538], [161] and [82].
[e] Ref. [155].
[f] From a simultaneous fitting to a straight line reported in refs. [155] and [538].
[g] $n/N\vert_{cB\Omega}$ and $D_{cB\Omega}$: calculated from the $cB\Omega$-model. The subscript "exp" indicates experimental values.

straight line of the measured B^S-values between 500 and 800 K and then calculate the B^S-values at higher temperatures by linear extrapolation. The isothermal bulk modulus (see table 9.6) has been obtained from B^S by using the specific heat data of Brooks [54] and the expansivity data extracted from Simmons and Balluffi's paper; in the table we include also the values of $\Delta L/L$ for each temperature, extracted from a linear interpolation of Simmons and Balluffi's measurements in the corresponding temperature range. The lattice parameter a (and Ω) has been obtained from $\Delta L/L$-values in conjunction with the value a (20°C) = 3.61473 Å quoted by Simmons and Balluffi.

The data included in table 9.6 permit the direct computation of s/h_{exp} for each temperature. We now proceed to its calculation for $T = 1023$ K, which is the mean temperature of the various measurements carried out in this material. A least squares fitting to a straight line of the B-values between 900 and 1300 K gives $B_0^{\text{SL}} = 1460$ kbar, $\mathrm{d}B/\mathrm{d}T|_P = -0.3244$ kbar/K and hence B is equal to 1128 kbar at 1023 K. Inserting these data into eq. (9.4) and using $\beta \simeq 68 \times 10^{-6}$ K^{-1} we get, for $T = 1023$ K, the value $s/h_{\text{exp}} = 1.8 \times 10^{-4}$ K^{-1} which gives the straight line drawn in fig. 9.7. By comparing it with the experimental values inserted in the figure, we actually see a striking agreement. It is worthwhile to remember that the line drawn in this figure has been obtained solely from elastic and expansivity data and compares favourably with defect parameters deduced from quite different techniques.

Evaluation of vacancy concentration in copper from the $cB\Omega$-model. In table 9.6 we have reproduced the experimental vacancy concentration for various temperatures; the values have been obtained from the relation

$$n/N = \exp(2.5) \exp(-1.28 \text{ eV}/kT), \tag{9.10}$$

which, according to Berger et al [154], describes in a unified way the results from differential dilatometry and resistivity measurements.

In order now to examine whether the $c^{\mathrm{f}}B\Omega$-model [eq. (9.5)] describes the correct vacancy concentration we have to determine the values of c^{f}. The most strict way is to determine it from a single measurement at the lower end of the temperature scale and then to examine whether (9.5) reproduces the correct n/N-values in the whole temperature range up to the melting point. The lowest temperature for which n/N has been determined is around 800 K. For this temperature eq. (9.10) leads to $n/N = 10.6 \times 10^{-8}$; by inserting this value into eq. (9.5) along with the corresponding values of B and Ω written in table 9.6 we get $c^{\mathrm{f}} = 0.1215$. Once this value of c^{f} is known, the $cB\Omega$-model gives the vacancy concentration through eq. (9.5) at various temperatures; the resulting values have been given in table 9.6 and inscribed as solid dots in fig. 9.8. For each temperature we have quoted the

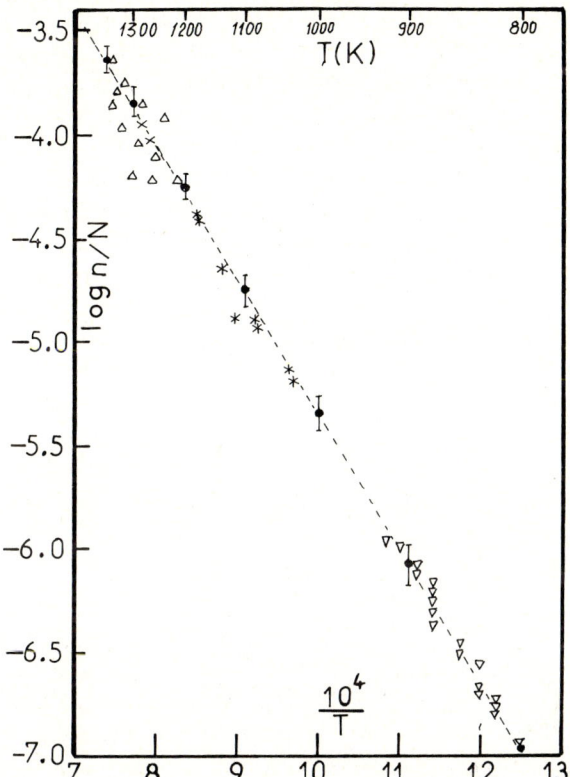

Fig. 9.8. Temperature dependence of vacancy concentration in copper; △: Simmons–Balluffi [151]; ▽: Berger–Ockers–Siegel [154]; ∗: Bourassa–Lengeler [152]; ●: calculated with $cB\Omega$-model; the exact values of the latter are given in table 9.6.

uncertainty of n/N by allowing for a plausible experimental error of ±1.5% in the B-value which has been used in our calculation; it is shown in the figure by bars. If one also considers the experimental error in the determination of n/N we see that, although we have started from the lower end of the scale and the concentration varies by ~ 3.5 orders of magnitude, the predicted values agree with the experimental ones within the experimental error. Obviously, if the relation $g^f = c^f B\Omega$ were incorrect by some percent then the coincidence observed in fig. 9.8 would have totally been destroyed.

Formation parameters from the $cB\Omega$-model. Once c^f has been determined from a single measurement, the values of s^f, g^f and h^f can be directly computed. For instance by using the elastic and expansivity data reported for 1000 and 1100 K we have $d(B\Omega)/dT \approx -3 \times 10^{-15}$ erg/grad and hence with $c^f = 0.1215$ eq. (9.1) gives $s^f = 2.64\ k$. Furthermore, we obtain g^f (1100 K) $= c^f B\Omega = 1.037$ eV and h^f (1100 K) $= c^f B\Omega + Ts^f = 1.287$ eV. The above values of s^f and h^f are in reasonable agreement with those included in table 9.5. Also our h^f-value agrees with those deduced from positron annihilation spectroscopy (PAS) [158–160] for the same temperature range. Several PAS investigations have yielded values of $h^f \approx 1.28$ eV; among them, according to Siegel [141], the value 1.29 ± 0.02 eV given by Triftshauser and McGervey [159] is the most straightforward, as far as concerns the assumptions made in their analysis. This latter value coincides with that predicted by the $cB\Omega$-model.

Calculation of the self-diffusion coefficient from the $cB\Omega$-model. The lowest temperature to which self-diffusion has been measured is $T_1 = 613.7$ K [161]. The value of the preexponential factor $fa^2\nu$ is obtained by using $f = 0.78$, $a = 3.63542$ Å (see table 9.6) and $\nu \approx \nu_D = 6.52 \times 10^{12}$/s. The value of D for $T = 613.7$ K is 1.63×10^{-18} cm^2/s. Inserting these data into eq. (9.7) gives $c^{act} = 0.2007$. Even if $fa^2\nu$ is in error by a factor of 2 the value of c^{act} changes only by 1.9%. Once c^{act} is known, eq. (9.7) immediately leads to the D-values at various temperatures. The D-values calculated in this way have been inserted in table 9.6 and plotted in fig. 9.9 along with the experimental results deduced from various investigators. It is really a striking fact that the calculated values, although extended over ten orders of magnitude, agree with the experimental ones within experimental error; notice that the uncertainty given in the calculated values comes from a plausible experimental error of 1.5% in the determination of the bulk modulus. If one takes $\nu = \tilde{\nu}$ with $h^m = 0.76$ eV one gets $c^{act} = 0.1963$ instead of 0.2007.

Let us now calculate the values of the activation parameters s^{act} and h^{act}. By proceeding in a similar way as in the case of the formation process – but now using $c^{act} = 0.2007$ instead of $c^f = 0.1215$ – we find (for 1100 K) $s^{act} = 4.36\ k$ and $h^{act} = 2.126$ eV. These values are again in excellent agreement with the experimental ones inserted in table 9.5. Furthermore, one can calculate the migration parameters from the difference of the (calculated) activation and formation parameters:

$$h^m = (2.126 \pm 1.5\%) \text{ eV} - (1.287 \pm 1.5\%) \text{ eV} = 0.84 \pm 0.05 \text{ eV},$$
$$s^m = (4.36 \pm 1.5\%)\ k - (2.64 \pm 1.5\%)\ k = 1.72 \pm 0.1\ k.$$

The calculated migration enthalpy agrees, within the uncertainty quoted, with the direct experimental value 0.76 ± 0.04 eV of Berger et al. [154]. Also the calculated s^m-value lies in the range of the values extracted from a comparison of the self-diffusion with the "formation" data (see table 9.5).

We finally turn to the vexing question about the origin of the curvature observed in the high temperature part of the self-diffusion plot. Some

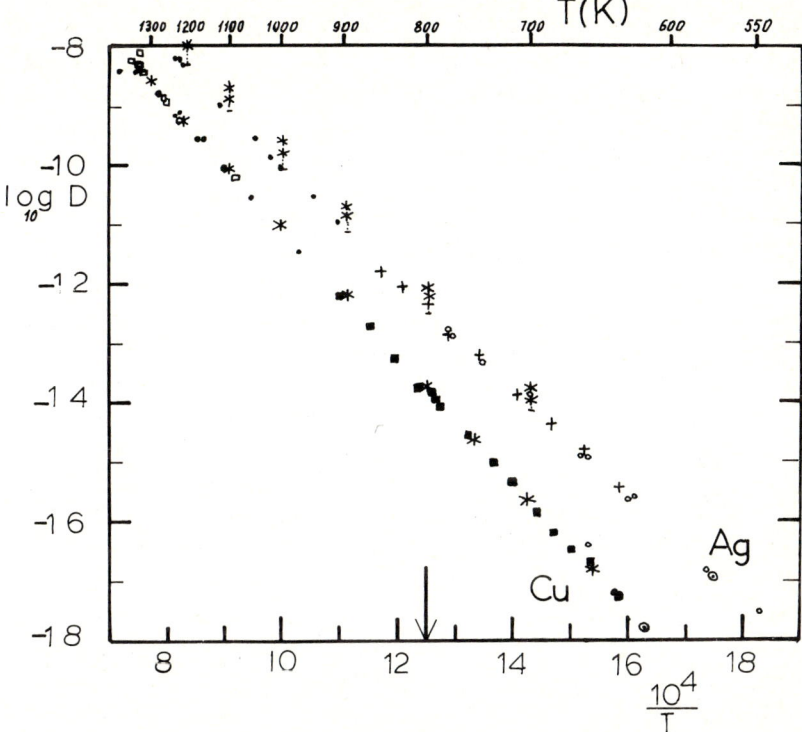

Fig. 9.9. Tracer self-diffusion in silver and copper; ●: Rothman and Peterson [538], Rothman, Peterson and Robinson [470]; □: Maier et al. [82]; ○: Lam et al. [166] for Ag, Lam et al. [161] for Cu; ×: Backus et al. [539]; ■: Bartdorff et al. [155]; ∗: calculated with $cB\Omega$-model. For Ag the upper and the lower asterisks correspond to $\nu = \nu_D$ and $\nu = \bar{\nu}$. The error bar corresponds to the upper asterisks and indicates the maximum correction from the temperature dependence of c^{act}. ⊙: starting point. The arrow indicates the highest temperature for which elastic data are available.

analyses accept that divacancies also contribute to the diffusion at temperatures higher than 1000 K. However, one should remember the following: when we calculated the D-values a c^{act}-value was used that had been determined from a measurement at 613.7 K; at this low temperature it is accepted by all workers that only monovacancies exist. However, this single c^{act}-value (i.e. this single measurement) enabled us to reproduce the whole diffusion curve up to the melting point, without having to make any further assumption, e.g. for an additional mechanism. It seems therefore likely that only monovacancies are operating in the whole temperature range; if divacancies exist they should contribute in a portion within the uncertainty of our computed values, introduced by the experimental error in B. However, one should be very cautious with the usual manners of "divacancy plus monovacancy" analyses, because they lead to unusually low s^f-values, which can be subjected to the same questions as those reported in §9.2.1 for aluminum.

Summarizing the discussion on copper one can state that the $cB\Omega$-model can reproduce all the defect data (differential dilatometry, self-diffusion, "resistivity", etc.) available to date *without* using any adjustable parameter.

9.2.3 Silver

This material has not been so extensively studied with the differential dilatometry technique as aluminum. Therefore its vacancy concentration is not exactly known at various temperatures. Simmons and Balluffi [162] have reported that at the melting point (1233 K) the quantity n/N is $1.7 \pm 0.5 \times 10^{-4}$ but have not attempted to derive values at lower temperatures. Applying relation (9.5) one gets, at $T = 1233$ K, the value:

$$g^f = 0.92 \, {}^{+0.04}_{-0.025} \, \text{eV}.$$

Simmons and Balluffi have estimated that $h^f = 1.09 \pm 0.1$ eV (see table 9.7) by assuming that $s^f = 1.5 \pm 0.5$ k. By using the PAS-technique

Table 9.7
Experimental and calculated enthalpies and entropies in silver.

Process	h (eV)	s (k-units)	Reference
Activation; self-diffusion at high temperatures	1.97 ± 0.1	4.74	[470]
$cB\Omega$-calculation	1.99 ± 0.04	4 ~ 6	
Activation; self-diffusion (effective values from all the temperature range)	1.86	3.7	*
$cB\Omega$-calculation	$1.94 \pm 4\%$	3.5 ~ 5.54	
Formation; estimated from differential dilatometry	1.09 ± 0.1	1.5 ± 0.5	[162]
Formation; combination of positron and differential dilatometry technique	~ 1.2	2.6 ± 0.3	see text
$cB\Omega$-calculation	1.2 ~ 1.36	2.7 ~ 4	

* These values correspond to a least squares fitting to a straight line of the data reported in refs. [470], [539] and [166].

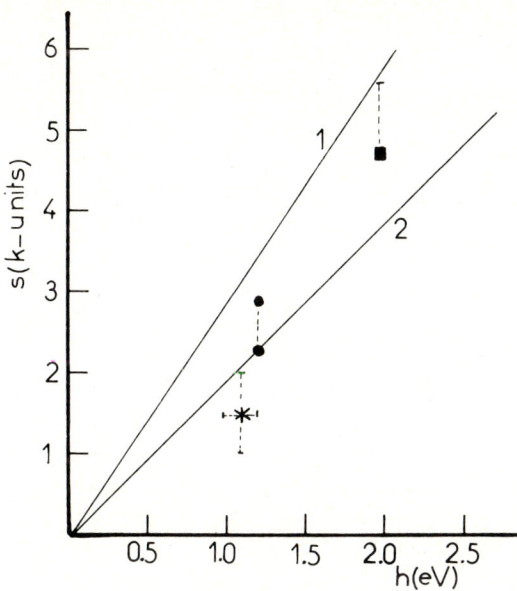

Fig. 9.10. Entropies and enthalpies in silver; ■: activation self-diffusion, Rothman et al. [470]; *: vacancy formation, Simmons and Balluffi [162]; ●: vacancy formation, see text. line 1: $cB\Omega$-model with c independent of temperature; line 2: $cB\Omega$-model with temperature dependent c, as discussed in the text.

McGervey and Triftshauser [163] have reported $h^f = 0.99 \pm 0.06$ eV but later [159] they gave 1.16 ± 0.02 eV. This value is in agreement with the recent result $h^f = 1.2$ eV of Campbell, Schulte and Jackman [164]. We therefore see that the recent PAS-measurements confirm the enthalpy value of Simmons and Balluffi but cast doubt on their pre-assumed $s^f = 1.5 \pm 0.5$ k. This is evident because combining the latest PAS-value $h^f = 1.16$ eV with the n/N-value given above one finds: $s^f = (2.87-2.26)$ k. This value 2.6 ± 0.3 k has been inserted in table 9.7 and in fig. 9.10 and seems to be more reliable because it is comparable to the value $(2.0-2.4)$ k directly measured by Simmons and Balluffi in the case of aluminum [135]. In the table and the figure we have also inserted the activation parameters obtained from the recent self-diffusion studies.

Let us now compare the above results with the predictions of the $cB\Omega$-model. Unfortunately, the bulk modulus has been measured only up to 800 K and furthermore its values are not as sure as those of aluminum. This can be seen from table 9.8 in which we have inserted two sets of B-values differing by some percents. The first set corresponds to the adiabatic bulk modulus measurements of Chang and Himmel [157]; these authors also give

Table 9.8
Calculated and experimental self-diffusion coefficients in silver.

T (K)	$\Delta L/L$ [a]	a [b] (Å)	Ω (Å³)	$B^{S,c}$ (kbar)	B^c (kbar)	$B^{S,d}$ (kbar)	B^d (kbar)	D_{calc} [e] (cm² s⁻¹)	D_{exp} (cm² s⁻¹)	D_{calc} [i] (cm² s⁻¹)
572.5	5.64	4.10065	17.23842	973	898	982	906	1.14×10^{-17}	1.14×10^{-17}	1.14×10^{-17}
700	8.42	4.11198	17.38178	940	851	956	864	$15.8^{+2.8}_{-7.8} \times 10^{-15}$	9.7×10^{-15} [f]	$13.4 \pm 8.7 \times 10^{-15}$
800	10.71	4.12132	17.50046	915	816	936	833	$8.75^{+12.2}_{-5.2} \times 10^{-13}$	$\sim 4.3 \times 10^{-13}$ [g]	$6.76^{+9}_{-4} \times 10^{-13}$
900	14.35	4.13617	17.69025	890	781	915	800	$1.94^{+2.05}_{-1.05} \times 10^{-11}$	$6.3 \pm 0.9 \times 10^{-12}$ [h]	$14^{+15}_{-7.5} \times 10^{-12}$
1000	16.94	4.14674	17.82623	865	746	895	768	$2.43^{+1.44}_{-1.22} \times 10^{-10}$	$8 \pm 1 \times 10^{-11}$ [h]	$16.5^{+15}_{-8} \times 10^{-11}$
1100	18.30	4.15225	17.89744	840	716	875	741	$18.46^{+14.4}_{-8.4} \times 10^{-10}$	$6.4 \pm 0.7 \times 10^{-10}$ [h]	$12^{+9}_{-5.3} \times 10^{-10}$
1200	21.17	4.16397	18.04945	815	681	854	708	$10.59^{+7}_{-4.4} \times 10^{-9}$	$3.6 \pm 0.4 \times 10^{-9}$ [h]	$6.6^{+4.2}_{-2.7} \times 10^{-9}$
1233			~ 18.09961				~ 697	$17.8^{+11.2}_{-7.1} \times 10^{-9}$	$5.9 \pm 0.6 \times 10^{-9}$ [h]	$10.95^{+6.71}_{-4.3} \times 10^{-9}$

[a] From ref. [162].
[b] From the values of $\Delta L/L_0$ and the lattice parameter 4.07765 Å quoted in ref. [162] for 20°C.
[c] Ref. [157].
[d] Ref. [165].
[e] The uncertainty given corresponds to a plausible error 2% in the value of c^{act} and another 2% in the experimental error of B. It does not include the "correction" due to the eventual temperature variation of c^{act}, which brings the calculated values in better agreement to the experimental ones.
[f] Ref. [166].
[g] Ref. [539].
[h] From the values of D_0 and h^{act} given in ref. [470].
[i] These values have been calculated from the $cB\Omega$-model by using $\bar{\nu}$ instead of ν_D (see text).

the isothermal bulk modulus up to 800 K whereas for the higher temperatures we have estimated it by making a linear extrapolation. The B^S-values of the second set have been extracted from the formulae given by Varshni [165] and the conversion to isothermal has been done with the same specific heat and expansivity data as used by Chang and Himmel. In the same table we have also inserted the values of $\Delta L/L$, a and Ω, extracted from the paper of Simmons and Balluffi [162].

From eq. (9.4) the value of s/h_{exp} for 909 K is found to be equal to 2.53×10^{-4} K^{-1}; it gives the straight line (1) drawn in fig. 9.10. An inspection of this figure suggests that the experimental values are consistent with a value of s/h_{exp} about 20–24% lower than that calculated above. However, this fact is not without an explanation. We shall return to this point later on.

We now proceed to the calculation of the self-diffusion coefficient from a single measurement, e.g. from the value $D = 1.14 \times 10^{-17}$ cm^2/s given by Lam, Rothman, Mehrer and Nowicki [166] for 572.5 K. The method has already been developed in §9.1. The term $fa^2\nu$ is equal to 5.83×10^{-3} cm^2/s if one uses $f = 0.78$, $a = 4.1$ Å (see table 9.8) and [167] $\nu = \nu_D = 4.45 \times 10^{12}$ s^{-1}. The same values have been used for the extraction of s^{act} given in table 9.7 from the published D_0-values. Inserting them into eq. (9.7) and considering that $B = 906$ kbar (see table 9.8) one gets $c^{act} = 0.1713$ with a plausible error of 2% due to the experimental error of B [157]; if one alternatively uses Chang and Himmel's value $B = 898$ kbar one gets $c^{act} = 0.1738$. By using the first value and Varshi's elastic data at various temperatures the D-values inserted in table 9.8 are obtained and shown as asterisks in fig. 9.9. For each value we give the uncertainty introduced by a total variation of 4% in the exponent of eq. (9.7), 2% from a plausible error in the value of c^{act} and another 2% in order to express the two possibilities of B in one value. Let us now compare these to the experimental self-diffusion coefficients which cover ten orders of magnitude (fig. 9.9). If one considers the uncertainty given, we observe an agreement up to 800 K, although D varies by more than four orders of magnitude. We recall that the bulk modulus has been directly measured only up to this temperature. At higher temperatures a systematic difference is observed, which, at the melting point, is about 50%, even if one considers the lower limit of the calculated value. The claim that this difference might be due to an additional mechanism must be precluded because the calculated value exceeds the experimental one. An obvious explanation might be an error of some percents due to the linear extrapolation of the B-values. A direct answer to this possibility can be given only when B-values will have been measured at higher temperatures. In view of this we proceed to the discussion of two other plausible sources of this difference.

In chapter 14 we shall emphasize that the statement of c^{act} or c^{f} being temperature independent is valid only as a first approximation. However, cases where we are dealing with a very large temperature range – as the present case where D varies by 10 orders of magnitude – justify a small error. But the $cB\Omega$-model has the advantage of permitting the computation of the limits of such an error by studying the temperature dependence of c. For the bounds of its thermal expansion coefficient β^c the following equation will be proven (see §14.4.2)

$$\delta - \left.\frac{\partial B}{\partial P}\right|_T < \frac{\beta^c}{\beta} < \delta - \frac{1}{2}\left(\left.\frac{\partial B}{\partial P}\right|_P + 1\right)$$

where $\delta \equiv -(\beta B)^{-1}(\partial B/\partial T)|_P$ is practically temperature independent. In the present case of Ag we have $(\mathrm{d}B/\mathrm{d}T)_P = -0.32$ kbar/K (at $T = 572.5$ K), $\beta \simeq 6.4 \times 10^{-5}$ grad^{-1}, $B = 906$ kbar and hence $\delta \simeq 5.5$. Furthermore the published values of $(\mathrm{d}B/\mathrm{d}P)_T$ (see table 10.1) lie between 6.5 and 6.78 so that we take an average value of 6.65. By inserting these values into the inequality we get:

$$-1.15 < \beta^c/\beta < 1.68.$$

By now turning to table 9.8 we see that the bulk volume of the crystal increases by $\sim 5\%$ from 572.5 K up to the melting point. Then the above inequality reveals that, for the same temperature range, the total variation of "c" lies between -5.5% and $+8.4\%$. By considering the upper bound and recalling that the value of c^{act} – determined at 572.5 K – is 0.1713, we find that at the melting point its upper limit is 0.1857 which gives from eq. (9.7) $D = 6.1^{+4.3}_{-2.6} \times 10^{-9}$ cm^2 s^{-1}. The latter value favourably agrees with the experimental one (see table 9.8). By employing a similar correction to the value of c^{act} we get the lower limit of the calculated D-value for each temperature. The error bars in fig. 9.9 show the lower limits due to the temperature variation of c^{act} in addition to the mentioned individual uncertainty of each D-value resulting from a plausible error of 4% in the exponent. An inspection of the figure shows that the experimental values lie actually in the predicted range.

Another plausible source of the differences observed in the high temperature region is the incorrect determination of the preexponential factor $fa^2\nu$. This possibility seems the most probable because no one guarantees that ν is exactly equal to ν_D. It may be incorrect by a factor of 2–3 or even more. However, the problem remains to investigate whether the error introduced by this approximation is in the desired direction. Simple physical expectations suggest that the vibration frequency of a self-diffusing atom in the jump direction towards the neighbouring vacancy should be smaller than the host frequency. This can also be more accurately seen from a simple model developed by Seeger and Mehrer [72] who suggest that $\nu = \tilde{\nu}$ (see eqs.

(7.38) or (9.8)). Inserting into eq. (9.8) the appropriate values ($a \simeq 4.1$ Å, $h^m \simeq 1.99 - 1.2 = 0.79$ eV and $M = 108/N$) one gets $\tilde{\nu} = 2.05 \times 10^{12}$ s^{-1}, which is about two times smaller than ν_D. Using this value of $\tilde{\nu}$, a preexponential factor $fa^2\tilde{\nu} = 2.69 \times 10^{-3}$ results instead of 5.83×10^{-3} cm^2/s. By inserting this value into eq. (9.7) – again at $T = 572.5$ K – one finds $c^{act} = 0.1674$ instead of 0.1713 (note that, although ν has been changed by a factor around 2, practically the same value of c^{act} results). Using these new values of $fa^2\nu$ and c^{act} we have recalculated – from the $cB\Omega$-model – the self-diffusion coefficient and the results are given in the last column of table 9.8. By comparing them to the experimental values one finds an agreement up to the melting point if one also considers the uncertainty quoted. We would like to stress the following: There is no doubt that the approximation of eq. (9.8) has a better theoretical justification than taking $\nu = \nu_D$; however, the latter is more useful for practical purposes, because when one starts to predict self-diffusion coefficients from a single diffusion measurement one usually does not know the value of h^m. But note that whatever approximation is used the value of c^{act} remains practically the same.

We shall now proceed to the calculation of the defect parameters from the $cB\Omega$-formulae; we start with the *activation parameters*. Using Varshni's [165] elastic data for temperatures between 700 and 800 K and $c^{act} = 0.1713$ we get from eq. (9.1):

$$s^{act} = -c^{act} \left.\frac{\partial(B\Omega)}{\partial T}\right|_P = 5.54 \, k.$$

Furthermore, by considering that for 700 K $g^{act} = c^{act} B\Omega = 1.608$ eV we find $h^{act} = 1.94$ eV. A similar calculation for the range 1100–1200 K gives $s^{act} \simeq 6 \, k$; by taking also into account that $g^{act} = 1.42$ eV we get $h^{act} = 1.99$ eV. We recall that these values might have an uncertainty of 4%. Their comparison to the experimental ones indicate a very good agreement as far as the enthalpy is concerned. The difference observed in the entropies is insignificant considering the large experimental error in their determination and considering also the fact that they have been extracted from the approximation $\nu = \nu_D$, which, as mentioned, might be incorrect. If one uses $\nu = \tilde{\nu}$ the experimental s^{act}-values inserted in the table must be increased by $0.8 \, k$ thus bringing a better agreement with the values calculated above.

It is of interest now to discuss the manner in which the temperature variation of c^{act} mentioned above might influence the values of the entropies calculated within the frame of the $cB\Omega$-model. In the last chapter, 14, we indicate that a more exact formula for the entropy exists:

$$s^{act} = -c^{act} \frac{d(B\Omega)}{dT} - \beta^c c^{act} B\Omega. \qquad (9.11)$$

By considering that the upper limit of β^c is 1.68β and taking into consideration that β is around 64×10^{-6} and 75×10^{-6} K^{-1} for 700 and 1100 K respectively, we find that the minimum values of the entropy which are predicted from the $cB\Omega$-model are:

$$s^{act}_{700\,K} \simeq 3.5\,k \quad \text{and} \quad s^{act}_{1100\,K} \simeq 4\,k.$$

We now find the experimental values of the activation entropy to be consistent (see table 9.7) with the above limits set by the $cB\Omega$-model.

We finally turn to the calculation of the *formation parameters*. Inserting into eq. (9.5) the limits of n/N determined by Simmons and Balluffi for $T = T_M$ one finds that c^f must lie between 0.1135 and 0.1217. By using the elastic data for 1200 and 1233 K we get from eq. (9.1) the values $s^f = 4.08$ and $4.37\,k$. As explained in the case of the self-diffusion process they do not take care of the eventual temperature variation of c^f. In order to study this effect we consider that near to the melting point β is around 8×10^{-5} K^{-1} whereas β^c is at the most $1.68\,\beta$ so that the maximum value of $\beta^c c^f B\Omega$ (see eq. 9.11) is around $1.44\,k$. This latter value provides the maximum error introduced in the calculation of s^f from the statement that c^f is temperature independent and must be subtracted from the values 4.08–$4.37\,k$ mentioned above (see eq. 9.11). We therefore conclude that the minimum formation entropy which can be extracted from the $cB\Omega$-model is in the range 2.6–2.9 k in agreement to the experimental data. Taking an average of $2.75\,k$ we see that the resulting enthalpy is, at least, $c^f B\Omega + Ts^f \simeq 1.21$ eV (recall the uncertainty of $\sim 4\%$) which agrees favourably with the recent experimental determination (see table 9.7) by means of the PAS technique.

Summarizing the above discussion we can say the following: although extending up to ten orders of magnitude the self-diffusion data of Ag can be reproduced from the $cB\Omega$-model even if one starts from the lower end of the scale. If the small deviations observed in the high temperature region are not due to an incorrect estimation of B or ν they can still be well accounted for from the known limits of the slight temperature variation of c^{act}. Furthermore, the recently published values of the formation and activation parameters also agree with those calculated from the $cB\Omega$-model. We finally recall that the current manner of analysing the defect data by assuming monovacancies and divacancies is subjected exactly to the same thermodynamical criticism as developed in the case of aluminum.

9.2.4 Lead

The differential dilatometry technique has been applied to Pb by Feder and Nowick; in their first paper [168] they concluded that the concentration

Table 9.9
Calculated and experimental entropies and enthalpies in lead.

Process	h (eV)	s (k-units)	Reference
Activation (self-diffusion)	1.09	5.3 (5.9)	[172] [a]
	1.13	5.94 (6.55)	b
$cB\Omega$-calculation	1.09	5.2	
Formation:			
differential dilatometry	0.49	0.7 ± 2.0	[169]
positrons technique	0.50 ± 0.03	–	[533]
	0.54 ± 0.02	–	[535]
$cB\Omega$-calculation	~ 0.59	~ 2.83	
Solubility in Pb of:			
Ni	0.48	1.8	[540]
Zn	0.42	2.6	[540]
Cd	0.19	0.8	[540]
Ag	0.55	4.5	c
Pt	0.53	3.0	[541]
Au	0.61	10.7	[542]

[a] The entropy has been extracted from the value $D_0 = 0.69$ cm^2/s by using $\nu_D = 1.8 \times 10^{12}$ s^{-1}. If alternatively $\tilde{\nu} = 9.8 \times 10^{11}$ (with $h^m = 0.5$ eV) is used one gets the entropy values included in parentheses.

[b] H.B. Vanfleet (private communication); by using $D_0 - 1.603$ cm^2/s and further taking either ν_D or $\tilde{\nu}$ as in note a.

[c] H.B. Vanfleet, reanalysis of the data of ref. [540] (private communication).

of vacancies at the melting point (600 K) is less than or equal to 1.5×10^{-4} whereas the formation enthalpy should be equal to or higher than 0.53 eV. These workers have improved their method [169] and gave the results included in the table 9.9 and in fig. 9.11. One should notice the large uncertainty in the entropy ($s^f = 0.7 \pm 2$ k); in view of the fact that negative values of s^f are precluded (the frequencies around the vacancy have to decrease) we shall only consider in the following the most probable values between 0.7 and 2.7 k. In the same table we have also included the results of positron annihilation spectroscopy which give a formation enthalpy in rough agreement with that of Feder and Nowick.

Self-diffusion measurements have been carried out from 447 K up to 595.6 K by Nachtrieb and Handler [170]; these early measurements were later confirmed by various investigators (for a compilation of a large number of references concerning self-diffusion and heterodiffusion in Pb we refer to an interesting paper by Vanfleet [171]). They are reproduced in fig. 9.12 as solid dots. The analysis however of these data has attracted much discussion. LeClaire [172] has suggested that the data can be described with

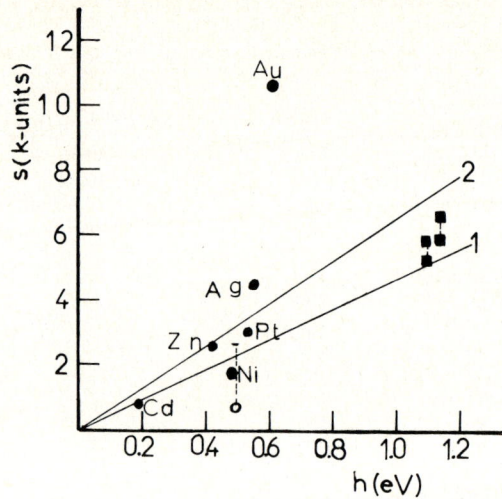

Fig. 9.11. Defect entropies and enthalpies in lead; ○: formation; ●: solubility; ■: self-diffusion; lines calculated from $cB\Omega$-model: (1) from elastic data mentioned in the text; (2) from information of Vanfleet.

$D = 0.69 \exp(-1.09 \text{ (eV)}/kT) \text{ cm}^2/\text{s}$. By considering that $f = 0.78$, $a = 5$ Å and taking $\nu = \nu_D \simeq 1.8 \times 10^{12} \text{ s}^{-1}$ [150] the preexponential factor leads to $s^{\text{act}} \simeq 5.3$ k: alternatively considering $\nu = \tilde{\nu} = 0.98 \times 10^{12} \text{ s}^{-1}$ (with $h^m = 0.5$ eV) the entropy s^{act} becomes slightly higher, e.g. 5.9 k. These values have been inserted into table 9.9 and fig. 9.11 along with those resulting from an analysis (over all the existing self-diffusion data) communicated to the authors by Vanfleet [173]. We have not included the values obtained from an analysis of Mehrer and Seeger [174] because they have also considered the pressure self-diffusion data of Hudson and Hoffman [175] which seem to suffer from a systematic error in the pressure calibration (Vanfleet, private communication).

We now proceed to the comparison of the existing "defect data" with the predictions of the $cB\Omega$-model. Adiabatic elastic data have been reported by Vold, Glicksman, Kammer and Cardinal [176] up to the melting point. These measurements are in agreement with those reported by Miller and Schuele [177] up to room temperature. For the conversion of the adiabatic bulk modulus to the isothermal one, we have used the expansivity data of Feder and Nowick and the specific heat measurements of Pochapsky [178]. By using at $T = 595.6$ K the values $\beta \simeq 10^{-4}$ K^{-1}, $(\partial B/\partial T)|_P = -0.22$ kbar/K, $B = 336.4$ kbar and further $B_0^{\text{SL}} \simeq 490$ kbar, we find $s/h_{\text{exp}} \simeq 4.1 \times 10^{-4}$ K^{-1}. This value gives the straight line, labelled as 1, in fig. 9.11; in the same figure we draw also the straight line 2 which corresponds to the

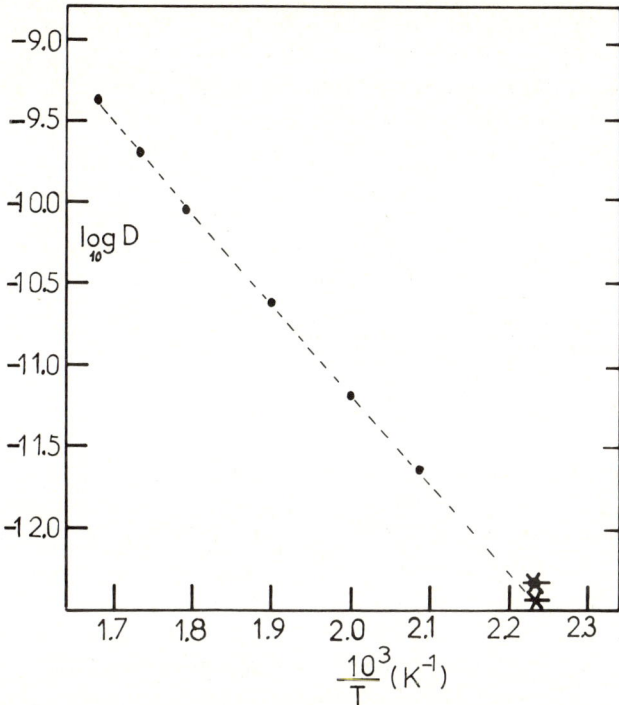

Fig. 9.12. Self-diffusion in lead; dots: experimental values; dashed line: $cB\Omega$-calculation with $c = 0.12466$. $*$: calculated with $\nu = \nu_D$ (higher point) and with $\nu \simeq \tilde{\nu}$ (lower point).

value $s/h_{exp} \simeq 5.66 \times 10^{-4}$ K^{-1} forwarded to the authors by Vanfleet [173]. We see that the formation and activation parameters actually scale with the $cB\Omega$-model if one considers the large experimental error in the formation entropy. In the same figure we have also included some solubility parameters of various impurities in Pb [178a]. Except for the point of Au the others seem to obey the $cB\Omega$-model. As will be seen later in fig. 11.13 a plot of $\ln D_0$ versus h^{act} for an extensive variety of impurities diffusing in lead gives also a very good straight line even for a slope only 14% higher than that corresponding to the s/h_{exp}-value predicted from the $cB\Omega$-model (see §11.2). Therefore we see that in this way a large amount of data (i.e. self-diffusion, differential dilatometry, solubility, heterodiffusion of various impurities) are correlated in a unified manner. This agreement becomes more striking if one considers the vexing questions (which are still open today [171]) as to the interpretation of the heterodiffusion of various impurities in lead. In other words, the model, without claiming any knowledge about the mechanism involved, correctly demands for all mechanisms the corresponding entropy and enthalpy to have a ratio equal to s/h_{exp} which is definitely predetermined from the matrix material.

We now turn to a more strict check of the $cB\Omega$-model; it examines whether it can reproduce the self-diffusion curve from the one end of the scale to the other. We start from the highest temperature $T = 595.6$ K for which Nachtrieb and Handler [170] measured $D = 4.09 \times 10^{-10}$ cm^2/s. Inserting this value into the eq. (9.7) and using the values (at $T = 595.6$ K) $f = 0.78$, $a = 5$ Å (and hence $\Omega = 31.285$ Å^3), $\nu = \nu_D$ and $B = 336.4$ kbar [176], we get $c^{act} = 0.12466$. If one alternatively uses $\nu = \tilde{\nu}$, the slightly smaller $c^{act} = 0.1197$ results. By now using this first value of c^{act} and the appropriate elastic and expansivity data for each temperature we calculate the D-values down to 447 K. The results are given as a dashed line in fig. 9.12; we see that it nicely describes all the diffusion data (dots) covering three orders of magnitude. Exactly the same agreement is observed if one inserts the values $\tilde{\nu}$ instead of ν_D and $c^{act} = 0.1197$ instead of 0.12466. For instance at the lower end of the scale (447 K) where $B = 369.6$ kbar, $\Omega = 30.8$ Å^3 depending on the pair ν, c^{act} adopted, we get $D = 4.6 \times 10^{-13}$ cm^2/s for the first case (i.e. with $\nu = \nu_D$) and $D = 3.6 \times 10^{-13}$ cm^2/s for the second. The latter D-value is shown as an asterisk in fig. 9.12, whereas the other happens to coincide with the solid dot at this temperature; their difference is undoubtedly within the experimental errors of the predicted D-value and the quantities used in our calculation.

We now proceed to the calculation of the values of the *activation parameters* from the $cB\Omega$-formulae. Using the data mentioned above for $T = 595.6$ K we get from eqs. (9.1) and (9.2): $s^{act} = 5.2\ k$ and $h^{act} = 1.09$ eV. In deriving the above results the value $c^{act} = 0.12466$ has been used; due to the fact that, as already mentioned, c^{act} may take a somewhat smaller value these results have an inherent uncertainty at least of 4%. A comparison with the experimental values inserted in table 9.9 indicates an excellent agreement.

The calculation of the *formation parameters* needs the determination of c^f; inserting at 595.6 K the value $n/N = 1.7 \times 10^{-4}$ (which has been measured by Feder and Nowick [169] at the melting point) we get $c^f = 0.0678$. Following the same procedure as for the activation process – but now using c^f instead of c^{act} we find: $h^f = 0.59$ eV and $s^f = 2.83\ k$. These values are comparable to the experimental cases, considering the large experimental error in the formation entropy.

Summarizing the present discussion in lead we can say that the $cB\Omega$-model without using any adjustable parameter not only reproduces the diffusion curve but also leads to parameters that correctly describe in a unified way a lot of other data obtained by quite different techniques.

9.3 bcc metals

The alkali-metals Na, Li and K are first considered because data for both formation and self-diffusion activation are available. Special attention is drawn to the question whether their self-diffusion plots can be reproduced from the temperature variation of their bulk properties. We further proceed to the examination of two bcc metals (W and Nb) which have melting temperatures and bulk moduli that are larger than those of alkali metals by one to two orders of magnitude. In this way the reliability of the connection of formation and activation Gibbs energy to the bulk properties can be checked in a very large temperature range.

The self-diffusion coefficient has been accurately measured also for other high-temperature bcc elements like α-iron. For the latter element due to the fact that the vacancy migration enthalpy is still a controversial problem we refer to ref. [178b]. We selected W for an extensive discussion because quenching experiments provide separate data for the formation and migration parameters. The capability of the $cB\Omega$-model to describe the temperature variation of the diffusion coefficient of carbon in iron [178c] in a very large temperature range is discussed in ch. 11.

9.3.1 Sodium

This metal is of interest because of a number of special qualities. Its Arrhenius plot for self-diffusion and its isothermal plot of log D versus P are strongly curved [75]. These qualities and the large amount of available elasticity measurements make it a crucial material for the check of the correctness of the $cB\Omega$-model. All facets will be extensively discussed.

Differential dilatometry in sodium was first applied by Sullivan and Weymouth [179], who obtained an unusual low formation enthalpy (0.14 eV). Later Feder and Charbnau [180] showed that this anomalous value was possible due to an hysteresis effect and published the new set of parameters included in table 9.10. They also found that the vacancy concentration at the melting point (97.9°C) is $7.5 \pm 0.4 \times 10^{-4}$. The macroscopic length was accurately determined with an interferometric method so that their error bars of $\Delta L/L_0$ were smaller than 10^{-5}; similarly, the errors in the relative lattice expansion $\Delta a/a_0$ measured by X-rays, were about 10^{-5}. Later, Ritter, Fritsch and Luescher [181] reported length data with an accuracy worse than that of Feder and Charbnau. On the other hand, Adlhart, Fritsch, Heidemann and Luescher [182] repeated the measurement of the lattice constant with the aid of a neutron back-scattering technique which allows the significant improvement of a three times greater accuracy with

Table 9.10
Entropies [a] and enthalpies in sodium.

Process	h (eV)	s (k-units)	Reference
Formation:			
differential dilatometry	0.42 ± 0.03	5.8 ± 1.1	[180]
	0.35 ± 0.02	3.7 ± 0.5	[182]
	0.354 ± 0.025	3.9 ± 0.7	b
	0.36 ± 0.03	4.0 ± 0.9	c
	0.306	2.2	d
$cB\Omega$-calculation; $T = 361$ K	0.332	3.27	
Activation:			
tracer self-diffusion	0.45	4.14	[185]
	0.44	3.63	e
	$\binom{0.37}{0.5}$	$\binom{0.25}{5.1}$	[184]
	$\binom{0.377}{0.517}$	$\binom{0.64}{5.5}$	f
quasi-elastic neutron scattering	0.43	3.42	g
	0.44	4.11	g
$cB\Omega$-calculation; $T = 361$ K	0.465	4.59	
$T \simeq 195$ K	0.393	1.83	

[a] The experimental activation entropies should increase by 0.64 k when the frequency $\tilde{\nu} = a^{-1}(h^m/M)^{1/2} \simeq 1.5 \times 10^{12}$ s^{-1} ($h^m \simeq 0.1$ eV) instead of ν_D is used.

[b] Quoted in ref. [183] by comparing their $\Delta a/a$-data to the volume data of ref. [180].

[c] Quoted in ref. [183] when their $\Delta a/a$-data are combined to the volume data of ref. [181].

[d] Quoted in ref. [183] when the "mean calculated value" of $h = 0.306$ eV of Ho [591] is adopted and allowing for the existence of divacancies.

[e] Ref. (75); The two sets of values correspond to a two-exponential fit of these data (Ref. [75] used the value $\nu = \nu_D = 3.33 \times 10^{12}$ s^{-1})

[f] These values correspond to a two-exponential fit of the data of ref. [75] made in ref. [189]; the entropies correspond to the same ν_D selected by Mundy in ref. [75].

[g] Ref. (187); The first set corresponds to polycrystalline sodium and the second to monocrystalline sodium.

respect to the earlier X-ray results. Adlhart, Fritsch and Luescher [183] then combined these recent lattice constant data with the earlier macroscopic length data and obtained the various sets of parameters included in table 9.10. The most important result of their paper is that interstitials have no essential part in the analysis of the volume data; furthermore, they show that the data are not in contradiction to an appreciable contribution of divacancies beyond single vacancies. This possible existence of divacancies complicates the situation so that an incontestable set of monovacancy parameters cannot yet be derived although the experimental differential dilatometry data have reached a high degree of accuracy. However, in table 9.10 (except in one case) the quoted parameters were obtained by making a least squares fitting to a straight line of $\ln(n/N)$ versus $1/T$. For the sake of comparison one case was added in which monovacancy parameters were derived by allowing the existence of divacancies. For the admissible fields of the single and divacancy formation parameters we refer to Adlhart et al. [183].

The self-diffusion has been studied by means of tracer techniques by various investigators [75,184,185]; a quasi-elastic neutron scattering technique has also been applied to self-diffusion by Göltz, Heidemann, Mehrer, Seeger and Wolf [186] and by Ait-Salem [187]. The various parameters obtained for self-diffusion are given in table 9.10 in which we have connected by parentheses these values extracted by each investigator under the assumption of the coexistence of two mechanisms. The others are effective parameters within the frame of a single mechanism and have been obtained mainly in the high temperature region up to the melting point. An inspection of the table indicates a reasonable consistency of the latter group although they were obtained by means of two different techniques in four independent investigations. This is important for the comparison with the $cB\Omega$-model (see below).

Mundy [75] detected a strong upward curvature in the Arrhenius plot (fig. 9.1) by extending the tracer measurements down to 194.4 K. The isotope effect data obtained by Mundy et al. [184] and by Mundy [75] show a gradual decrease with increasing temperature. These results attracted wide interest and their explanation remains today a point of contradicting views. By following Seeger's proposals Mundy [75] suggested that the coexistence of monovacancies and divacancies might provide a plausible explanation. On these grounds he performed a two-exponential fit of his data and arrived at the temperature independent pairs of parameter given in table 9.10. Therefore these two sets of (h, s)-parameters must be thought of as providing the two contributions D_{1V}, D_{2V}, e.g. [188] of monovacancies and divacancies to the diffusion process. These parameters indicate that at 194.4 K $D_{1V} = 1.45 \times 10^{-12}$ cm^2 s^{-1} and $D_{2V} = 8.4 \times 10^{-14}$ cm^2 s^{-1}, which means that the monovacancy mechanism is almost exclusively responsible for the diffusion process at low temperatures. On the other hand, at $T = 368.7$ K (i.e. close to the melting point) one finds that $D_{2V}/D_{1V} = 2.2$. This kind of analysis has been recently refined by Mehrer [62] who proposes: $D_{1V} = 3.76 \times 10^{-3} \exp(-0.365 \text{ eV}/kT)$ cm^2/s and $D_{2V} = [0.292 \exp(-0.457 \text{ eV}/kT) + 8 \exp(-0.575 \text{ eV}/kT)] f_{2V}$, where the two activation enthalpies in the divacancy contribution pertain to the two saddle points between the various divacancy configurations and where f_{2V} is the correlation factor of the divacancy (which is not just a temperature independent constant). According to this analysis the contribution of monovacancies to the whole diffusion coefficient at 368.2 K is only around 24%.

Feit [189] has alternatively suggested that the behaviour of Na is consistent with a picture involving sodium atoms at two interstitial sites (of nearly equal energy) beyond the vacancies. However, in view of the finding of Adlhart et al. [183] this explanation is not supported. Fano and Jacucci's [190] calculations suggest that second-near-neighbour jumps occur at high temperatures and contribute more than one fourth of the atomic migration processes at the melting point. One should stress however that although these double jumps may explain the sharp upward curvature in the immediate vicinity of the melting point they cannot account for [186,188] the curvature in the whole temperature range.

Gilder and Lazarus [37] have proposed another explanation of the curvature over the entire temperature range within the frame of a single mechanism (monovacancy) with temperature dependent parameters. It is the most plausible one as it does not need any ad hoc assumption; the temperature dependence of h and s is in principle allowed from thermodynamics. To which extent these temperature dependence can account for the whole curvature has to be quantitatively assessed. Audit and Gilder [74] have performed a careful microscopic calculation which showed that the formation volume increases significantly with temperature; thermodynamics then demand that both h^f and s^f should also quickly increase with increasing T. This pioneering calculation also showed that β^f increases with temperature and consistently exceeds β by more than a factor of ten, i.e. $11.7 < \beta^f/\beta < 14.7$. The compatibility of this calculation with the $cB\Omega$-model has been shown by the authors in a recent paper [191]. Audit and Gilder have concluded that "the neglect of the temperature increase of v^f in an analysis of diffusion data is unrealistic even when it is assumed that at least two diffusion mechanismes are operative". Two further points should be noted: their previous calculation indicates that in the temperature range 195–370 K the formation enthalpy should increase by 0.04 eV whereas the explanation of the diffusion curvature requires a total increase of the activation enthalpy by 0.95 eV. In view of the fact that the temperature dependence of the migration volume v^m has not yet been calculated it is still not clear if Audit and Gilder's numerical result can account for the whole curvature. The second point which should be noticed is that according to Mehrer's opinion [62] the volume v^f should decrease (and not increase) with temperature. However, this latter proposal has not been well justified and therefore cannot be considered as casting doubt on Audit and Gilder's results.

Another point that seems to be still open is the study of the influence of anharmonicity on the temperature dependence of the isotope effect. Some preliminary calculations by Achar [77] and Franklin [192] have shown this influence to be small but Gilder and Lazarus' [37] proposals indicate the opposite. Much work still remains to be done towards this direction because there is a direct experimental confirmation that in bcc-Zr which also has a strongly curved diffusion plot [193] the isotope effect increases with temperature, thus showing that divacancies cannot explain the curvature.

From the above brief review of the main proposals for the explanation of the curvature it becomes clear that quite contradictory views exist. The only point on which all the workers agree and which is experimentally confirmed is that in the low temperature region ($\sim$ 195 K) only monovacancies are operating. This point of contact is all that is needed for a convincing description of the tracer self-diffusion curve in terms of the $cB\Omega$-model; it will be attempted below. We discuss only the existing expansivity and elastic data of sodium which are needed for our applications. The conversion of the adiabatic bulk modulus to the isothermal one has been done with the help of the specific heat data reported in table 1 of Martin's paper [194]. The high temperature expansivity has been taken from the paper of Adlhart et al. [182]. In table 9.11 we give for the temperatures 195 and 295 K the adiabatic and isothermal bulk modulus measured by various workers by means of various techniques. As noticed by Martinson [195] the degree of over-all agreement between the entries of this table is somewhat disappointing in view of the care with which all the experiments appear to have been done. It should be noticed that there is no single elasticity experiment covering the whole temperature range (194.4 K up to melting point) of interest. We face the difficulty to select low temperature data (195 K up to RT) and to join them with data obtained above room temperature. As noticed by Fritsch, Geipel and Prasetyo [196] the systematic errors involved may be as high as $\sim$ 3% which however play an important role in the application of the $cB\Omega$-model. Care has therefore been taken for the selection of a consistent set of elastic data. In order to lower the systematic errors we take the average of the B-values reported for the dry-ice point and room temperature (see table 9.11). It is noticeable that at 195 K the mean value of 67.4 kbar is close enough to the value of 67.2 kbar obtained by Beecroft and Swenson [197] by a direct method for the isothermal bulk modulus. The other values come from the adiabatic bulk modulus obtained from the ultrasonics technique and are converted to isothermal in the standard thermodynamical manner. Exactly the same observation holds for the temperature of 299 K for which the mean value of 62.94 kbar practically coincides with the directly measured value of 63 kbar by Beecroft and Swenson as corrected later by Monfort and Swenson [197]. Taking now the linear interpolation (i.e. $(\partial B/\partial T)_P = -0.0429$ kbar/K) of these two mean values at 195 and 299 K we obtain the B-values of the last column of table 9.12 from 194.4 up to 313 K. Above this temperature the values of this column come from a least squares fitting to a straight line of the measured

B-values (in their high temperature region) by Fritsch, Nohman, Korpium and Lüscher [198]; the experimental method of the latter workers has the advantage of directly giving the isothermal bulk modulus. The fitting of their data shows that at high temperatures $(dB/dT)_P = -0.0843$ kbar/K, which is appreciably greater – as expected – than the value -0.043 kbar/K of the low temperature data. For more recent elastic data, see §A.4.

Computation of the self-diffusion curve. For reasons already explained we shall start our calculation by using the experimental value of $D = 1.59 \times 10^{-12}$ cm^2/s at the lowest temperature (194.4 K) of Mundy's measurements. At this temperature Killean and Lischer [150] give $\Theta_D = 140$ K (and hence $\nu_D = 2.9 \times 10^{12}$ s^{-1}) whereas the density measurements of Diederich and Trivisonno [199] lead to $a = 4.255$ Å (i.e. $\Omega = 38.517$ Å^3). By using also $f = 0.727$ and $\nu = \nu_D$ we find $fa^2\nu = 3.84 \times 10^{-3}$ cm^2 s^{-1}. Then by applying

Table 9.11
Adiabatic and isothermal bulk modulus of sodium at 195 and 299 K.

T (K)	B^S (kbar)	B_T (kbar)
195	72.5 [a]	~ 70 [a]
	68.2 [b]	65 [b]
	71.9 [c]	67.2 [d]
		average value: 67.4
299	68.8 [a]	64.1 [a]
	67.6 [e]	63 [e,d]
	66.1 [f]	61.6 [f]
	67.67 ± 0.6 [g]	63.05 [g]
		average value: 62.94

[a] Ref. [195]; these values are consistent with those reported in ref. [201].
[b] Ref. [199].
[c] The value of B^S has been quoted in ref. [195] as extracted from ref. [197] by applying some appropriate corrections in the pressure scale and using $B^S/B = 1.0362$.
[d] Ref. [197].
[e] Ref. [197].
[f] Daniels [592].
[g] The B^S-value comes from eq. (a) of table 3 of ref. [196]; the B-value is extracted in the usual thermodynamic manner.

Table 9.12
Elastic and expansivity data of sodium.

T (K)	$\Omega^{a,b}$ (10^{-24} cm^3)	β^a (10^{-4} grad^{-1})	B^c (kbars)	T (K)	$\Omega^{a,b}$ (10^{-24} cm^3)	β^a (10^{-4} grad^{-1})	B^c (kbars)
194.4	38.517		67.426	345	39.8399	2.2326	59.57
208	38.63		66.84	352.9	39.9107	2.2593	58.90
222.7	38.76		66.21	357	39.9478	2.2731	58.55
235.5	38.88		65.66	357.2	39.9496	2.2738	58.54
247.5	38.98		65.15	360.6	39.9806	2.2852	58.25
258.7	39.08		64.67	362.6	39.9989	2.2920	58.08
273	39.23		64.05	362.7	39.9998	2.2923	58.08
285.1	39.32		63.53	367.5	40.0440	2.3085	57.67
298.2	39.4405	2.0737	62.97	368.7	40.0551	2.3125	57.57
307.7	39.5189	2.1061	61.80	370.2	40.069	2.3175	57.44
308	39.5214	2.1071	61.78	288	39.34		63.41
313	39.5632	2.1241	61.40	364.5	40.0163	2.2984	57.92
327.9	39.6901	2.1747	61.01				

[a] The values of Ω and β above RT have been calculated from the analytical expressions given in ref. [182].
[b] Ref. [199]; the Ω-value at 195 K has been calculated from the corresponding density $\rho = 0.991$ g/cm^3. Then the Ω-values up to RT have been extracted from a linear interpolation of this value (at 195 K) and that obtained for 308 K with the manner explained in note a.
[c] For temperatures up to 313 K the B-values come from linear interpolation of the values 67.4 kbar and 62.94 kbar given in table 9.11 for 195 and 299 K respectively. At higher temperatures the B-values are from a least squares fitting to a straight line of the high temperature isothermal bulk modulus data of ref. [198].

Table 9.13
Calculated and experimental self-diffusion in sodium.

T (K)	D_{exp} (cm^2/s)	D_{calc} (cm^2/s)	$\dfrac{D_{calc} - D_{exp}}{D_{exp}}$ (%)	T (K)	D_{exp} (cm^2/s)	D_{calc} (cm^2/s)	$\dfrac{D_{calc} - D_{exp}}{D_{exp}}$ (%)
194.4	1.59×10^{-12}	1.59×10^{-12}	0	345	5.79×10^{-8}	$5.7^{+1.4}_{-1.1} \times 10^{-8}$	-1.5
208	6.97×10^{-12}	$7.52^{+3.63}_{-2.48} \times 10^{-12}$	$+7.9$	352.9	8.31×10^{-8}	$8.16^{+1.92}_{-1.58} \times 10^{-8}$	-1.8
222.7	2.65×10^{-11}	$3.17^{+1.4}_{-1} \times 10^{-11}$	$+19.6$	357	9.81×10^{-8}	$9.74^{+2.24}_{-1.86} \times 10^{-8}$	-0.7
235.5	8.69×10^{-11}	$9.55^{+3.9}_{-2.8} \times 10^{-11}$	$+9.9$	357.2	9.94×10^{-8}	$9.81^{+2.26}_{-1.9} \times 10^{-8}$	-1.3
247.5	2.08×10^{-10}	$2.43^{+0.94}_{-0.68} \times 10^{-10}$	$+16.8$	360.6	1.12×10^{-7}	$1.13^{+0.26}_{-0.2} \times 10^{-7}$	$+0.9$
258.7	5.20×10^{-10}	$5.39^{+1.95}_{-1.46} \times 10^{-10}$	$+3.7$	362.6	1.21×10^{-7}	$1.23^{+0.27}_{-0.23} \times 10^{-7}$	$+1.7$
273	1.33×10^{-9}	$1.34^{+0.33}_{-0.34} \times 10^{-9}$	$+0.8$	362.7	1.27×10^{-7}	$1.24^{+0.27}_{-0.23} \times 10^{-7}$	-2.4
285.1	2.63×10^{-9}	$2.74^{+0.88}_{-0.67} \times 10^{-9}$	$+4.2$	367.5	1.51×10^{-7}	$1.50^{+0.33}_{-0.28} \times 10^{-7}$	-0.7
298.2	5.81×10^{-9}	$5.52^{+1.7}_{-1.3} \times 10^{-9}$	-5	368.7	1.64×10^{-7}	$1.58^{+0.34}_{-0.3} \times 10^{-7}$	-3.7
307.7	9.66×10^{-9}	$10.39^{+3}_{-2.35} \times 10^{-9}$	$+7.5$	370.2	1.86×10^{-7}	$1.67^{+0.36}_{-0.3} \times 10^{-7}$	-10
308	1.02×10^{-8}	$1.05^{+0.31}_{-0.23} \times 10^{-8}$	$+2.9$	288	3.23×10^{-9}	$3.22^{+1.22}_{-0.78} \times 10^{-9}$	0
313	1.33×10^{-8}	$1.38^{+0.38}_{-0.31} \times 10^{-8}$	$+3.8$	364.5	1.34×10^{-7}	$1.33^{+0.3}_{-0.24} \times 10^{-7}$	-0.7
327.9	2.65×10^{-8}	$2.54^{+0.66}_{-0.54} \times 10^{-8}$	-4.2				

eq. (9.7) for $T = 194.4$ K we find $c^{act} = 0.223$ when the value $B = 67.426$ kbar of the last column of table 9.12 is used. Due to the fact that in this temperature range the "mean" value of B used for the calculation of c^{act} differs by 4% from the "extreme" B-value reported by some investigators (see table 9.11) the calculated value of c^{act} has an inherent uncertainty of the same amount. Once c^{act} is known the D-values at various temperatures can be calculated up to the melting point. The results are given in table 9.13 for those temperatures for which D has been directly measured. For each value we give the uncertainty resulting from a plausible experimental error of 2% in the value of $B\Omega$ used for each temperature. Although, as we have seen, this error is greater ($\sim 3\%$) a comparison of the calculated D-values with the experimental ones indicates a very good agreement. In order to get an idea about the extent of this agreement we give in the last column the percentage deviation of the directly computed D-value from the experimental one by assuming that the latter has no experimental error whatsoever. This was done in spite of the fact that the experimental errors of the D-values reported by Mundy [75] lie between 1 and 3%. We see that at three temperatures the difference exceeds 10%; but a 1% increase of the B-value used in our calculation is enough to eliminate the difference. In view of the fact that the measured B-values have, in this temperature range, an experimental error of 3% the agreement observed in table 9.13 should be regarded as excellent.

We proceed now to the examination of the correlation between the various defect entropies and enthalpies. Due to the strongly curved plot of sodium we cannot use the approximate expression of eq. (9.4) but we use – regardless of the process – the accurate expression (9.3). By inserting the appropriate data for 360.6 K, i.e. $\beta = 2.29 \times 10^{-4}$, $B = 58.25$ kbar [198] and $dB/dT = -0.084$ kbar/K, the $cB\Omega$-model gives

$$\left.\frac{s}{h}\right|_{cB\Omega} = 8.46 \times 10^{-4} \text{ K}^{-1}.$$

The above value gives the straight line drawn in fig. 9.13. In the same figure we have inserted the corresponding self-diffusion and formation parameters obtained from the same temperature region. The agreement of the activation parameters with the general behaviour predicted from the $cB\Omega$-model is obvious although they have been extracted from a variety of experiments. Furthermore, the formation parameters obey the $cB\Omega$-scheme within the experimental error in spite of the large uncertainty, especially for the formation entropy. The above agreement is noticeable because in the usual cases of fcc metals the activation enthalpy exceeds the formation one by a factor of around 2, whereas in the present case the ratio is appreciably smaller. According to the $cB\Omega$-model the corresponding entropies s^{act} and

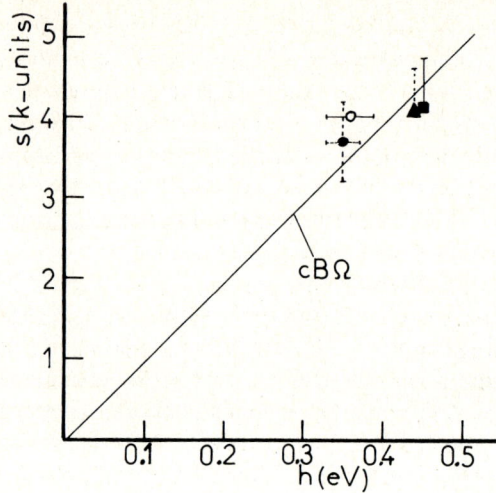

Fig. 9.13. Defect entropies and enthalpies in sodium; ●: formation parameters, Adlhart et al. [182]; ○: formation parameters, $\Delta a/a$-data of [183] combined to the volume data of [181]; ■: self diffusion activation, Nachtrieb et al. [185]; ▲: self-diffusion activation Ait-Salem [187]; the error bars of the activation entropies correspond to $\bar{\nu}$ instead of ν_D; line: $cB\Omega$-calculation for 360.3 K.

s^f should also not differ much; this behaviour is not only qualitatively verified but is further enhanced by the fact that the experimental ratio s/h is just equal to that predetermined from the bulk properties β, B and $(\partial B/\partial T)|_P$.

We now proceed to the determination of the values of the (self-diffusion) activation parameters. Using the data mentioned above for 360.6 K and further recalling that $c^{act} = 0.223$ and $\Omega = 40$ Å^3 we find from eqs. (9.1) and (9.2):

$$s^{act} = -c^{act}\Omega\left(\beta B + \left.\frac{\partial B}{\partial T}\right|_P\right) = 4.59\ k$$

and

$$h^{act} = c^{act}\Omega\left(B - T\beta B - T\left.\frac{\partial B}{\partial T}\right|_P\right) = 0.465\ \text{eV}.$$

We remind that even if there were no experimental errors in the quantities used in the above calculation the values just given would have an uncertainty of 4% due to the plausible error of the value of c^{act}. Repeating the same calculation for $T = 370.2$ K we get $s^{act} = 4.6\ k$ and $h^{act} = 0.47$ eV. These values strongly change in the low temperature region. For instance, if one repeats the same calculation for 194.4 K and uses the values: $\beta = 2 \times 10^{-4}$ K^{-1} [197], $B = 67.426$ kbar, $(\partial B/\partial T)_P = -0.0429$ kbar/K and $\Omega =$

38.517 Å^3 he finds $s^{act} = 1.83\ k$ and $h^{act} = 0.393$ eV, with again a plausible uncertainty of at least 4% for the reasons explained above. These values being appreciably smaller than those calculated above at 370.2 K directly show that the $cB\Omega$-model can account quantitatively for the curvature of the diffusion plot. The difference of the activation enthalpies at the melting point and at $T = 194.4$ K ($0.47 \pm 4\%$ and $0.39 \pm 4\%$ eV) gives in effect the required increase of 0.95 eV [74] (see above) of h^{act} in order to explain the curvature within the frame of a single mechanism.

We now proceed to the determination of the values of the formation parameters. As noticed by Adlhart et al. [183] the concentration of vacant sites at the melting point is $\sim 7.8 \times 10^{-4}$. Accepting that this value corresponds only to monovacancies and that the elastic data of table 9.12 for $T = 370.2$ K (i.e. $B = 57.44$ kbar) are appropriate for the melting point (370.9 K) we get from eq. (9.5) the value $c^f = 0.159$. This value must have an uncertainty of at least around 1.4% due to the experimental uncertainty of B in this temperature range [198]. Following the same procedure as in the activation process – but now inserting c^f instead of c^{act} – we find at $T = 360.6$ K: $s^f = 3.27\ k$ and $h^f = 0.332$ eV. In spite of their uncertainty these values are comparable to the experimental ones inserted in table 9.10. Combining these calculated formation values with the activation ones, one obtains: $s^m \simeq 1.3\ k$ and $h^m \simeq 0.13$ eV with plausible uncertainty of 6%. In the absence of any direct determination of the migration parameters we simply note that they coincide with those resulting from the difference of the experimental activation self-diffusion and formation parameters. This very low calculated value of h^m explains the failure of some experimental efforts to study vacancy kinetics with quenching techniques in Na.

The curvature of the isothermal plots $\ln D$ *versus* P. The pressure experiment of Mundy [75] at 288 and 364.5 K (these temperatures are studied in the two last lines of tables 9.12 and 9.13) showed a significant curvature of the corresponding $\ln D$ versus P plots. For reasons of completeness we shall now attempt a direct explanation of this curvature within the frame of the $cB\Omega$-model. This will be done only for the isotherm $T = 288$ K because at $T = 364.5$ K no elastic data under pressure exist.

Gilder and Lazarus [37] have unmistakably shown (see also the analysis of $\ln D$ versus P plots in §7.3.4) that the curvature at 288 K can be completely explained if the compressibility of the activation volume is equal to $\kappa^{act} = 33 \pm 5 \times 10^{-3}$ kbar^{-1}. It is a matter of simple calculation to see whether this value is predicted from the $cB\Omega$-model. Equation (8.31) suggests that:

$$\frac{\kappa^{act}}{\kappa} = 1 - B\frac{d^2B}{dP^2}\left(\frac{dB}{dP} - 1\right)^{-1}.$$

Grover, Getting and Kennedy [200] have measured dB/dP close to RT and found $B\, d^2B/dP^2 \simeq -dB/dP$; in view of the fact that dB/dP is 3.7–4 [197,200,201] the last term of the above equation is around 1.35. Note that this term is not sensitive to the selection of dB/dP. Therefore the $cB\Omega$-model suggests that κ^{act}/κ is 2.35. By considering that at $T = 288$ K, $B \equiv 1/\kappa = 64.45 \pm 3\%$ kbar (see table 9.12) we get $\kappa^{act} = 36.5 \times 10^{-3}$ kbar^{-1}. This value is just equal to that reported by Gilder and Lazarus' analysis. We therefore see that the $cB\Omega$-model can quantitatively account for the curvature of the isothermal diffusion plot at $T = 288$ K (see fig. 9.14) by using purely bulk quantities (i.e. κ, dB/dP, d^2B/dP^2) *without* having to borrow anything from the defect data. In other words, the curvature observed in the isothermal $\ln D$ versus P plot is strictly predetermined from the elastic data of the host material. At this point we suggest the interesting experiment of studying the pressure dependence of the diffusion coefficient of an impurity in sodium. According to the $cB\Omega$-model the isothermal plot $\ln D$ versus P must again show a curvature that can be described with the same compressibility, i.e. the same value of κ^{act} as in the case of self-diffusion.

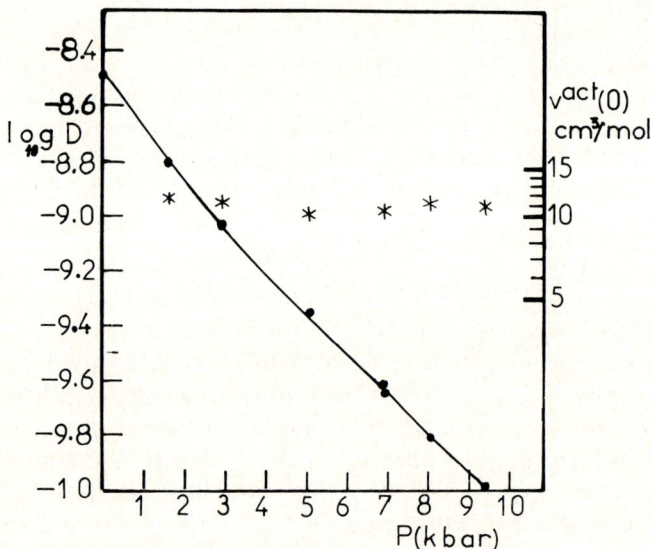

Fig. 9.14. Pressure variation of D in sodium for $T = 288$ K; ●: D-values Mundy [75]; *: $v^{act}(0)$-values Varotsos et al. [36]; the strong curvature of D, according to the $cB\Omega$-model, is described by $v^{act} = v^{act}(0) \exp(-\int_0^P \kappa^{act} dP)$ because $v^{act}(0) = -kT(d \ln D/dP) \exp\int_0^P \kappa^{act} dP + kT\kappa \exp\int_0^P \kappa^{act} dP$ remains unchanged with pressure if $\kappa^{act} = 36.5 \times 10^{-3}$ kbar^{-1} (see $cB\Omega$-model in ref. [36]).

Calculation of the activation and formation volume. The calculation of v^{act} and v^f of sodium within the frame of the $cB\Omega$-model is of high interest because in its proof (see ch. 14) this theory demands a correction of c when pressure is applied to highly compressible materials like Na. According to this proof the compressibility κ^c of c is at most roughly equal to the bulk compressibility. In the present case at 288 K we have $\kappa^c \simeq 1/64.45 \text{ kbar}^{-1} \pm 3\%$. This value holds irrespective if c refers to an activation or a formation process. Allowing for a pressure dependence of c the differentiation of $g = cB\Omega$ with respect to pressure gives

$$v = c\Omega\left(\frac{dB}{dP} - 1\right) - cB\Omega\kappa^c. \tag{9.12}$$

This relation * can give v^{act} from the known value of c^{act} (or v^f from c^f). We would like to stress here that eq. (9.12) is not a result of any model but is a pure thermodynamic result without the smallest approximation.

As mentioned before the experimental value of dB/dP varies between 3.7 and 4. In the following we shall take the value 3.95 ± 0.2 of Monfort and Swenson [197], which lies between those reported by Grover, Getting and Kennedy [200] (4.06 ± 0.07) and by Ho and Ruoff [201] (3.7). Introducing into eq. (9.12) $c^{act} = 0.223 \pm 4\%$, appropriate values of B and Ω given in table 9.12 and $\kappa^c = 1/64.45 \text{ kbar}^{-1}$ we get

$$v^{act} = 0.66 \begin{smallmatrix} +0.04 \\ -0.05 \end{smallmatrix} \Omega - 0.22\Omega = 0.44 \begin{smallmatrix} +0.04 \\ -0.05 \end{smallmatrix} \Omega.$$

In the errors we give only the uncertainty resulting from that of dB/dP; there is another 4% uncertainty due to c^{act} as already discussed. The experimental value is $v^{act} = 0.467 \pm 0.008\Omega$ [75] in excellent agreement with the calculated one **.

Now the calculation of v^f is straightforward. It is the same as for the activation volume but using c^f instead of c^{act}. By recalling that $c^f = 0.159$ we get:

$$v^f = 0.47 \pm 0.03\Omega - 0.16\Omega = 0.31 \pm 0.03\Omega,$$

i.e. 71% of the value of v^{act}. The migration volume results from the difference of v^{act} and v^f, i.e. $v^m = 0.13\Omega$ but with an uncertainty of $\sim 0.07\Omega$. An experimental value of v^m is still lacking.

Summary. From the above discussion of the applications of the $cB\Omega$-model to sodium we can say the following:

(1) The quantities s^f/h^f and s^{act}/h^{act} derived from measurements in the high temperature region with three quite independent techniques (differen-

---

* Notice that from this relation a more accurate expression for κ^{act} can be derived.
** In reality what the $cB\Omega$-model suggests is that the accurate value of v^{act} lies between 0.44 and 0.66 (see ch. 14).

tial dilatometry, tracer self-diffusion and quasi-elastic neutron study of self-diffusion) are all equal to the value predetermined from the bulk properties β, B and $(\partial B/\partial T)_P$.

(2) The experimental values of the formation and activation parameters coincide with those calculated from the $cB\Omega$-model.

(3) By using only the D-measurement at 194.4 K (where all workers agree that only monovacancies exist) and the existing elastic and expansivity data we can reproduce the whole diffusion curve up to the melting point. In other words, the curvature of the Arrhenius plot can be quantitatively explained from the increase of h^{act} predicted by the $cB\Omega$-model.

(4) By using solely bulk elastic data the $cB\Omega$-model can quantitatively account for the curvature of the isothermal $\ln D$ versus P plot at 288 K.

(5) The experimental activation volume, at 288 K, agrees with that calculated from the $cB\Omega$-model; for its determination the *only* information taken from defect data was the D-value at $T = 194.4$ K.

9.3.2 Lithium

The differential dilatometry technique has been applied by Feder [202] but the formation parameters extracted show a large uncertainty. Positron anihilation spectroscopy has been insensitive to vacancies, as in all alkali metals [142]. On the other hand there is a large amount of experimental data concerning self-diffusion in this material. In addition to older NMR studies of self-diffusion (D^{SD}) a number of more recent measurements of this kind on pure Li is available [203–208]. The resulting parameters have been compiled in table 9.14 and are plotted in fig. 9.15. In the same table we have also inserted the activation parameters resulting from tracer self-diffusion studies (D^T) of Lodding, Mundy and Ott [209]. The comparison of NMR to tracer studies showed that below 40°C the ratio D^T/D^{SD} approaches the value 0.727 which is just the correlation factor for a monovacancy mechanism. Above 40°C this ratio decreases continuously approaching the value of ~ 0.4 at the melting point. This experimental fact has been considered by Mehrer [62] as evidence for a strong divacancy contribution at high temperature although alternative explanations of this behaviour have been proposed [37].

We now proceed to the comparison of the above data with the predictions of the $cB\Omega$-model. Elastic data are available only up to room temperature and are given in table 9.15. A comparison of the B-values reported for 100, 150 and 295 K by various investigators shows a scatter of a few percent. We have therefore decided to use their average, which has been added to the table. The Ω-values, resulting from the density of Li reported

Table 9.14
Calculated and experimental entropies and enthalpies in lithium.

Process	h (eV)	s (k-units)	Reference
Formation:			
differential dilatometry	0.34 ± 0.04	0.9 ± 0.8	[202]
$cB\Omega$-calculation	0.36 ± 5%	1.3 ± 5%	
Activation (self-diffusion) in ^{7}Li: [a]			
NMR	0.52 ± 0.01	$1.94^{+0.16}_{-0.18}$ ($2.37^{+0.16}_{-0.18}$)	[203]
	0.52 ± 0.01	1.49 (1.91)	[204]
	0.52	–	[205]
	0.52	–	[206], [207]
	0.56 ± 0.02	–	[208]
tracer	0.56 ± 0.01	2.95 (3.38)	[209]
$cB\Omega$-calculation	0.53–0.52	1.9–1.87	

[a] The activation entropy has been extracted from the published D_0-values by using $a \simeq 3.5$ Å and $\nu \simeq \nu_D = 7 \times 10^{12}$ s^{-1}; the values given in parentheses are obtained when the value $\nu \simeq \tilde{\nu} = 4.6 \times 10^{12}$ s^{-1} (with $h^m = 0.19$ eV) is used instead of ν_D. Note that *only* in the tracer measurements f has been taken equal to 0.727.

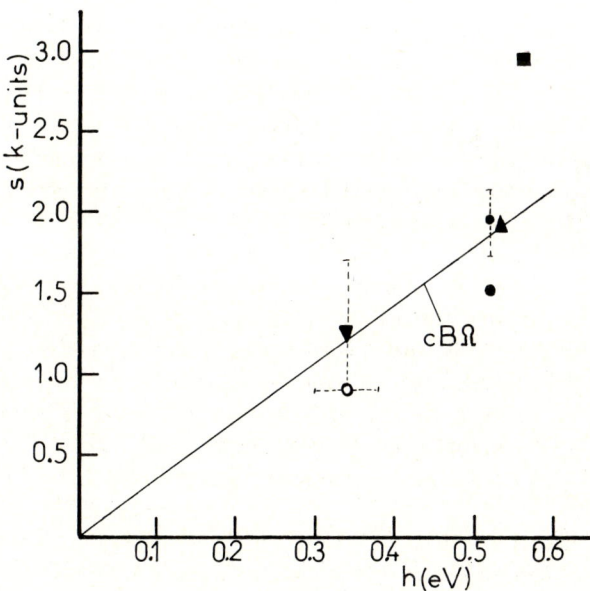

Fig. 9.15. Defect entropies and enthalpies of lithium; formation: ○: Feder [202]; ▼: $cB\Omega$-model; activation: ●: NMR [203–208]; ■: tracer [209]; ▲: $cB\Omega$-model.

Table 9.15
Experimental B-values of lithium.

T (K)	B [a] (kbar)	B [b] (kbar)	B [c] (kbar)	B_{average} (kbar)	Ω [d] (Å^3)
100	124.3 (125.3)	126.5	122	124.3	
150	122.4 (123.2)	–	121	121.7	
295	115.8 (115.7)	115	112	114.3	21.8
195				119.4 [e]	21.5

[a] Ref. [47]; the values given in parentheses correspond to ^{7}Li while the others are for ^{6}Li. In taking the averages we have used the values of ^{6}Li which however differ only slightly from the other values.
[b] Ref. [210].
[c] Ref. [211].
[d] From the density values given in ref. [47] for ^{7}Li.
[e] By linear interpolation of the "average" B-values at 150 and 295 K.

by Felice, Trivisonno and Schuele [47] have also been added. The thermal volume expansion coefficient at room temperature is around 1.4×10^{-4} K^{-1}. Two values of dB/dT have been reported; Day and Ruoff [210] suggest -6×10^{-2} kbar/K whereas Swenson [211] gives -5×10^{-2} kbar/K; in view of this small difference their mean value of -5.5×10^{-2} kbar/K will be used. Finally, Felice et al. [47] suggest that $B_0^{\text{SL}} = 129.5$ kbar. Inserting the above values into eq. (9.4) and using $\exp \int_0^{RT} \beta \, dT \simeq 1.02$ we get, at room temperature, $s/h_{\text{exp}} = 3.1 \times 10^{-4}$ K^{-1} which gives the straight line drawn in Fig. 9.15. The uncertainty is $\sim 10\%$, due to the scatter of the experimental values mentioned above. The agreement is satisfactory in view of the fact that the activation entropy has a large uncertainty and furthermore depends on the frequency factor selected for its extraction from the published D_0-values.

We shall now examine whether the $cB\Omega$-model reproduces the self-diffusion curve. For this application we start from 195 K, which is the lowest temperature at which a self-diffusion coefficient has been reported and examine if it can predict the corresponding D-values up to 295 K, which is the highest temperature at which elastic data are directly available. In the last column of table 9.16 we have inserted the experimental values of D^{SD} from 195 up to 295 K. They are calculated from the explicit formula suggested by Messer and Noack [204] as fitting the published (uncorrelated) self-diffusion data. Inserting the data of 195 K into eq. (9.7) we get the value $c^{\text{act}} = 0.3091$ when the values [47] $\nu \simeq \nu_D \simeq 7 \times 10^{12}$ s^{-1} and $a = 3.5$ Å are used; alternatively, if one inserts the value $\nu = \tilde{\nu} = 4.6 \times 10^{12}$ s^{-1} (with $h^{\text{m}} \simeq 0.19$ eV) one finds $c^{\text{act}} \simeq 0.3046$. Note that f has now been set

Table 9.16
Calculated and experimental (uncorrelated) self-diffusion in lithium.

T (K)	B [a] (kbar)	Ω [a] (Å^3)	$D^{SD}_{cB\Omega\text{-calc.}}$ [b] (cm^2/s)	$D^{SD}_{cB\Omega\text{-calc.}}$ [c] (cm^2/s)	$D^{SD}_{\text{exper.}}$ [d] (cm^2/s)
195	119.4	21.5	1.34×10^{-15}	1.34×10^{-15}	1.34×10^{-15}
220	118.1	21.57	$4.7^{+4.1}_{-1.9} \times 10^{-14}$	$4.5^{+2.8}_{-1.8} \times 10^{-14}$	4.6×10^{-14}
245	116.8	21.65	$7.8^{+4.5}_{-2.8} \times 10^{-13}$	$7.1^{+4}_{-2.6} \times 10^{-13}$	7.9×10^{-13}
270	115.6	21.73	$7.7^{+3.8}_{-3.8} \times 10^{-12}$	$6.8^{+3.3}_{-2.3} \times 10^{-12}$	8.3×10^{-12}
295	114.3	21.8	$5.2^{+2.4}_{-1.6} \times 10^{-11}$	$4.5^{+2}_{-1.4} \times 10^{-11}$	5.1×10^{-11}

[a] From linear interpolation of the experimental values B and Ω given for 150 and 295 K in table 9.15. The quantity $B\Omega$ may have an uncertainty of 2%, which reflects the "errors" given in columns 4 and 5. Attention is drawn to the point that the Ω-values of Felice et al. [47] differ from those of Feder [202]; however, in view of the fact that we allow for a ~ 2% error in $B\Omega$ this difference does not change our conclusions.

[b] From eq. (9.7) $D = a^2 \nu \exp(-c^{\text{act}} B\Omega/kT)$ by using $a = 3.5$ Å, $\nu = \nu_D = 7 \times 10^{12}$ s^{-1} and $c^{\text{act}} = 0.3091$ (see the text).

[c] From the same relation as in footnote [b] by using $\nu = \tilde{\nu} = 4.6 \times 10^{12}$ s^{-1} and $c^{\text{act}} = 0.3046$ (see the text).

[d] Extracted from the explicit formula: $D^{SD} = D_{10} \exp(-Q_1/RT)[1 + D_{21} \exp(-Q_2/RT)]$ with $D_{10} = 0.038$ cm^2/s, $Q_1 = 12$ kcal/mole, $D_{21} = 250$ cm^2/s, and $Q_2 = 4$ kcal/mole, as suggested in ref. [204].

equal to 1 because the diffusion is uncorrelated; once c^{act} is known, eq. (9.7) in conjunction with the elastic and expansivity data can give the value of D at any temperature and pressure. In table 9.16 we give the corresponding values of B and Ω that result from a linear interpolation of the experimental B (average) and Ω-values reported for 150 and 295 K. The direct insertion of these data into eq. (9.7) gives the "calculated" D^{SD}-values inserted in columns 4 and 5 of the table. In the fourth column we give the values that result from the use of the set $(\nu_D, c^{act} = 0.3091)$ whereas for those of the fifth column the set $(4.6 \times 10^{12}\ s^{-1}, c^{act} = 0.3046)$ has been employed for the sake of comparison. As we have already mentioned for Cu, the first set had to be used in view of the fact that the quantity h^m – required for the determination of $\tilde{\nu}$ – is usually unknown. The two sets give values that agree within the uncertainty quoted. A comparison with the experimental values indicates that eq. (9.7) reproduces the self-diffusion curve up to 295 K, which, as already mentioned, is the highest temperature with measured elastic data (see fig. 9.17). The question whether the $cB\Omega$-model can describe the diffusion curve up to the melting point cannot be answered due to the lack of direct elastic data in this temperature region. A measurement of B in this region is highly desirable because it would answer the question whether the observed curvature of the diffusion curve is due to an additional mechanism (e.g. divacancies) or to a nonlinear decrease of B which may occur when the melting point is approached. (After the completion of the main part of the text, B-values have been reported up to 350 K; see §A.4 and §A.11, note 17.)

We now proceed to the determination of the self-diffusion activation parameters in terms of $cB\Omega$-formulae. By inserting the appropriate data given in table 9.16 for 295 K into eqs. (9.1) and (9.2) with $c^{act} = 0.3091$ we get $h^{act} = 0.53$ eV and $s^{act} = 1.9\ k$. If one alternatively uses $c^{act} = 0.3046$ the calculated values become $h^{act} \simeq 0.52$ eV and $s^{act} \simeq 1.87\ k$. A comparison with the experimental parameters inserted in table 9.14 actually indicates a striking agreement (fig. 9.16).

We now calculate the formation parameters: At the melting point (453.4 K) the concentration $n/N = 4.5 \times 10^{-4}$ of vacancies has been measured by Feder [202]. A rough estimation of B_M at the melting point can be made by assuming that B decreases linearly with temperature up to the melting point; then by using the B-value given in table 9.16 for the temperature of 295 K and recalling that $(\partial B/\partial T)_P \approx -5.5 \times 10^{-2}$ kbar/K we get $B_M \simeq 105.6$ kbar. Furthermore, using the lattice parameter (3.50814 Å) of Feder given for 20°C and the value $\Delta L/L \simeq 45 \times 10^{-4}$ corresponding to the melting point we get $\Omega_M \simeq 21.9$ Å^3. Inserting these values into eq. (9.5) we directly get the approximate value $c^f \simeq 0.2085$. Working in a similar manner as for the activation process – but now inserting $c^f = 0.2085$ instead of c^{act}

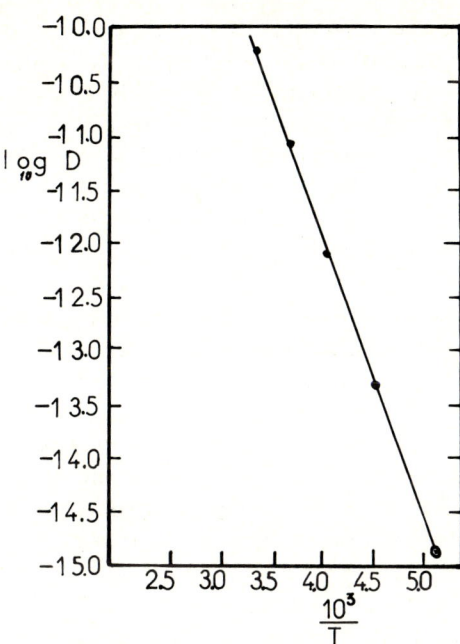

Fig. 9.16. Self-diffusion in lithium; ●: experimental values [204]; line: $cB\Omega$-calculation. The exact values are given in table 9.16.

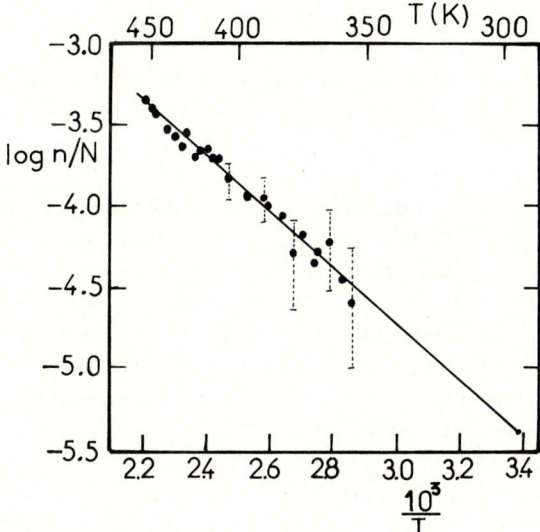

Fig. 9.17. Vacancy concentration in lithium; ●: experimental values from fig. 2 of ref. [202]; line: $cB\Omega$-model.

– we get for $T = 295$, from eqs. (9.2) and (9.1): $h^f = 0.36$ eV and $s^f = 1.3\ k$. In spite of the fact that the above estimation of c^f may have an error of at least 5% the calculated h^f, s^f values are in satisfactory agreement with those obtained from the differential dilatometry technique (table 9.14).

In fig. 9.17 we have inserted in the form of a line the values of n/N resulting from eq. (9.5) when one uses the value $c^f = 0.2085$ and assumes, as above, that the isothermal bulk modulus decreases linearly from 295 K up to the melting point. In the same figure we have inserted for the sake of comparison the experimental points directly measured by Feder [202]. For a few of them the experimental error is shown; the agreement is satisfactory when one considers the experimental error of the measured points and the approximations involved in the drawing of the line.

Summarizing the present discussion in lithium we can say that the $cB\Omega$-model reproduces the self-diffusion curve up to the temperature (295 K) to which direct elastic data are available and gives formation and activation enthalpies and entropies that are in close agreement to the experimental ones.

9.3.3 Potassium

This metal has not directly been studied by the differential dilatometry technique. An attempt was made by Pokorny [211a] to evade the lack for such experiments by combining the very accurate measurements of the lattice constant by Stetter, Adlhart, Fritsch, Steichele and Lescher [211b] with measurements of the linear thermal expansivity of the bulk material by Schouten and Swenson [215]. The result gave a concave curvature for the density of defects (in the logarithmic scale) versus temperature. This was interpreted as if both interstitials and vacancies were present. However, it is generally accepted that the necessary accuracy can only be achieved if the two types of experiments are carried out on the same sample. As potassium, furthermore, is insensitive to the positron technique, the knowledge of its vacancy formation parameters must be considered as poor. An indication of their values comes from the specific heat measurements of MacDonald [212] and Kraftmakher and Strelkov [213]. They have shown that the formation enthalpy is 0.39–0.4 eV whereas the entropy is around 2.55 k (table 9.17). On the other hand self-diffusion has been well studied by Mundy, Miller and Porte [214] with the tracer technique. The study was made in a large temperature range (220.8–334.8 K) and detected a curved diffusion plot. Although the curvature is slight and not detectable in a logarithmic plot, they determined a systematic deviation from a straight line. In view of this curvature the data were fitted by a sum of two exponentials: $D = 1.0\exp(-11220/RT) + 0.05\exp(-8890/RT)$ cm^2/s.

Table 9.17
Calculated and experimental entropies and enthalpies in potassium.

Process	h (eV)	s (k-units)	Reference
Formation	0.39–0.4	~ 2.55	[212,213]
Activation:			
tracer self-diffusion: low T	0.386	2.56	[214] [a]
$cB\Omega$-calculation: low T	0.382	2.52	
$cB\Omega$-calculation: high T	0.428	4.32	

[a] Low temperature values (see text).

An inspection of this formula indicates that in the low temperature range only the second term is present; its preexponential factor (0.05) and exponent indicate that in this temperature range the activation enthalpy and entropy are 0.386 eV and 2.56 k when the values $f = 0.727$, $a = 5.3$ Å and $\nu_D = 1.9 \times 10^{12}$ s^{-1} are employed. These values correspond to the point inserted in fig. 9.18. Close to the melting point the first term is 60% of the second and corresponds to an enthalpy of 0.49 eV and the entropy 5.5 k. It is now obvious that if the first term actually describes an additional mechanism the values 0.49 eV and 5.5 k are characteristic of this mechanism; on the other hand, if solely one mechanism is responsible for the diffusion process then it must have – close to the melting point – parameters that are smaller than 0.49 eV and 5.5 k but, of course, higher than those (0.386 eV, 2.6 k) corresponding to the low temperature range. The situation is analogous to that of sodium.

We now turn to the comparison of the above data with the prediction of the $cB\Omega$-model. Fortunately expansivity data have been given by Schouten and Swenson [215] and elastic data by Fritsch and Bube [216] up to the melting point. These data are shown in table 9.18 at those temperatures at which Mundy et al. have measured the (tracer) diffusion coefficients. The values given have been extracted by interpolation from ref. [215] where B^S, β, C_P and (the molar volume) V are listed for a large number of temperatures. The conversion of B^S to B (up to 288 K) has been made by means of the standard thermodynamical formula and using the specific heat and expansivity data inserted in the table. At temperatures higher than 288 K the B-values have been reported in ref. [216] to have an uncertainty of ± 0.7 kbar.

We start with the calculation of s/h_{exp} at 220.8 K. The low temperature data inserted in the table indicate that $(\partial B/\partial T)_P \simeq -0.028$ kbar/K and $B_0^{\text{SL}} \simeq 39$ kbar. By using also in eq. (9.4) the values $\beta = 2.052 \times 10^{-4}$ K^{-1}, $B = 32.68$ kbar and $\exp \int_0^{220.8} \beta \, dT \simeq 1.02$, we get $s/h_{\text{exp}} = 5.57 \times 10^{-4}$ K^{-1}

Table 9.18
Experimental and calculated (tracer) self-diffusion in potassium.

T (K)	V^a (cm³/mole)	C_P^a (J/mole K)	β^a (10^{-4} grad^{-1})	$B^{S\,a}$ (kbar)	B^b (kbar)	$D_{cB\Omega}$-calc. (cm²/s)	$D_{exper.}^c$ (cm²/s)
220.8	44.787	27.386	2.052	34.39	32.68	9.02×10^{-11}	$9.02 \pm 0.2 \times 10^{-11}$
220.8	44.787	27.386	2.052	34.39	32.68		$8.57 \pm 0.2 \times 10^{-11}$
229.7	44.872	27.582	2.080	34.25	32.45	$2.0^{+0.8}_{-0.6} \times 10^{-10}$	$1.92 \pm 0.03 \times 10^{-10}$
237.1	44.943	27.745	2.103	34.14	32.27	$3.6^{+1.4}_{-1.1} \times 10^{-10}$	$3.77 \pm 0.05 \times 10^{-10}$
237.5	44.947	27.754	2.105	34.14	32.26	$3.8^{+1.3}_{-1.1} \times 10^{-10}$	$4.07 \pm 0.02 \times 10^{-10}$
246.4	45.035	27.957	2.193	34.04	31.96	$7.5^{+2.7}_{-2} \times 10^{-10}$	$7.73 \pm 0.07 \times 10^{-10}$
255.8	45.128	28.174	2.233	33.94	31.75	$1.4^{+0.5}_{-0.35} \times 10^{-9}$	$1.51 \pm 0.03 \times 10^{-9}$
273	45.304	28.655	2.282	33.71	31.33	$4.1^{+1.3}_{-1} \times 10^{-9}$	$4.82 \pm 0.10 \times 10^{-9}$
288	45.460	29.135	2.338	33.52	30.97	$0.91^{+0.27}_{-0.21} \times 10^{-8}$	$1.15 \pm 0.01 \times 10^{-8}$
304.7	45.641	29.839	2.412		30.2 ± 0.7	$2.5^{+0.8}_{-0.6} \times 10^{-8}$	$2.97 \pm 0.02 \times 10^{-8}$
319.8	45.810	30.743	2.492		29.5 ± 0.7	$5.5^{+1.6}_{-1.3} \times 10^{-8}$	$6.64 \pm 0.07 \times 10^{-8}$
320.2	45.814				29.5 ± 0.7	$5.5^{+1.7}_{-1.3} \times 10^{-8}$	$6.58 \pm 0.1 \times 10^{-8}$
334.8	~45.979				28.9 ± 0.7	$1.1^{+0.4}_{-0.25} \times 10^{-7}$	$1.27 \pm 0.01 \times 10^{-7}$

[a] Ref. [215].
[b] The B-values up to 288 K result from ref. [215] by applying the thermodynamic correction to the B^S-values inserted in the preceding column. From there on the B-values are taken from fig. 2 of ref. [216].
[c] ref. [214].

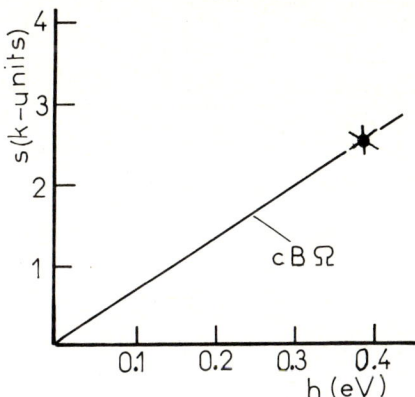

Fig. 9.18. Low-temperature activation entropy and enthalpy in potassium; ●: Mundy et al. [214]; it coincides with the formation parameter (see text); ∗: $cB\Omega$-calculation; line: $cB\Omega$-model.

which gives the straight line drawn in fig. 9.18. We see that it describes very well the low temperature activation parameters of Mundy et al. Furthermore, it describes the formation parameters mentioned above; they lead to a point that coincides with that shown in the figure for the activation parameters.

We proceed to the computation of the diffusion curve from the $cB\Omega$-model. We start from the value $D = 9.02 \times 10^{-11}$ cm^2/s reported by Mundy et al. for the lowest temperature 220.8 K of their measurements. Introducing this value into eq. (9.7) and using $\nu = \nu_D$ and $f = 0.727$ as above, we get $c^{act} = 0.22$. This value of c^{act} in conjunction with the isothermal elastic data and the values of Ω given in table 9.18 leads, with the help of eq. (9.7) to the calculated D-values inserted in the table. For each temperature up to 288 K we give the uncertainty resulting from an experimental error of 2% in the value of $B\Omega$; at higher temperatures we give only the uncertainty resulting from the experimental error quoted in ref. [216]. A comparison of the calculated D-values to the experimental ones (fig. 9.19) indicates an agreement up to the melting point, if one also considers the uncertainties given. This is quite interesting because the diffusion plot shows a curvature which is *totally* accounted for by the $cB\Omega$-model, although we have started the calculation from the lowest temperature at which "the second mechanism" was absent. The fact that the $cB\Omega$-model predicts the curvature is also apparent from the following example in which we see that it predicts values of h^{act} and s^{act} that increase with temperature.

The high temperature B-values show that $(\partial B/\partial T)_P \simeq -0.043$ kbar/K. Applying now eqs. (9.1) and (9.2) we get for 220.8 K: $h^{act} \simeq 0.382$ eV and

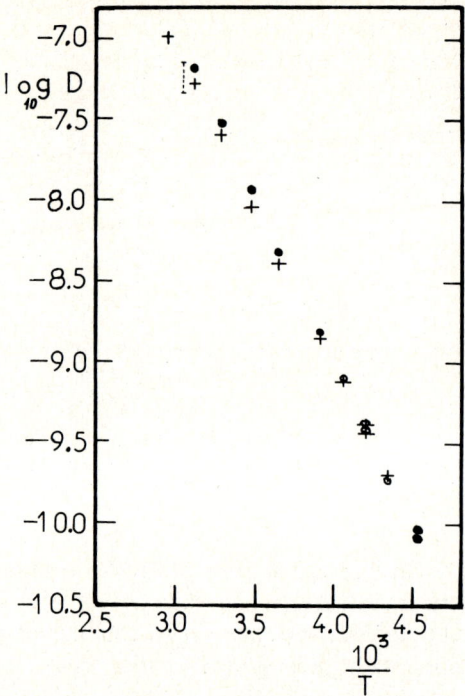

Fig. 9.19. Self-diffusion in potassium. ●: experimental. +: $cB\Omega$-model; error bars indicated. The exact values are given in table 9.18.

$s^{act} \simeq 2.52\ k$, and for 319.8 K: $h^{act} \simeq 0.428$ eV and $s^{act} \simeq 4.32\ k$. We observe that the low temperature parameters coincide with those of Mundy et al. [214] in the same temperature range. Furthermore the calculated values at 319.8 K are actually between the bounds expected from the analysis of the data of Mundy et al. in terms of two exponentials.

The present example of potassium provides one of the most strict checks of the validity of the $cB\Omega$-model. Remember that potassium is a highly compressible material having a bulk modulus two orders of magnitude smaller than that of tungsten which is also a bcc metal. The present example shows that when precise elastic and expansivity data are available up to the melting point, these data can lead to a reproduction of the whole self-diffusion curve by means of the $cB\Omega$-model, without using any adjustable parameter.

9.3.4 Tungsten

Differential dilatometry experiments have not yet been performed on this

Table 9.19
Experimental and calculated entropies and enthalpies in tungsten.

Process	h (eV)	s (k-units)	Reference
Formation:			
quenching	3.67 ± 0.2	2.3 (2.4 ~ 3.4)	a a
positron spectroscopy	4.0 ± 0.3	–	b
$cB\Omega$-calculation	3.85 ~ 3.4	3.2 ~ 2.8	
Migration (quenching)	1.78 ± 0.1	–	a
Activation (self-diffusion)	5.45	3	c
	5.82	4.75 ~ 5.82	d
	~ 5.2	4.6	[221]
	~ 6.1	5.86	e
$cB\Omega$-calculation	5.88	4.88	

[a] Ref. [218]; quenching experiment in conjunction to transmission electron microscopy studies; see also the text for the entropies given in parentheses.
[b] Ref. [217]; positron annihilation technique. Earlier PAS studies have led to a value of ~ 3.5 eV (see the references in [217]).
[c] Ref. [220] the values given correspond to an analysis in terms of two exponentials.
[d] From the data of ref. [220] by analysing them in terms of one mechanism (see the text).
[e] Ref. [222]; the entropy (from refs. [221] and [222]) has been extracted by taking $\nu = \nu_D$ (see the text).

metal (T_M = 3483 [217] ~ 3495 K [218]). Quenching-in resistivity experiments in conjunction to transmission electron microscopy investigations have reached a satisfactory degree of perfection and have been reported by Rasch, Siegel and Schultz [218]. However, as they mention, some reservation should be made on the accuracy of their formation parameters (given in table 9.19 and plotted in fig. 9.20) because the temperature calibration of the tungsten resistivity is at present not unique. Furthermore, Maier, Peo, Saile, Schaefer and Seeger [217] have reported a rather complete study for various bcc refractory metals by means of the positron annihilation technique. Gupta and Siegel [219] have suggested that some aspects of the analysis of Maier et al. should be reconsidered. However, it should be noticed that the PAS-formation enthalpy [217] (4.0 ± 0.3 eV) in W is in reasonable agreement with that (3.67 ± 0.2 eV) deduced from "quenching" studies [218]. For the formation entropy no accurate data are at present available apart of an estimate given by Rasch et al. [218].

Self-diffusion studies have been completed by the detailed measurements of Mundy, Rothman, Lam, Hoff and Nowicki [220] in an extensive temperature range (1700–3400 K); these workers have observed a curved plot in

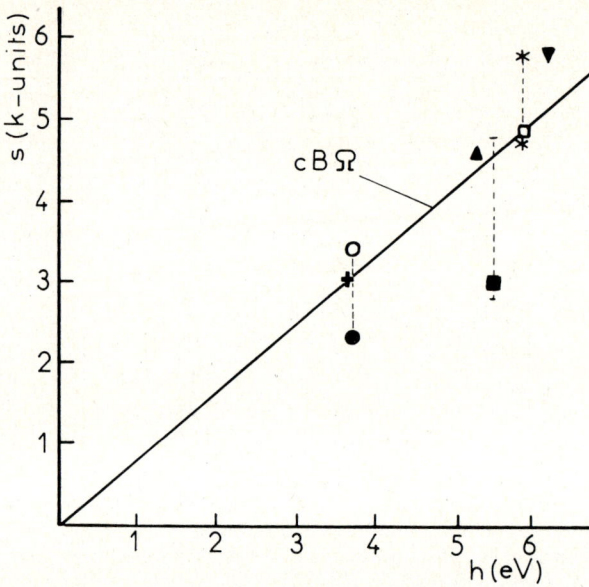

Fig. 9.20. Defect entropies and enthalpies of tungsten; formation: ●: Rasch et al. [218]; ○: extracted from data of Rasch et al. [218]; +: mean value $cB\Omega$-calculation; activation; ▲: Danneberg [221]; ▼: Korolev et al. [222]; ■: Mundy et al. [220]; *: extracted from Mundy et al. [220]; □: $cB\Omega$-model. Line: $cB\Omega$-model.

contrast to previous suggestions. Mundy et al. have described their data in terms of two exponentials but they also noticed that the statistical fit of their data was equally good when analysed in terms of a single mechanism with temperature dependent parameters. In table 9.20 we reproduce the directly measured values of the (tracer) diffusion coefficient at some temperatures along with the values computed from their formula:

$$D = D_1 + D_2 = 0.04 \exp(-5.54 \text{ (eV)}/kT)$$
$$+ 46 \exp(-6.9 \text{ (eV)}/kT) \text{ (cm}^2 \text{ s}^{-1}).$$

According to them, the first term describes the contribution of monovacancies and the preexponential factor $0.04^{+0.17}_{-0.1}$ gives $s^{\text{act}} = 3$ k shown in fig. 9.20 when the frequency factor $\nu = \tilde{\nu} = 2.5 \times 10^{12}$ s^{-1} (with $h^m = 1.8$ eV) is used. It should be stressed however that when their data are analysed within the frame of a single mechanism the resulting entropy value (even in their low temperature range) must be higher by ~ 2 k – or even more – depending on the frequency factor which is employed. In the same figure we have inserted the activation parameters resulting from the data of Danneberg [221] and those of Korolev, Pavlinov and Gavrilyuk [222]; the activation entropy has been extracted from their D_0-values by assuming that

Table 9.20
Calculated and experimental self-diffusion coefficients in tungsten

T (K)	B (kbar)	Ω (Å^3)	$D_{\text{calc.}}$ (cm^2/s)	$D_{\text{exper.}}^{\text{a}}$ (cm^2/s)	$D_{\text{exper.}}^{\text{b}}$ (cm^2/s)	D_1^{b} (cm^2/s)	D_2^{b} (cm^2/s)
1705	2682	16.124	3.17×10^{-18}	3.17×10^{-18}	3.34×10^{-18}	3.15×10^{-18}	1.88×10^{-19}
1809.6	2654	16.16	$3.2^{+2.8}_{-1.6} \times 10^{-17}$	3.89×10^{-17}	2.97×10^{-17}	2.69×10^{-17}	2.84×10^{-18}
1943.6	2615	16.20	$4.4^{+3.5}_{-2} \times 10^{-16}$	4.30×10^{-16}	3.59×10^{-16}	2.99×10^{-16}	6.0×10^{-17}
2078.9	2587	16.25	$3.9^{+2.8}_{-1.7} \times 10^{-15}$	2.70×10^{-15}	3.35×10^{-15}	2.48×10^{-15}	8.75×10^{-16}
2202.7	2545	16.29	$2.7^{+1.8}_{-1.1} \times 10^{-14}$	1.85×10^{-14}	2.13×10^{-14}	1.37×10^{-14}	7.62×10^{-15}
2315.2	2521	16.34	$11.1^{+7}_{-4.3} \times 10^{-14}$	8.78×10^{-14}	9.99×10^{-14}	5.53×10^{-14}	4.45×10^{-14}
2630.2	2418	16.46	$4.4^{+2.24}_{-1.5} \times 10^{-12}$	$2.39 \sim 7.22 \times 10^{-12}$	4.25×10^{-12}	1.45×10^{-12}	2.8×10^{-12}
2895.2	2343	16.57	$4.8^{+2}_{-1.5} \times 10^{-11}$	3.99×10^{-11}	5.84×10^{-11}	1.31×10^{-11}	4.53×10^{-11}
3099.1	2299	16.68	$2^{+0.8}_{-0.58} \times 10^{-10}$	2.51×10^{-10}	3.34×10^{-10}	5.5×10^{-11}	2.79×10^{-10}
3199.8	2277	16.73	$3.8^{+2.4}_{-1.5} \times 10^{-10}$	6.32×10^{-10}	7.3×10^{-10}	1×10^{-10}	6.3×10^{-10}
3242.3	2268	16.75	$5^{+3}_{-2} \times 10^{-10}$	1.2×10^{-9}	1.0×10^{-9}	1.4×10^{-10}	8.74×10^{-10}
3408.9	2223	16.81	$1.4^{+0.8}_{-0.5} \times 10^{-9}$	3.08×10^{-9}	3.27×10^{-9}	3.5×10^{-10}	2.9×10^{-9}
(3493?)	(2090)	(16.99)					

[a] Directly measured values.
[b] From the fitting formula given by Mundy et al. [220]; see text.

$\nu = \nu_\mathrm{D} = 7.27 \times 10^{12}$ s^{-1} (i.e. $\Theta_\mathrm{D} = 348$ K [150]) as usual. Note that the latter parameters should be more representative than the "monovacancy parameters" of Mundy et al. for a single mechanism process as their D_0-values have been directly determined from a least squares fitting to a straight line of their data. These values also seem to be consistent with the D-value calculated from Mundy's et al. [220] data for 1705 K and 2500 K, which give a slope close to 5.82 eV and $D_0 = 0.618$ cm^2/s. This D_0-value leads to $s^\mathrm{act} = 4.75\,k$ or $5.82\,k$ if the values ν_D or $\tilde{\nu}$ respectively are used. These values have been inserted as asterisks in fig. 9.20 for the sake of comparison.

We further examine whether the above results are in accord with the $cB\Omega$-model. The adiabatic bulk modulus has been directly measured only up to 2073 K by Lowrie and Gonas [223] and expressed with a polynomial. We shall assume that their formula is correct up to 3400 K. In order to convert the adiabatic values to the isothermal ones we use the specific heat data of Morizur, Radenac and Cretenet [224] which unfortunately extend only up to 2800 K. The thermal expansion coefficient has been taken from the Metals Handbook [225] but again the formula given does not hold at very high temperatures. Therefore, in the conversion we used a linear extrapolation of the C_P and β-values up to 3400 K. The B-values obtained in this way have been inserted in table 9.20 along with the corresponding Ω-values; the results corresponding to temperatures appreciably higher than 2800 K have been separated in order to emphasize that they are only rough estimates.

We start with the calculation of s/h_exp; for $T = 1705$ K the data mentioned above give $B = 2682$ kbar, $\beta = 18 \times 10^{-6}$ K^{-1}, $(\partial B/\partial T)_P = -0.268$ kbar/K, $B_0^\mathrm{SL} = 3140$ kbar and $\exp \int_0^{1705} \beta \mathrm{d}T \simeq 1.024$. Inserting these values into eq. (9.4) we get $s/h_\mathrm{exp} = 7.2 \times 10^{-5}$ K^{-1} which leads to the straight line drawn in fig. 9.20. In spite of the scatter we see that the general trend of the experimental data is consistent with the expectation of the $cB\Omega$-model. The small deviation observed in the formation parameters is not without explanation. Transmission electron microscopy studies [218] showed that the concentration of vacancies at the melting point 3695 K is $2 \pm 1 \times 10^{-4}$. Applying eq. (9.5) we find that g^f should lie between 2.934 eV and 2.584 eV (the first corresponds to $n/N = 10^{-4}$ and the other to 3×10^{-4}). The Arrhenius plot of the quenched-in resistivity has shown that $h^\mathrm{f} = 3.67 \pm 0.2$ eV. Combining this h^f-value with the values of g^f reported above we find that s^f should lie between 2.4 and 3.4 k and that therefore the mean value $s/h = 6.8 \times 10^{-5}$ K^{-1} should result. Rasch et al. [218] have reported only the smallest s^f-value because they used the lower limit $n/N = 10^{-4}$ although the value 3.4 k is equally acceptable. We see now that the average of the above s/h-values coincides with the prediction of the $cB\Omega$-model.

We now calculate each formation parameter separately from the $cB\Omega$-formulae. In order to determine c^f we need a single value of n/N, e.g. that measured at the melting point. Inserting the above n/N-value into eq. (9.5) we get $c^f = 0.125$ (when $n/N = 10^{-4}$) or $c^f = 0.11$ when $n/N = 3 \times 10^{-4}$. Using the elastic and the expansivity data mentioned above for $T = 1705$ K we get from eqs. (9.1) and (9.2): $h^f = 3.85$ or 3.4 eV and $s^f = 3.2$ or 2.8 k. The first of the values corresponds to $n/N = 10^{-4}$, the second to $n/N = 3 \times 10^{-4}$. These parameters are really in excellent agreement with the experimental ones. However, one should recall that they are derived by assuming a linear extrapolation of B and Ω (these values are set in table 9.20 between parentheses). As we expect B to decrease nonlinearly, close to the melting point the values h^f and s^f might lie somewhat higher.

The calculation of the activation self-diffusion parameters from the $cB\Omega$-formulae can be done in a similar fashion. In order to determine c^{act} we need a single diffusion measurement, e.g. at $T = 1705$ K, for which Mundy et al. have reported the value: $D = 3.17 \times 10^{-18}$ cm^2/s. By inserting it into eq. (9.7) we get $c^{act} = 0.19078$. For its calculation we have used $a = 3.183$ Å, $f = 0.727$ and $\nu = \nu_D$. If alternatively one uses $\nu = \tilde{\nu}$ one gets $c^{act} = 0.185$. Using the same procedure as in the formation process (i.e. if we use c^{act} instead of c^f) we find: $h^{act} = 5.88$ eV and $s^{act} = 4.88$ k, in agreement with all the experimental values inserted in fig. 9.20 except for the "monovacancy values" (5.45 eV, 3 k) of Mundy et al. which turn into 5.8 eV, 4.75–5.82 k when their data are analysed within the frame of a single mechanism.

We finally proceed to the computation of the self-diffusion curve within the frame of the $cB\Omega$-model. We have already used a single D-measurement, i.e. that of the lowest temperature (1705 K) of Mundy et al. [220] and obtained $c^{act} = 0.19078$. Inserting this value into eq. (9.7) the computation of D for the other temperatures is straightforward; the results are given in table 9.20. We have allowed for a 2% uncertainty in the value of B and calculated the resulting error bars of D_{calc}. Only at the higher temperatures – separated by the blank line in the table – have we allowed for a 3% uncertainty (see below). In the same table we have intentionally reproduced the two contributions D_1 and D_2 which – according to Mundy et al. – might correspond to two mechanisms. An inspection of the table up to 2895.2 K – where, as mentioned above, the elastic and expansivity data can be regarded as reliable – indicates that the calculated values agree with the experimental ones if the experimental errors are considered. Note that in this range the value of D already varies by *seven* orders of magnitude (fig. 9.21).

At this point the following comment can be made: the D-values up to 2895.2 K have succesfully been reproduced within the frame of the $cB\Omega$-

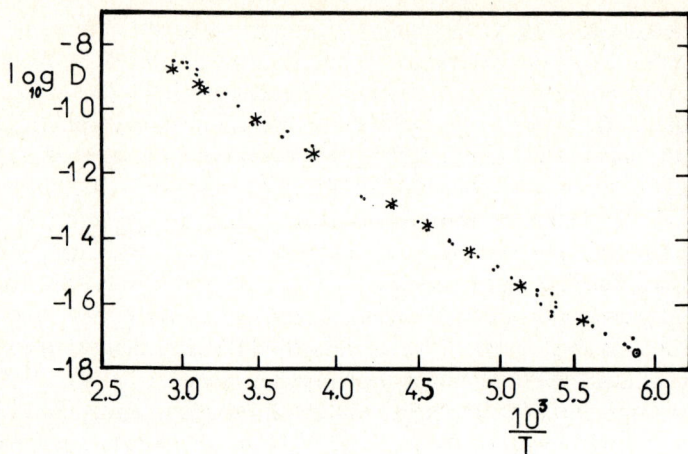

Fig. 9.21. Self-diffusion in tungsten. $*$: calculated values ($cB\Omega$); $\bullet$: experimental values (table 9.20).

model from the c^{act}-value determined from 1705 K without having to consider a second mechanism operating. On the other hand according to the analysis of Mundy et al., which considers two mechanisms, the calculation of the two contributions D_1 and D_2 given in the same table indicates that at 1705 K only the D_1-mechanism is operating whereas at 2895.2 K the D_2-mechanism dominates. It therefore seems that the curvature can be described by one mechanism solely from the elastic and expansivity data.

We finally turn to the inspection of temperatures higher than 2895.2 K. Although up to 3200 K the calculated values agree with the experimental ones within the error bars, at the two highest temperatures they differ by ~ 40% at least. This is not unexpected; in this temperature range β and C_P should show a strong upward curvature whereas in our calculation we have assumed only a linear increase. Such an upward curvature will create an excessive decrease of B – according to the thermodynamical relation $B = B^S/(1 + \gamma\beta T)$ – even if B^S continues to decrease linearly up to the melting point and γ does not increase in this temperature range (one also expects that B^S will show an excessive decrease). Such a nonlinear decrease of B will result in an increase of the calculated D-value, thus bringing, in any case, better agreement with the experimental data.

Summarizing the present considerations on tungsten we can say that the $cB\Omega$-model reproduces successfully – *without* any adjustable parameter – the self-diffusion data up to the temperature to which precise elastic and expansivity data are available. It further leads to formation and activation parametes that are in favourable agreement with the existing experimental estimates.

9.3.5 Niobium

Although self-diffusion in Nb has often been measured [226] we shall restrict the present discussion to the experiments of Einziger, Mundy and Hoff [227] because they cover temperatures in a very extended range (1354 to 2695 K). The adiabatic elastic constants at high temperature have been measured by Talmor, Walker and Steinemann [228]. We have extracted values from their fig. 2 and corrected them in order to obtain the isothermal bulk modulus. The correction needed specific heat data [225] and lattice constants [225,229]. In the absence of suitable high temperature data the trend of C_P and the thermal expansion coefficient was linearly extrapolated. In the preexponential factor we set $f = 0.727$ and introduce $\nu = \nu_D$ corresponding to a Debye temperature of 275 K. The data at the lowest temperature (1354 K) were taken as a starting point. Inserting the corresponding [227] values $T = 1354$ K, $D = 3.84 \times 10^{-16}$ cm^2/s, $B = 1496.7$ kbar and $\Omega = 18.41 \times 10^{-24}$ cm^3 one gets $c^{act} = 0.2042$. The diffusion coefficient can then be calculated for any higher temperature by inserting the necessary values of B and Ω into eq. (9.7). The result is shown in fig. 9.22 as a full line. In the same figure the experimental values of Einziger et al. [227]

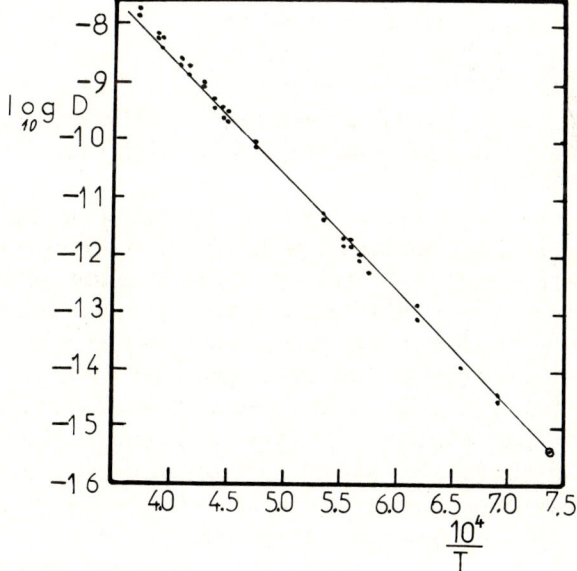

Fig. 9.22. Self-diffusion in niobium (tracer); ●: experimental values from ref. [227]; line: $cB\Omega$-model.

are shown as dots. One sees that at the highest temperature (2695 K), although the diffusion coefficient is nearly *8 orders* of magnitude larger, its calculated value 0.931×10^{-8} cm^2/s, is *only* 16% below one of the experimental points (1.13×10^{-8} cm^2/s). This is a very satisfactory agreement, because, although the elastic constants have been measured up to the highest temperature, they usually have an error of a few percent, and furthermore, the diffusion experiments indicate the possibility of errors in the temperature. A change of only 1% in the bulk modulus and of 8 K in the temperature is sufficient to bring $D_{2695\,K}$ very near to the mean experimental value.

9.4 Hexagonal and tetragonal metals

In this paragraph we discuss one hexagonal (Zn) and one tetragonal (Sn) crystal for both of which vacancy formation parameters have been determined by similatenous measurements of length and lattice parameters in two directions. Due to the anisotropy of the diffusion coefficient in different directions these materials provide an opportunity to examine how far the proposed connection between g^{act} and $B\Omega$ correctly describes the temperature variation of the diffusion coefficient for each direction separately.

9.4.1 Zinc

For this hexagonal crystal the length and the lattice parameter changes along the *a*- and the *c*-axis have been recently measured by Balzer and Sigvaldason [230]. The measurements extend from 296.1 K up to 683 K (i.e. only ten degrees lower than the melting temperature 692.4 K) where the concentration of monovacancies was found to be $4.9 \pm 0.3 \times 10^{-4}$. The formation parameters obtained from these data have been inserted in table 9.21. In the same table we have inserted the parameters given by an earlier study of Current and Gilder [231]; they performed only lattice parameter studies – along both axes – and have combined their data with earlier length measurements [232]. This combination led to a somewhat lower vacancy concentration at the melting point; furthermore, this analysis gave a comparable formation enthalpy to that deduced by Balzer and Sigvaldason but an appreciably lower formation entropy. Pathak and Desai [232a] obtained a value $h^f = 0.47$ eV by using only the temperature variation of the X-ray lattice parameter. For reasons discussed in ref. [230] we shall consider in the following (see fig. 9.23) only the formation parameters published by Balzer and Sigvaldason. Note that the positron technique (see table 9.21)

Table 9.21
Entropies and enthalpies in zinc.

Process	h (eV)	s (k-units)	Reference
Formation:			
differential dilatometry	0.53 ± 0.07	1.1 ± 1	a
	(0.52 ± 0.05)	(0.5 ± 0.3)	b
positrons technique	0.53 ± 0.02	–	[142]
$cB\Omega$-calculation	0.555	1.69	
Activation (tracer self-diffusion)			
c-axis	0.994 ± 0.0065	2.63	c
from self-diffusion coefficient D_h	1.019 ± 0.0065	3.53	c, d
$cB\Omega$-calculation (c-axis)	0.965	2.94	
$cB\Omega$-calculation for self-diffusion coefficient D_h	0.995	3.03	

[a] Ref. [230]; these values are preferable above those of ref. [231].
[b] Ref. [231]; these values are included in parentheses for the reasons explained in the text.
[c] Ref. [234]; see text.
[d] Ref. [238]; these workers have analysed the data of ref. [234] and gave $D_0 = 0.186^{+0.037}_{-0.031}$ and the enthalpy value given in the table. The entropy is extracted from the D_0-value and approximating the quantity $\frac{3}{2}f_3 a^2 \nu_h$ by $f_3 \approx 0.75$ (which may differ from the real one only by some percents), $\nu_h = \nu_D$ and taking $a = 2.6648$ Å from ref. [230].

gave an h^f-value in reasonable agreement with the values reported from differential dilatometry studies. The "quenching studies" by Simon, Vostry, Hillairet and Vajda [233] have led to a formation enthalpy of 0.45 ± 0.05

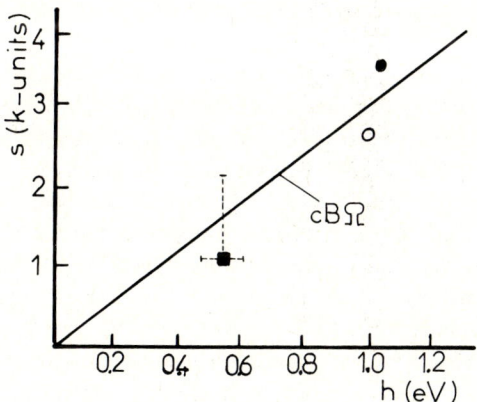

Fig. 9.23. Entropies and enthalpies of zinc; ■: formation, Balzer and Sigvaldason [230]; ○: self-diffusion along c-axis, Peterson and Rothman [234]; ●: from self-diffusion coefficient D_b; line: $cB\Omega$-model.

eV; this value has not been inserted into table 9.21 in spite of the fact that it agrees within the error bars with those obtained by the other techniques. As Balzer and Sigvaldason suggested probably "cold-work" problems have here caused a slight reduction of the h^f-value.

The (tracer) self-diffusion and the isotopic effect have been extensively studied by Peterson and Rothman [234] in addition to earlier studies of Shirn, Wajda and Huntington [235]. In the following we shall only consider the values of Peterson and Rothman as the most recent but one should notice that they are in rough agreement with the earlier data. The self-diffusion in Zn is anisotropic and the measured values D_c and $D_{\perp c}$ refer to diffusion along the c-axis and perpendicular to it. According to Buescher, Gilder and Shea [236] the diffusion along the c-axis is described by:

$$D_c = \tfrac{3}{4}c^2 f_1 \nu_c \exp(-g_c/kT), \tag{9.13}$$

where f_1, ν_c and g_c are the correlation factor, attempt frequency and activation Gibbs energy for the jumps along c-axis and c is the corresponding lattice parameter. Although f_1 depends only slightly on temperature [234] we shall take the value $f_1 = 0.75$ whereas for ν_c we shall take $\nu_c = \nu_D = 6.626 \times 10^{12}$ s^{-1}; the latter value comes from $\Theta = 328$ K as recommended by Ledbetter [237]. Using $c = 4.9456$ Å [230] we find for the preexponential factor 9.4×10^{-3} cm^2 s^{-1}. Combining this value with $D_{0,c} = 0.13$ cm^2/s, published by Peterson and Rothman, we get $s^{act} = 2.63\ k$ with a plausible uncertainty of $1\ k$ due mainly to the arbitrary selection of ν_c.

The value of $D_{\perp c}$ has two components arising from jumps along the basal plane (D_b) and from the horizontal component of the jumps made out of the basal plane. The first component is given by [236]

$$D_b = \tfrac{3}{2}a^2 f_3 \nu_b \exp(-g_b/kT), \tag{9.14}$$

where f_3, ν_b and g_b stand for the correlation factor, attempt frequency and activation Gibbs energy for the diffusion purely on the basal plane (the other component has a different correlation factor f_2). Component D_b has been extracted by Chhabildas and Gilder [238] from the $D_{\perp c}$-values measured by Peterson and Rothman [234] and is described with an activation enthalpy of 1.019 ± 0.0065 eV and a preexponential factor $D_{0,b} = 0.186 \pm 0.04$ cm^2/s. The latter value gives $s^{act} = 3.53 \pm 0.2\ k$ (see table 9.21 and fig. 9.23) when combined with $a = 2.6648$ Å, $\nu_b = \nu_D$ and $f_3 = 0.75$. The total diffusion $D_{\perp c}$ measured perpendicularly to the c-axis is therefore described by a sum of two exponentials – with nearly equal exponents – and hence we shall not further discuss it in our applications; this is intentionally done because if D_c and D_b are correctly described from the $cB\Omega$-model the coefficient $D_{\perp c}$ has also to be correct.

In table 9.22 we give the necessary expansivity and elastic data required

Table 9.22
Self-diffusion and various properties of zinc.

T (K)	C_p[a] (J/mole K)	a[b] (Å)	c[b] (Å)	Ω[b] (Å^3)	β[b] (10^{-6} grad^{-1})	B^S[c] (kbar)	B[d] (kbar)	$D_c^{\text{exper.}}$[e] (cm^2/s)	$D_b^{\text{exper.}}$[e] (cm^2/s)	c_c^{act}	c_b^{act}
512.1	28.25	2.673	5.012	15.506	~98	580	530	5.9×10^{-11}	1.76×10^{-11}	0.163	0.168
513.5	28.27	2.674	5.013	15.521	~98	580	530		1.87×10^{-11}		0.168
655.5	30.99	2.682	5.054	15.742	~104	559	498.5	6.12×10^{-9}	2.74×10^{-9}	0.164	0.167
670.1	31.40	2.683	5.058	15.766	~105	557	495.4	9.04×10^{-9}	4.06×10^{-9}	0.164	0.167
684.3	31.85	2.684	5.061	15.787	~106	555	492.3	1.29×10^{-8}	5.86×10^{-9}	0.164	0.167
691	32.12	2.684	5.063	15.793	~106	554	491.4	1.55×10^{-8}	6.92×10^{-9}	0.164	0.167

[a] Kramer and Nölting [545]; the detailed data have been kindly forwarded to the authors by Prof. J. Nölting.
[b] Ref. [231]; the values of a and c (and hence Ω) are obtained by interpolation of these data. The values of a and c are only rough estimates from their data and may be in error by some percent (they correspond to the lattice expansivity and not to the macroscopic one); an accurate extraction of β is not necessary because it is needed only for the conversion of B^S to B and the B^S-value has already an experimental error of the order of ~1%.
[c] Ref. [230]; by analysing the high temperature adiabatic elastic data of Alers and Neighbours [543] they gave the values $B_0^{SL} = 655$ kbar and $dB^S/dT = -0.146$ kbar/K; the latter values give those inserted in the table.
[d] From the B^S-values by applying the standard thermodynamical correction and using the specific heat and the expansivity data given in the table.
[e] direct experimental values (tracer technique) reported in ref. [234]; the values of D_b from the measured values D_a, and D_c have been determined with the parameters reported in ref. [238].

for the application of the $cB\Omega$-formulae. The conversion of B^S to B has been done in the standard thermodynamical manner by using the C_P and β-values inserted in the table. Using the data of 655.5 K and considering that $(dB/dT)_P \simeq -0.21$ kbar/K we find from eq. (9.3): $s/h = 2.63 \times 10^{-4}$ K^{-1} with a plausible uncertainty of $\sim 5\%$; this value gives the straight line drawn in fig. 9.23 and agrees with the experimental points of one considers the experimental error.

We shall now examine whether the temperature dependence of D_c and D_b obeys the $cB\Omega$ scheme. As we have already mentioned one can determine c_c^{act} (resp. c_b^{act}) from a single D_c (resp. D_b) experimental value by applying

$$D_c = \tfrac{3}{4} c^2 f_1 \nu_c \exp\left(-c_c^{act} B\Omega/kT\right), \tag{9.15}$$

respectively

$$D_b = \tfrac{3}{2} a^2 f_3 \nu_b \exp\left(-c_b^{act} B\Omega/kT\right). \tag{9.16}$$

Once c_c^{act} (resp. c_b^{act}) is known, D_c (resp. D_b) can be computed at any temperature with the help of elastic and expansivity data given in the table. That this method can adequately describe the diffusion curves is obvious from the following strict check: inserting the experimental D_c and D_b values into eqs. (9.15) and (9.16) we have calculated the values of c_c^{act} and c_b^{act} for various temperatures; the results are given in table 9.22. If one takes into account that the B-values used in our calculation have an experimental uncertainty of at least 1% an inspection of these results indicates that c_c^{act} (resp. c_b^{act}) actually remains constant in the whole temperature range studied up to date, although the diffusivity changes by 500. This method of checking the validity of the $cB\Omega$-model is unambiguous as it directly proves the basic interconnection of the expansivity and elastic data with the measured values D_a and D_b without involving any adjustable or empirical parameter. If the $cB\Omega$-model were incorrect – even by a small factor – the constancy of c_c^{act} (resp. c_b^{act}) over the whole temperature range would have been destroyed. An interesting observation is that for both directions a and b the temperature dependence of the diffusion coefficient is directly described by the bulk modulus and *not* from the corresponding linear compressibility.

In order now to present the diffusivity in zinc in the same way as for other materials we proceed to the computation of the self-diffusion Arrhenius plots for D_c and D_b. The experimental values of D_c and D_b at 512.1 K give $c_c^{act} = 0.163$ and $c_b^{act} = 0.168$ (see table 9.22) when inserted into eqs. (9.15) and (9.16). In conjunction with the elastic data the value of c_c^{act} leads to the calculated D_c-values represented by the solid line of fig. 9.24. As expected –

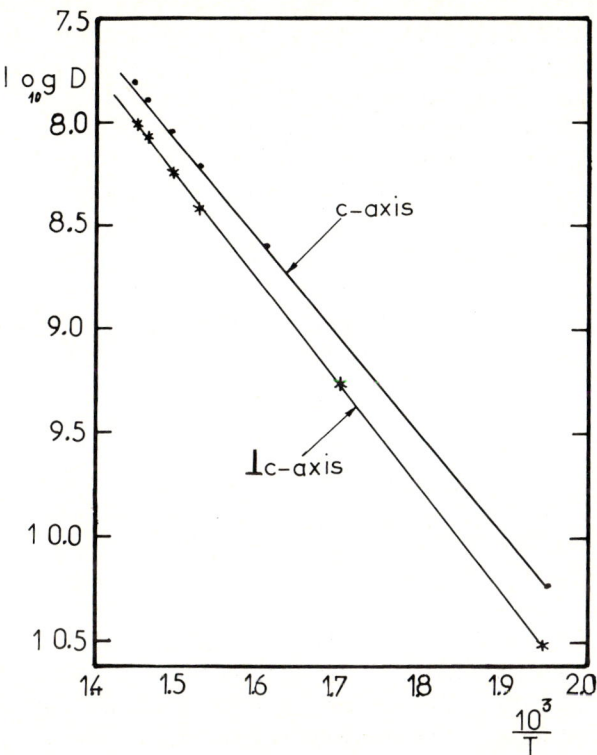

Fig. 9.24. Self-diffusion in zinc; ● and *: experimental values, Peterson and Rothman [234]; lines: $cB\Omega$-model.

due to the constancy of c_c^{act} checked above – this line passes through the experimental points; the small deviation observed in the high temperature region is not real because if we allow for an experimental error of 1% in the B-values the predicted line passes exactly through the experimental points. Similarly the value of c_b^{act} leads – through eq. (9.16) – to the determination of D_b for each temperature; then adding the term $\frac{1}{2}f_2 a^2 \nu_D \exp(-c_c^{act} B\Omega/kT)$ as indicated by Buescher et al. [236] we find the $D_{\perp c}$-values represented from the other solid line shown in fig 9.24. We see that this line accurately describes the D-values measured in the direction perpendicular to the c-axis.

The calculation of the activation parameters is straightforward. Inserting the value $c_c^{act} = 0.163$ into eq. (9.1) and (9.2) and using the elastic and expansivity data mentioned above for $T = 655.5$ K we find $s_c^{act} \simeq 2.94\,k$ and $h_c^{act} = 0.965$ eV with a plausible error of 5%, mainly due to the experimental uncertainty of dB/dT. These values are in fine accordance with the experimental ones inserted in table 9.21, especially if one recalls that the extraction of the experimental s^{act}-value suffers from the approximation $\nu_c \simeq \nu_D$. In a similar way by using $c_b^{\text{act}} = 0.168$ – instead of c_c^{act} – we find $s_b^{\text{act}} \simeq 3.03\ k$ and $h_b^{\text{act}} = 0.995$ eV with the same uncertainty of 5% as above. These values are again in very good agreement with the experimental ones.

We proceed to the determination of the formation parameters; assuming that the values of B and Ω given in table 9.22 for 691 K are also appropriate for the melting point (692.4 K), the insertion of the value [230] $n/N = 4.9 \times 10^{-4}$, for $T = T_M$, into eq. (9.5) gives: $c^f = 0.0937$. Once c^f is known one can easily compute the concentration and the formation parameters of vacancies at any temperature. The values of h^f and s^f are found by inserting the value of c^f into eqs. (9.1) and (9.2) for $T = 655.5$ K. Inserting the elastic and expansivity data mentioned above we get $h^f \simeq 0.555$ eV and $s^f = 1.69\ k$ with an uncertainty similar to that of the activation parameters. A comparison with the experimental values of Balzer and Sigvaldason shows a satisfactory agreement if one considers the experimental error.

A reliable check for the activation volumes cannot be made because the published values of dB/dP show a very large scatter; Vaidya and Kennedy [239] reported values between 2.715 and 4.76; on the other hand Swartz and Elbaum [240] gave $dB/dP \simeq 6.4$.

Summarizing the present discussion on zinc we can say that the $cB\Omega$-model reproduces the diffusion coefficients measured for both directions. Furthermore, *without* using any adjustable parameter, it leads to formation and self-diffusion activation parameters that agree with the experimental ones for both axes.

9.4.2 White tin

Simultaneous measurements of length and lattice parameters have been performed by Balzer and Sigvaldason [241] on tetragonal Sn single crystals in the *a*- and *c*-directions from room temperature up to 497 K (i.e. about 8° below the melting point 504.8 K). The thermal expansion was found to be anisotropic but for all temperatures the relative changes of the length and the lattice parameter were equal within their experimental error (3×10^{-5}). Due to this fact they concluded that even close to the melting point the vacancy concentration must be below their experimental error, i.e. $n/N \leqslant 3 \times 10^{-5}$ and, by assuming that $h = 0.50$ eV, reported that the upper limit of their formation entropy is around $1.1 \pm 1\ k$ (see table 9.23) and the asterisk of fig. 9.25). One should stress however the following: the "mean" activation enthalpy h^{act} from self-diffusion studies along and perpendicular to the *c*-axis is around 1.05–1.1 eV. In view of the fact that the formation enthalpy in metals is usually equal to or slightly above half the self-diffusion

Table 9.23
Entropies and enthalpies in tetragonal tin.

Process	h (eV)	s (k-units)	Reference
Formation:			
length measurements	0.52 ±0.05	4.5 ±1	a
differential dilatometry	0.50	1.1±1	b
differential dilatometry	(0.56)	(2.48)	b
differential dilatometry	(0.57)	(2.72)	b
$cB\Omega$-calculating	0.55 ~ 0.56	2.52	
Activation (self-diffusion):			
⊥ c-axis	1.094 ± 0.04	$6.64^{+0.09}_{-0.1}$	
∥ c-axis	1.11 ± 0.05	$7.51^{+0.33}_{-0.49}$	c
⊥ c-axis	1.01 ± 0.02	$4.61^{+0.3}_{-0.45}$	
∥ c-axis	1.11 ± 0.035	$7.55^{+0.56}_{-1.5}$	d
⊥ c-axis	1.085	5.75	
∥ c-axis	1.055	6.9	e
$cB\Omega$-calculation ⊥ c-axis	1.02	~ 4.64	
cB-calculation ∥ c-axis	1.00	~ 4.55	
Formation:			
positron technique	0.50–0.56	–	[242a, b]
resistivity technique	0.51 ± 0.05	–	[242]

[a] Ref. (243) The entropy is unusually high, see the text.
[b] Ref. (241) For the values given in parentheses, see the text.
[c] Ref. (246) measurements between 160–223°C.
[d] Meakin and Klokholm [544]; Measurements between 178 and 222°C.
[e] Quoted in Lazarus [395].

activation enthalpy we see that a formation enthalpy around 0.57 eV is quite plausible. This h^f-value does not contradict the resistivity measurements of Desorbo [242] who obtained $h^f = 0.51 \pm 0.05$ eV, because the latter value has been obtained after very rapid quenching which freezes-in the vacancies but also creates large cold-work effects; these are difficult to separate from the vacancy-induced resistivity changes. Furthermore, notice that if the value 0.56 eV deduced from the positron technique by Dedoussis, Charalambous and Chardalas [242a] and by Mackenzie and Lichtenberger [242b] is accepted, a concentration of 3×10^{-5} at $T = 497$ K gives $s^f = 2.48$ k; alternatively, if the highest formation enthalpy 0.57 eV is chosen an entropy of 2.72 k results. These points have also been inserted in the figure.

We finally discuss the formation parameters $(0.52 \pm 0.05$ eV$)$, $(4.5 \pm 1$ $k)$ reported by Current [243]; he performed only length-change measurements and estimated the vacancy contribution to the crystal volume by fitting a quadratic curve for the thermal expansion in the low-T region and then extrapolating this theoretical curve to high temperatures. The difference of

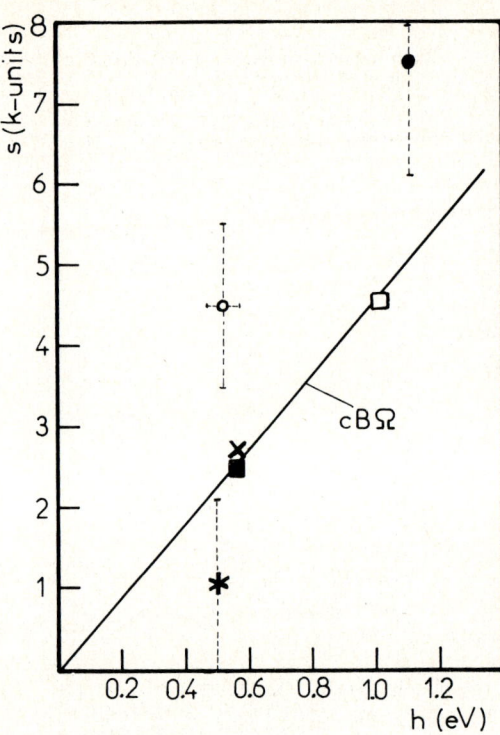

Fig. 9.25. Entropies and enthalpies of tin; ○: Current [243]; ∗: Balzer and Sigvaldason [241], $h^f = 0.5$ eV; ■: $h^f = 0.56$ eV, [241]; ×: $h^f = 0.57$ eV, [241]; ●: activation parallel to c-axis; □: activation normal to c-axis; line: $cB\Omega$-model.

the high-T measured thermal expansion from the one extrapolated from low T was then attributed exclusively to the creation of vacancies. This led to the very high concentration of vacancies $(7 \pm 2) \times 10^{-4}$ at T_M and to the parameters inserted in table 9.23 and fig. 9.25. This concentration is more than one order of magnitude higher than the upper limit found by Balzer and Sigvaldason. The difference is not without an explanation: Current's extrapolation from the low-T to the high-T region may be not appropriate because the lattice expansivity close to the melting point might show a strong upward curvature (greater than that estimated by Current) due to the high anharmonicity of this material in this temperature region. In such a case the real difference between the bulk expansivity and the lattice expansivity would be appreciably smaller than that estimated by Current, thus leading to appreciably lower values of n/N. This would drastically decrease the formation entropy reported by Current while probably changing only slightly the formation enthalpy. These conclusions are in favour of the more recent results of Balzer and Sigvaldason and should be regarded as highly reliable despite their great experimental uncertainty of 3×10^{-5}.

The (tracer) self-diffusion of tetragonal Sn has been studied by various investigators, both along and normal to the c-axis in the temperature region 423–496 K, i.e. almost up to the temperature limit of the differential dilatometry. The published activation enthalpies and entropies have been inserted in table 9.23. The entropies have been extracted from the published D_0-values by writing $D_0 = D_0' \exp(s^{\rm act}/k)$, where

$$D'_{0,c} = c^2 \nu_{\rm D} = 4.2 \times 10^{-3} \text{ cm}^2/\text{s}, \tag{9.17}$$

$$D'_{0,\perp c} = a^2 \nu_{\rm D} = 1.4 \times 10^{-2} \text{ cm}^2/\text{s}, \tag{9.18}$$

and taking a and c from table 9.24 and $\nu_{\rm D} = 4.16 \times 10^{12}$ s^{-1} from $\Theta_{\rm D} = 200$ K [156]. As the attempt frequencies for both directions have been set equal to $\nu_{\rm D}$ – which might be incorrect – we have intentionally deleted in the above formulae some appropriate numerical factors due to the geometry of the lattice; therefore the "true" activation entropies may differ from those given in the table by 1–2 k.

We now proceed to the application of the $cB\Omega$-formulae. The data given in table 9.24 suggest that $(\partial B/\partial T)_P = -0.27$ kbar/K. Therefore the application of eq. (9.3) gives, at $T = 497$ K, the value $s/h = 3.93 \times 10^{-4}$ K^{-1}. The denominator of the above equation is equal to 586.5 kbar, which is only 3% smaller than the value $B_0^{\rm SL} = 606$ kbar resulting from the elastic data of Kammer, Cardinal, Vold and Glicksman [244]. If one uses alternatively the approximate expression (9.4) the ratio decreases by 4%. The former (exact) value gives the straight line drawn in fig. 9.25. Considering the above remarks concerning the formation parameters the agreement of the prediction of the $cB\Omega$-model with the published values should be regarded as encouraging, specially if one recalls that the entropy of Current [243] has to be, any how, substantially lowered relative to the position shown in the figure.

We now proceed to the computation of the self-diffusion curve [245]. We introduce the preexponential factors (9.17) and (9.18) into eq. (9.7) and calculate $c^{\rm act}$ for the lowest temperature ($T = 423$ K) for which diffusivity has been measured. Using the experimental D-values and the elastic data of Table 9.24 we get: $c^{\rm act}_{\|c} = 0.100$ and $c^{\rm act}_{\perp c} = 0.102$ with a plausible uncertainty of 2–4%, mainly due to the experimental error of B and the approximation of the preexponential factors mentioned above. Once $c^{\rm act}$ is known, eq. (9.7) gives the D-values quoted in table 9.24 and indicated in fig. 9.26 in the form of lines. We actually observe an agreement of the "experimental lines" with those calculated. In order to realise the degree to which they agree we give in the same table at the highest temperature of the measurements the

Table 9.24
Calculated and experimental self-diffusion in tetragonal tin.

Quantity	$T = 423$ K	$T = 497$ K	Reference
B^S (kbar)	532	519	[244]
B (kbar)	492	472	a
β (10^{-6} grad^{-1})	~ 80	~ 84	b
C_P (J/K mole)	28.71	29.78	c
Ω (Å^3)	27.12	27.29	d
D_c exper. (cm^2/s)	4.52×10^{-13}	4.22×10^{-11}	e
D_c calc. (cm^2/s)		$3^{+1.27}_{-0.9} \times 10^{-11}$	
$D_{\perp c}$ exper. (cm^2/s)	1.0×10^{-12}	$8.1 \sim 9.6 \times 10^{-11}$	e, f
$D_{\perp c}$ $cB\Omega$-calc. (cm^2/s)		$6.8^{+3.1}_{-2} \times 10^{-11}$	

^a From ref. [244] by applying the thermodynamical correction and using the corresponding values of β and C_P given in the table.
^b Ref. [241]; by extracting the values of $\Delta L/L$ from figs. 1 and 2 of ref. [241].
^c Kramer and Nölting [545]; these values have been calculated from the relation $C_P = 22.62 + 0.0144\, T$ (J K^{-1} mole^{-1}) kindly forwarded to the authors by J. Nölting (Univ. Göttingen).
^d From $\Delta a/a$ and $\Delta c/c$ of figs. 1 and 2 of ref. [241] and using, for $T = 293$ K, the values $a = 5.8197$ Å and $c = 3.1749$ Å.
^e Ref. (246); by applying the relations $D_c = 7.7 \exp[-1.11 \text{ (eV)}/kT]$ and $D_a = 10.7 \exp[-1.094 \text{ (eV)}/kT]$ of ref. [246].
^f Meakin and Klokholm [544], by applying their relation $D_a = 1.4 \exp[-1.01 \text{ (eV)}/kT]$.

calculated and the experimental values. The uncertainty quoted in the calculated values and indicated in the figure comes from a (minimum) error of 2% of the exponent of the D-expression as mentioned above.

An inspection of the table indicates that, although D_c (or $D_{\perp c}$) varies from $T = 423$ K to $T = 497$ K by about two orders of magnitude, the calculated values agree with the experimental ones within the uncertainty given. This is significant because at first sight one should expect the diffusivity along e.g. the c-axis to be correlated to the corresponding linear compressibility. However, we find that the temperature dependence of the diffusion coefficient is well deduced for both directions from the bulk modulus, i.e. from the compressibility of the bulk. This is not unexpected in view of the thermodynamical origin of the $cB\Omega$-model, in which only bulk quantities appear (see ch. 14).

We first proceed to the computation of the *activation* parameters at $T = 497$ K. Applying eqs. (9.1) and (9.2) and considering the elastic and expansivity data of table 9.24 (recall that $(\partial B/\partial T)_P = -0.338$ kbar/K) we get:

with $c_c^{\text{act}} = 0.100$: $s^{\text{act}} = 4.55\, k$ and $h^{\text{act}} = 1.00$ eV,
with $c_{\perp c}^{\text{act}} = 0.102$: $s^{\text{act}} = 4.64\, k$ and $h^{\text{act}} = 1.02$ eV,

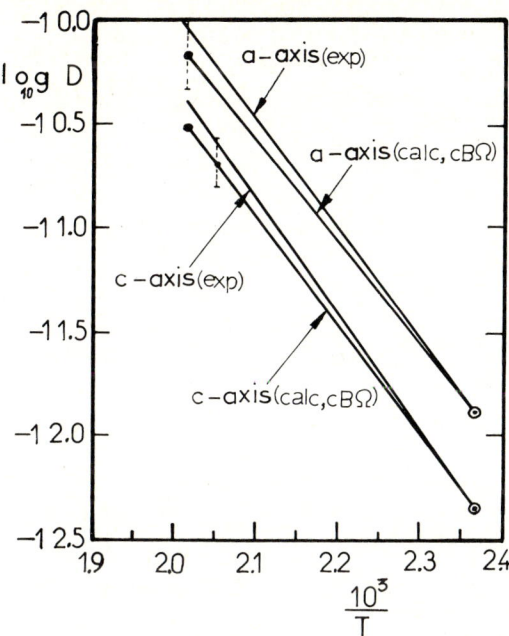

Fig. 9.26. Self-diffusion in tin. The experimental lines correspond to Coston and Nachtrieb [246]; the experimental results of Meakin and Klokholm [544] are closer to the calculated lines; the error bars indicate the rotation of the lines due to experimental uncertainty of the bulk modulus.

with an uncertainty of 2–4% due to the plausible error in the c-values as mentioned above. These values have been inserted in table 9.23; a comparison to the experimental values indicates a satisfactory agreement for both axes if one further bears in mind that the experimental entropies, as already stated, have a further inherent uncertainty of 1–2 k. Note that the $cB\Omega$-model correctly predicts them to be comparable for both axes.

We turn to the calculation of *formation* parameters from $cB\Omega$-formulae. In order to determine c^f we need a single measurement of n/N. However, as we have seen this concentration is not known with accuracy even close to the melting point. Nevertheless we shall intentionally proceed to some estimates. Inserting, at $T = 497$ K, the upper value $n/N = 3 \times 10^{-5}$ of Balzer and Sigvaldason into eq. (9.5) one immediately gets $c^f = 0.0555$. On the other hand, if at the same temperature a value 1×10^{-4} is accepted we get $c^f = 0.04906$. The values of h^f and s^f are calculated as in the case of the activation process by introducing c^f instead of c^{act}; the result at $T = 497$ K with $c^f = 0.0555$ is $h^f = 0.555$ eV and $s^f = 2.52$ k. In spite of the large experimental uncertainty of the formation parameters the above values, especially the enthalpy, are in reasonable agreement with those compatible with the data of Balzer and Sigvaldason [241].

We finally proceed to the pressure experiments on tetragonal tin. Coston and Nachtrieb [246] have studied D as a function of pressure in both cases, parallel and perpendicular to the c-axis. They found an activation volume of 5.3 ± 0.3 cm^3/mole which was isotropic and practically temperature independent within their temperature range and experimental uncertainties. By comparing it to the molar volume (~ 16.3 cm^3/mole) we find an unusual small ratio $v^{\text{act}}/\Omega = 0.325 \pm 0.019$ for both directions. Let us now investigate if these findings are in accordance with the $cB\Omega$-model. We have already seen that the c^{act}-values for both directions are, within 2%, equal; therefore the $cB\Omega$-model predicts (see eq. 8.19, taking c^{act} for c^{f}) that the activation volumes for the two directions should also be equal within 2%, as experiments show. In order now to compute from eq. (8.19) the value of v^{act} we need the value of dB/dP. Vaidya and Kennedy [239] obtain $(\partial B/\partial P)_T = 3.65$ by fitting their measurements to the Murnaghan equation but from the equation $dB/dP = 2\gamma + \frac{1}{3}$ they report $(dB/dP)_T = 4.76$. Taking their average 4.2 and inserting it into eq. (8.19) we find for both axes $v^{\text{act}} = 0.32$ Ω, in excellent agreement with the experimental value (recall that $c_a^{\text{act}} = 0.102$ and $c_c^{\text{act}} = 0.100$).

Summarizing the present discussion on tetragonal tin we may state that the $cB\Omega$-model accounts both for the formation parameters as well as for the temperature and pressure dependence of the self-diffusion along both axes.

Fast diffusion in Sn. The noble metals (Cu, Ag, Au) are fast diffusers in tetragonal (crystalline) tin. By following Warburton and Turnbull [247] we can say that their diffusivities range from 1 to 10 orders of magnitude greater than the self-diffusivity of tin [246] depending on the element, the temperature and the crystallographic direction. The major question therefore remains whether the $cB\Omega$-model can reproduce these fast heterodiffusivities; this is extremely important because they exhibit a striking anisotropy being $1\frac{1}{2}$ to $2\frac{1}{2}$ orders of magnitude higher along the c-axis than in orthogonal directions. Furthermore this behaviour sharply contrasts with that of normal diffusion (e.g. self-diffusion) because, although anisotropy was present, the diffusivity in directions transverse to the c-axis was larger only by a factor of two.

Dyson [248] has described the diffusivities of *silver* with the relations (in cm^2/s)

$$D_{\parallel c} = 7.1 \times 10^{-3} \exp[-12.3 \pm 0.4 \text{ (kcal)}/RT] \text{ cm}^2 \text{ s}^{-1},$$

$$D_{\perp c} = 1.8 \times 10^{-1} \exp[-18.4 \pm 0.5 \text{ (kcal)}/RT] \text{ cm}^2 \text{ s}^{-1}.$$

The resulting D-values for 423 and 497 K are given in table 9.25.

Table 9.25
Calculated and experimental fast diffusion of silver and gold in tin. [a]

Process	T (K)	$D_{\text{exper.}}$ (cm²/s)	$D_{cB\Omega\text{-calc.}}$ (cm²/s)	$h^{\text{act}}_{\text{exper.}}$ (eV)	$h^{\text{act}}_{\text{calc.}}$ (eV)
Diffusion of Ag into Sn:					
$\parallel c$-axis	423	3.11×10^{-9}	–	0.534 ± 0.017	
	497	$2.75^{+1.37}_{-0.92} \times 10^{-8}$	$3.9^{+1}_{-0.8} \times 10^{-8}$	0.534 ± 0.017	0.62
$\perp c$-axis	423	5.53×10^{-11}	–	0.799 ± 0.022	
	497	$1.44^{+0.95}_{-0.57} \times 10^{-9}$	$1.76^{+0.64}_{-0.48} \times 10^{-9}$	0.799 ± 0.022	0.85
Diffusion of Au into Sn:					
$\parallel c$-axis	423	1.19×10^{-8}	–	0.477 ± 0.017	
	497	$8.4^{+4.2}_{-2.8} \times 10^{-8}$	$11.1^{+2.5}_{-2} \times 10^{-8}$	0.477 ± 0.017	0.55
$\perp c$-axis	423	1.13×10^{-10}	–	0.768 ± 0.022	
	497	$2.61^{+1.7}_{-1} \times 10^{-9}$	$3^{+1}_{-0.8} \times 10^{-9}$	0.768 ± 0.022	0.80

[a] Experimental D-values from the fitting by Dyson [248]. For more accurate (calculated) enthalpies, see table 9.26.

In order to examine whether the $cB\Omega$-model describes these diffusivity curves we determine c^{act} – along each direction – from a single D-value, e.g. at $T = 423$ K. We shall use for the attempt frequencies $\nu_{\parallel c}$ and $\nu_{\perp c}$ the rough approximation $\nu_\parallel = \nu_{\perp c} = \nu_{\text{D,Sn}}(m_{\text{Ag}}/m_{\text{Sn}})^{-1/2} = 4.38 \times 10^{12}$ s^{-1}, where $\nu_{\text{D,Sn}}$ is the Debye frequency of tin. Writing

$$D_{\parallel c} = c^2 \nu_{\parallel c} \exp(-c^{\text{act}}_{\parallel c} B\Omega / kT) \qquad (9.19)$$

and [245]

$$D_{\perp c} = a^2 \nu_{\perp c} \exp(-c^{\text{act}}_{\perp c} B\Omega / kT) \qquad (9.20)$$

and inserting the experimental values for $T = 423$ K, $D_{\parallel c} = 3.11 \times 10^{-9}$ cm²/s and $D_{\perp c} = 5.53 \times 10^{-11}$ cm²/s we find: $c^{\text{act}}_{\parallel c} = 0.062$ and $c^{\text{act}}_{\perp c} = 0.0849$. Once these are known, eqs. (9.19) and (9.20) can give the corresponding diffusion coefficients at any temperature. For instance, by calculating $D_{\parallel c}$ and $D_{\perp c}$ for $T = 497$ K we find values that agree with the experimental ones (see table 9.25). The interesting fact is that in this temperature range (423–497 K) $D_{\parallel c}$ has increased by about one order of magnitude in contrast to $D_{\perp c}$, which changed faster i.e. roughly by two orders of magnitude. The success of the $cB\Omega$-model consists in managing to determine correctly the decidedly larger anisotropy of the diffusivity at 497 K.

Inserting the c^{act} values into the $cB\Omega$-formulae we find for 497 K: $h^{\text{act}}_{\parallel c} = 0.62$ eV and $h^{\text{act}}_{\perp c} = 0.85$ eV. The first value differs by 14% from the experimental ones whereas the deviation of the second is appreciably smaller (see table 9.25). These differences are well understood and cannot be considered as an adequacy of the $cB\Omega$-model because of the various simplifications. The insertion of $\nu_{\parallel c} = \nu_{\perp c} = 4.38 \times 10^{12}$ s^{-1} and the deletion of the correlation factor f, which in non-cubic crystals is not a single geometrical constant, and of numerical factors due to the geometry of the lattice (they undoubtedly differ for the two directions due to the difference in dimensionality) may easily introduce into the preexponential factor an error of a factor up to 5. This will immediately reflect an error of about 12% in the determination of $c_{\parallel c}^{act}$ and hence an equal error to the calculated value of $h_{\parallel c}^{act}$ (see also below).

The Dyson et al. [249] measurements on the diffusivity of *gold* have shown that $D_{\parallel c} = 5.8 \times 10^{-3} \exp[-11.0 \pm 0.4 \text{ (kcal)}/RT]$ and $D_{\perp c} = 1.6 \times 10^{-1} \exp[-17.7 \pm 0.5 \text{ (kcal)}/RT]$, which give for 423 and 497 K the values inserted in table 9.25. In a similar fashion, as for silver, we approximate: $\nu_{\parallel c} = \nu_{\perp c} = 3.25 \times 10^{12}$ s^{-1}; then from the D-values at $T = 423$ K we find $c_{\parallel c}^{act} = 0.0548$ and $c_{\perp c}^{act} = 0.08046$. These can be used to calculate the diffusion constants given in the same table for $T = 497$ K; they are in satisfactory agreement with the experimental ones. Note that again the anisotropic diffusion behaviour is sufficiently well accounted for from the $cB\Omega$-model. The determination of the activation enthalpies and the uncertainties involved in their determination are similar as in the case of silver.

Because of the uncertainty in ν an unambiguous determination of c^{act} requires in principle the knowledge of D at two temperatures; by writing:

$$D = A \exp(-c^{act} B\Omega/kT), \tag{9.21}$$

we have included in A the attempt frequency, the correlation factor and the geometrical factor. As a first approximation we accept A as temperature independent and then determine A and c^{act} for each direction (see table 9.26) from the D-values at $T = 423$ and 497 K. As expected these more accurate values of c^{act} differ by some percent from those determined by an

Table 9.26
Calculated activation parameters for the fast diffusion of silver and gold in tin.

Quantity	Diffusion of Ag in Sn		Diffusion of Au in Sn	
	$\parallel c$	$\perp c$	$\parallel c$	$\perp c$
c^{act}	0.05348	0.08	0.04795	0.07704
A (cm^2/s)	6.30×10^{-4}	4.78×10^{-3}	6.82×10^{-4}	4.98×10^{-3}
h^{act} (eV)	0.535	0.8	0.48	0.77
s^{act} (k-units)	2.435	3.643	2.183	3.51
v^{act}/Ω	0.171	0.256	0.153	0.247
$a^2 \nu_D \sqrt{m_{Sn}/m_i}$ (cm^2/s)		1.47×10^{-2}		1.09×10^{-2}
$c^2 \nu_D \sqrt{m_{Sn}/m_i}$ (cm^2/s)	4.42×10^{-3}		3.28×10^{-3}	

arbitrary selection of the preexponential factor (e.g. $\nu = \nu_D$, etc.) and lead to activation enthalpies that coincide with the experimental ones. The values of A obtained in this way differ from the preexponential factors of eqs. (9.19) and (9.20) by a factor of about 5.

The following question could be asked: when the D-values at two temperatures are known, the quantities D_0 and h^{act} can be immediately determined by writing $D = D_0 \exp(-h^{act}/kT)$; therefore, what is the advantage of determining A and c^{act} from eq. (9.13) (i.e. by plotting $\ln D$ versus $B\Omega/kT$) instead of D_0 and h^{act}? We report three basic advantages:

(1) The values (D_0, h^{act}) can account only for a linear Arrhenius plot in contrast to (A, c^{act}), which can also describe a curved Arrhenius plot by accepting a nonlinear temperature dependence of B.

(2) The knowledge of D_0 can tell us nothing about the true activation entropy s^{act}. Even if we know the other numerical factors included in D_0, we cannot determine s^{act} from the product $\nu_{att} \exp(s^{act}/k)$ because of the absence of a good estimate of the attempt frequency (for the definition of the latter, see ch. 5). It is true that a direct determination of ν_{att} might be achieved by a further experiment (e.g. inelastic relaxation, etc.) but this is not possible for s^{act}. However, once c^{act} is known the activation entropy can be calculated from the $cB\Omega$-formula (9.1). The values of s^{act} derived in this manner are given in table 9.26. The separate knowledge of ν and s^{act} helps in understanding the kind of mechanism responsible for the diffusion process.

(3) the third advantage is the capability of determining the pressure variation of the diffusion coefficient (i.e. the corresponding activation volume). Obviously the knowledge of (D_0, h^{act}) – as determined above – does not help toward the determination of v^{act}; on the other hand, c^{act} can immediately lead to v^{act} through eq. (8.19). In this way we have calculated the activation volumes for each direction by accepting the value $dB/dP = 4.2$, which has already been used (see table 9.26).

A quite interesting fact now emerges from the inspection of these results: the calculated activation volume along the c-axis is appreciably lower (50% for Ag and 60% for Au than perpendicularly to it. The result contrasts sharply with the self-diffusion behaviour, for which we have seen that the activation volumes along the two directions are equal. The experimental verification of this predicted anisotropy is highly desirable.

Closing this discussion on Sn we emphasize that the above manner of analysis of the diffusion curve (i.e. the determination of A and c^{act} instead of D_0 and h^{act}) is suggested for any material in view of the advantages discussed above. This point has also been discussed in §7.3.

9.5 Noble gas solids

For these materials the formation parameters were obtained with the Simmons–Ballufi method and the activation parameters (mainly) with NMR-techniques. The check of eq. (9.4) is of special value because the enthalpies lie about two orders of magnitude lower than those of alkali halides.

We shall see below that the validity of the $cB\Omega$-model as far as the relation between s and h is concerned can accurately be checked only for krypton. This is so because in Ne, Ar and Xe the experimental uncertainties in the isothermal elastic data and in the defect parameters are large so that a direct comparison of the experimental values of the ratio s/h and that calculated from the $cB\Omega$-model is not equally reliable. Recent results for He are discussed in §A.9.

9.5.1 Neon

In fig. 9.27 we plot the formation parameters reported by Schoknecht [52] by means of the usual Simmons–Balluffi technique and the activation (self-diffusion) parameters of Sirovich and Norberg [250] obtained by the NMR-technique. Note that the latter activation parameters also agree with those reported earlier by Henry and Norberg [251], also from the NMR-technique. The numerical values are given in table 9.27. In the figure we see that the representative points of formation and activation processes

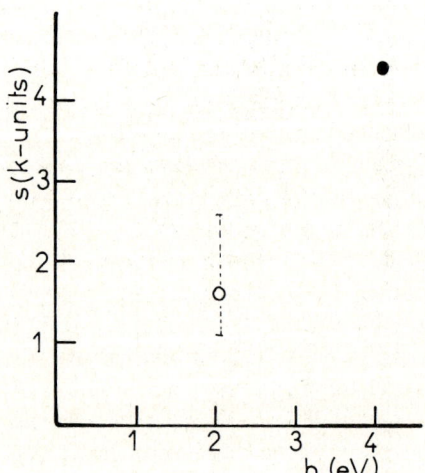

Fig. 9.27. Defect entropies and enthalpies of neon; ○: formation; ●: self-diffusion activation;

Table 9.27
Activation entropies and enthalpies in noble gas solids (self-diffusion).[a]

Noble gas solid	h^{act} (eV)	s^{act} (k-units)	Reference
Ne	0.041	4.4	[250]
Ar	0.156	3.8	[257]
	0.171	4.6	[256] [b]
	0.165	9.2	[256] [c]
Kr	0.208	7.2	[262]
	0.217	7.0	[261]
Xe	0.321	8.1	[264]

[a] See also ref. [266], p. 1220; the formation parameters are given in table 10.5.
[b] Thin film results.
[c] Results from tracer exchange in large crystals.

are compatible with a straight line through the origin of the axes although derived from quite different techniques. This agrees with the prediction of eq. (9.3) that the ratio s/h must be the same for various processes and only depends on bulk properties. We now turn to the comparison of the calculated s/h_{exp}-values with those obtained from experiments. The calculation of s/h from the $cB\Omega$-model requires the knowledge of $(dB/dT)_P$. Unfortunately the isothermal bulk modulus data show a large scatter as noticed by Korpiun and Luescher [252]. The data of Anderson, Fugate and Swenson [253] obtained for 13.5 and 19.9 K by the piston displacement technique as analysed by Birch [254] give for $(\partial B/\partial T)_P$ a value between -0.31 and -0.32 kbar K^{-1}; this value differs by a factor of two or more from that extracted for the same temperature region from B-data of Batchelder, Losee and Simmons [255], which have been obtained by X-ray diffraction. In view of this large uncertainty we do not proceed with the calculation of s/h.

9.5.2 Argon

The activation parameters obtained by Berne, Boate and De Paz [256] or by Parker, Glyde and Smith [257], both obtained by tracer exchange techniques, are consistent when extracted from thin films (see table 9.27 and fig. 9.28). However, large crystals give approximately the same activation enthalpy but an activation entropy larger by a factor of two (the point corresponding to the large-crystal results has not been inserted in the figure). The formation parameters in Ar are also not known with apprecia-

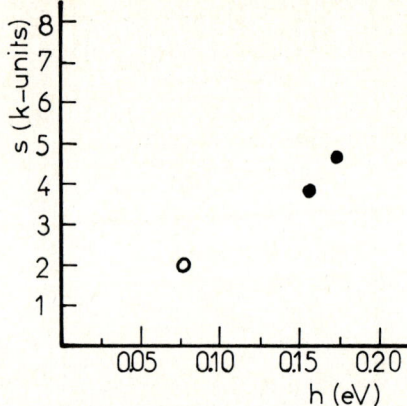

Fig. 9.28. Defect entropies and enthalpies of argon; ○: formation; ●: self-diffusion activation.

ble accuracy. Only an upper limit for the concentration of vacancies can be set from the Simmons–Balluffi measurements of Schwalbe [258]. Macrander [53] has recently reported that the following sets of the values (h, s) are all consistent with the results of Schwalbe: (0.062 eV, 0 k), (0.078 eV, 2 k) and (0.093 eV, 4 k). Of course these sets lead to the same approximative value of g, but cast considerable doubt on the exact value of the ratio $s/h_{\exp}$. In fig. 9.28 the most probable set (0.078 eV, 2 k) has been inserted. In view of these large uncertainties it is worthless to proceed to a comparison with the calculated value of s/h. However, as shown in the figure, the most probable sets of s and h (one for formation and two for activation) may give a straight line which passes fairly well through the origin of the axes.

9.5.3 Krypton

Birch [254] has re-analysed the data of Anderson and Swenson [259] in two ways, both giving the value $(\partial B/\partial T)_P = -0.16$ kbar K^{-1} for the temperatures between 100 and 110 K. The average value of B for 110 K given in these two ways is 15.5 kbar, whereas β is equal to 16.4×10^{-4} K^{-1} [51]. We have selected this set of B-values because recent measurements of Korpiun, Kampfer and Luescher [260] showed, for 77 K, $B = 21.5 \pm 0.6$ kbar, in excellent agreement with the analysis of Birch which indicates a value of between 21.4 and 21.5 kbar for the same temperature. The high temperature data of Birch show that B_0^{SL} is around 36 kbar, whereas Korpiun and Coufal [51] give 41.5 kbar. By using $b = 1.1$ and the average of the mentioned B_0^{SL}-values, eq. (9.4) gives for 110 K the value $s/h = 3.83 \times$

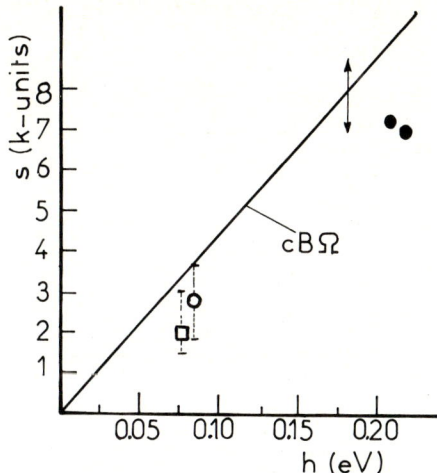

Fig. 9.29. Defect entropies and enthalpies of krypton; ○: formation, [51]; □: formation, [263]; ●: self-diffusion activation, [261,262], see also table 9.27; line: calculated from the $cB\Omega$-model; the arrow indicates the variation of the calculated line due to the experimental uncertainty of B_0^{SL}.

10^{-3} K^{-1}. In fig. 9.29 a line has been drawn with this slope. It is evident from the figure that the formation parameters reported by Korpiun and Coufal [51], the activation parameters (NMR-technique) of Cowgill and Norberg [261] and those of Chadwick and Morrison [262] are not far from the $cB\Omega$-scheme. Note further that the formation parameters $h = 0.077$ eV and $s = 2^{+1}_{-0.5}$ k reported by Losee and Simmons [263], also included in the figure, agree with the prediction of the $cB\Omega$-model within the experimental errors.

9.5.4 Xenon

The activation parameters have been obtained by means of NMR-techniques by Yen and Norberg [264], whereas the formation parameters are only estimates obtained from the application of the classical law of corresponding states to the values observed in Kr. These values are given in fig. 9.30. The high-temperature elastic data (100–150 K) of Anderson and Swenson [259] as re-analysed by Birch [254] give the mean values (from the two ways of analysis), $B = 14.7$ kbar, $dB/dT = -0.167$ kbar/K and $B_0^{SL} = 42$ kbar. The value of β for 159 K [265] is $\sim 1.3 \times 10^{-4}$ K^{-1} and b is 1.1. When inserted into eq. (9.4) these values give $s/h_{exp} = 4.3 \times 10^{-3}$ K^{-1} which gives the straight line plotted in fig. 9.30. Taking into consideration

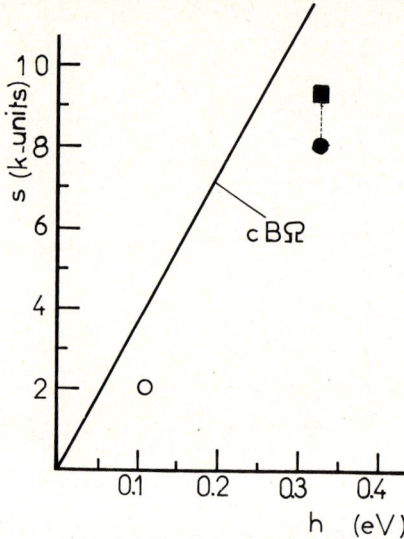

Fig. 9.30. Defect entropies and enthalpies of xenon; ○: formation; ●: self-diffusion activation with $\nu = \nu_D$; ■: self-diffusion activation with $\nu = \tilde{\nu}$; line: $cB\Omega$-model.

the experimental error one sees that both formation and activation parameters systematically deviate by a factor of two. This deviation may not be real because as already mentioned the formation entropy is only a very rough estimate whereas the activation entropy results from a D_0-value (which, as discussed by Chadwick and Glyde [266], has a very large uncertainty) with $\nu \simeq \nu_D$; if one alternatively uses $\nu = \tilde{\nu}$ with $h^m \approx 0.2$ eV, the entropy s^{act} should increase by 1.3 k. Despite these uncertainties one should consider still another plausible reason; as discussed in the last chapter a possible temperature dependence of c – when properly employed – should turn the straight line in the figure to a smaller slope, thus leading to a better agreement with the experimental data. However, in view of the great uncertainties it is not worthwhile to proceed to such a correction until more accurate experimental values are established.

9.6 Alkali halides

9.6.1 Sodium fluoride

Miller and Smith [267] give at 300 K $B^S = 482$ kbar and $(\partial B^S/\partial T)_P = -0.178$ kbar/K from which we get $B_0^{SL} = 535.4$ kbar. By converting to isothermal values we get for 300 K $B = 456.9$ kbar and $(\partial B/\partial T)_P = -0.26$ kbar/K, whereas $\beta = 1.08 \times 10^{-4}$ K^{-1} and $b \simeq 1.02$. From these, eq. (9.4)

§9.6]   Alkali halides   247

Table 9.28
Defect entropies and enthalpies in NaF.

Process	h (eV)	s (k-units)	Reference
Schottky formation	2.56	9.86	a
	2.42	7.54	b
Cation vacancy migration	0.97	0.93	a
	0.95	0.93	b
Anion vacancy migration	1.54	6.5	a
	1.46	11.7	b
Bound cation vacancy, dipole orientation with ITC-technique			
Ca^{2+}	0.86	3.7 ± 1	c
Ca^{2+}	0.86	4.25 ± 1	d
Mg^{2+}, nn	0.78	2.1 ± 0.6	e
Mg^{2+}, nnn	0.83	2.6 ± 0.6	f

^a Ref. [270]; From the fitting of the ionic conductivity curve without considering long range electrostatic interactions.
^b Values given in [270] when electrostatic interactions are included in the analysis of the conductivity curve.
^c Ref. [268] The entropy value has been extracted by assuming $\nu \simeq \nu_D$ (note that in [268] the value $1.86\,k$ has been reported; this differs from the "real value" given in the table by $\ln 2\pi$).
^d Ref. [268], assuming $\nu = \nu_{TO}$.
^e Ref. [115a], nn neighbours and assuming $\nu = \nu_{TO}$.
^f Ref. [115a], nnn neighbours.

gives for 300 K $s/h_{exp} = 4 \times 10^{-4}$ K^{-1}. The defect parameters which describe the motion of the cation vacancy bound to a Ca^{2+}-ion at the same temperature region [268] give (see table 9.28) $s/h_{exp} = 3.7 - 4.2 \times 10^{-4}$ K^{-1} in excellent agreement with the value resulting from the $cB\Omega$-model.

Assuming that B decreases linearly with T we get for 900 K the value $B = 300.9$ kbar. The expansivity data quoted by Lagu and Dayal [269] give for 900 K $\beta \simeq 1.35 \times 10^{-4}$ K^{-1} and $b \simeq 1.08$. Equation (9.4) then gives $s/h_{exp} = 4.425 \times 10^{-4}$ K^{-1}. A straight line with this slope and the other data have been plotted in fig. 9.31. The maximum error of s/h_{exp} can be estimated to be $\pm 20\%$, due mainly to the extraction of $(\partial B/\partial T)_P$ from the corresponding adiabatic value. In the same figure we see the experimental values of s/h for various processes which are not far from the values predicted from the $cB\Omega$-model. However, it should be noted that the defect parameters given by Bauer and Whitmore [270] must be accepted with some caution. This is due to the following two reasons: (1) these values have been extracted from a fitting to the conductivity curves of pure NaF and NaF doped with Ca^{+2}; but as stated by Beniere, Chemla and Beniere [69] a

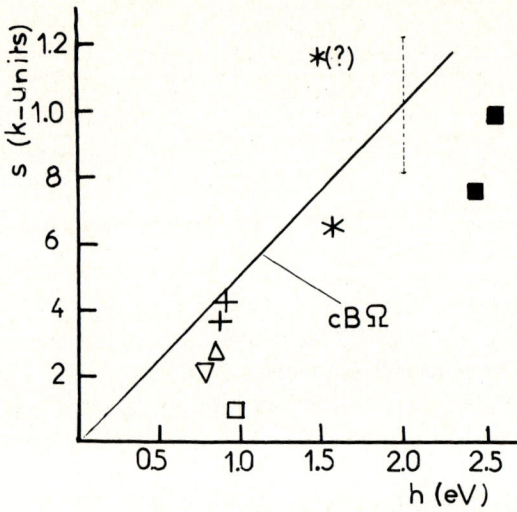

Fig. 9.31. Defect enthalpies and entropies of sodium fluoride; ■: Schottky formation; □: cation vacancy migration; ∗: anion vacancy migration; +: Ca^{2+} dipole orientation; ▽: nn Mg^{2+} dipole orientation; △: nnn Mg^{2+} dipole orientation; line: $cB\Omega$-model.

consistent set of defect parameters needs a simultaneous fitting not only of the conductivity of a pure crystal and a crystal doped with various amounts of divalent cations but also of diffusion data. For example we remember that Jacobs and Pantelis [271] – with a procedure similar to that of Bauer and Whitmore – obtain the unreliable value of 10 k for the entropy of anion vacancy migration in KCl instead of the value 3.20 k of ref. [69]. (2) Bauer and Whitmore have observed nonrandom deviations between their experimental results and the theory which was employed for the analysis of their data. They attempted to explain these deviations by considering the usual long-range Coulomb interactions between isolated defects as suggested by Lidiard [272]. However, the nonrandom deviations persisted over the entire temperature range under observation so that Lidiard's suggestion obviously cannot explain the effect. Before going into detail we report a recent neutron scattering study of Yoshizawa and Hirakawa [273] who stated that anharmonicity of lattice vibrations is extremely important in NaF. Therefore, the defect parameters, which are closely related to the lattice vibrations, should be strongly influenced by anharmonicity and should therefore change in a detectable way with temperature. Repeating the calculation of s/h_{exp} close to 1200 K and assuming that B continues to decrease linearly with T up to this temperature we have the values $B = 222.9$ kbar, $\beta \simeq 1.55 \times 10^{-4}$ K^{-1} and $b = 1.12$. We get for $T = 1200$ K a value $s/h_{\text{exp}} = 4.716 \times 10^{-4}$ K^{-1}, i.e. a value of 18% greater than that correspond-

Table 9.29
Minimum variations of the defect parameters of NaF with temperature. [a]

T	300 K	600 K	900 K	1200 K
s/h_0 (10^{-4} K^{-1})	4.187	4.416	4.632	4.936
h/h_0	1.037	1.0423	1.052	1.081
$s/h_{\exp}$ (10^{-4} K^{-1})	4	4.219	4.425	4.716

[a] B is assumed to decrease linearly with T. The ratios refer both to formation and migration.

ing to 300 K. This means that s should increase by 18% from 300 to 1200 K. Of course, this is only valid if B continues to decrease linearly up to 1200 K; however, there is recent experimental evidence from Jones [274] and Hart [275] that B decreases, faster than linearly thus giving a greater value for $(\partial B/\partial T)_P$. According to eq. (9.4) this results in an increase of s by *more than* 18%.

Similarly eq. (9.2) indicates that the defect enthalpy should increase with temperature. In order to see what temperature dependence of the defect parameters can be derived from the elastic constants with the help of the $cB\Omega$-model we have calculated from eq. (8.9), (8.13) and (8.17) the contents of table 9.29 and plotted the results in fig. 9.32. For 600 K the values inserted are $B = 378.9$ kbar, $\beta = 1.2 \times 10^{-4}$ K^{-1} and $b \simeq 1.05$. For B_0 we have used the value 511.5 kbar which results from $B_0^{SL} = 535.4$ kbar by assuming that in NaF the ratio B_0^{SL}/B_0 is approximately the same as that of NaCl (1.047).

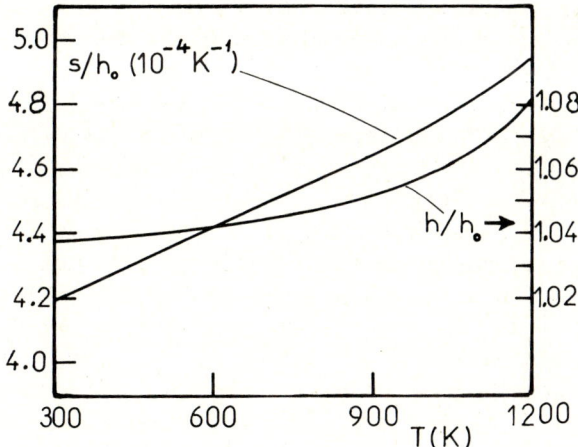

Fig. 9.32. Minimum variation of the defect parameters of NaF with temperature, predicted from the $cB\Omega$-model (when B decreases linearly with T). The path of the curves is similar to those published long ago [92] for KCl.

In order to study the temperature dependence of g we restrict ourselves to the process of the formation of Schottky defects and take as a representative value for the formation enthalpy at $T = 0$, $h_0^f = 2.3$ eV. Table 9.30 gives the resulting numerical values of s^f, h^f and g^f. The following remarks could be useful: (1) the calculated s-values are comparable to the experimental values 7.54–9.86 k reported in ref. [270]. A direct comparison cannot be made due to the uncertainty of the assumed value $h_0 = 2.3$ eV which may be in error by 10% and to a lesser extent to the lack of accurate experimental values of B. (2) An inspection of the g^f-values resulting from the temperature dependent h- and s-values which are given in table 9.30 shows that g^f decreases faster than linearly. This is easily seen by comparing the g^f-values computed directly for each temperature with the quantities g^{SL} also given in the table. The latter are the result of a calculation in which the s- and h-values are assumed to remain constant above 300 K, i.e. from the formula $g^{SL}(T) = h_{300K} - Ts_{300K}$. The fact of this excessive decrease of g^f reflects an upward curvature of the conductivity plot and is in agreement with the trend of the experimental data. It would be interesting to check whether the predicted temperature variation of the defect parameters can quantitatively account for the curvature. Due, however, to the lack of direct elastic data at high temperatures this is not possible for NaF. The results included in the table can be thought as giving the *minimum* temperature dependence of the defect parameters because, as already mentioned, they have been derived under the assumption that B decreases linearly up to 1200 K. However, a semi-quantitative argument is possible from these results. For 1200 K, $g^{SL}(1200) - g(1200) = 1.526 - 1.474 = 0.052$ eV, so the deviation of the conductivity plot from linearity is expected to be of the order $\exp[0.052 \text{ (eV)}/2kT] \simeq 30\%$. A further curvature comes from the temperature variation of the migration parameters which can be calculated in the same manner. The value 30% is not only beyond the experimental error (1 to 2%) but it also has the same order of magnitude as the deviation estimated from experimental data.

Table 9.30
Schottky formation parameters of NaF by using $h_0^f = 2.3$ eV.

Quantity	300 K	600 K	900 K	1200 K
s (k-units), from eq. (8.9)	8.302	8.756	9.184	9.787
h (eV), from eq. (8.13)	2.385	2.397	2.420	2.486
g (eV) calc. from h, s	2.170	1.944	1.706	1.474
g^{SL} (eV)	2.170	1.956	1.741	1.526
g_{exp} (eV) [a]	2.225	2.030	1.835	1.640

[a] By using the values $h_{exp} = 2.42$ eV and $s_{exp} = 7.54$ k given in ref. [270]; see table 9.28.

9.6.2 Potassium chloride

Smith and Cain [276] give the values $\beta = 1.11 \times 10^{-4}$ K^{-1} and $B = 173.5$ kbar at 295 K in agreement with the values reported by Bartels and Schuele [277]. A linear extrapolation of the values given for the isothermal bulk modulus by Bartels and Schuele for 195 and 295 K yields $B_0^{SL} = 217.35$ kbar and $(\partial B/\partial T)_P = -0.11$ kbar/K, whereas for RT, $b \simeq 1.02$. A linear extrapolation of the adiabatic values at 195 and 295 K gives $B_0^{SL} \simeq 206$ kbar which differs from the above value by 5.5%; this is not easily explained because in the quasi-harmonic approximation the difference should be approximately zero [278]. When inserted into eq. (9.4) these values give $s/h_{exp} = 4.26 \times 10^{-4}$ K^{-1}, which is the slope of the line drawn in fig. 9.33. Assuming that B decreases linearly with T we obtain for 900 K, $B \simeq 107$ kbar, whereas the expansivity data of Srivastava and Merchant [279] give $\beta = 1.66 \times 10^{-4}$ K^{-1} and $b \simeq 1.07$. These values give $s/h_{exp} = 4.54 \times 10^{-4}$ K^{-1}, which is about 7% higher than that found for 300 K. Recent values of defect parameters for various processes have been collected in table 9.31 and are indicated in the fig. 9.33. Their general trend follows the prediction of the $cB\Omega$-model, i.e., that s is proportional to h for various processes. The

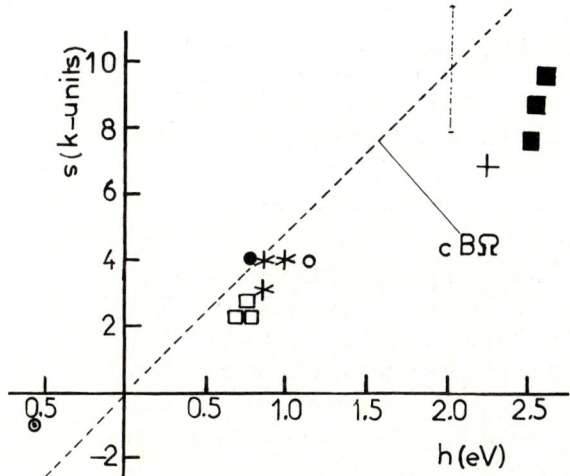

Fig. 9.33. Defect entropies and enthalpies of potassium chloride; ○: migration of SO$_4^{2-}$; +: solubility parameters of SO$_4^{2-}$; ■: formation parameters (Schottky); ●: solubility parameters of Mn^{2+}; □: cation migration parameters; *: anion migration parameters; ⊙: association; The error bars indicate the variation of the line calculated from the $cB\Omega$-model due to the experimental uncertainty of the elastic data.

Table 9.31
Defect entropies and enthalpies in KCl.

Process	h (eV)	s (k-units)	Reference
Solubility of SO_4^{2-}	2.24	6.9	[546]
Solubility of Mn^{2+}	0.79	4.2	[547]
Schottky formation	2.54	8.99	[69]
	2.50	7.5–7.9	[89]
	2.59	9.61	[280]
Cation vacancy migration	0.68	2.4	[89]
	0.73	2.7	[280]
	0.73	2.4	[69]
Anion vacancy migration	0.85	3.2	[69]
	0.99	4.14	[280]
	0.85	4.1	[89]
Migration of SO_4^{2-}	1.12	3.9	[89]
Association of "Sr^{2+} + cation vacancy"	−0.56	−1.08	[69]

following remarks could be useful: the data reported by Chandra and Rolfe [280] or by Beniere et al. [69] seem to agree better with the calculated s/h_{exp}-value. The parameters reported by Acuna and Jacobs [89,95] show a larger deviation. This is not surprising because their analysis makes the following arbitrary assumptions: (1) they fix some parameters which have been calculated microscopically (with an unpredicted error) and then they determine the others from fitting to the experimental data of Beniere et al. [69], etc. (2) They assume that both cation and anion Frenkel defects are operative, a fact for which there is no experimental evidence. Of course, the introduction of these additional defects was made basically for the explanation of the curvature observed in the conductivity plots by assuming temperature independent defect parameters (a third arbitrariness). However, an explanation of the curvature can be achieved, without assuming additional defects, by allowing for a temperature dependence of the enthalpies and entropies. This explanation seems to be currently accepted because it has been directly experimentally verified in a number of cases [281] in AgCl and AgBr. Thermodynamics indicate that when g falls faster than linearly, an upward curvature of the conductivity (or diffusion) plot can never be explained – in the frame of a single mechanism – with an enthalpy that decreases with temperature, although, as already mentioned, such proposals have appeared many times in the literature [96,97] (see also §A.11, note 4).

9.6.3 Sodium chloride

The highest temperature for which direct elastic data are available is 800 K. The data of Spetzler, Sammis and O'Connell [282] for 800 K give: $B = 157.39$ kbar, $\beta = 1.719 \times 10^{-4}$ K^{-1}, $(\partial B/\partial T)_P = -0.1586$ kbar K^{-1}, $B_0^{SL} = 284.7$ kbar and $b \simeq 1.1$. Inserting these values into eq. (9.4) we get $s/h_{\text{exp}} = 5.08 \times 10^{-4}$ K^{-1} which corresponds to the slope of the straight line plotted in fig. 9.34 in which the parameters h and s have also been marked for a number of processes [69,283–287]. We again see a rough agreement between the predictions of the $cB\Omega$-model and these more recent data if one also considers the error bars in the drawing of the straight line. An interesting point is that published by Beniere and Rokbani [286] concerning the solubility parameters of Y^{3+}; it introduces two cation vacancies in the lattice and is found to have the same s/h-value as other defects obtained in quite different types of experiments. Furthermore, the association parameters of Ca^{2+} or Sr^{2+} with a cation vacancy, although they

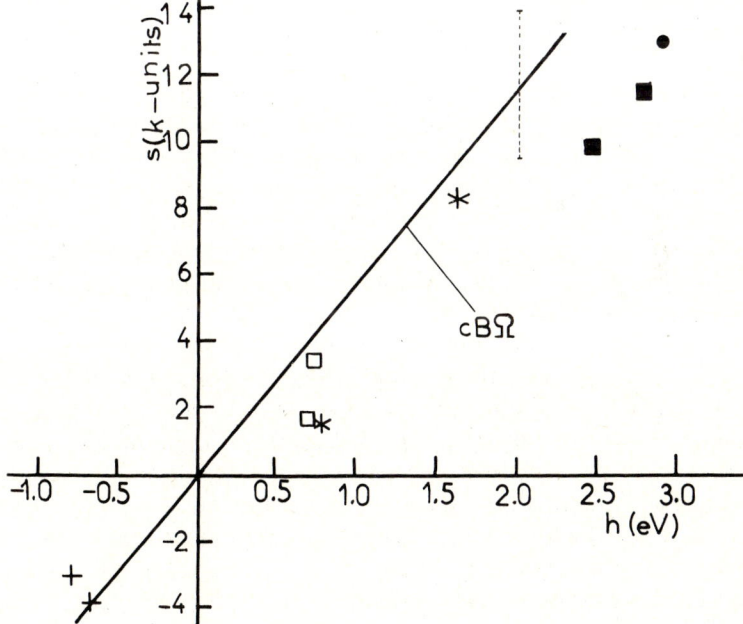

Fig. 9.34. Defect entropies and enthalpies of sodium chloride; +: association; ●: solubility parameters of Y^{3+}; ■: formation parameters (Schottky); □: cation migration; ∗: anion migration; line: $cB\Omega$-model for 800 K. The error bars of the straight line correspond to a plausible error of 20% in the calculated s/h_{exp}-value, mainly due to the uncertainty of the value of $(\partial B/\partial T)_P$.

Table 9.32
Defect entropies and enthalpies in NaCl.

Process	h (eV)	s (k-units)	Reference
Solubility of Y^{3+}	2.9	13	[286]
Schottky formation	2.44	9.8	[69]
	2.745	11.5	[283]
Cation vacancy migration	0.69	1.64	[69]
	0.715	3.3	[283]
Anion vacancy migration	0.77	1.38	[69]
	1.616	8.25	[283]
Association of "Ca^{2+} + cation vacancy"	−0.752	−3.12	[287]
Association of "Sr^{2+} + cation vacancy"	−0.671	−3.89	[287]

lie on the negative part of the scales, fit well within the scheme of the $cB\Omega$-model for thermally excited processes. All experimental values are given in table 9.32.

The following important remark must be made: in table 9.32 one notices a large scatter of the anion vacancy parameters. The enthalpy value of 0.77 eV recently reported by Beniere et al. [69] is about one half of that reported by Nadler and Rossel [283] or of the value 1.4–1.75 eV reported earlier by Alnatt, Pantelis and Sime [284]. It was believed that the value of Beniere et al. [69] is more reliable because it was obtained by a simultaneous fitting of both the conductivity and diffusivity data, whereas the others have been obtained only from conductivity data. Moreover, Cook and Dryden [285] have reported the preparation of single crystals of NaCl doped with sulphide ions and they showed that, after quenching from high temperature, the first stage of aggregation of the sulphide ions is the formation of a sulphide ion – anion vacancy pair. By measuring the ionic conductivity they were then able to determine the enthalpy of migration of anion vacancies in a temperature range where other energy parameters do not have to be considered. The resulting value is $h^{m-} = 1.24 \pm 0.05$ eV (well above the Beniere et al.-value) which is almost two times greater than h^{m+}. This result casts doubt on the shell-model calculations of Catlow, Corish, Diller, Jacobs and Norgett [90], who claimed that the migration enthalpies of anion and cation vacancies in alkali halides with NaCl-structure differ only very little. The basic advantage of Cook and Dryden's experiment is that it was the first one made on NaCl doped with divalent anions. Beniere et al. did not include data on conductivity of NaCl doped with divalent anions as they did for KCl. At the moment there is no decisive answer to resolve this discrepancy. Some comments which might help towards this direction are given below. There is no doubt that for KCl the defect data of Beniere et al. are more complete (due to their inclusion of $KCl + SO_4^{2-}$) than for NaCl. For the latter the parameters given by Beniere et al. are in good agreement with the predictions of the $cB\Omega$-model if one considers the lower error bar of the straight line in fig. 9.34. In the case of KCl the formation parameters of ref. [69] are also within the expected region. The solubility parameters of ref. [286] and the association parameters of Mashida and Fredericks [287] for both Ca^{2+} and Sr^{2+} as well as the experiments with sulfide ions [285,288] agree with the calculated s/h_{exp}-values. Only the migration parameters of ref. [69] seem to deviate from the $cB\Omega$-model in the sense that the anion vacancy which has the higher enthalpy cannot have the smaller entropy. In view of the enthalpy value of ref. [285] this seems to be uncertain. We cannot further discriminate between these two possibilities. Measurements of anion parameters with the procedure of ref. [285] are desirable [288]. Last, the following remark should be added; although the idea of temperature dependent parameters seems to be now generally accepted, a re-analysis of experimental data in the light of this suggestion has not yet been reported. Probably if the experimental data of Beniere et al. are re-analysed by allowing for temperature dependent enthalpies and entropies the anion migration parameters might result in a quite different range, thus showing that the discrepancy between Beniere et al. and Cook and Dryden [285] is only formal, being due only to the kind of analysis. Alternatively another possibility to clarify the above situation is the measurement of the ionic thermocurrent of $NaCl + S^{2-}$ in order to detect the migration enthalpy of the bound anion vacancy which may be comparable to that of the free anion vacancy.

9.6.4 Potassium bromide

Smith and Cain [276] give for 295 K the values $\beta = 1.16 \times 10^{-4}$ K^{-1} and $B = 146.4$ kbar. By using the temperature derivative -0.0715 kbar K^{-1} of the adiabatic bulk modulus of Barsch and Chang [289] and the adiabatic modulus 153.0 kbar K^{-1} of Smith and Cain [276] for 295 K one gets $B_0^{SL} \simeq 175$ kbar. In conjunction with the B-value given by Smith and Cain for 295 K this leads to $(\partial B/\partial T)_P = -0.097$ kbar K^{-1}. Using $b = 1.02$ the above data, when inserted into eq. (9.4), give for 295 K the value $s/h_{exp} = 4.66 \times 10^{-4}$ K^{-1} which leads to the straight line drawn in fig. 9.35. The probable error of the calculated value of s/h_{exp} which is indicated in the figure with bars is around 20% and is due mainly to the uncertainty of the

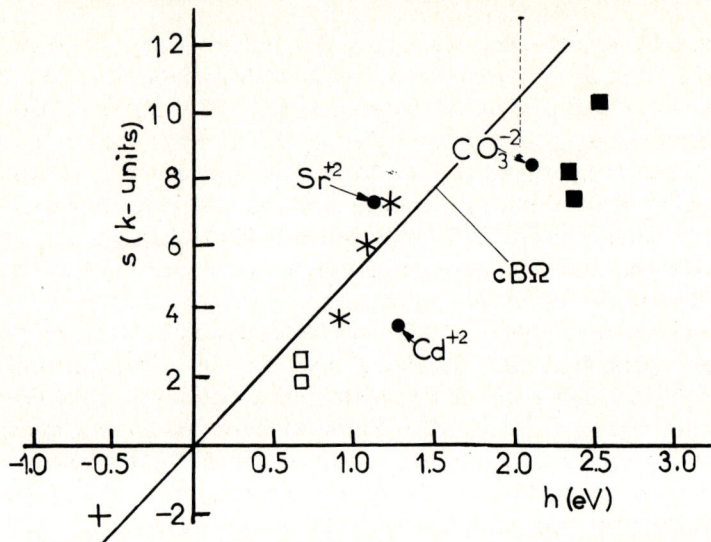

Fig. 9.35. Defect entropies and enthalpies of potassium bromide; ●: solubility; ■: Schottky formation; □: cation migration; *: anion migration; +: association; line: $cB\Omega$-model.

value of dB/dT. Therefore it is not worthwhile to repeat the calculation for higher temperatures because s/h_{exp} then usually exceeds its RT-value by about the same percentage. The various experimental values of h and s are

Table 9.33
Defect entropies and enthalpies in KBr.

Process	h (eV)	s (k-units)	Reference
Solubility of Cd^{2+}	1.28	3.6	[548]
Solubility of Sr^{2+}	1.14	7.3	[548]
Solubility of CO_3^{2-}	2.12	8.4	[548]
Schottky formation	2.53	10.3	[549]
	2.37	7.2	[550]
	2.33	8.1	[457] [a]
Cation vacancy migration	0.65	1.9	[549]
	0.67	2.5	[550]
Anion vacancy migration	1.22	7.3	[549]
	0.9	3.8	[550]
	1.08	6.3	[457] [a]
Association of "Ca^{2+} + cation vacancy"	−0.61	−2.1	[549]

[a] Dislocation model.

given in table 9.33. In view of the large uncertainty of the determination of the anion migration parameters (discussed for the previous materials) one can say that the general trend of the experimental data is compatible with the $cB\Omega$-model especially when the lower error bar of the straight line of fig. 9.35 is considered. Note also that association parameters of "Ca^{2+} + cation vacancy", though both negative, scale with other defect parameters. The solubility parameters show a significant scatter but such experiments have a large experimental uncertainty [287]. A repetition of these measurements is highly desirable.

The general remarks made for the analysis of the defect data in NaCl are also valid here. However, one should note that the values of the defect parameters of KBr are less reliable than those in the case of KCl. This is so because all the data included in fig. 9.35 come from the analysis only of conductivity data, whereas in the case of NaCl and KCl, diffusion data were simultaneously included in the fitting procedure.

9.6.5 Sodium iodide

From the expansivity data of Rapp and Merchant [290] we get at 570 K the values $\beta = 1.79 \times 10^{-4}$ K^{-1} and $b \simeq 1.1$. The elastic data of Cyrbu, Ul'yanow and Botaki [291] give $B_0^{SL} \simeq 180$ kbar and further, when converted to isothermal values at 570 K, $B \simeq 123$ kbar and $(\partial B/\partial T)_P \simeq -10^8$ dyn cm^{-2} K^{-1}. Equation (9.4) then gives $s/h_{exp} \simeq 4.3 \times 10^{-4}$ K^{-1} with a plausible uncertainty of $\pm 20\%$. These data are consistent with those published by Barsch and Schull [292]: $B_0^{SL} = 183.95$ kbar and $dB^S/dT = -0.073$ kbar K^{-1}. After correction the latter value gives $dB/dT \simeq -0.1$ kbar K^{-1}. The straight line in fig. 9.36 corresponds to the value of s/h_{exp} just mentioned. In the same figure the contents of table 9.34 are inserted. Only recently obtained defect parameters are considered, because as discussed by Kostopoulos, Reddy and Beniere [293], the analysis of the earlier defect data of Wylde [294] and Pershits and Veisman [295] had various shortcomings. Furthermore, recent ITC-experiments of Triantis and Kostopoulos [296] on NaI doped with small amounts (~ 5 ppm) of Ca^{2+} give for the bound cation vacancy motion the values $h^m = 0.56$ eV and $s^m = 2.4 \pm 0.5$ k which give $s^m/h^m = 3.7 \times 10^{-4}$ $K^{-1} \pm 20\%$ in good agreement with the $cB\Omega$-value of s/h_{exp}. The migration entropies included in the table have been obtained by assuming that the preexponential frequency in eq. (9.7) is equal to ν_{TO}; this approximation might however have introduced a small systematic shift in the experimental entropy values. In view of this uncertainty, the experimental and the calculated values of s/h_{exp} are in agreement.

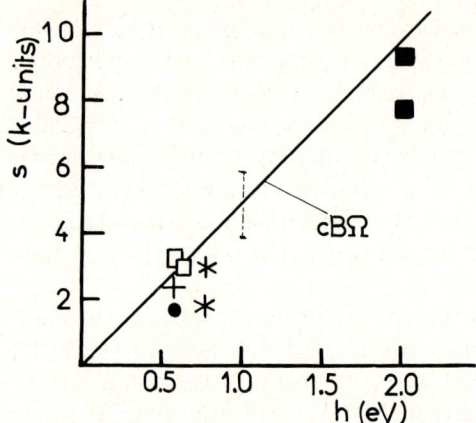

Fig. 9.36. Defect entropies and enthalpies of sodium iodide; ■: Schottky formation; *: anion migration; □: cation migration (two values); ●: bound cation migration; +: reorientation of Ca^{2+}-cation vacancy dipole; line: $cB\Omega$-model for 570 K. The error bars correspond to a plausible uncertainty of the calculated s/h_{exp}-value.

Another important point should be mentioned: at temperatures close to the melting point (i.e. about 50 K from T_M) a strong upward curvature has been observed by Kostopoulos, Beniere and Reddy [293,297] in both the conductivity and the (anion and action) diffusion plots. This experimental result verifies an earlier prediction of the authors [92] that at temperature close to the melting point anharmonicity becomes stronger and hence the

Table 9.34
Defect entpopies and enthalpies in NaI.

Process	h (eV)	s (k-units)	Reference
Schottky formation	2.02	9.33	a
	2.00	7.64	b
Cation vacancy migration	0.60	3.02	a
	0.58	3.23	b
Anion vacancy migration	0.75	1.68	a
	0.77	2.91	b
Bound cation vacancy migration; ITC technique	0.59	1.7	[120e]

[a] Ref. [551], parameters from the analysis of the conductivity and diffusion data without considering the strong curved part of the plots.
[b] Ref. [551], analysis includes long range electrostatic interactions.

bulk modulus should decrease faster than linearly. According to §9.1 this leads to an excessive decrease of g^f and therefore to an excessive creation of Schottky defects and an excessive increase of the mobility of both anion and cation vacancies. The plots of $\ln(\sigma T)$ and $\ln D^+$ (or $\ln D^-$) versus $1/T$ start showing in such cases an upward bend as explained in §7.3. According to the $cB\Omega$-model this should happen for all three almost at the same temperature, exactly as experimentally observed [293]. In this material there are the following properties which invalidate alternate explanations of the curvature suggested by other authors:

(1) the existence of vacancy pairs cannot explain the curvature, because it contributes only to diffusion and not to conductivity, whereas the curvatures of the plot of $\ln(\sigma T)$ or of $\ln D$ versus $1/T$ are strikingly similar; furthermore, the correlation factor is close to 0.78 [293] which means that the contribution of vacancy pairs to the diffusivity should be negligible.

(2) The curvature observed in the conductivity plot cannot be attributed only to the coexistence of cation and anion vacancies; this results from the fact that the separate study of diffusion of cations and anions shows the cation vacancies to be appreciably more mobile than the anions.

(3) If Frenkel defects [89] or some interstitialcy mechanism [90] were responsible for this pronounced curvature, they should displace the correlation factor far from the value of 0.78 which is characteristic only for the single vacancy mechanism.

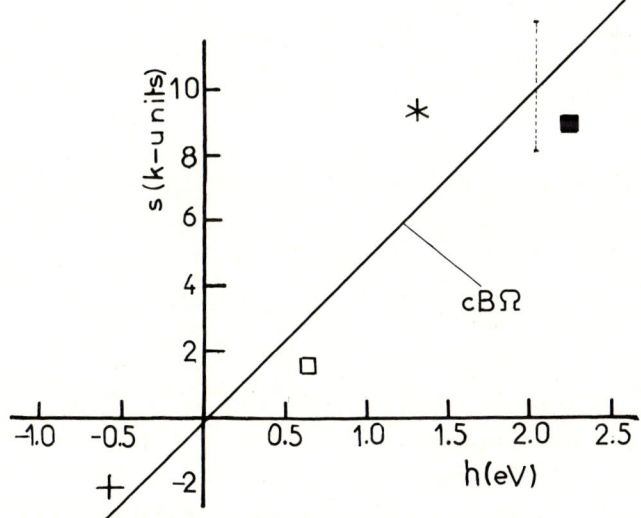

Fig. 9.37. Defect enthalpies and entropies of potassium iodide; *: anion migration; □: cation migration; ■: Schottky formation; +: association.

(4) The long-range electrostatic interactions (LDH-interactions) cannot explain the effect, because, as noticed by Kostopoulos, Beniere and Reddy [293,297] their inclusion leads only to an insignificant curvature.

9.6.6 *Potassium iodide*

Smith and Cain [276] give for 295 K $B = 115.1$ kbar and $\beta = 1.23 \times 10^{-4}$ K^{-1}. The values $dB^S/dT = -0.051$ kbar K^{-1} [289] and $B^S = 121.5$ kbar [276] for 295 K give $B_0^{SL} = 136.5$ kbar which in conjunction with the value of B for 295 K gives $dB/dT = -0.0725$ kbar K^{-1}. Inserting these data into eq. (9.4) and using $\exp \int_0^{RT} \beta dT \simeq 1.02$ we find $s/h_{\text{exp}} \simeq 4.36 \times 10^{-4}$ K^{-1} with a maximum uncertainty of 20% due again to the estimation of dB/dT. This value of s/h_{exp} gives the line drawn in fig. 9.37. The use of the value $dB^S/dT = -0.0594$ kbar K^{-1} given by Barsch and Schull [292] would increase it by 5%. Due again to the large uncertainty of the value of s/h_{exp} for RT, we shall not repeat the calculation at higher temperatures. The error bar in the figure is indicative of the expected variation of the line, due to the uncertainty of 20% of the calculated s/h_{exp}-value. An inspection of the figure shows that the defect parameters obey the $cB\Omega$-model if one considers the uncertainty in the drawing of the straight line. Table 9.35 contains the defect parameters measured by Chandra and Rolfe [298].

9.7 Silver halides

Silver chloride and silver bromide exhibit quite interesting properties in the high temperature region [299–307]; their ionic conductivity is found to be several times larger just below the melting point than would be expected from extrapolation of the low temperature data. For details we refer to the

Table 9.35
Defect entropies and enthalpies in KI.

Process	h (eV)	s (k-units)
Schottky formation	2.21	8.88
Cation vacancy migration	0.63	1.58
Anion vacancy migration	1.29	9.34
Association of "Sr^{2+} + cation vacancy"	−0.54	−2.2

review article by Friauf [65]. The most interesting fact emerging from the study of these materials is the high temperature conductivity (or diffusion) curvature which cannot be explained within the framework of the conventional Lidiard–Debye–Hückel (LDH) model. Proposals [302] for an additional interstitialcy mechanism have also proved to be unsuccessful [303]. The definite answer as to the origin of this curvature came from the pioneering experiment of Batra and Slifkin [305,306]. By studying the diffusion of sodium tracer in these materials they have definitely proved that the major part of the curvature is due to the temperature dependence of the formation parameters of cation Frenkel defects. In general, AgCl and AgBr exhibit a similar behaviour, their only difference being the various stronger "anomalies" in the latter [307]. Therefore we proceed to the explanation of the properties only of AgBr. As we shall see it provides a characteristic example for the application of the $cB\Omega$-model because, fortunately, its elastic constants have been measured up to the melting point. The explanation of all the "anomalies" reported in the literature is quite straightforward if one also carefully considers thermodynamics.

We simply note that a diffusion study in AgF has just been reported by Raaen, Svare and Fjeldly [308] by using the NMR-technique.

Silver bromide. We start the discussion from the linear part of the conductivity plot. Aboagye and Friauf (AF) [307] have obtained the migration and formation parameters inserted in fig. 9.38. In the same figure we have also inserted the migration parameters for the Na^+-motion reported by

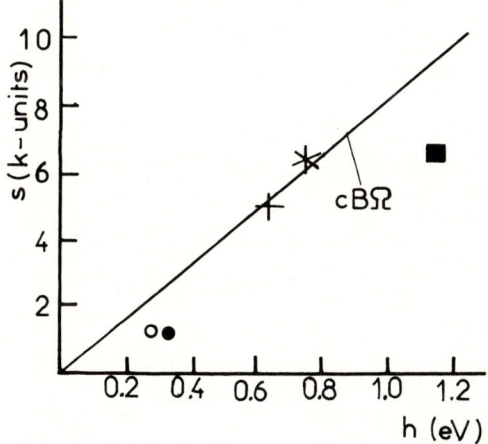

Fig. 9.38. Defect entropies and enthalpies of silver bromide; +: migration of Na^+ into a cation vacancy, Batra and Slifkin [305]; *: migration of Rb^+ into cation vacancy, Cardegna and Laskar [308a]; ■: formation of a cation Frenkel defect, Aboagye and Friauf [307]; ●: cation vacancy motion, [307]; ○: silver interstitial motion (mean values of ref. [307]).

Batra and Slifkin (BS) [305] and for Rb$^+$-motion by Cardegna and Laskar [308a]. Let us now examine whether these values are consistent with the prediction of the $cB\Omega$-model. The values of B measured by Loje and Schuele [309] for 300 and 195 K are 377 kbar and 419.5 kbar respectively. These values give $(dB/dT)_P = -0.4$ kbar/K and $B_0^{SL} = 498.4$ kbar. Using also the value $\beta = 1.05 \times 10^{-4}$ reported by Strelkow [310] for 300 K we immediately get from eq. (9.4) $s/h_{\mathrm{exp}} \simeq 7.4 \times 10^{-4}$ K^{-1} which gives the straight line drawn in the figure. We observe that the parameters of Batra and Slifkin (BS) nicely agree with the $cB\Omega$-model whereas the AF-values seem to slightly deviate from this scheme. This is in favour of the $cB\Omega$-model because as noticed by BS the curvature starts at lower temperatures than Aboagye and Friauf thought. In other words the latter workers have included in their analysis with temperature independent parameters a part of the curved region; therefore their parameters are not representative of the low temperature linear part of the conductivity region (see also below).

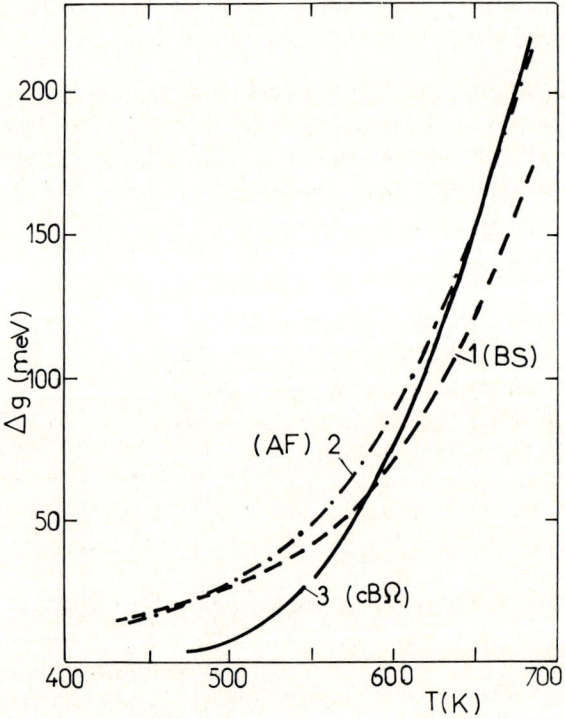

Fig. 9.39. Calculated (solid line) and experimental excessive decrease (dashed lines, AF = ref. [307], BS = ref. [305]) of g^f in AgBr.

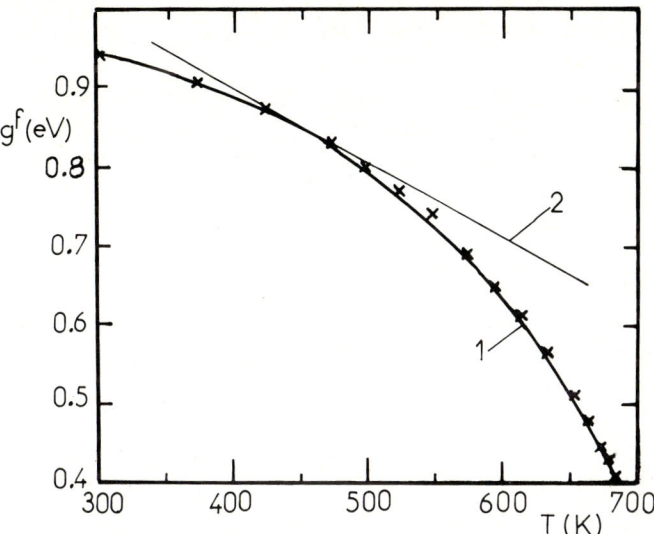

Fig. 9.40. Curve 1: calculated temperature dependence of g^f in AgBr from $cB\Omega$-model; curve 2: see text.

We now turn to the "anomaly" observed in the high temperature region. It is mainly due to the excessive decrease Δg^f of the Gibbs formation energy of cation Frenkel defects in respect to the normal linear decrease. In curve 2 of fig. 9.39 we plotted the Δg^f-values, as reported by AF, which are necessary for an explanation of their conductivity data. They assumed that the useful linear range lies between 170 and 245°C; Obviously, if they would have used a smaller "useful linear region" their low temperature parameters would have been different, thus leading to different Δg^f-values. Actually, BS have reported the somewhat smaller Δg^f-values of curve 1 which have to be considered to be more reliable.

A quite useful remark can now be made. The fact that AF have overestimated the Δg^f-values means that they – their value $h^f = 1.14$ eV being correct [305] – have underestimated the formation entropy. This fact is in accordance with the prediction of the $cB\Omega$-model which suggests a value of s^f close to 9–9.5 k, as can be extracted from fig. 9.38, instead of the point $s^f = 6.55$ k reported by AF.

We now proceed to a further treatment of the $cB\Omega$-model in order to examine whether it describes quantitatively the strong curvature observed. The points in fig. 9.40 are the g^f-values that result from an application of eq. (8.18); in this application we have used the h^f-value reported by AF and the expansivity measurements of Lawn [311]. The elastic constants measured by Tannhauser, Bruner and Lawson [312] from RT up to the melting point show an excessive fall of B near the melting point. The points of curve 1 were obtained by a fit of a polynomial of third degree [313]. Obviously, the $cB\Omega$-model predicts a strong nonlinear decrease of g^f versus T; furthermore it shows, in agreement with the observation of BS, that this nonlinear decrease starts earlier than the end of the useful linear region 443–518 K reported by AF. However, in order to have the same "system of reference" with the analysis performed by AF we employed a least square fitting to the points of fig. 9.40 between 420 K and 500 K and have drawn the straight line labelled with 2. The values obtained from this Straight line (SL) are labelled as g^{SL} and their difference with the actual g^f-values of curve 1 give the Δg-values inserted in fig. 9.39 (curve 3). A comparison of this curve with the experimental curves 1 and 2 indicates a fine agreement if one considers the experimental errors of the values used in the application of the $cB\Omega$-model. Especially the (adiabatic) elastic data have a large experimental error because they have been extracted from ultrasonic experiments close to the melting point. The $cB\Omega$-model can therefore account quantitatively for the strong curvature of AgBr especially as the experiment of BS excludes any additional mechanism.

On intuitive grounds it has been repeatedly proposed in the literature that the curved plot of the silver halides is the result of a fast decrease of the formation enthalpy due to the softening of the lattice close to the melting point. This proposal has lately found a theoretical support from three publications of Harwell's group [97,314]. It was stated that new accurate potentials were discovered which lead to formation enthalpies for Schottky and Frenkel defects that quickly decrease with temperature. Then by considering a temperature independent formation entropy the curvature observed in conductivity and diffusion plots could be explained. Equation (3.25), however, shows that such a description of the experimental data with a temperature independent entropy is thermodynamically inconsistent. Furthermore, in §3.6 it was proved by means of thermodynamics that a *temperature-decreasing enthalpy leads to a downward curvature and not to an upward one*. We therefore conclude that the only consistent explanation of the curvature observed in AgBr is that proposed on the basis of the $cB\Omega$-model; it predicts (see §8.2) a nonlinear decrease of g^f, thus on thermodynamical grounds leading to a simultaneous increase of the formation enthalpy and entropy. This result is also consistent with the specific heat measurements of Kanzaki [315], Kobayashi [316] and Bremer and Nölting [316a]. The plot of $\ln(T^2\Delta C_P)$ versus $1/T$ shows an upward bend which from the thermodynamics of specific heats indicates a simultaneous increase of h^f and s^f (however, see §7.2 for the shortcomings of the analysis of the specific heat data in a T-region with considerable anharmonicity).

We finally turn to the upward curvature observed in the bromine diffusion plots of AgBr. This has been described in the literature [65] as a "double anomaly" within the following scheme: "The excessive increase of cation Frenkel defects in silver halides near the melting point should suppress the number of anion vacancies through the mass action law. Therefore a downward curvature in the anion diffusion plot would be expected in contrast to experiments". However, the above scheme is questionable if one carefully considers the mass action law for the simultaneous presence of Frenkel and Schottky disorder in conjunction with the influence of anharmonicity. The anion diffusion takes place through the anion vacancies (for simplicity we do not consider here any contribution from vacancy pairs); their concentration x_a is given by [see eq. (4.49)]

$$x_a = \frac{1}{\sqrt{2}} \exp\left[-\left(g^{f,S} - \tfrac{1}{2}g^{f,F}\right)/kT\right], \tag{9.22}$$

where $g^{f,S}$ and $g^{f,F}$ denote the formation Gibbs energies for Schottky and cation Frenkel defects respectively. If we assume for a moment that $g^{f,S}$ decreases only linearly with temperature, then – in view of the fact that $g^{f,F}$ decreases faster than linearly – the difference $g^{f,S} - (g^{f,F}/2)$ is an increasing function of T; therefore the bromine plot $\ln D$ versus $1/T$ has to show a downward bend which contradicts the observed behaviour. In order to explain the upward curvature the difference $g^{f,S} - (g^{f,F}/2)$ has to decrease faster than linearly which – in view of the form of $g^{f,F}$ versus T – means that $g^{f,S}$ decreases also faster than linearly. This latter behaviour cannot be a priori precluded and hence the form of the anion diffusion plot cannot be in any event considered as having to be "anomalous". Let us now turn to the $cB\Omega$-model. The Gibbs energies $g^{f,F}$ and $g^{f,S}$ can be written as

$$g^{f,F} = c^{f,F}B\Omega, \tag{9.23}$$

$$g^{f,S} = c^{f,S}B\Omega. \tag{9.24}$$

These equations indicate that $g^{f,S}$ and $g^{f,F}$ both decrease faster than linearly in the high temperature region because of the nonlinear decrease of B already mentioned. The excessive decreases $\Delta g^{f,F}$ and $\Delta g^{f,S}$ of the energies $g^{f,F}$ and $g^{f,S}$ in respect to their linear extrapolation from the low temperature region are given by:

$$\Delta g^{f,F} = c^{f,F}(B^{SL} - B)\Omega \tag{9.25}$$

and

$$\Delta g^{f,S} = c^{f,S}(B^{SL} - B)\Omega, \tag{9.26}$$

where B^{SL} is the linear extrapolation of the graph $B = f(T)$ from the low temperature to the high temperature region ($B^{SL} > B$; see fig. 9.2). As cation Frenkel disorder predominates (this is generaly accepted) we have:

$$c^{f,F} < c^{f,S}. \tag{9.27}$$

By considering now the inequality (9.27) we see that eqs. (9.25) and (9.26) imply $\Delta g^{f,S} > \Delta g^{f,F}$ at every temperature and hence the difference $g^{f,S} - (g^{f,F}/2)$ – which regulates x_a through eq. (9.22) – has to decrease faster than linearly. In the high temperature region the $cB\Omega$-model thus indicates an excessive increase of the concentration of anion vacancies and hence explains the upward curvature of the bromine diffusion plot even if there were no contributions from vacancy pairs. A quantitative account of this curvature on the $cB\Omega$-basic can be done in the manner described by ref. [318].

A last remark should be added. In all the above discussions only the Δg^f-values resulting from the nonlinear decrease of g^f have been considered. In the conduction and diffusion experiments, according to the $cB\Omega$-model, another contribution to the excessive decrease of g^{act} comes from the nonlinear decrease of the Gibbs migration energy g^m:

$$g^m = c^m B \Omega \tag{9.28}$$

and hence

$$\Delta g^m = c^m (B^{SL} - B)\Omega, \tag{9.29}$$

where g^m refers either to cation vacancy or to interstitial motion. As g^m is always smaller than g^f a comparison of eq. (9.28) to (9.23) and (9.24) indicates that c^m is smaller than $c^{f,F}$ or $c^{f,S}$ by a factor larger than three; this factor will be seen later in table 12.5. Then the comparison of eq. (9.29) to (9.25) and (9.26) indicates that the major source of the curvature in the Arrhenius plots comes from the formation energies and not from the migration ones. This latter conclusion is in fundamental agreement with the observations of Batra and Slifkin [305,306].

Summarizing the above considerations one sees that the $cB\Omega$-model can naturally explain all the observed "anomalies" (conductivity curvature, sodium diffusion curvature and bromine diffusion curvature) whereas the other conventional theories fail to do so.

Determination of $g^{f,S}$ for silver halides. A useful prediction can be made within the frame of the $cB\Omega$-model on the determination of the lower limits of $g^{f,S}$ for each temperature; this could not be extracted from the experimental data. The only point known from the experiments of Fouchaux and Simmons [319,320] on AgCl is that the formation enthalpy must be larger than 1.45 eV whereas the formation entropy should exceed $7\,k$. Equations (9.23) and (9.24) give $g^{f,F}/g^{f,S} = c^{f,F}/c^{f,S}$. Writing eq. (8.19) for both types of defects one gets

$$g^{f,F}/g^{f,S} = v^{f,F}/v^{f,S}, \tag{9.30}$$

where $v^{f,F}$ and $v^{f,S}$ denote the formation volumes per (cation) Frenkel defect and Schottky defect respectively. Abey and Tomizuka [321] have reported

$v^{f,F} = 16.7$ cm³/mole which, as expected, is appreciably smaller than the molecular volume $2\Omega = 25.7$ cm³/mole. But we know from the experiments of Lazarus and coworkers [322,323] that in ionic solids $v^{f,S}$ characteristically exceeds the molecular volume. Accepting this for AgCl we have from eq. (9.30) for any temperature:

$$g^{f,S} > g^{f,F} 2\Omega / v^{f,F}$$

or

$$g^{f,S} > 1.54 g^{f,F}.$$

Considering that $h^{f,F} = 1.45$ eV and $s^{f,F} = 9.4\ k$ [307] we obtain, e.g. for 400 K, $g^{f,F} \simeq 1.126$ eV and hence $g^{f,S} > 1.734$ eV. This latter result is consistent with the limits of Fouchaux and Simmons [319] but strongly invalidates the microscopic proposal [97] that for AgCl $h^{f,S} = 1.505$ eV and $s^{f,S} = 8\ k$ (which give for 400 K, $g^{f,S} = 1.23$ eV).

Repeating the above calculation for AgBr ($v^{f,F} = 14$ cm³/mole, $2\Omega = 29$ cm³/mole) [324] one concludes that

$$g^{f,S} > 2 g^{f,F}$$

and hence approximately:

$$h^{f,S} > 2 h^{f,F}.$$

The last inequality is useful in the following sense: in the high temperature region $h^{f,F}$ takes large values (larger than 1.13 eV reported by AF) for the linear region because of the upward curvature of conductivity or Na^+-diffusion plots; therefore $h^{f,S}$ should take still larger values thus explaining the unusually large slope observed by Batra and Slifkin [71] in the high temperature region of the bromine diffusion plot.

9.8 Summary of the progress in the study of point defects in ionic crystals

During the last two decades considerable progress has been made in the study of point defects of alkali and silver halides, both in the experimental and in the theoretical field.

Referring to experimental studies a *breakthrough* was achieved in two directions which had a very strong impact on our knowledge in this field: (1) the diffusion studies of Slifkin and coworkers definitely proved that the physical origin of the high-temperature conductivity (or diffusion) curvature in silver halides is the temperature dependence of the formation enthalpy and entropy, in agreement to the suggestions [92,93] stressing the role of anharmonicity in the study of point defects; (2) the conductivity studies at various pressures of Lazarus and co-workers (see §10.4.2) definitely showed that the relaxation volume of Schottky defects in alkali halides is *positive*.

Concerning the theoretical studies, earlier microscopic calculations consistently disagreed with the experimental findings described above; more precisely: these early calculations led (1) to an origin of the curvature which was attributed either to the operation of an additional mechanism (in alkali halides [90]) or to a *decrease* [96,97] of the formation energy u^f (or enthalpy h^f) with temperature, whereas thermodynamics indicates [352] an enthalpy *increasing* with temperature, and (2) to *negative* relaxation volumes [325–327]. Since 1979 it was suggested [630] that the distinction between "isochoric" and "isobaric" defect parameters had been overlooked in the above calculations and that the hitherto published microscopic energies do *not* correspond to the energy u^f (or the enthalpy h^f) but to a quite different physical quantity, i.e. to u^* (see §A.8). Later, calculations have actually considered the above distinction and hence the disagreement between theory and experiments disappeared (see also §A.11, note 4).

10 | THE $cB\Omega$-MODEL: DEFECT VOLUME AND GIBBS ENERGY

10.1 Introduction

According to the $cB\Omega$-model the defect Gibbs energy g is directly connected to the corresponding defect volume v through eq. (8.22):

$$g = Bv\left(\frac{dB}{dP} - 1\right)^{-1}. \tag{10.1}$$

This relation has to be valid for formation, migration or activation processes. The interesting point is that the right-hand side of this equation comes from isothermal pressure experiments whereas the other comes usually from isobaric heating experiments.

The use of the above connection is obvious; it can directly lead to the determination of the defect volume which is usually extracted from pressure experiments with a considerable uncertainty (about 10%). The knowledge of the value of v is useful because, for instance, the ratio v^f/Ω can in some cases directly show whether Schottky or Frenkel disorder predominates in a crystal; but more generally v^{act} shows how the diffusion (or conductivity) varies upon compression.

One important point should be noticed here: the above connection between g and v is valid regardless if the quantity c involved in the $cB\Omega$-formulae is explicitly *temperature independent* or not. Equation (10.1) can be proved to be valid under the *only* restriction that the compressibility κ^d of the defect volume cannot exceed the bulk compressibility κ by a large factor. The latter restriction is generally accepted and is well justified from microscopic concepts (see ch. 14). Therefore eq. (10.1) is expected to be valid for all cases. Inversely, if in a given case the above equation is found not to be fulfilled this directly means that κ^d considerably exceeds κ, e.g. by one order of magnitude.

Such a violation seems extremely unlikely to occur for formation processes of the usual defects i.e. a vacancy or interstitial. Equation (10.1) may be used as a helpful criterion for the correctness of microscopic calculations, this being so because at $T = 0$ K it reveals that a microscopic calculation leading to a correct value of the formation enthalpy h_0^f (at $T = 0$ K) has to lead simultaneously to a correct formation volume. However, as known, there are numerous calculations in the literature which claim that they succeed to calculate h_0^f but fail to give even the correct sign of v^f [325–327]. In such cases the self-consistency of the calculation and the correctness of the potentials used can be checked by computing the relaxations of the atoms (or ions) around the vacancy (or the defect) and examining whether they lead to a value which is connected to the formation enthalpy through eq. (10.1). For recent developments, see §A.8.

In this chapter we check the validity of eq. (10.1) for various categories of solids. In each case we usually compute the right-hand side – which comes from pressure experiments – and compare it to the experimental value of g which is usually obtained from a quite different experiment. Obviously, in this check no adjustable parameters are present so that in essence a basic interconnection of defect data with elastic data emerges.

Another point which is checked in this chapter is the question if the quantity v/g $[= B^{-1}((dB/dP) - 1)]$ is independent of the type of the process, as predicted by the $cB\Omega$-model. For this check we plot, whenever possible, the values of v for various processes versus the corresponding values of g. Not only do we observe a linear dependence but further we *also* find that the gradient of the straight line is just that predetermined from bulk properties as predicted by the $cB\Omega$-model.

Connection of v/g with the mean Grüneisen constant. The relation (10.1) can give the value of v/g as long as the quantities B and $(dB/dP)_T$ are known. However, in some cases the latter quantity is unknown so that we indicate a slightly different method for the calculation of v/g.

Slater [328] has initially proved the relation

$$\gamma = \frac{1}{2} \left. \frac{dB}{dP} \right|_T - \frac{1}{6};$$

it has been later corrected by Dugdale and MacDonald [329] and by Shanker, Gupta and Sharma [330] to

$$\gamma = \frac{1}{2} \left. \frac{dB}{dP} \right|_T - \frac{1}{2}, \qquad (10.2)$$

from which one can estimate the quantity $(dB/dP)_T$:

$$\left. \frac{dB}{dP} \right|_T = 2\gamma + 1 \qquad (10.3)$$

and hence eq. (10.1) can be written as

$$\frac{v}{g} = \frac{2\gamma}{B}. \tag{10.4}$$

The Grüneisen constant γ, as introduced by Slater, is an average quantity as he has obtained it – within the Debye approximation of monoatomic solids – by assuming that all the modes vary upon compression in a similar way:

$$-\frac{d \ln v^i}{d \ln V} \equiv \gamma.$$

Attention is drawn to the point that eq. (10.1) – and hence eq. (10.4) – is valid as long as the quantity c is considered as exactly pressure independent. In the case of a possible pressure dependence eq. (10.1) is not completely valid anymore; we can easily prove [646] that it has to be changed into

$$v = \frac{g}{B}\left(\frac{dB}{dP} - 1 - \frac{\kappa^c}{\kappa}\right), \tag{10.5}$$

or

$$v = \frac{2g}{B}\left(\gamma - \tfrac{1}{2}\frac{\kappa^c}{\kappa}\right), \tag{10.6}$$

where κ^c is the "compressibility" of c defined by

$$\kappa^c \equiv -\frac{1}{c}\left.\frac{dc}{dP}\right|_T. \tag{10.6a}$$

Equations (10.5) and (10.6) can be used for the determination of the bounds of v/g because in ch. 14 it is shown that $|\kappa^c|/\kappa$ lies only between zero and unity. Equations (10.1) and (10.4), which are sufficiently accurate for most cases, result from eqs. (10.5) and (10.6) by setting $\kappa^c = 0$.

Prediction of the sign of v in terms of the Grüneisen constant. Equation (10.6) indicates that the defect volume is negative provided that $\gamma < \tfrac{1}{2}\kappa^c/\kappa$. Taking the lower limit -1 of κ^c/κ we find as the necessary condition for a negative value of v the inequality $\gamma < -1/2$. However, there is the possibility of v being negative for positive values of γ but only up to $1/2$. Materials having negative activation volumes are B4-AgI, δ-cerium, γ-cerium, ε-plutonium, etc. They will be discussed in this chapter.

Connection of v/g with melting properties. Cornet [331] and Languille et al. [332] proved by making use of the classical Lindemann law that close to the melting point T_M the value of the mean Grüneisen constant is given by

$$\gamma = \tfrac{1}{3} + \tfrac{1}{2}B\frac{\partial \ln T_M}{\partial P}. \tag{10.7}$$

It is noticeable that in deriving the above expression no empirical coefficient

Table 10.1
Self-diffusion; comparison of $B_0^{act}[(dB/dP)-1]^{-1}$ with the experimental values of g^{act} for fcc metals.

Metal	T (K)	B (kbar)	dB/dP	$v_{exper.}$ (cm^3/mole)	$B_0(dB/dP-1)^{-1}$ (eV)	h (eV)	s (k-units)	g (eV)
Pb	595.6	369.6 [a]	7.08 [b]	13 [c]	0.82	1.09 [d,e]	5.3 [f]	0.82
	595.6			0.64Ω ± 10% [g]	0.77	1.09 [d,e]	5.3 [f]	0.82
Ag	630.7 [h]	960.5 [i]	6.5 [j]	9.2 [k,l]	1.67	1.86 [m]	3.65 [m]	1.66
	630.7		6.78 [n]	9.2 [k,l]	1.59	1.86 [m]	3.65 [m]	1.66
	300	1007 [o]	~6.5	9.2 [k,l]	1.75			1.76
Cu	613.7 [p]	1267 [q]	5.87 [r]	6.5 [l]	1.75	2.08 [s]	3.5 [s]	1.90
	613.7 [p]	1267 [q]	5.7 [t]	6.5 [l]	1.81	2.12 [u]	4.15 [u]	1.90
	300	1370	5.7 [t]	~6.5 [l]	1.97	2.12 [u]	4.15 [u]	2.01
Al	700	632.6 [v]	5.31 [w]	0.71Ω ± 18% [x]	1.12	1.476 [y]	5.25–5.82 [y]	1.12 ~ 1.16

[a] Vold, Glicksman, Kammer and Cardinal [176]
[b] H.B. Vanfleet (private communication); high temperature value (T = 600 K).
[c] Nachtrieb, Resing and Rice [552]; activation volume from self-diffusion under pressure.
[d] Nachtrieb and Handler [170]; tracer measurements between 479–595 K as reanalysed by LeClaire [172].
[e] LeClaire [172].
[f] From the value $D_0 = 0.69$ cm^2/s given in ref. [172] by using $v_D = 1.8 \times 10^{12}$ s^{-1}, $a = 4.94$ Å and $f = 0.78$.
[g] Hudson and Hoffman [175]; activation volume from self-diffusion under pressure; this value seems to be somewhat smaller than the real one due to the pressure calibration (H.B. Vanfleet, private communication).
[h] The calculation has been made at this temperature for the following reasons: (1) it is the lowest temperature at which self-diffusion has been measured (see ref. [539]). (2) the value of dB/dP has been measured at RT and is assumed to be roughly valid at 630.7 K, (3) the value of B has been measured only up to 800 K.

[i] Varshni [165]; he quotes adiabatic values. Correction to the isothermal value has been made by using the specific heat and expansivity data reported by Vollmer and Kolhaas [553].

[j] Daniels and Smith [554]; they give the value $dB^S/dP = 6.18$ which has to be increased by $\sim 5\%$ in order to give dB/dP.

[k] Tomizuka, Lowell and Lawson [555]; activation volume at high T for self-diffusion under pressure.

[l] Beyeler and Adda [147]; activation volume at high T for self-diffusion under pressure.

[m] Backus, Bakker and Mehrer [539]; the values of the activation enthalpy and entropy are obtained from a least squares fitting to a straight line of the self-diffusion data. In order to extract the activation entropy we have used the values $a = 4.08$ Å, $\nu_D = 4.7 \times 10^{12}$ s^{-1}, $f = 0.78$; we notice that when one uses the values $D_0 = 0.046$ and $h = 1.76$ eV (Peterson [73]) one obtains a g-value comparable to that given in the table.

[n] Fisher, Manghnani and Katahara [343]; this value has been derived from shock wave data.

[o] Kittel [556].

[p] This is the lowest temperature at which self-diffusion data exist (Lam et al. [161]); the selection of this temperature has been made for the same reasons as mentioned in footnote h.

[q] Chang and Himmel [157]; the correction from B^S to B has been made by using the specific heat data of Brooks [54].

[r] By increasing by $\sim 5\%$ the value $dB^S/dP = 5.95$ at RT reported in note j above; this value practically agrees with the recent results of Chiarodo, Green, Spain and Bolsaitis [557].

[s] Bartdorff, Neumann and Reimers [155]; these workers have made a least squares fit of the low-temperature self-diffusion data reported by Lam, Rothman, and Nowicki [161] and by Maier, Bassani and Schuele [82]; the activation entropy is calculated from the value $D_0 = 0.216$ cm^2/s by using $a = 3.61$ Å, $\nu_D = 6.52 \times 10^{12}$ s^{-1} and $f = 0.78$.

[t] By increasing by $\sim 5\%$ of the value $dB^S/dP = 5.44$ reported by Hiki and Granato [558].

[u] Values reported in ref. [155] when one makes a simultaneous least squares fit to a straight line of the data reported by Lam et al. [161], Maier et al. [82] Bartdorff et al. [155] and by Rothman and Peterson [538].

[v] Tallon and Wolfenden [137].

[w] RT-value reported by Schmunk and Smith [559].

[x] Engardt and Barnes [560]; activation volume for self-diffusion with the spin-echo technique of NMR in the temperature range 673–723 K.

[y] Lundy and Murdock [146]. The entropy is obtained from $D_0 = 1.71$ cm^2/s by using either $\nu_D = 6.8 \times 10^{12}$ s^{-1} or $\bar{\nu} = 3.83 \times 10^{12}$ s^{-1} with $h^m = 0.7$ eV.

in Lindemann's law had to be assumed which could introduce some uncertainty; it directly resulted from the relation $\nu \propto \sqrt{T_M/(M\Omega^{2/3})}$ which has a good justification (see ch. 13).

Inserting eq. (10.7) into eq. (10.6) we get:

$$\frac{v}{g} = \frac{2}{B}\left(\frac{1}{3} + \tfrac{1}{2}B\frac{\partial \ln T_M}{\partial P} - \tfrac{1}{2}\frac{\kappa^c}{\kappa}\right) \tag{10.8}$$

or by considering the Clausius–Clapeyron law:

$$\frac{v}{g} = \frac{2}{B}\left(\frac{1}{3} + \tfrac{1}{2}B\frac{\Delta H}{\Delta V} - \tfrac{1}{2}\frac{\kappa^c}{\kappa}\right), \tag{10.9}$$

where ΔH is the latent heat of fusion and $\Delta V \equiv V_{\text{liquid}} - V_{\text{solid}}$.

Due to the fact that the bounds of $|\kappa^c|/\kappa$ are narrow enough, eqs. (10.8) and (10.9) can determine v/g with sufficient accuracy considering that the experimental errors of v are usually very large. However, as explained in ch. 14, the approximation $\kappa^c = 0$ is usually very good.

General thermodynamics demand that the quantities g and B have to be positive. Therefore the sign of v depends on the sign of terms in parentheses of eqs. (10.8) and (10.9). Hence the above equations have the important property of predicting the sign of v as a function of bulk properties only, which are usually well known.

A final remark should be added: some materials (e.g. δ-Pu) contract upon irradiation. Equation (10.8) might be useful for the prediction of these properties because v can refer to any type of isolated defects.

The insertion of eq. (8.18) into eq. (10.8) gives

$$\frac{v}{h_{\text{exp}}} = \frac{2b}{B_0^{\text{SL}}}\left(\frac{1}{3} + \tfrac{1}{2}B\frac{\partial \ln T_M}{\partial P} - \frac{1}{2}\frac{\kappa^c}{\kappa}\right), \tag{10.10}$$

which, to a *rough* approximation, becomes

$$\frac{v}{h_{\text{exp}}} \simeq \frac{\partial \ln T_M}{\partial P} \tag{10.11}$$

or

$$\frac{v}{h_{\text{exp}}} \simeq \frac{\Delta H}{\Delta V}. \tag{10.12}$$

This relation is the famous empirical law of the "corresponding states" of Nachtrieb and coworkers.

Table 10.2
Calculated and experimental activation (self-diffusion) volumes in fcc metals.

Metal	$v_{\text{exper.}}$ (cm^3/mole)	$v_{\text{calc.}}$ (cm^3/mole)	$v/g|_{\text{exper.}}$ 10^{-11} cm^3/erg	$B^{-1}(\mathrm{d}B/\mathrm{d}P - 1)$ [a] (10^{-11} cm^3/erg)
Pb	13	13	1.645	1.645
Ag	9.2	9.16 ~ 9.24	0.57 ~ 0.61	0.57 ~ 0.6
Cu	6.5	6.5 ~ 6.8	0.36	0.37 ~ 0.39
Al	7.3	7.3 ~ 7.6	0.65 ~ 0.68	0.68

[a] The data are from table 10.1, given at high temperatures.

10.2 Metals

10.2.1 fcc metals

In table 10.1 we give all the necessary data for the check of eq. (10.1) for various fcc metals. The quantities g and v refer to the self-diffusion activation process; the latter quantity has a large experimental uncertainty, around 10% or even more. Both for high and low temperatures a satisfactory agreement is observed. This agreement inversely allows the calculation of v^{act} from experimental values of g^{act} with the help of relation (10.1). The results of this calculation for high temperatures have been inserted into table 10.2. This agreement is significant because it shows that one can actually find v^{act} from g^{act}, which is useful because v^{m} is extremely difficult to calculate [35] and furthermore the calculation of v^{f} is connected with considerable difficulties and depends on the specific model employed [143,333–336]. Figure 10.1 shows that v^{act} has no direct connection with g at the temperatures where self-diffusion is measured whereas in fig. 10.2 their connection through the $cB\Omega$-model is directly evident.

Summarizing the present discussion in fcc metals we can say that the inspection of tables 10.1 and 10.2 shows that actually the quantity $Bv^{\text{act}}[(\mathrm{d}B/\mathrm{d}P) - 1]^{-1}$ is comparable to the activation Gibbs energy g^{act}.

10.2.2 bcc metals

The case of sodium has been extensively discussed in ch. 9 because of its special features and the volume of experimental data available. In this chapter we shall examine the validity of relations (10.1) and (10.5) for other bcc metals as K, β-Ti, δ-Ce, γ-Yb and ε-Pu, which, as known, have (except possibly K) some peculiar properties. For instance K, β-Ti and γ-Yb have *positive* self-diffusion activation volumes in contrast to δ-Ce and ε-Pu. It is

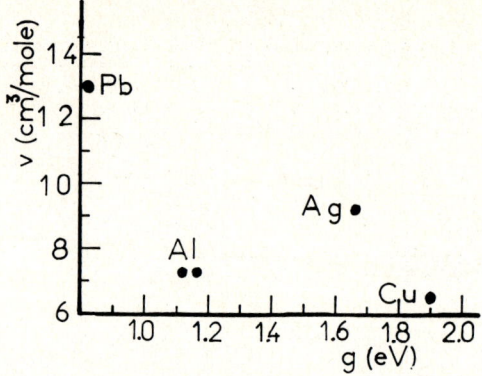

Fig. 10.1. Connection of v and g of fcc-metals from self-diffusion experiments (compare with fig. 10.2).

therefore of major importance to see whether this behaviour is in accordance with the $cB\Omega$-model. In fig. 10.3 the validity of eq. (10.1) in demonstrated for K, β-Ti and γ-Yb; the data used are those discussed in the following.

Potassium. Schouten and Swenson [215] have found $B \simeq 31$ kbar close to RT whereas dB/dP is around [200,337] $3.9 \sim 4$. Kohler and Ruoff [338] give $v^{act} \simeq 26.0$ cm^3/mole ($\simeq 0.59\Omega$) from creep measurements. Therefore the quantity $Bv/[(dB/dP) - 1]$ (i.e. by taking eq. 10.1) gives: $g^{act} = 0.284$

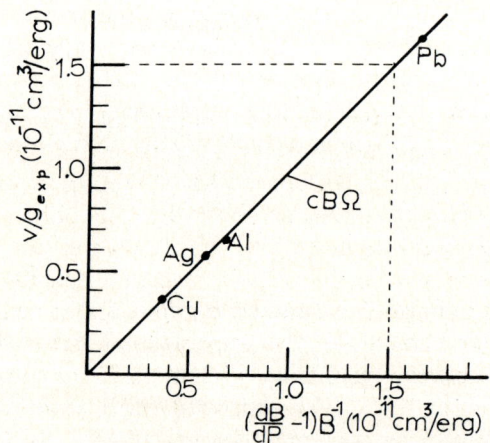

Fig. 10.2. Demonstration of the correctness of eq. (10.1) for fcc-metals. Line: $cB\Omega$-model.

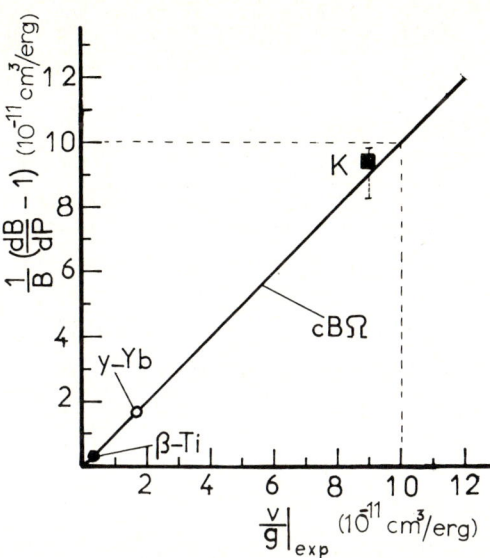

Fig. 10.3. Demonstration of the correctness of eq. (10.1) for bcc-metals. Line: $cB\Omega$-model.

eV ± 10%. Mundy et al. [214] found $h^{act} \approx 0.4$ eV whereas their value of $D_0 = 0.16$ cm^2 s^{-1} gives $s^{act} \simeq 3.74\ k$ by using $a = 5.225$ Å, $\nu = \nu_D = 1.9 \times 10^{12}$ s^{-1} and $f = 0.727$. An experimental value of $g^{act} \simeq 0.3$ eV therefore results for room temperature, which compares favourably with the value 0.284 eV predicted from eq. (10.1). A more precise investigation (analogous to that of §9.3.1) reveals that the experimental value of g^{act} should be slightly smaller than 0.3 eV thus leading to a better agreement with the calculated value.

β-titanium. The activation volume for self-diffusion has been obtained by Jeffery [339] with a considerably uncertainty, i.e. $v^{act} = 3.6 \pm 1$ cm^3/mole = $0.33 \pm 0.1 \Omega$ at $T \simeq 1273$ K. However, he has noticed that "one may, without greatly stretching either the imagination or the data, obtain a value ranging from 2 to 8 cm^3/mole, but in any case v^{act} is positive". Jeffery's result that v^{act} is positive anyhow invalidates earlier measurements [340] which gave a negative sign. An intuitive suggestion concerning the reason why initial measurements in Ti showed negative activation volumes is given by Lazarus [341]. According to Fisher and Dever [342] the isothermal bulk modulus, at $T \simeq 1273$ K, is around 909 kbar whereas dB/dP should be around 3.34 [343]. In view of the fact that v^{act} has a considerable uncertainty one can use as a first approximation $g^{act} \simeq h^{act} = 1.43$ eV. This is a value for low temperatures ($\sim$ 1273 K); due to the fact that the Arrhenius plot is strongly curved it can be safely used only for the low temperature part.

Let us now examine the validity of eq. (10.1). Inserting the above values of $g^{act} \simeq h^{act}$, B and dB/dP we get $v^{act} = 3.6$ cm^3/mole, in excellent agreement with the experimental value of 3.6 ± 1 cm^3/mole. If the value $s^{act} = -3\ k$ reported by Walsoe de Reca and Libanati [344] is used, one finds at $T = 1273$ K in conjunction with $h^{act} \simeq 1.43$ eV the value $g^{act} = 1.76$ eV and hence $v^{act} \simeq 4.4$ cm^3/mole which again agrees with the experimental v^{act}-value considering the large experimental error. To the best of our knowledge microscopic calculations of v^{act} have not been reported to date.

γ-*ytterbium*. Studying the self-diffusion between 1003 and 1073 K and for pressures between 4.5 and 10 kbars Fromont [345] found that the activation volume is 0.59 Ω or 15.7 ± 0.4 cm^3/mole. The activation enthalpy is $h^{act} = 1.255$ eV which with a typical value of $s^{act} \simeq 3\ k$ gives at the highest temperature of the measurements $g^{act} \simeq 0.978$ eV and therefore $v/g = 1.67 \times 10^{-11}$ cm^3/erg.

We now turn to the value of v/g predicted by the $cB\Omega$-model. Fromont suggested $\gamma \simeq 1.13$ and $B \simeq 136$ kbar; eq. (10.4) then gives $v/g = 1.66 \times 10^{-11}$ erg/cm^3 which is in excellent agreement with the experimental one.

One further point should be mentioned. We shall see below that some other bcc metals of the same group have negative activation volumes. The positive value of v^{act} for γ-Yb was anyhow predicted from the $cB\Omega$-model; this is evident from eq. (10.6) which can yield the order of magnitude of the lower bound of v^{act}. For instance by setting $(|\kappa^c|/\kappa)_{max} \simeq 1$ and $\gamma = 1.13$ one definitely gets a positive value for v^{act}.

δ-*cerium (bcc) and ϵ-plutonium*. The metal cerium has a stable fcc structure from 283 K to 993 K. It then transforms to the bcc form (δ-cerium) which subsists up to the melting point ($\sim$ 1023 K) [346]. δ-cerium and ϵ-plutonium have the outstanding property that they increase their density upon transformation from a bcc metallic phase to the liquid phase. It may not be fortuitous that both of them also show negative self-diffusion activation volumes, i.e. self-diffusion coefficients that increase upon compression. It is therefore of high interest to examine whether these negative activation volumes are predictable from the $cB\Omega$-model. It should be preliminarily noticed that their Arrhenius plots are linear in contrast to all the other bcc metals for which curved Arrhenius plots have been detected. This exception might be due to the fact that the bcc phase of cerium and plutonium extends only over a narrow range of temperatures thus creating difficulties in detecting a curvature.

The self-diffusion measurements of Dariel, Dayan and Languille [346] showed that for δ-cerium $h^{act} = 0.93 \pm 0.03$ eV and (by taking $\nu = \nu_D$)

§10.2]   Metals   279

$s^{act} = 1.7 \pm 0.35$ k; therefore at $T = 1000$ K one gets $g^{act} = 0.78 \pm 0.06$ eV. A little later Languille, Dariel, Calais and Coqblin [347] gave somewhat different values i.e. $h^{act} = 0.88$ eV and $s^{act} = 1.51$ k which however lead, at $T = 1000$ K, practically to the same value of $g^{act} = 0.75$ eV. Studying the self-diffusion under pressure Languille, Calais and Fromont [332] found that the activation volume is around -2 cm^3/mole i.e. -0.1 Ω with an uncertainty of, at least, around 10%. Therefore the experimental value of v/g is around -2.7×10^{-12} erg/cm^3 with an uncertainty around 15%.

We now turn to the prediction of the $cB\Omega$-model concerning δ-cerium for which no elastic data exist in the literature. As a rough approximation we shall take the RT-value $B \simeq 200$ kbar as suggested by Languille et al. [347] and further the $cB\Omega$-formulae of v/g in function of the melting properties. The quantity $\partial \ln T_M/\partial P$ according to Languille et al. [332] is negative and around -5×10^{-12} erg/cm^3. Therefore eq. (10.8) gives the bounds of the ratio v/g predicted from the $cB\Omega$-model: $v/g = -1.7 \times 10^{-12}$ erg/cm^3 when $\kappa^c = 0$ and $v/g = -6.7 \times 10^{-12}$ erg/cm^3 when $(\kappa^c/\kappa)_{max} = 1$. We therefore see that in spite of the rough approximation the experimental value of $v/g = -2.7 \times 10^{-12}$ erg/cm^3 lies between the limiting *negative* values predicted by the $cB\Omega$-model.

We finally turn our attention to ε-Pu for which again the elastic data under pressure are unknown to us. Cornet's self-diffusion studies [331] give at $T = 800$ K the value $g^{act} = 0.63 \pm 0.04$ eV whereas the earlier studies of Dupuy [348] give the same g^{act}-value but with an uncertainty of $\sim 30\%$. Furthermore, Cornet gives for the activation volume of monovacancy self-diffusion $v^{act} = -4.9$ cm^3/mole or -0.336 Ω. Therefore the experimental value v/g is around -8×10^{-12} cm^3/erg. We now proceed to the bounds of v/g predicted from the $cB\Omega$-model. Due to the lack of reliable elastic experimental data we take the value $B \simeq 55$ kbar as suggested by Cornet who also gives $\partial \ln T_M/\partial P \simeq -7 \times 10^{-12}$ erg/cm^3. Inserting these values into eq. (10.8) we get: $v/g = 5 \times 10^{-12}$ cm^3/erg when $\kappa^c = 0$ and $v/g = -13 \times 10^{-12}$ cm^3/erg when $(\kappa^c/\kappa)_{max} = 1$. In spite of the fact that the bounds predicted from the $cB\Omega$-model are not so narrow in the present case, we see that the experimental value of v/g is well within these bounds.

10.2.3 Tetragonal metals

In table 10.3 we give all the appropriate data for the check of eq. (10.1) for the two tetragonal metals In and Sn. As expected in the case of indium the activation volume 8.1 ± 0.4 cm^3/mole obtained from self-diffusion studies is higher than the monovacancy formation volume 6.1 ± 0.2 cm^3/mole measured by the positron technique at various pressures. Their

Table 10.3
Comparison of $Bv[(dB/dP)-1]^{-1}$ with the experimental values of g for tetragonal metals.

Metal	T (K)	B (kbar)	dB/dP	v (cm³/mole)	$Bv(dB/dP-1)^{-1}$ (eV)	h (eV)	s (k-units)	$g_{\text{exper.}}$ (eV)
In	~400	370 [a]	6 [b]	6.1±0.2 [c]	0.47 [d]	0.45–0.55 [e]	2 [f]	0.38–0.48
	~400	370 [a]		8.1±0.4 [g]	0.62	0.81±0.015 [h]	3–4 [i]	0.66–0.70
Sn	~450	480 [j]	3.65 [k]	5.3±0.5 [l]	1.0	1.05 [m]	~3 [i]	0.94
			4.76 [n]		0.70			

[a] Kittel [561]; for RT the value $B = 411$ kbar is reported and therefore we decrease by ~10% in order to get the B-value at 400 K.
[b] Guinan and Steinberg [562]; value at RT.
[c] This value corresponds to the formation volume and has been reported by Dickman, Jeffery and Gustafson [563]; it has been obtained by positron-annihilation measurement at 14 kbars by heating from RT up to 498 K (which is the melting point of indium at 14 kbars).
[d] This value must be somewhat smaller due to the small temperature increase of dB/dP.
[e] Seeger [142]; formation enthalpy obtained by positron technique.
[f] Typical value of formation monovacancy entropy.
[g] Ott and Norden-Ott [564]; activation volume from self-diffusion under pressure.
[h] Dickey [565]; activation enthalpy from self-diffusion.
[i] Typical values for self-diffusion activation entropy.
[j] From the value 554 kbar reported in ref. [562] for RT by decreasing by about ~15%; this value agrees, within a few percent, with the high temperature measurements of Kammer, Cardinal, Vold and Glicksman [244]
[k] Vaidya and Kennedy [239]; by fitting their measurements to the Murnaghan equation.
[l] Nachtrieb and Coston [566]; activation volume from self-diffusion under pressure.
[m] Huang and Huntington [567]; activation enthalpy for self-diffusion.
[n] Quoted in ref. [239] when they use the equation $dB/dP = 2\gamma + \frac{1}{3}$.

Table 10.4
Calculated and experimental defect volumes in tetragonal indium. [a]

g (eV)	$v_{cB\Omega\text{-calc.}}$ (cm³/mole)	$v_{\text{exper.}}$ (cm³/mole)	Process
~ 0.48	6.25	6.1 ± 0.2	Monovacancy formation
~ 0.66	8.6	8.1 ± 0.4	Activation (self-diffusion)
~ 0.18	2.3	2 ± 0.6	Vacancy migration

[a] For references of g and v, see table 10.3.

difference gives a migration volume of 2 ± 0.6 cm³/mole, which is, at most, a little smaller than half the formation volume.

An inspection of table 10.3 indicates that the quantity $Bv[(dB/dP) - 1]^{-1}$ actually agrees with the experimental g-value in spite of the great experimental uncertainties. The degree of success of the $cB\Omega$-model in calculating the activation volume in Sn, for both axes, has been extensively treated in the previous chapter and hence we shall not further discuss it here. We turn to indium, for which, as already mentioned the various defect volumes are precisely known. In order to better realise the validity of eq. (10.1) in this material we give once more in table 10.4 the experimental values of the formation, activation and migration volumes; the latter values are compared with those resulting from the equation $v = (g/B)[(dB/dP) - 1]$ by using the elastic data of table 10.3. The agreement is significant considering the noncubic structure of the material.

Note that indium is one of the few metals for which v^f is precisely known; this allows the verification of the ratio v/g actually being equal to the bulk quantity $B^{-1}[(dB/dP) - 1]$ irrespective of the process (see fig. 10.4). To the best of our knowledge no microscopic calculations for any of these volumes exist.

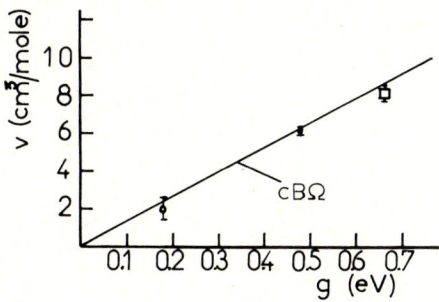

Fig. 10.4. Correlation of the defect volumes with the Gibbs energies in indium; ○: migration; ●: formation; □: activation of the self-diffusion; line: $cB\Omega$-model.

Table 10.5
Comparison of $Bv^f[(dB/dP)-1]^{-1}$ with the experimental values of g^f for noble gas solids.[a]

Noble gas solid	T (K)	B (kbar)	dB/dP	$Bv^f(dB/dP-1)^{-1}$ (eV)	h^f (eV)	s^f (k-units)	$g^f_{exper.}$ (eV)
Ne	20	8.26[b]	8.50[b]	0.0157	0.0206 ± 0.0026[c]	$1.6^{+1}_{-0.5}$[c]	$0.0178^{+0.0035}_{-0.0043}$
	13.5	10.1[b]	7.8[b]	0.0208			$0.0187^{+0.0032}_{-0.0037}$
Ar	77	14.1[d]	8.4[d]	0.048	0.062–0.093[e]	0–4[e]	0.062–0.066
	60	18.6[d]	7.8[d]	0.067			0.062–0.072
Kr	110	15.8[d]	7.6[d]	0.074	$0.086^{+0.0086}_{-0.0086}$[f]	$2.8^{+0.8}_{-0.9}$[f]	$0.067^{+0.009}_{-0.009}$
	100	17.3[d]	7.6[d]	0.080			$0.069^{+0.009}_{-0.009}$
Xe	159	14.8[d]	8.8[d]	0.076	0.107[g]	~2	0.079

[a] Macrander [53] measured the formation volumes in Kr and Ar and found that they are approximately equal to the atomic volume (with an uncertainty of 20%); therefore in this table v^f has been taken equal to Ω and hence the calculated values of g have an inherent uncertainty of, at least, 20%. Further notice that in the case of Kr the g^f-values obtained from refs. [51] and [263] are practically the same.
[b] Birch [254]; Birch analysed the results of Anderson, Fugate and Swenson [253].
[c] Schoknecht [52].
[d] Anderson and Swenson [568].
[e] These values are estimates and are consistent with the data of Schwalbe [258]; see Macrander [53] for details.
[f] Korpiun and Coufal [51].
[g] This value has been estimated by Losee and Simmons [263] from the law of corresponding states.

10.3 Noble gas solids

Table 10.5 contains all the appropriate quantities for the check of eq. (10.1) for the formation process in noble gas solids (NGS). In all cases the second member of this equation has been calculated by setting $v^f \simeq \Omega$. It can be compared to the experimental g^f-values obtained from the differential dilatometry technique for all NGS except Xe and He. The comparison indicates an agreement considering that the relation $v^f \simeq \Omega$ may have an inherent uncertainty of even 20% [53] (see also fig. 10.5). The latter experimental fact of $v^f \simeq \Omega$ has been established for krypton from a constant-volume X-ray study of Macrander [53]. In the same study the ratio v^f/Ω for argon has been found to be somewhat less than unity but with a considerable uncertainty depending on the choice of h^f and s^f. For Ne and Xe no direct experimental values of v^f have been reported to date and therefore we have used the same value as for Kr, i.e. $v \simeq \Omega$, by bearing in mind, as mentioned, an uncertainty of 20%. The earlier values $v^f/\Omega = 1.5^{+0.3}_{-0.1}$ for Kr by Korpiun and Coufal [51] or $v^f/\Omega = 1.5^{+0.3}_{-0.9}$ by Schoknecht [52] have not been considered because, as discussed by Macrander, they come from bulk data and x-ray data obtained in separate experiments, thereby allowing the possibility of nonidentifiable errors. For the case of ^{3}He, see §A.9.

We now turn to the point which according to Chadwick and Glyde [349] remains the outstanding question in the calculation of defects in NGS.

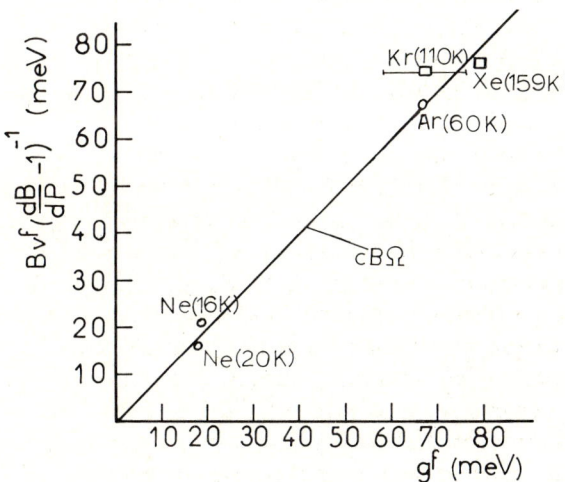

Fig. 10.5. Demonstration of eq. (10.1) for noble gas solids; the data are from table 10.5. Line: $cB\Omega$-model.

Numerous calculations have shown that in krypton the calculated value of the formation enthalpy and hence the value of g^f differs by at least 22% from the experimental one. On the other hand, as is clear from table 10.5, the quantity $g^f\{\equiv Bv^f/[(dB/dP)-1]\}$ calculated from the $cB\Omega$-model nicely agrees with the experimental g^f within the experimental errors mentioned. This is important because it ensures the compatibility between Macrander's value $v^f \simeq \Omega$ and the experimental value of g^f; on the other hand the opinion has been expressed that the incorrect value of h^f (and g^f) is due to a bad estimation of the relaxations around the vacancy. In view of the present result it becomes clear the this cannot be the cause so that the disagreement will have to be attributed elsewhere.

Another important remark should be added. The NGS belong to the class of the highly compressible materials. In ch. 14 it is explained that for this class of materials a possible pressure dependence of c, in the expression $g = cB\Omega$, has to be considered if one desires a very accurate result. This pressure dependence turns the expression (10.1) into (10.5). In view of the fact that κ^c is of the order of κ and that dB/dP is close to 8 (see table 10.5) one immediately sees that the possible pressure dependence of c might change v^f at most by 15%. Unfortunately the experimental values of v^f in NGS, except in the case of Kr, are known with an uncertainty greater than 15% so that a deviation of eq. (10.1), if any, from the experiments cannot be judged. It is however useful to note that future more accurate experiments should give a value of v^f lying between $(g/B)[(dB/dP)-1]$ and $(g/B)\{[(dB/dP)-1]-\kappa^c/\kappa\}$ where κ^c/κ is at most around unity.

10.4 Ionic crystals

10.4.1 Ionic crystals with CsCl-structure

In table 10.6 we give all the appropriate data for the check of eq. (10.1) for Schottky formation, cation vacancy migration and anion vacancy migration in TlBr and in CsCl. We observe that irrespective of the process the quantity $Bv[(dB/dP)-1]^{-1}$ generally agrees with the experimental value of g if one considers the experimental uncertainty. Note that in these materials the entropies (formation and migration) are not known with accuracy so that rough estimates of them have been used; however, an error in s does not change the value of g more than a few percent. It might be useful to remark that for these materials it has been experimentally verified [350] that the anion vacancies are more mobile than the cation vacancies, i.e. $g^{m-} < g^{m+}$ and further $h^{m-} < h^{m+}$. In view of the fact that the $cB\Omega$-model demand s/h to be a bulk property irrespective of the process, the entropy

Table 10.6

Comparison of $Bv[(dB/dP)-1]^{-1}$ with the experimental values of g for formation and migration processes in crystals with CsCl-structure.

Crystal	T (K)	B (kbar)	dB/dP	v (cm³/mole)	$Bv(dB/dP-1)^{-1}$ (eV)	h (eV)	s (k-units)	$g_{\text{exper.}}$ (eV)	Process
TlBr	700	135 [a]	~8.2 [b]	44.4 [c]	0.86±10%	1.10 [c]	~5 [d]	0.8	Schottky formation
	300	213 [e]	7.5 [b]	13.7 [f]	0.47±10%	0.56 [c]	~2 [d]	0.51	Cation vacancy migration
	300	213 [e]	7.5 [b]	6.8 [g]	0.23±10%	~0.25 [c]	~1 [d]	0.22	Anion vacancy migration
CsCl	700	123 [h]	6.9 [i]	80–87 [c]	1.73–1.88	2.1–2.3 [c]	~5 [d]	1.8–2	Schottky formation
	300	173 [j]	6.2 [i]	18 [f]	0.62±10%	0.62 [c]	~2 [d]	0.57	Cation vacancy migration
	300	173 [j]	6.2 [i]	5.5 [g]	0.19±10%	0.2 [c]	~1 [d]	0.17	Anion vacancy migration [k]
	300	173 [j]	6.2 [i]	9.0 [g]	0.31±10%	0.3 [c]	~1 [d]	0.27	Anion vacancy migration [k]

[a] Landolt–Börnstein tables [569]; the value $B^S = 150$ kbar at 700 K has been decreased by ~10% to give the isothermal one.
[b] In ref. [569] the value $dB^S/dP = 7.15$ for RT is given which gives $dB/dP \approx 7.5$ for RT. The value for 700 K has been estimated by assuming that $d^2B/(dTdP)$ in TlBr is the same as that given for NaCl by Spetzler et al. [282].
[c] Samara [442]; the formation volume per Schottky defect is obtained from ionic conductivity under pressure by combining the measurement of the extrinsic and intrinsic region.
[d] Estimated value because in ref. [442] the entropies are not given.
[e] By applying a correction of 5% to the value $B^S = 224$ kbar given in ref. [569] for 300 K.
[f] Ref. [442]; migration volume for cation vacancy motion.
[g] Ref. [442]; migration volume for anion vacancy motion.
[h] From the value $B^S = 136$ kbar given in ref. [569] for 700 K by applying a correction of 10%.
[i] Gard, Puri and Verma [570]; they give for 300 K the value $dB^S/dP = 5.9$ which gives $dB/dP \approx 6.2$; by following the same arguments as in footnote b the value $dB^S/dP = 6.9$ results for $T = 700$ K. In ref. [569] for RT the value $dB^S/dP = 5.7$ is given which is practically the same as that given by Gard et al.
[j] From the value $B^S = 182$ kbar given in ref. [569] by applying a correction of 5%.
[k] In ref. [442] two alternative values are given for the enthalpy and the volume of the anion vacancy migration.

for the anion migration should be smaller than that of the cation, in sharp contrast to alkali halides with NaCl-structure. An experimental verification of this result is desirable.

As mentioned above, in both materials the quantity $Bv[(dB/dP) - 1]^{-1}$ agrees, irrespective of the process, with the corresponding value of g within the experimental error, which is $\sim 10\%$ (especially due to the experimental uncertainty of the volume v). However, a careful consideration of the $cB\Omega$-model (see ch. 14) predicts that if in future more accurate experiments will be carried out some deviation between these two values would not be surprising. For instance, a careful ITC-experiment under various pressures could show that $Bv[(dB/dP) - 1]^{-1}$ slightly differs from g where now v and g refer to the reorientation process of a dipole. This might be expected from a small pressure variation of the quantity c. More precisely, by considering that the absolute value of the compressibility κ^c of c may be at most around κ we see that c can change at most by $\exp\int_0^P \kappa \, dP - 1$ at a pressure P, i.e. in CsCl (for $T = 300$ K) by 0.58% per kbar. For the usual pressure ranges of up to 5 kbars it means that from 0 to 5 kbars c may change at most up to 3% which of course lies within the current experimental error.

Summarizing the present discussion on TlBr and CsCl one can say that within the current experimental accuracy the quantity $Bv[(dB/dP) - 1]^{-1}$ is found to be equal to g irrespective of the process. This is also demon-

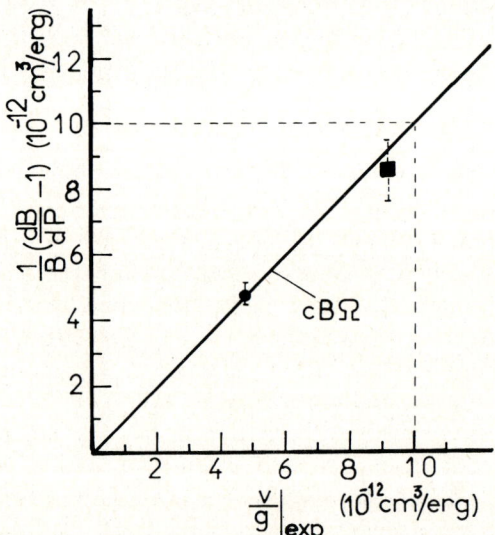

Fig. 10.6. Demonstration of $v/g = [(dB/dP) - 1]B^{-1}$ for TlBr; ■: formation at 700 K; ●: cation or anion vacancy migration at 300 K. Line: $cB\Omega$-model.

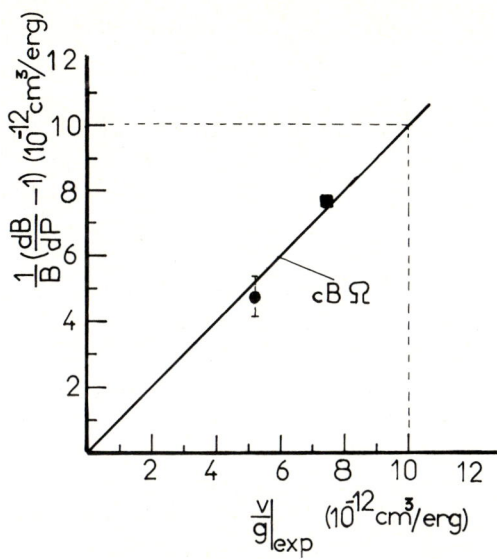

Fig. 10.7. Demonstration of $v/g = [(dB/dP) - 1]B^{-1}$ for CsCl; ■: formation at 700 K; ●: cation or anion vacancy migration at 300 K.

strated in figs. 10.6 and 10.7 where we see that the bulk quantity $B^{-1}[(dB/dP) - 1]$ actually gives straight lines with a slope equal to unity as predicted from the $cB\Omega$-model when plotted as a function of the experimental quantity v/g. The data given above are taken from table 10.6

10.4.2 Alkali halides with NaCl-structure

These materials have raised considerable interest during the last 15 years. Many microscopic calculations [325,327] have shown that a strong contraction around the Schottky defect (cation and anion vacancy) should be expected, i.e. a large negative relaxation volume. When added to the molecular volume 2Ω this leads either to a small positive formation volume appreciably smaller than 2Ω [325] or to a negative formation volume [327]. On the other hand, careful experiments carried out by Lazarus and co-workers [322,323], which have been confirmed independently through the quite different technique used by Spalt, Lohstöter and Peisl [351], have definitely shown an outward relaxation around the defect and hence a formation volume appreciably larger than 2Ω in all alkali halides studied to date. The origin of this sharp contrast between theory and experiment has been extensively discussed by Varotsos et al. [36]: the microscopic calcula-

Table 10.7

Comparison of $g_{calc} \equiv Bv[(dB/dP)-1]^{-1}$ with the experimental values of g for formation and migration [a] processes for crystals with NaCl-structure.

Crystal	T (K)	B (kbar)	dB/dP	v (cm³/mole)	$g_{calc.}$ [eq. (10.1)] (eV)	h (eV)	s (k-units)	$g_{exper.}$ (eV)	Process
NaCl	800	157 [b]	6.13 [b]	55±9 [c]	1.75	2.44 [d]	9.8 [d]	1.76 [d]	Formation
	300	237 [b]	5.35 [b]	9.5 [e]	0.54±10%	0.66 [f] – 0.69 [d]	3.17 [f] – 1.64 [d]	0.58 [f] – 0.65 [d]	Cation migration
	300	240 [g]	5.38 [g]	9.5 [e]	0.54				Cation migration
	300 at 5 kbar [h]	264	4.95 [i]	7±1 [c]	0.49±0.7				Cation migration
KCl	800	132 [j]	5.46 [k]	61±9 [c]	1.87	2.54 [d] – 2.5 [l]	8.99 [d] – 7.7 [l]	1.92 [d] – 1.96	Formation
	300	175 [k]	5.46 [k]	10 [e]	0.41	0.68 [l]	2.4 [l]	0.62 [l]	Cation migration
NaBr	800	158 [m]	5.44 [g]	46±5 [c]	1.7±0.2	1.72 [n]	–	~1.5	Formation
KBr	800	117.5 [j]	5.47 [g]	54±9 [c]	$1.47^{+0.25}_{-0.25}$	2.37 [o]	8.6 [o]	1.77	Formation
						2.09 – 2.33 [p]	4.3 – 8.1 [p]	1.79 – 1.77	Formation

288   The cBΩ-model: Defect volume and Gibbs energy   [§10.4]

§10.4]   Ionic crystals   289

a The migration parameters for NaBr and KBr are not given in the table because, as mentioned, the low pressure (<1 kbar) migration volumes in these materials have not been reported by Yoon and Lazarus [322]; the v^m-values given in ref. [322] correspond at pressures between 2–7 kbars where we cannot estimate the value of dB/dP due to the lack of $\partial^2 B/\partial P^2$-values.
b Ref. [282].
c Ref. [322]; formation volume for Schottky defects from ionic conductivity measurements under pressure.
d Ref. [69].
e Biermann [353]; migration volume of cation vacancy. Ionic conductivity measurement under pressure in the low pressure region 0 ~ 0.5 kbar. In ref. [322] somewhat lower values are given but they have been extracted from pressures higher than 1 kbar and therefore they are consistently lower than those given by Biermann (see footnote h).
f Allnatt and Pantelis [571].
g Roberts and Smith [376]
h All the values given in this line correspond to $p = 5$ kbar; this has been done because the value 7 ± 1 cm^3/mole has been obtained in ref. [322] for this pressure range.
i By using the zero pressure values $dB/dP = 5.35$ and $(\partial^2 B/\partial P^2)_{P=0} = -8 \times 10^{-2}$ kbar^{-1} extracted from the data of ref. [282].
j From the B^S-value given at 800 K by Slagle and McKinstry [371].
k Ref. [376]; this value corresponds to 295 K.
l Acuna and Jacobs [89].
m Estimated from the value $B = 199$ kbar given in ref. [376] for R.T.
n Flynn [572].
o Crawford and Slifkin [573].
p Brown and Jacobs [574].

tions have been carried out within the frame of a harmonic approximation [649] and they disregard the vibrational contribution; on the other hand it has been shown that anharmonicity and vibrational effects play a major role and that their correct inclusion leads to positive values of the formation volumes in agreement to the experimental data [36].

In table 10.7 we give all the data required for the check of eq. (10.1) for Schottky formation and cation migration in some alkali halides. It should be stated in advance that at ambient pressure NaCl and KCl are the best studied and their defect parameters (h, s) have been obtained by a simultaneous fitting of their conductivity and diffusion data. The values of (h, s) of NaBr and KBr must be considered as rough estimates as they have been derived solely from an analysis of the conductivity data.

Sodium chloride. In this material the quantity $Bv^f[(dB/dP) - 1]^{-1}$ agrees very well with the experimental g^f-value (see table 10.7). This means that at high temperatures v^f can be accurately calculated from g^f and the elastic data; at $T = 0$ one has from eq. (10.1):

$$v_0^f = \frac{h_0^f}{B_0} \left\{ \left.\frac{dB}{dP}\right|_{T=0} - 1 \right\}$$

or equivalently by considering eq. (8.15)

$$v_0^f \simeq \frac{h_{\text{exp}}^f}{B_0^{\text{SL}}} \left\{ \left.\frac{dB}{dP}\right|_{T=0} - 1 \right\}.$$

Setting [282] $B_0^{\text{SL}} = 284.7$ kbar, $(\partial B/\partial P)_{T=0} = 4.88$ and taking from table 10.7 the value $h_{\text{exp}} = 2.44$ eV we find $v_0^f = 32$ cm^3/mole. This value is appreciably smaller than 55 cm^3/mole corresponding to 800 K but still exceeds the molecular volume $2\Omega_0 = 43$ Å^3 at $T = 0$. Therefore any microscopic calculation which claims to have correctly found the formation enthalpy, at $T = 0$ should also give a formation volume larger than $2\Omega_0$ (i.e. a positive relaxation volume). Furthermore, the present results show that the significant increase of the formation volume with temperature is in agreement with the experimental data (see table 10.8). In order to realise the temperature dependence of v^f, we calculate it from the relation $v^f = (g^f/B)[(dB/dP) - 1]$ for various temperatures. The value of v^f at $T = 936$ K is about twice that at absolute zero. In the same table we give the v^f-values for a possible pressure dependent c^f, e.g. with $\kappa^c \simeq \kappa$. They are calculated from relation (10.5); this gives the minimum values of v^f that can be extracted from the $cB\Omega$-model by allowing for a possible pressure dependence of c^f. An inspection of the table indicates that the experimental v^f-values truly lie between the two limiting values calculated from the $cB\Omega$-model.

Table 10.8
Calculated and experimental formation volume for Schottky defects in sodium chloride.

T	$\mathrm{d}B/\mathrm{d}P$ [a]	B [a] (kbar)	g [b] (eV)	$v_{\text{calc.}}^{\text{f}}$ [c] (cm³/mole)	$v_{\text{calc.}}^{\text{f}}$ [d] (cm³/mole)	$v_{\text{exper.}}^{\text{f}}$ (cm³/mole)
0	4.88	~270	~2.32	~32	~24	–
300	5.35	238.4	2.19	38.5	29.6	–
800	6.14	157.4	1.76	55.4	44.6	55 ± 9 [e]
936	6.35	135.4	1.65	~63	51	56 ± 10% [f]

[a] All the elastic data come from ref. [282].
[b] From the values $h^{\text{exper.}} = 2.44$ eV and $s^{\text{exper.}} = 9.8\,k$ given in ref. [69]; the value at $T = 0$ is obtained from $h_0 = h^{\text{exper.}} B_0/B_0^{\text{SL}}$ where $B_0 = 270$ kbar and $B_0^{\text{SL}} \simeq 284.7$ kbar are from ref. [282].
[c] Calculated from relation (10.1) which assumes that c is pressure independent.
[d] Calculated from the thermodynamical relation (10.5) with $\kappa^c \simeq \kappa$; the values given represent the lower limit of v^{f} (see §10.1).
[e] Ref. [322].
[f] Ref. [323]; this paper gives for $T = 936$ K, $v^{\text{act}} = 35$ cm³/mole which in conjunction to the value $v^{\text{m}} = 7 \pm 1$ cm³/mole given in ref. [322] for the cation migration process gives $v^{\text{f}} \simeq 56$ cm³/mole with an uncertainty around 10%.

We can certify that the rate of increase of the formation volume with temperature is appreciably higher than that of the bulk volume. The v^{f}-values calculated for 800 and 936 K lead to a thermal expansion coefficient β^{f} of the formation volume which is around 10^{-3} K^{-1} and therefore exceeds the bulk one ($\beta \simeq 2 \times 10^{-4}$ K^{-1}) by a factor of around 5. (Notice that the ratio β^{f}/β depends on temperature and that at 200 K it must be appreciably higher than 5). This fast temperature dependence of v^{f}, i.e. the high value of β^{f}, reveals two important points:

(a) From thermodynamics the temperature variation of the formation enthalpy [see eq. (6.45)] is directly connected to the ratio β^{f}/β. A large value of β^{f}/β demands, on thermodynamical grounds, a significant value of $(\partial h^{\text{f}}/\partial T)_P$ which again leads to an upward curvature of the conductivity plot. It is this point to which we not turn: in order to convince the reader about the major role of $(\partial h^{\text{f}}/\partial T)_P$ in the curvature plot we give the following example: consider relation (6.45)

$$\left.\frac{\mathrm{d}h^{\text{f}}}{\mathrm{d}T}\right|_P = c_V^0 + \frac{Tv^{\text{f}}\beta^{0^2}}{\kappa^0}\left(2\frac{\beta^{\text{f}}}{\beta^0} - \frac{\kappa^{\text{f}}}{\kappa^0}\right)$$

in which – as explained by Gilder and Lazarus [37] – c_V^0 is appreciably smaller than $(\partial h^{\text{f}}/\partial T)_P$ and to a first approximation can be disregarded. Even if we choose the extreme case of $\beta^{\text{f}} = \beta^0$ and $\kappa^{\text{f}} = \kappa^0$, the insertion of the typical values $v^f = 56$ cm^3/mole, $\beta = 2 \times 10^{-4}$ K^{-1} and $B = 130$ kbar give (for $T = 1000$ K) $(\partial h/\partial T)_P \simeq 3.5\ k$. This corresponds to an increase of h^f by 0.06 eV for a temperature increase of 200 K. By choosing $\beta^f/\beta = 5$ and further $\kappa^f/\kappa = 5$ (see below), h^f increases by 0.3 eV which gives $(\partial h/\partial T)_P \simeq 17.5\ k$. This increase can account for the curvature observed in the conductivity plot without having to introduce arbitrary mechanisms or any other kind of defects beyond Schottky defects. Microscopic calculations [96,97] which claim that the enthalpy h^f (or u^f) decreases with temperature and that this decrease can explain an upwards curvature of the conductivity plot are questionable; this is so because pure thermodynamics reveal [352] that a decrease of h^f with temperature leads to a downward and *not* to an upward curvature of the conductivity plot. (see §A.11, note 4 and §A.8).

(b) The inequality $\beta^f \gg \beta$ can be directly verified in cases where β^f refers to the change of the macroscopic volume when a divalent (or trivalent) cation or anion is added; the experimental technique may be similar to that recommended later for the case of AgBr. It should be noticed that Raab and Peisl [131] have already verified the above inequality in close agreement with eq. (8.26) of the $cB\Omega$-model for the expansion coefficient of the volume corresponding to the addition of an F-center to KCl.

It is interesting to examine now whether the calculated value of β^{act}, which should be equal to β^f, agrees with the experimental data. An experimental value of β^{act} can be extracted from the paper of Martin Lazarus and Mitchell [323] as follows: For low pressures the activation volume at 936 K and 1004 K is 35 cm^3/mole and 38 cm^3/mole ($\pm 10\%$) respectively. These two values give a mean thermal expansion coefficient of $1.3^{+3.6}_{-1.6} \times 10^{-3}$ K^{-1} which is comparable to the β^f-value of 10^{-3} K^{-1} calculated above from the $cB\Omega$-model.

We finally turn our attention to the curvature in the form of a "knee" observed [322, 323] in the isothermal plots $\ln \sigma$ (or $\ln D$) versus P. This can be directly explained from the fact that the compressibility κ^{act} of the activation volume is not zero but has a value comparable to that predicted from the $cB\Omega$-model. In table 10.9 we give the values of κ^{act}, at $T = 936$ K, for pressures up to 6 kbar. They come from the $cB\Omega$-formula [eq. (8.31)]. The bulk compressibility κ for $P = 0$ at the same temperature is $\sim 7.4 \times 10^{-3}$ kbar^{-1} (see table 10.8), i.e. 3 to 4 times smaller than the κ^{act}-values given in table 10.9. Recalling that $v(P)^{act} = v(0)^{act} \exp(\int_0^P - \kappa^{act} dP)$, an inspection of the table indicates that beyond 3 kbar the factor $\exp(\int_0^P - \kappa^{act} dP)$ starts deviating from unity by more than 10%; the experimental error of the determination of the activation volume is about 10% and hence the $cB\Omega$-theory suggests that curvature should be detectable only for pressure higher than 3 kbar. Turning now to the experimental data we see that Martin, Lazarus and Mitchell [323] found a "knee" in the $\ln D$ versus P plots at

Table 10.9
Compressibility of the activation volume of NaCl for $T = 936$ K. [a]

P (kbar)	κ^{act} (10^{-2} kbar^{-1})	$\exp \int_0^P \kappa^{act} dP$
0	2.5	1
2	2.6	1.05
3	2.7	1.08
4	2.8	1.11
6	3.0	1.18

[a] Data taken from ref. [36].

936 K and 1004 K approximately at the pressure predicted above; the same is observed in the isothermal $\ln \sigma$ versus P plots of Yoon and Lazarus [322] for the intrinsic conductivity. In order now to examine whether the curvature agrees quantitatively with the $cB\Omega$-model we consider the following. A plot of $\ln D$ versus P shows the activation volumes $v^{act} = 38 \pm 10\%$ for ambient pressure and $26 \pm 10\%$ cm^3/mole at 6 kbar [323]. These two values imply an "average" compressibility $\kappa^{act} = 5.26^{+2.1}_{-3.0} \times 10^{-2}$ kbar^{-1} which agrees – within the error bars – with that calculated in table 10.9 from the $cB\Omega$-model.

We now proceed to the check of eq. (10.1) for the case of the cation migration process. From table 10.7 we see that the calculated quantity $Bv^m[(dB/dP) - 1]^{-1}$, although agreeing within the error bars ($\pm 10\%$ mainly due to v^m) with the experimental g^m-value, shows a trend to be slightly smaller. There are three plausible reasons:

(a) the published experimental values of v^m do not exactly correspond to zero pressure; for instance the value 7 ± 1 cm^3/mole reported by Yoon and Lazarus [322] has been derived from a least squares fitting of the $\ln \sigma$-values above 1 kbar; therefore this value of v^m is smaller than the value for

Table 10.10
Calculated and experimental migration volumes in NaCl and KCl ($T = 300$ K).

	g^m [c] (eV)	$v^m_{calc.}$ [a] (cm^3/mole)	$v^m_{calc.}$ [b] (cm^3/mole)	$v^m_{exper.}$ [c] (cm^3/mole)
NaCl	0.58 ~ 0.65	10.3–11.5	7.9–8.9	7 ± 1 to $9 \pm 10\%$
KCl	~ 0.62	15.2	11.8	10

[a] From eq. (10.1).
[b] From eq. (10.5) with $\kappa^c/\kappa \simeq 1$.
[c] See table 10.7

ambient pressure. On the other hand, had they considered pressures smaller than 1 kbar they would doubtlessly have obtained a higher migration volume, thus leading to a better agreement between $Bv^m[(dB/dP)-1]^{-1}$ and g^m_{exp}. This explanation is further strengthened by the fact that Biermann [353], who measured a lower pressure-region (0–0.5 kbar), obtained a value of 9 cm^3/mole, which is 10–20% higher than that of Yoon and Lazarus. This leads to the conclusion that the low pressure measurements of Biermann [353] are actually compatible with those of Yoon and Lazarus (pressures up to 6 kbar) which are anyhow more accurate.

(b) The experimental g^m-values are based on the published s^m-values. Apart from their large experimental uncertainty (see table 10.7) the latter have the shortcoming that they are based on the arbitrary assumption that the attempt frequency ν is equal to the Debye frequency ν_D (recall that the published s^f-values do not suffer from this arbitrarity). It is quite possible that ν is smaller than ν_D – which is roughly equal to ν_{TO} in NaCl – by a factor 2–4, thus leading to an increase of the published s^m-value by 0.6 to 1.4 k. Then the experimental g^m-values should decrease by 0.02–0.04 eV, thus leading to a better agreement with $Bv^m[(dB/dP)-1]^{-1}$.

(c) As explained in ch. 14, c^m may depend slightly on pressure to an extent that would give its compressibility a value comparable to the bulk compressibility, i.e. $\kappa^c \simeq \kappa$. In order to envisage this effect we now proceed inversely to the computation of the migration volume. In table 10.10 we first calculate v^m from eq. (10.1), in which c^m is accepted as a constant, and then give in the next column the same quantity from eq. (10.5), allowing a pressure dependence of c^m that would correspond to $\kappa^c/\kappa = 1$ [see eq. (10.5)]. A comparison with the experimental values indicates a satisfactory agreement considering the large experimental errors and the remarks developed above in (a) and (b). This agreement is significant if one remembers that microscopic calculations *still fail* to give even a rough estimate of v^m.

KCl, NaBr, KBr. The discussion of these materials is exactly analogous to that of NaCl. We see in table 10.7 that the quantity $Bv^f[(dB/dP)-1]^{-1}$ favourably compares with the experimental values of g^f (this agreement becomes even better if one allows for a small pressure dependence of c^f corresponding to $\kappa^c/\kappa \simeq 1$). Therefore the formation volumes v^f in these materials can be correctly computed from g^f and the elastic data; note that the microscopic calculations cannot even give a reliable estimate of v^f. In the case of the migration process, at first sight, a disagreement appears in the case of KCl. However, this can be explained if one recalls the remarks (a), (b) and (c) for NaCl. For this reason in table 10.10 we also give the calculated v^m-value for KCl in a similar fashion as for NaCl. We see that the calculated value $v^m = 11.8$ cm^3/mole is comparable to the experimental

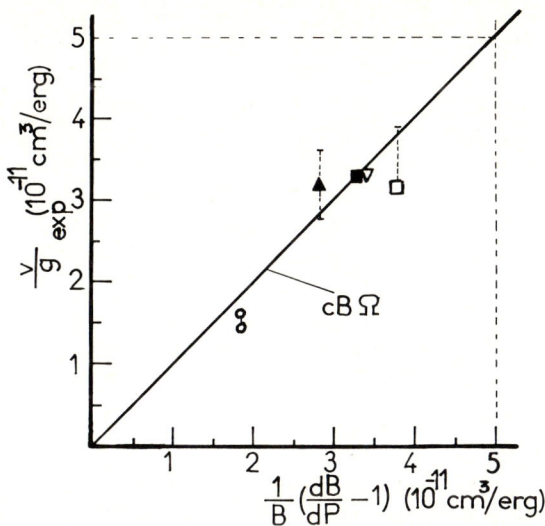

Fig. 10.8. Correlation of v/g with bulk properties for alkali halides; ■: Schottky formation in NaCl (at 800 K); □: KBr; ▽: KCl; ▲: NaBr; ○: cation vacancy migration in NaCl at 300 K; the data are taken from table 10.7; line: predicted by $cB\Omega$-model.

one (10 cm³/mole) bearing in mind the experimental uncertainties and especially remark (c) made for NaCl. For recent calculations, see §A.8.

Summarizing the present discussion in alkali halides we can say that actually the formation (and migration) volume in alkali halides is connected to the Gibbs energy and the elastic data in the way predicted from the $cB\Omega$-model (see fig. 10.8). The temperature and pressure dependence of the defect volume observed in the experiments can be accounted for by the same model.

10.4.3 Silver halides

In table 10.11 we give all the data required for the calculation of the quantity $Bv^f[(\mathrm{d}B/\mathrm{d}P)-1]^{-1}$ where v^f is the formation volume per cation Frenkel defect, in AgCl and AgBr. In AgCl the computation of this quantity gives 1.22 eV with an uncertainty of around 10% mainly due to the experimental error of v^f. This value is in agreement with that of g^f (1.21 eV) for the same temperature, resulting from the parameters h^f and s^f reported from the analysis of conductivity data. In view of the complexity of this analysis, the real values of the parameters h^f and s^f may be different from those given in table 10.11 but, anyhow, this cannot change the value for g^f

Table 10.11
Comparison of $Bv[(dB/dP)-1]^{-1}$ with experimental values of g for formation (unless otherwise cited) in silver halides.

	T (K)	B (kbar)	dB/dP	v (cm³/mole)	$Bv(dB/dP-1)^{-1}$ (eV)	h (eV)	s (k-units)	$g_{exper.}$ (eV)
AgCl	295	417 [a]	6.9 [b]	16.7 [c]	1.22	1.45 [d]	9.4 [d]	1.21
AgBr	295	378 [a]	7.16 [e]	14±1.5 [f]	0.89	1.13 [d] – 1.07 [f]	6.55 [d] – 6.9 [f]	0.96–0.89
	295	378 [a]	7.16 [e]	16 [g]	1.01			
	295	378 [a]	7.16 [e]	13 [h]	0.83			
	295	378 [a]	7.49 [b]	14 [f]	0.85			
B4 – AgI	295	240 [i]	~0 [j]	−3 [k]	0.75	0.80–0.78 [k]	—	~0.78 [l]
	395	~240	−2 [j]	−9.3 [k]	~0.77	0.80–0.78 [k]	—	~0.78 [l]

[a] Cain, [381].
[b] Loje and Schuele [309]; for AgCl they report for the pressure derivative of B^S the value 6.57 which has to be increased by ~5% in order to get dB/dP.
[c] Formation volume of cation Frenkel defect measured by Abey and Tomizuka [321]; conductivity under pressure.
[d] Aboagye and Friauf [307].
[e] By increasing the value $dB_S/dP = 6.82$ reported in ref. [381] by ~5%.
[f] Lansiart and Beyeler [324]; formation volume for cation Frenkel defect (conductivity measurements under pressure between 285 and 500 K).
[g] Kurnick [393]; formation volume for cation Frenkel defect at high T.
[h] Lazarus [575]; formation volume for cation Frenkel defect at high T.
[i] Shaw [576].
[j] Shaw [577].
[k] Ref. [355]; activation volume for RT and for conduction along c-axis.
[l] Estimated value; due to the fact that activation entropy is not reported in ref. [355] we have taken $g^{act} \simeq h^{act}$; it is expected that g^{act} must be lower than h^{act} by some percentage thus giving a better agreement between calculation and experiment.

Ionic crystals

by more than some percent. Therefore, the agreement between g^f_{exper} and $Bv^f[(dB/dP)-1]^{-1}$ can in any event not be considered fortuitous and it directly leads to the conclusion that v^f can be correctly calculated from the value of g^f and the corresponding elastic data. Until 1979 microscopic calculations could not give even a rough estimate of v^f; it was then proved [630] that the distinction between "isobaric" and "isochoric" defect parameters plays a prominent role in defect calculations, a fact that has been totally disregarded in earlier publications in ionic crystals [97,314,325–327]. Subsequent calculations have actually considered this fact (see §A.8, and §A.11, note 4).

Exactly the same comments hold for AgBr. One sees in table 10.11 that the quantity $Bv^f[(dB/dP)-1]^{-1}$ takes values between 0.83 and 1.01 eV. If one again considers the experimental error one finds that the calculated values agree with the experimental ones which lie between 0.89 and 0.96 eV. In essence this agreement means that v^f is correctly connected to g^f and to the elastic data through the $cB\Omega$-model. To the best of our knowledge a microscopic calculation of v^f in AgBr has not been published until 1980.

We now proceed to the interesting case of B4-AgI [354]. Attention is drawn to the point that for this material all the defect parameters given in the table do not correspond to a pure formation process but to an activation one involving both formation and migration. What makes this material extremely interesting is the fact that $dB/dP = 0$ close to RT and -2 around 395 K; furthermore, its experimental activation volume is negative varying from -3 cm^3/mole at RT to -9.5 cm^3/mole around 395 K. This unusual behaviour, verified during the last years by Allen and Lazarus [355] provides a very strict check for the $cB\Omega$-model; the fact that a negative value of v^{act} is accompanied by a value of dB/dP smaller than unity (provided that g^{act} is positive) had been qualitatively predicted from the $cB\Omega$-model through eq. (10.1). It is apparent (see table 10.11) that there is also quantitative agreement between g^{act} and $Bv^{\text{act}}[(dB/dP)-1]^{-1}$ for both temperatures 295 K and 395 K.

The success for B4-AgI can also be visualised in the following way: eq. (8.26)

$$\beta^{\text{act}} = \beta + \left(\frac{dB}{dP} - 1\right)^{-1} \frac{d}{dT}\left(\frac{dB}{dP}\right)$$

contains only bulk properties and reveals that even if the bulk expansion coefficient is practically zero (as in the case of B4-AgI) the corresponding coefficient β^{act} of the activation volume may take appreciably large values. Considering that in this material $(dB/dP)_{\text{RT}} \simeq 0$ and $(dB/dP)_{395\text{K}} \simeq -2$ we get $d^2B/dTdP \simeq -2 \times 10^{-2}$ K^{-1}; furthermore, by recalling that $\beta \simeq 0$ the relation under discussion gives the quite large value of $\beta^{\text{act}} \simeq 6.7 \times 10^{-3}$

K^{-1} at $T = 395$ K. This value of β^{act} calculated purely from bulk properties with the $cB\Omega$-model is in excellent agreement with the value 6.5×10^{-3} K^{-1} extracted from fig. 11 of Allen and Lazarus [355].

Some experiments can now be suggested, based on the idea that this value of the expansion coefficient of the "defect volume" has to be independent of the type of defect: (a) The activation volume $v^{act,i}$ of the diffusion of an (e.g. monovalent) ion (i) in B4-AgI should depend on temperature and give a thermal expansion coefficient $\beta^{act,i}$ equal to that calculated above. (b) The introduction of small amounts of divalent cations will create a number of vacancies and hence an increase of the volume of the crystal. Then the difference of the volumes of the undoped and the doped crystal must be temperature dependent in a way predetermined by the value of the expansion coefficient reported above; this dependence can be measured with an interferometric technique. Similar experiments are suggested for AgBr and AgCl; especially the first material is expected to have a large value of the expansion coefficient β^d of the defect volume. For AgBr these experiments should better be carried out in the temperature region where the conductivity plot is strongly curved upwards (in this region one again expects $\beta^d \gg \beta$).

We finally turn to the validity of the eq. (10.1) where now g and v refer to the migration process. In table 10.12 we have collected all the appropriate data required for checking this equation for AgCl and AgBr. One sees that for two migration processes (cation vacancy resp. silver interstitial motion) appropriate defect parameters have been published. In AgCl the quantity $Bv^m[(dB/dP) - 1]^{-1}$ for the cation vacancy motion nicely agrees with the experimental g^m-value, whereas for the cation interstitial motion it has a value $0.23 \pm 10\%$ eV which is almost the average of the two published g^m-values. The scatter of the latter is not so surprising if one again recalls the complexity of the analysis of the conductivity data in silver halides. In AgBr the agreement is again very good for both migration processes. We are therefore led to the conclusion that the migration volumes in silver halides are correctly connected to g^m and to the bulk elastic data through the $cB\Omega$-model. On the other hand microscopic calculations for the migration energies of cation vacancy, anion vacancy and cation interstitial in AgCl have been published by Jacobs, Corish and Catlow [647]. They found that these energies decrease as the temperature increases and *compared*, for $T = 250°C$, the calculated values with the experimental migration enthalpies. Their calculation, however, did not consider the above mentioned distinction between "isobaric" and "isochoric" defect parameters and therefore this comparison should be repeated after considering the corrections indicated in ref. [630] and briefly discussed in §A.8.

General remarks. Eq. (10.1) reveals that the ratio v/g, irrespective of the

Table 10.12
Comparison of $Bv[(dB/dP)-1]^{-1}$ with the experimental values of g for migration for silver halides.[e]

	T (K)	B (kbar)	dB/dP	v (cm³/mole)	$Bv(dB/dP-1)^{-1}$ (eV)	h (eV)	s (k-units)	$g_{\text{exper.}}$ (eV)
AgCl, cation vacancy motion	295	417	6.9	4.7 [a]	0.34 ± 10%	0.288 [b]	−0.5 [b]	0.30
						0.275 [c]	−0.64 [c]	0.29
AgCl, silver interstitial motion	295	417	6.9	3.25 [a]	0.23 ± 10%	0.308 [b]	−0.147 [b]	0.31
						0.104 [c]	−3.24 [c]	0.19
AgBr, cation vacancy motion	295	378	7.49	5.5 ± 0.5 [d]	0.31 ± 0.03	0.325 [c]	1.16 [c]	0.30
			7.16	5.5 ± 0.5 [d]	0.33 ± 0.03			
AgBr, silver interstitial motion	295	378	7.49	3.6 ± 0.4 [d]	0.22 ± 0.02	0.278 [c]	1.35 [c]	0.24
			7.16	3.6 ± 0.4 [d]	0.23 ± 0.03			

[a] Abey and Tomizuka [321]; conductivity measurements under pressure.
[b] Corish and Jacobs [302].
[c] Aboagye and Friauf [307].
[d] Lansiart and Beyeler [324]; conductivity measurements under pressure (285–500 K).
[e] For the elastic data used, see the references of table 10.11.

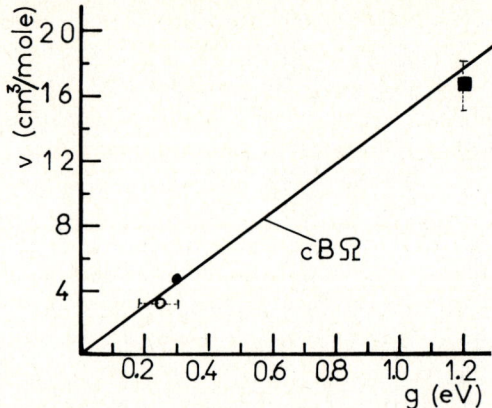

Fig. 10.9. Relation between v and g for variation defects in AgCl; ■: formation at cation Frenkel defects; ●: migration of cation vacancies; ○: cation interstitial migration. The data are from tables 10.11 and 10.12.

process, has to be equal to the bulk property $B^{-1}[(dB/dP)-1]$. In figs. 10.9 and 10.10 we plot the various values of v versus the corresponding values of g for AgCl and AgBr. In both materials we actually observe a very good linear correlation if one also considers the experimental errors. Furthermore in these figures we have drawn straight lines with a gradient equal to $B^{-1}[(dB/dP)-1]$ which is predicted from the $cB\Omega$-model. In both materials the agreement is obvious.

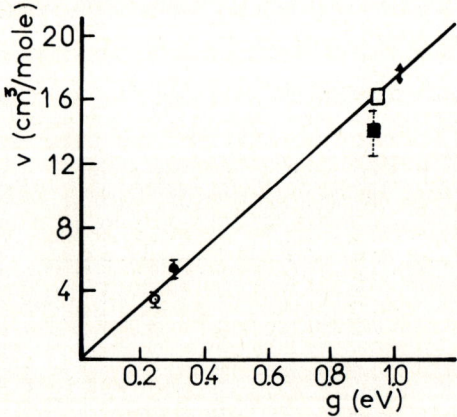

Fig. 10.10. Relation between v and g for defects in AgBr; ■: cation Frenkel formation, Lansiart and Beyeler [324]; □: cation Frenkel formation, Kurnick [393]; ●: cation vacancy migration; ○: cation interstitial migration. The data for migration come from ref. [324]. For details, see tables 10.11 and 10.12.

§10.4]   *Ionic crystals*   301

The elastic data used for the drawing of the lines are those given in table 10.11. In the case of AgBr the value $dB/dP = 7.325$ has been used which is just the average of the two published experimental values. The arrows indicate points through which the line would pass if each of these direct experimental dB/dP-values were used.

Silver halides provide one of the best examples of the validity of the $cB\Omega$-model as to the proportionality of v with g. This is so because their elastic data are precisely known but especially because the volume of formation exceeds that of the cation interstitial motion by a large factor (i.e. around 4–5). We expect future pressure experiments measuring the activation (or migration) volume of various ions diffusing in these materials to give points that lie on the lines of figs. 10.9 and 10.10. Recent pressure experiments in AgCl and AgBr are reported in ref. [648].

The coincidence of representative points with the line helps in determining the minimum value of g^f of Schottky defects in silver halides (because $v^{f,S} \gg 2\Omega$), an important factor in the analysis of anion diffusion experiments (see §9.7).

Summarizing the present discussion in silver halides (AgCl, AgBr and B4-AgI) we can say that all the published data support the validity of the relation eq. (10.1) irrespective of the process (see figs. 10.9, 10.10 and 10.11).

10.4.4 Lead fluoride

The conductivity under pressure has been studied by Samara [356] and

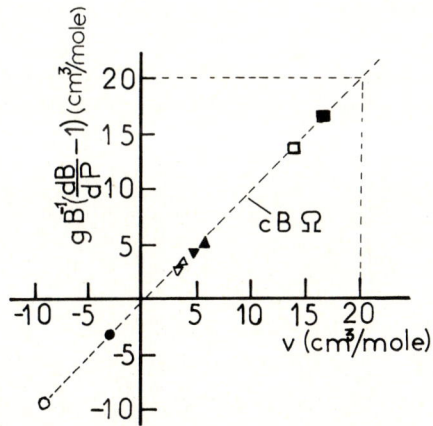

Fig. 10.11. Demonstration of eq. (10.1) for silver halides; ■: cation Frenkel formation in AgCl; □: cation Frenkel formation in AgBr; ▲: cation vacancy motion in AgBr; ▼: cation vacancy motion in AgCl; △: cation interstitial motion in AgBr; ▽: cation interstitial motion in AgCl; ●: activation in B4-AgI (295 K); ○: activation in B4-AgI (395 K); the data are from tables 10.11 and 10.12

Table 10.13
Correlation of the defect volume and the Gibbs energy in cubic PbF_2.[c]

T (K)	B[a] (kbar)	dB/dP[a]	h[b] (eV)	s[b] (k-units)	g[b] (eV)	$v/g\|_{exper.}$[b] (10^{-12} cm^3/erg)	$B^{-1}(dB/dP-1)$[a] (10^{-12} cm^3/erg)	Process[b]
300	630	7.13	0.26 ± 0.02	~2	0.21	9.4	9.7	Migration of fluorine vacancy
400			0.52 ± 0.03	~3	0.44	8.3	~9.7	Migration of fluorine interstitial
340			0.73	~4	0.61	10.9	~9.7	Activation in stage II (i.e. anion Frenkel pair formation and fluorine vacancy migration)

[a] Ref. [358]; the elastic data refer to 296 K. These data are also used for the approximate calculation of the quantity $B^{-1}[(dB/dP)-1]$ at the higher temperatures 340 and 400 K; the error introduced with this approximation is small.

[b] Ref. [356]; the entropies given are only estimates whereas the numerical values of the volumes are 1.9, 3.5 and 4 cm^3/mole respectively for the three processes given in the table.

[c] The values of B and dB/dP given in this table correspond to adiabatic measurements; however, they differ only slightly from the isothermal values for the temperatures under discussion. After the completion of the main part of the book new point defect parameters [650] and elastic data [651] appeared. The latter lead to a B-value that coincides with that given in the table. Table 4 of ref. [650] leads to $g^m = 0.20$ and 0.346 eV for the fluorine vacancy and interstitial motion, respectively, at $T = 300$ K. These g^m-values give: $v/g|_{exper.} = 9.9$ and 10.5, instead of 9.4 and 8.3 given in the above table. Obviously the new values are in *better* agreement with the value 9.7 (in units of 10^{-12} cm^3/erg) predicted from the $cB\Omega$-model.

by Oberschmidt and Lazarus [357] whereas the elastic data under pressure have been reported by Rimai and Sladek [358]. These adiabatic data have been used in table 10.13 for checking the validity of eq. (10.1) for various processes, i.e. migration of a fluorine vacancy (vac), migration of a fluorine interstitial (int) and activation in stage II, which according to Samara involves anion Frenkel pair formation and fluorine vacancy migration. For this check we give in table 10.13 the experimental value v/g which, according to the $cB\Omega$-model, is equal to the bulk quantity $B^{-1}[(dB/dP)-1]$. The experimental value v/g has an experimental uncertainty of at least 20%; 10% arising from the experimental v-value, 5–10% from the h-value and an additional error from the estimation of the entropies, which unfortunately have not been reported. An inspection of the table clearly indicates an agreement between v/g and $B^{-1}[(dB/dP)-1]$ if one considers the experimental errors; this agreement is significant for two reasons:

(a) it indicates that v can be correctly calculated from the g-value and the elastic data, irrespective the process; (for recent microscopic calculations of v, see §A.11, note 6.)

(b) in the past suggestions have been forwarded connecting the ratio v/g for the migration process to a combination $\tilde{C}$ of the elastic constants appreciably different from the bulk modulus (see ch. 14). Using this combination a value of v/g results appreciably smaller than the experimental one (e.g. by a factor of 4) thus invalidating the use of $\tilde{C}$ (see §14.1).

The above applications refer to cubic PbF_2. But when it transforms into the orthorhombic structure the migration volumes for vacancy fluorine motion and interstitial fluorine motion are found to be roughly equal in contrast to the cubic phase where they differ by a factor of two [356]. This behaviour is again in striking agreement with eq. (10.1) because in the orthorombic structure the experimental enthalpies for the above two processes do not differ much ($h^{m,vac} = 0.36 \pm 0.02$ eV and $h^{m,int} = 0.41 \pm 0.02$ eV) and hence the corresponding Gibbs energies $g^{m,vac}$ and $g^{m,int}$ are roughly equal. Because of eq. (10.1) the latter equality demands comparable migration volumes, in accordance to the experimental data. The $cB\Omega$-model is also compatible [644,663] with the measurements just published [642] for CdF_2.

10.5 Correlation between the defect volume and the enthalpy

We have already discussed in the previous chapter the validity of the relation $g = Bv[(dB/dP)-1]^{-1}$ or equivalently $g = cB\Omega$ with

$$c \equiv \frac{v}{\Omega}\left(\frac{dB}{dP}-1\right)^{-1}; \quad (10.13)$$

in this form it is valid regardless of any temperature variation of c and it is absolutely based on the fact that c is solely (practically) pressure independent [646]. If one further considers a temperature independent c it can be calculated from $c = g/B\Omega$ or at absolute zero from

$$c = \frac{h_0}{B_0 \Omega_0}. \tag{10.14}$$

Combining eqs. (10.13) and (10.14) one immediately gets

$$v = \frac{h_0}{B_0 \Omega_0} \left\{ \left. \frac{dB}{dP} \right|_T - 1 \right\} \Omega.$$

Considering the approximate relation (8.15)

$$h_0 \simeq h_{\exp} B_0 / B_0^{SL} \tag{10.15}$$

and $\Omega/\Omega_0 = b$, the previous relation reads

$$v \simeq \frac{h_{\exp} b}{B_0^{SL}} \left\{ \left. \frac{dB}{dP} \right|_T - 1 \right\}. \tag{10.16}$$

This equation assumes that c is simultaneously temperature and pressure independent and furthermore that the Arrhenius plot is almost linear because only in the latter case is eq. (10.15) valid. Therefore eq. (10.16) is clearly not equally general as eq. (10.1) so that it needs an experimental confirmation for the various types of processes, formation, migration and activation. For materials that exhibit a curved Arrhenius plot we will use experimental parameters ($h_{\exp}$, v) only extracted from the linear part. In fig. 10.12 we plotted the experimental v-values (for various processes) as a function of $h_{\exp}$ for a large body of materials; the references are given in

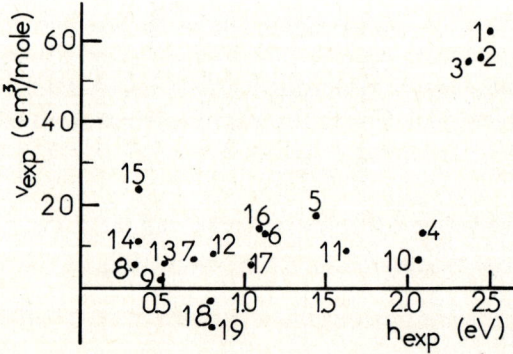

Fig. 10.12. Experimental defect volume versus defect enthalpy (compare with fig. 10.13; the numbers refer to table 10.14.

Table 10.14
Various defect volumes in solids

Dots of figs. 10.12 and 10.13	Kind of defect	Solids	Data for enthalpy [ref.]	Data for volume [ref.]	Data for elastic constants [ref.]
1, 2, 3	Formation of Schottky defect	KCl, NaCl, KBr	[36]	[322]	[36]
4	Formation of Schottky defect	LiF	[578]	[578]	[579]
5, 6	Formation of (cation) Frenkel defect	AgCl, AgBr	[307]	[580]	[381]
7	Migration of cation vacancy	NaCl	[69]	[322]	[581]
8	Migration of cation vacancy	AgBr	[307,324]	[324]	[309]
9	Migration of anion bound vacancy	CaF_2	[106,603]	[106,603]	[429]
10	Activation of monovacancy	Cu	[582]	[147]	[554,558]
11	Activation of monovacancy	Ag	[583]	[147,555]	[554,584]
12	Activation of monovacancy	In	[565]	[564]	[239]
13	Formation of monovacancy	In	[585]	[563]	[239,562]
14	Activation of monovacancy	Na	[586]	[586]	[587]
15	Activation of monovacancy	K	[36]	[36]	[36]
16	Activation of monovacancy	Pb	[36]	[36]	[36]
17	Activation of monovacancy	Sn	[567]	[566]	[239]
18	Activation of silver ion	B4-AgI at RT	[588]	[588]	[577,589,590]
19	Activation of silver ion	B4-AgI at 400 K	[588]	[588]	[577,589,590]

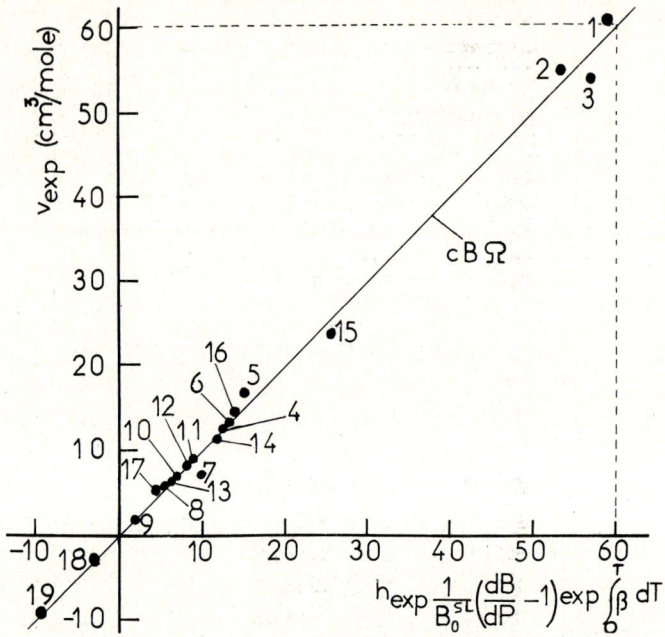

Fig. 10.13. Demonstration of eq. (10.16); the numbers refer to table 10.14; the ordinates and abscissae are given in the same units.

table 10.14. A significant scatter in the figure is obvious and it seems that the v-values (covering approximately three orders of magnitude and having positive and negative values) do not have any direct connection with the enthalpies, which vary by about two orders of magnitude. Turning now to eq. (10.16) we see that the product of the abcissae of fig. 10.12 and the bulk quantity $(b/B_0^{SL})[(dB/dP)-1]$ must become equal to the ordinate. Figure 10.13 shows the result of this procedure. Almost all points in this figure lie on the straight line with a gradient equal to unity, thus justifying the validity of eq. (10.16). It should be added that a large number of the available experimental points has not been included in the figures in order not to overload them or because they lie beyond the region of the coordinates (see also fig. A.2).

A final remark implies the following: a combination of eqs. (10.3) and (10.7) directly leads to a connection of $(dB/dP)-1$ with the quantity $d \ln T_M/dP$; the insertion of this result into eq. (10.16) gives a direct connection of the ratio v/h_{exper} to the melting properties. This connection has long been noticed as an important empirical fact.

11 THE $cB\Omega$-MODEL: HETERODIFFUSION

11.1 Temperature dependence of heterodiffusion

The same considerations as for self-diffusion (§§ 7.3 and 9.1) can be applied to the heterodiffusion coefficient. The quantities β, B and Ω depend only on the matrix material, while the constants c and f depend also on on the diffusant and the type of the diffusion process. In order to calculate c from eq. (7.52) one has to know the frequency ν^i of the diffusing atom (i) for the passage through the saddle-point; lacking any better approximation we determine it by comparison to the frequency ν^m of an atom of the matrix material (m) after considering the influence of their masses. To a rough approximation [359] one can set

$$\nu^i/\nu^m = (m^m/m^i)^{1/2}, \tag{11.1}$$

In the present calculation we assume that ν^m is equal to the Debye frequency ν_D of the matrix material and we furthermore take it as temperature independent. The $cB\Omega$-model therefore gives

$$D^i = fa^2\nu_D(m^m/m^i)^{1/2} \exp(-cB\Omega/kT). \tag{11.2}$$

The constant c is now calculated from the D^i-value at a given temperature from

$$c = (-kT/B\Omega) \ln(D^i_{\text{corr}}/fa^2\nu_D), \tag{11.3}$$

where

$$D^i_{\text{corr}} = D^i(m^i/m^m)^{1/2}. \tag{11.4}$$

Once c has been determined, one can calculate the value of the diffusion coefficient for any other temperature. We proceed to two examples:

Zirconium diffusing in niobium. The elasticity data etc. of the matrix material (Nb) have already been mentioned [228]. The factor f has been set equal to 0.727 although it is not quite sure that it is exclusively a case of vacancy diffusion. At the lowest temperature (1900 K) measured by Einziger and Mundy two values of D have been given [360] 4.51×10^{-11} and 3.82×10^{-11} cm^2/s. The resulting values of c from eq. (11.3) are 0.1867 and 0.1884. By using the first couple of values, the calculation for the highest

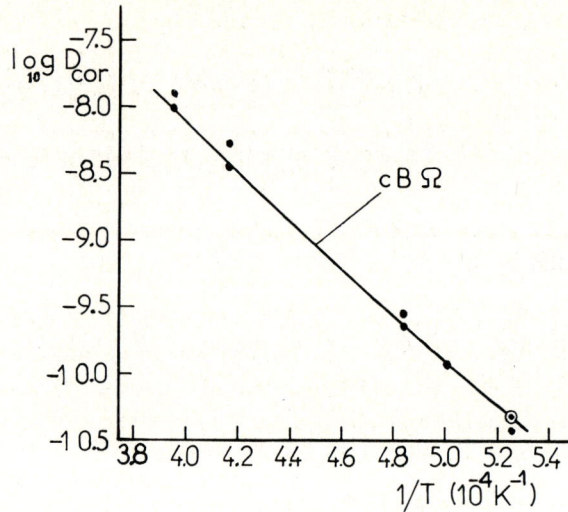

Fig. 11.1. Heterodiffusion of Zr in Nb. The line represents the values calculated from the $cB\Omega$-model whereas the dots denote experimental points from ref. [360].

temperature (2523 K) gives $D = 9.79 \times 10^{-9}$ cm^2/s, while the second couple gives 8.71×10^{-9} cm^2/s. A comparison of the experimental and the calculated results is possible in fig. 11.1 where the plot of log D_{corr} in function of $1/T$ is given for $c = 0.1884$. The points are the experimental values. A decrease of the isothermal bulk modulus at 1900 K by 1% suffices to bring the plot between the two available experimental points (9.6 and 12.5×10^{-9} cm^2/s).

Carbon diffusing in α-Fe [360a]. A compilation of diffusion coefficients obtained by various methods and in a wide range of temperatures has been given by Lord and Beshers [361]. In order to determine the constant c we use the corrected experimental diffusion constant $D^i_{\text{corr.1}} = 9.26 \times 10^{-7}$ cm^2 s^{-1} at $T_1 = 1058$ K and the bulk modulus $B_1 = 1.275 \times 10^{12}$ dyn cm^{-2} which results from Dever's [362] measurements after correction to isothermal values. For f we used the value $1/6$ by assuming that the diffusion proceeds by interstitials in octahedral sites [362]. The value of ν_D results from $\Theta_D = 428$ K as calculated for high temperatures by Killean and Lisher [150]. The application of eq. (11.3) gives $c = 0.06766$. We now proceed to the calculation of D at a low temperature T_2 according to the $cB\Omega$-model. Of course, the values of a, B and Ω appropriate to this temperature have to be considered. Using for $T_2 = 237$ K $B_2 = 1.67 \times 10^{12}$ dyn/cm^2 and $\Theta_2 = 470$ K, we obtain $D_2 = 4.01 \times 10^{-20}$ cm^2 s^{-1} which agrees with the experimental value 4.1×10^{-20} cm^2 s^{-1}. This exceedingly good accuracy naturally is fortuitous, because errors of 1% and more are present in the value of B.

Note that in this case the $cB\Omega$-model has calculated D_2 from D_1 (which is 2×10^{13} times larger) purely by comparing the anharmonic bulk qualities of the matrix material. Details can be found elsewhere [360a] together with an application to Sb diffusion in Cu.

A large number of heterodiffusion parameters can be found in ref. [362a].

11.2 Correlation between the diffusion coefficients of atoms diffusing in a given matrix

The $cB\Omega$-model leads to a correlation between the diffusion constants and the activation enthalpies of all elements diffusing with the same mechanism in a given matrix. Equation (9.4) can be written in the form

$$s = Fh_{\exp}, \qquad (11.5)$$

where

$$F \equiv -b(\beta B + dB/dT)/B_0^{\mathrm{SL}}. \qquad (11.6)$$

The factor F is a function of temperature that depends only on the host material and not on the diffusing atom. For almost linear diffusion plots and within the region of the diffusion experiments it usually increases with temperature by some percent. Equation (7.40b) can be written $D = fa^2\nu \exp(s/k)\exp(-h_{\exp}/kT)$, where s and $h_{\exp}$ are the activation parameters for diffusion. Introducing eq. (11.5) into this equation one obtains for the diffusion of atoms (i)

$$D^i = f^i a^2 \nu^i \exp(h_{\exp}^i F/k) \exp(-h_{\exp}^i/kT) \qquad (11.7)$$

where a and F are qualities of the matrix, while f, ν and h depend on the diffusing element. Due to eqs. (11.1) and (11.4), equation (11.7) simplifies to

$$\ln D_{\mathrm{corr}}^i = \ln(f^m a^2 \nu^m) - (T^{-1} - F)h_{\exp}^i k^{-1} \qquad (11.8)$$

when various atoms (i) are diffusing in the same matrix (m), if the self-diffusion follows the same mechanism as the heterodiffusion and if the attempt frequencies ν^i and ν^m follow eq. (11.1). One sees that for a given temperature a linear connection emerges between $\ln D_{\mathrm{corr}}^i$ and $h_{\exp}^i$. A plot of the above quantities for various diffusants in a given matrix should therefore give a straight line with a slope equal to $-(T^{-1} - F)/k$ and an intercept equal to $\ln(f^m a^2 \nu^m)$. For a different temperature all quantities change – D^i changes strongly and F, h^i to a lesser extent – but the linear connection will still hold. In case that various foreign atoms diffuse with the same mechanism but different than that of self-diffusion, eq. (11.8) still holds by replacing f^m with another factor f^i.

We investigate here the applicability of eq. (11.8) for a number of solids. We emphasize however that the use of eq. (11.8) *cannot* be considered as a *serious check* for the reliability of $cB\Omega$-model; this is so because T^{-1} is usually appreciably larger than F and hence the value of the slope of $\ln D^i_{corr}$ is practically governed by T^{-1} whereas F has only a minor influence. On the other hand the contents of § 11.2.4 can be considered as a reliable and significant check of the capability of the $cB\Omega$-model to describe the diffusion data. In view of the large inaccuracy of this investigation no attempt was made to classify the foreign atoms diffusing in the same element into categories – as demanded from eq. (11.8) – according to their mechanism.

11.2.1 fcc metals

Copper. A large number of experimental results on the heterodiffusion in copper have been compiled by Butrymowicz, Manning and Read [363]. Beyond those they recommend, we have considered publications by the group of Gorbachev, Klotsman, Rabovskii, Talinskii and Timofeev [364] and a few others. In the following considerations we accepted most of the published data without critical evaluation. Only the results for Mn, P and Pt have been rejected because they fall completely outside the expected region. Figure 11.2 contains the experimental values of $\log D^i_{corr}$ as function of h^i_{exp} for 1023 K. In the same figure points for self-diffusion in Cu have been added. The straight line is a least square fit to a linear law. Its slope and the intercept in the figure are -3.602 eV^{-1} and -2.85 with a correlation factor of 0.93. For a plot of natural logarithms the slope is -8.29 eV^{-1}. This experimental slope can be compared to the expression $(T^{-1} - F)/k$. Using the elastic data mentioned in ref. [157] and § 9.2.2 one gets the value 9.24 eV^{-1} which is 11% larger than the experimental one.

The same procedure was applied to the temperatures 1173 and 1323 K. The resulting slopes (for natural logarithms) were -6.81 and -5.62 eV^{-1} with the correlation factors 0.91 and 0.87. As the measurements of elastic data do not extend beyond 800 K, the computation of $(T^{-1} - F)/k$ can only be carried out under the assumption that B falls linearly with temperature. The calculated slopes for the two temperatures then have the values -7.77 and -6.64 eV^{-1} which are around 18% larger than the experimental slopes.

Silver. The results of a similar investigation for Ag at 909 K are shown in fig. 11.3. The straight line in the figure has a slope of -3.73 eV^{-1} and an intercept -3.88 with a correlation factor 0.88. These values have disregarded the point for Ru which resulted from experiments made exclusively at high temperatures; a slight curvature might have given a larger activation

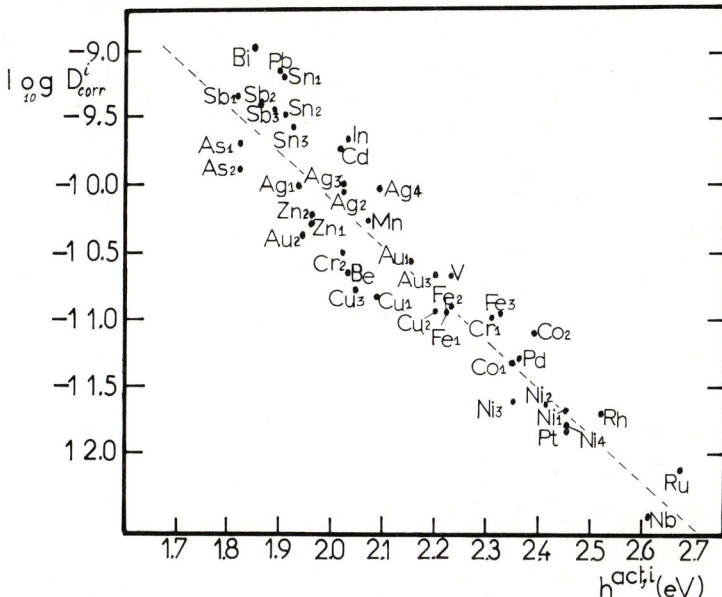

Fig. 11.2. Heterodiffusion in Cu ($T = 1023$ K); Ag1, As1, As2, Au1, Au2, Be, Co1, Co2, Cr1, Cu1, Cu2, Cu3, Fe1, Nb, Ni1, Ni2, Pd, Pt, Rh, Ru, Sb1, Zn1: ref. [363]; Ag2, Au3, Bi, Cd, In1, Pb, Sb3, Sn3: ref. [364]; Ag3: Shih and Stark [458]; Ag4: Krautheim et al. [459]; In: Krautheim et al. [461]; Sb2, Sn2: Krautheim et al. [460]; Cr2, Mn, V: Hoshino et al. [462]; Fe2: Wenzl et al. [463]; Fe3: Sen et al. [464]; Ni2: Seran [465]; Ni3: Dutt et al. [466]; Sn1: Sen et al. [467]; Zn: Dutt and Sen [468].

entropy than the one that corresponds to 909 K. The values published for the heterodiffusion of Pb in Ag have large error bars; for the calculation of the slope we have chosen the lowest value. Using the bulk modulus mentioned in § 9.2.4, the $cB\Omega$-model gives for the slope -9.83 eV^{-1} which agrees within 15% with the value 8.6 eV^{-1} of the slope for natural logarithms. The difference may be due to the fact that the elements plotted in fig. 11.3 do not all diffuse with the same mechanism.

Aluminum. The data for the diffusion in Al at 850 K are plotted in fig. 11.4. This figure could not contain the diffusion of Zr ($h = 2.51$ eV, log $D_{\text{corr}} = -11.49$). The slope and intercept of the best fit to a straight line are -2.88 eV^{-1} and -4.074 with a correlation factor of 0.918. Using the elastic data of Tallon and Wolfenden [137] one obtains $(T^{-1} - F)/k = 10.17$ eV^{-1} which is 55% larger than the experimental value 6.625 eV^{-1}. This difference is well understandable considering the scatter of the experimental data (e.g.

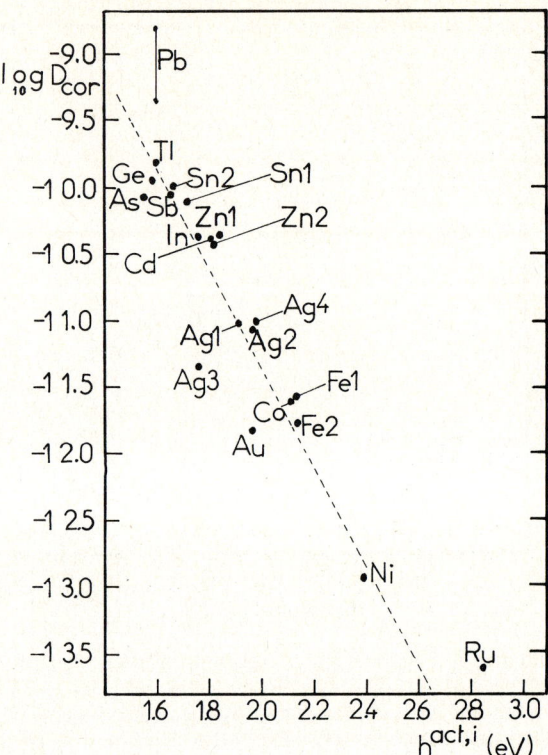

Fig. 11.3. Heterodiffusion in Ag ($T = 909$ K); Ag1: Tomizuka and Sonder [469]; Ag2: Rothman et al. [470]; Ag3: Bihr et al. [471]; Ag4: Gas and Bernardini, [472]; As: Hehenkam and Wubbenhorst [473]; Cd: Tomizuka and Slifkin [474]; Co: Bernardini et al. [475]; Fe: Bernardini and Cabane [476]; Fe2: Bharati et al. [477]; Cd, In, Sn1: Tomizuka and Slifkin [474]; Ge: Hoffman [478]; Ni: Sen et al. [464]; Pb: Sen et al. [467]; Ru: Pierce and Lazarus [479]; Au: Jaumot Jr. and Sawatzky [480]; Sb: Sonder et al. [481]; Sn2: Gas and Bernardini [472]; Tl: Hoffman [478]; Zn1: Dutt and Sen [468]; Zn2: Sawatzky and Jaumot [482].

the points for Au) and the fact that the elements do not all diffuse in Al with the same process. The latter is evident from the large variation of the enthalpy values.

Lead. The plot between log D_{corr} and h at $T = 600$ K is shown in fig. 11.5. The slope and intercept are -4.704 eV^{-1} and -3.85 with a correlation factor of 0.98. The slope in a plot of natural logarithms is 10.83 eV^{-1}. Using the data mentioned in ref. [176] and in § 9.2.4 the application of eq. (11.8) gives $(T^{-1} - F)/k = 14.4$ eV^{-1}. The good correlation factor of 0.98 would suggest that only one mechanism is involved. However, this is not the case as became clear from a variety of experiments [171].

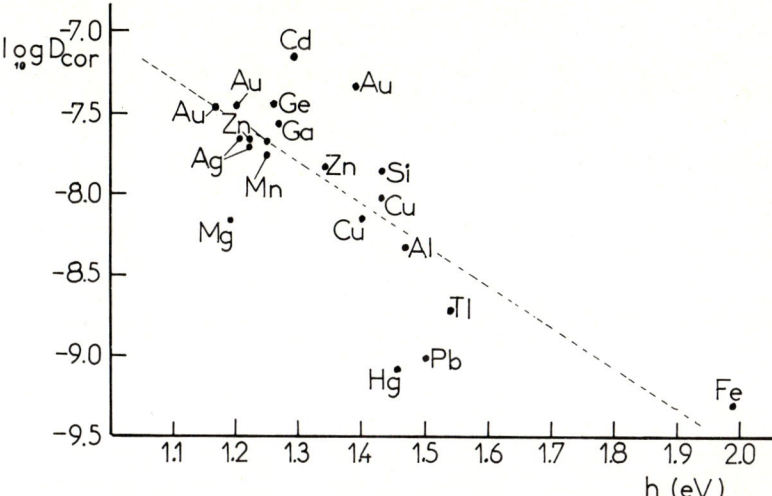

Fig. 11.4. Heterodiffusion in Al ($T = 850$ K); Au, Cu, Mg, Zn, Si: Hirano and Fujikawa [483]; Al, Mn: Lundy and Murdock [146]; Au, Ag, Cd, Fe: Alexander and Slifkin [484]; Ge, Ga, Zn, Cu, Ag, Au: Peterson and Rothman [485]; Hg, Tl, Pb: Sawayanagi and Hasiguti [486]; Zn: Hilliard et al. [487].

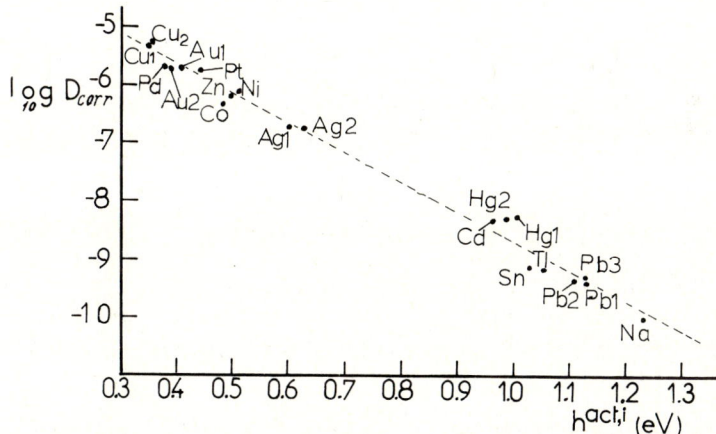

Fig. 11.5. Heterodiffusion in Pb ($T = 600$ K); Ag1: ref. [173]; Ag2: Dyson et al. [488]; Au1: Decker et al. [489]; Au2: Warburton [490]; Cd: Vanfleet et al. [491]; Co: Kusunoki and Nishikawa [492]; Cu1: ref. [173]; Cu2: Dyson et al. [488]; Hg1: Warburton [493]; Hg2: Vanfleet et al. [491]; Na: Owens and Turnbull [494]; Ni: ref. [173]; Pb1: Warburton [493]; Pb2: Resing and Nachtrieb [495]; Pb3: Baker and Gilder [496]; Pd: ref. [173]; Pt: ref. [173]; Sn1: Decker et al. [497]; Sn2: Donovan-Vojtovic et al. [498]; Tl: Resing and Nachtrieb [495]; Zn: Decker et al. [359].

11.2.2 bcc metals

Niobium. The heterodiffusion in niobium is shown in fig. 11.6 for the temperature 2000 K. The point for diffusing tungsten ($h = 6.76$ eV, log $D_{\text{corr}} = -20.1$) could not be included in the figure, but it has been taken into consideration in making the least square fit. In the figure the slope and intercept are -1.99 eV^{-1} and -2.33 with a correlation factor 0.85. The elastic data for Nb have been mentioned in § 9.3.5 For 2000 K we obtained $B = 1.37 \times 10^{12}$ dyn/cm^2 and $dB/dT = -1.86 \times 10^8$ dyn cm^{-2} K^{-1}. By using these data in eq. (11.6) a slope of -4.79 eV^{-1} resulted which agrees within 4% with the experimental value -4.59 for the slope of the natural logarithms. We must state here that this good agreement exclusively results

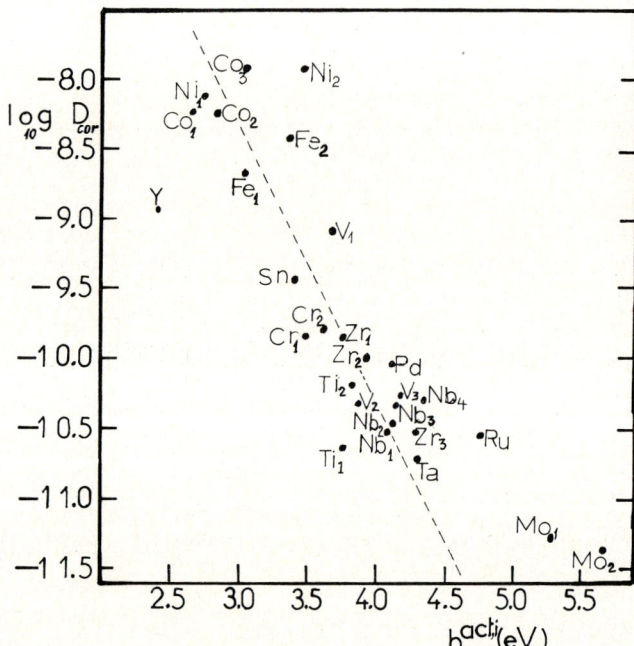

Fig. 11.6. Heterodiffusion in Nb ($T = 2000$ K); Co1: Pelleg [499]; Co2: Ablitzer [226]; Co3: Peart et al. [500]; Cr1: Pelleg [501]; Cr2: Pelleg [502]; Fe1: Ablitzer [503]; Fe2: Peart et al. [500]; Mo1: Roux and Vignes [504]; Mo2: [505]; Nb1: Einziger et al. [227]; Nb2: Ablitzer [226]; Nb3: Lundy et al. [506]; Nb4: Pelleg and Lindberg [507]; Ni1: Ablitzer [226]; Ni2: Agarwala and Hirano [508]; Pd, Ru: Sathyaraj et al. [509]; Sn: Askill [510]; Ta: Lundy et al. [506]; Ti1: Pelleg [501]; Ti2: Roux [511]; V1: Agarwala et al. [512]; V2: Roux and Vignes [504]; V3: ref. [505]; W. Roux and Vignes [504] (this point – coordinates (6.76, −20,1) – could not be contained in the figure); Y: Gornyy and Altovskiy [513]; Zr1: Roux and Vignes [504]; Zr2: Einziger and Mundy [360]; Zr3: ref. [505].

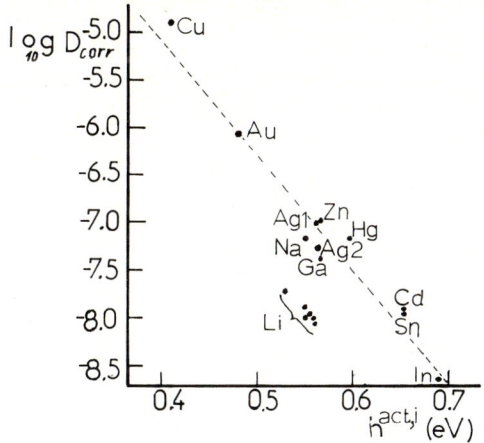

Fig. 11.7. Heterodiffusion in Li ($T = 393$ K); Ag1: Ott and Norden [514]; Ag2: Mundy and McFall [515]; Au: Ott [516]; Bi, Pb, Sb, Sn: Ott et al. [517] (the values of Bi (2.05, -11.9), Pb (1.09, -11.1) and Sb (1.8, -12.3) lie outside the dimensions of the figure); Cd, Hg, Ga: Ott [518]; Cu: Ott [519]; In: Ott [520]; Li: Lodding et al. [209], Ott et al. [521], ref. [204]; Na: Mundy et al. [522]; Zn: Mundy et al. [523].

from the use of the measurements for W. If the latter point is ignored, the value 4.59 of the slope and the intercept turn into 2.65 and 12.4 with a correlation factor of 0.84.

Lithium. The results for the hetero-diffusion in Li are shown in fig. 11.7 at $T = 393$ K. The points for Pb, Sb and Bi lie beyond experimental uncertainties outside any linear connection of the diffusion constants with h and cannot be plotted within the scale of this figure. Within the frame of the $cB\Omega$-model their behaviour must be attributed to a cooperation of several diffusion mechanisms of equal activation enthalpy. In view of the fact that the diffusion in bcc elements is often controversial, we cannot make any comments. Nine other elements give a linear connection with a slope and intercept of -11.3 eV^{-1} and -0.74 with a correlation factor of 0.97. The corresponding slope of the natural logarithms is -25.91 eV^{-1} which coincides with the value 25.92 expected from eq. (11.8) after using the elastic constants and the expansivity data mentioned in § 9.3.2. The published values for self-diffusion give the group of points that is shown in the figure; the coincidence between the calculated and the experimental values of the slope of the line is due to the fact that $T^{-1} \gg F$.

γ-uranium. Using the compilation of diffusion parameters given by Warburton and Turnbull [247], values of the "corrected" diffusion coefficient were calculated for 1100 K. A straight line was actually obtained (fig.

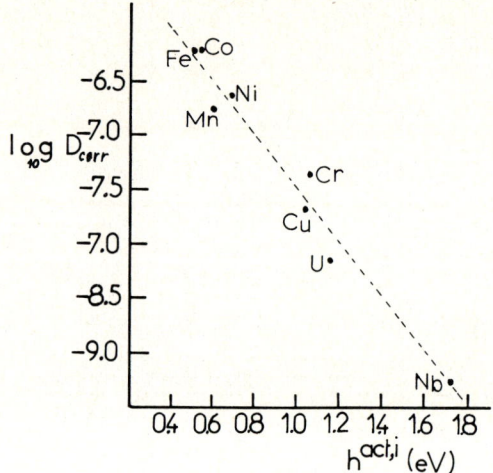

Fig. 11.8. Heterodiffusion in γ-uranium ($T = 1100$ K). All points: ref. [247].

11.8) with a slope and an intercept equal to -2.51 eV^{-1} and -4.99 with a correlation factor of 0.98. The elastic constants have not been measured so that numerical comparison with eq. (11.8) is not possible for uranium; however, the linearity observed is a good indication.

β-titanium. It is interesting to extend the study of the linearity between log D_{corr} and h to matrices that show strongly bent Arrhenius plots, as for example β-Ti. As the enthalpy in such cases is strongly temperature dependent, one has to choose a temperature and determine the value of the activation enthalpy at that temperature for each diffusant from the slope of the corresponding Arrhenius plot. From a compilation by Askill and Gibbs [365] of such values of D_0 and h, valid separately for low and high temperatures and after certain corrections by Hood [366], the diffusion coefficients D_{corr} have been calculated for $T = 1250$ and 1800 K and plotted in figs. 11.9 and 11.10. The slopes on the log scale are 3.11 eV^{-1} and 2.31 eV^{-1} with correlation factors 0.70 and 0.92. The intercepts are -3.99 and -1.85. As the correlation at 1250 K seems unsatisfactory, we calculated the slopes after separating the elements according to their solubility [366]. Materials with low solubility (marked with large dots) then gave a slope and intercept of -3.39 eV^{-1} and -3.09 with the much more satisfactory correlation factor 0.92. The rest of the diffusing elements are indicated with small dots; all form a group of lower diffusion coefficients. There seems to be no connection to a linear law. The lack of sufficient elasticity data for

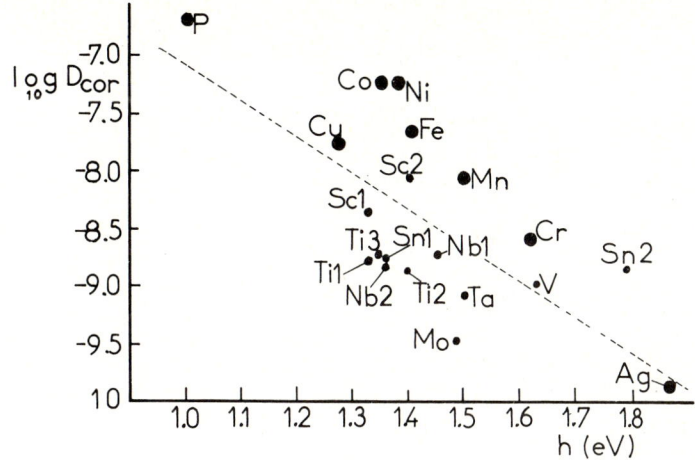

Fig. 11.9. Heterodiffusion in β-titanium ($T = 1250$ K); Ag and Sc2 for low T: Askill [524]; Nb2 and Ti3 for low T: Pontau and Lazarus [525]; V and Ti2: Murdock et al. [526]; Ta: Askill [527]; Sn2: Jackson and Lazarus [367]; all other points: ref. [366]; line: best fit through all points.

β-Ti prevents the calculation of the corresponding bulk quantity $(T^{-1} - F)/k$ for the two cases.

Writing eq. (9.2) for a high (II) and a low (I) temperature and dividing them one finds that h_{II}/h_{I} is a bulk property and should therefore have the

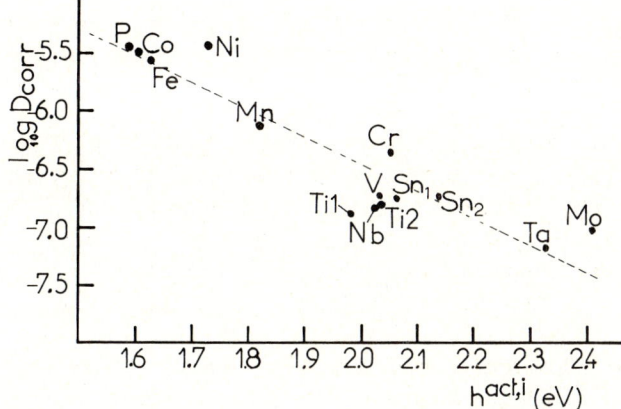

Fig. 11.10. Heterodiffusion in β-titanium ($T = 1800$ K); Sn2: Jackson and Lazarus [367]; V and Ti2: Murdock et al. [526]; Ta: Askill [527]; all other points: ref. [366]; line: best fit through all points.

Table 11.1.
Comparison of data of β-Ti for $T_I = 1250$ K and $T_{II} = 1800$ K.

Diffusant	h_{II}/h_I	$c \times 10^2$	$B_{II}(10^{11}\,\text{dyn/cm}^2)$
P	1.67	11.2	4.01
Sn1	1.51	16.2	3.89
Nb1	1.40	16.5	3.83
Mo	1.62	18.0	3.55
Cr	1.26	15.8	3.64
Mn	1.21	14.6	3.73
Fe	1.15	13.4	3.48
Ni	1.26	12.5	3.56
Co	1.19	12.5	3.65
Ti1	1.45	16.3	3.87
Nb2	–	16.4	4.16
Ti3	–	16.1	4.27
V	1.25	16.7	3.72
Ti2	1.45	16.3	3.51
Ta	1.55	17.0	4.03
Sn	1.20	16.4	3.81
Mean value	$1.37 \pm 3\%$		$3.82 \pm 1.6\%$

same value for all diffusants irrespective of the mechanism with which they diffuse as long as the same mechanism occurs at both temperatures. Because of the method with which h_I and h_{II} were determined, the ratio h_{II}/h_I might be in error by 15%. The results for a variety of diffusants are given in table 11.1, the mean value is $1.37 \pm 3\%$, and therefore there is no single diffusant for which h_{II}/h_I is definitely outside the error bars.

A determination of the numerical value of h_{II}/h_I from the $cB\Omega$-model is not possible; it would need elastic data for both temperatures. However, measurements for 1800 K are lacking. On the other hand, the considerations can be reversed and h_{II}/h_I used to anticipate the value of B_{II}. This is possible because the adiabatic bulk modulus $B^S = 8.96 \times 10^{11}$ dyn cm^{-2} has been measured by Fisher and Dever [342], just above the transformation temperature to the β-phase; from this one obtains the isothermal value $B_I = 9.09 \times 10^{11}$ dyn cm^{-2}. Combining this value with the values of h_{II}, the constant c has been determined for each diffusant. As we are at a loss as to the error in these values, we give three decimals, although they will not all be of significance. The bulk modulus for $T_{II} = 1800$ K can now be calculated from

$$-B_{II} = \frac{kT_{II}}{c\Omega_{II}} \ln\left(\frac{D_{\text{corr},II}}{fa_{II}^2 \nu}\right).$$

Their values are given in the last column. The mean value resulting from this column is $B_{II} = 3.82 \times 10^{11}$ dyn cm$^{-2} \pm 1.6\%$.

A strong curvature of the Arrhenius plots is a common feature for almost all atoms diffusing in β-Ti (and β-Zr).

β-zirconium. Upon transforming from the α-phase to the β-phase this metal is known to produce embryos of a so-called Ω-phase, mainly when alloyed. As this phase persists well above the transition temperature, (1135 K), a study was carried out at the high temperature of 1900 K. The published experimental values of log D_{corr} for each diffusant at various temperatures near 1900 K were used in order to determine a corresponding value of h. The points for the diffusants Zr [193], Nb [367a], Ta [527] and V [512] give a very irregular picture. A fit to a linear law gave a correlation factor of 0.44. The reason for which eq. (11.8) completely fails for β-Zr is not clear, perhaps it has to do with the concentration of Ω-embryos not being the same in the various samples. The activation enthalpies lie between 2 and 3 eV, while corrected diffusion coefficients for the diffusants investigated do not differ by more than 30%.

Various atoms diffusing in the same matrix. It is interesting to note that the various heterodiffusion curves for a given matrix do usually not intersect [365]. An intersection of the curves for the diffusants (i) and (j) would mean that for a certain temperature the quantity

$$\ln D^i - \ln D^j = -\frac{(c^i - c^j)B\Omega}{kT}$$

would be equal to zero, whereas for other temperatures it would have positive or negative values. The latter fact, however, gives $c^i \neq c^j$ thus making an intersection impossible, provided that the mechanism is the same for all temperatures. A crossing can become possible in spite of $c^i - c^j \neq 0$, only when the quantities f and ν are not the same for both diffusion curves. It is not expected [367] that the quantity f would differ by a large factor; on the other hand the attempt frequencies might easily differ sufficiently to make a crossing possible, because a large difference in the preexponential factor shifts the Arrhenius curve vertically and thus can lead to an intersection.

11.2.3 Ionic crystals

Sodium chloride. Diffusion has been measured for a number of anions and cations. The results for 850 K are given in fig. 11.11. As all h_{exp}-values coincide with 2 eV within 15%, experimental inaccuracies can horizontally displace the relative position of the points in the figure to a considerable

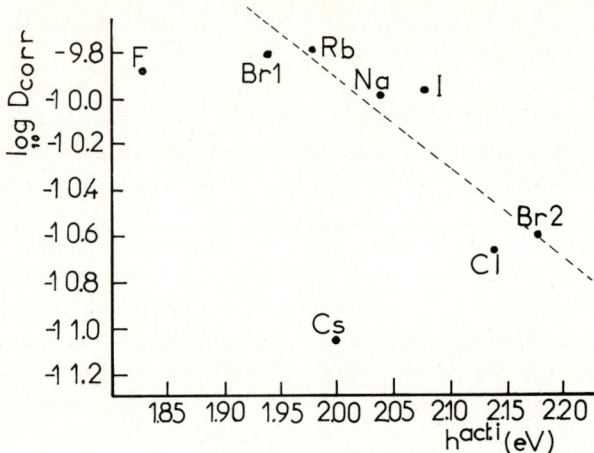

Fig. 11.11. Heterodiffusion in NaCl ($T = 850$ K); Na: Rothman et al. [528]; Rb, Cs: Beniere et al. [529a]; Br, I: Dobrovinskaya et al. [529]; Br2, Cl, I, F: Beniere et al. [530].

extent. Furthermore, the diffusion coefficients do not differ by more than a factor of 10, so that a large error in the slope and the correlation factor could be expected. One sees that Cs^+ and F^- fall clearly away from the other points. The remaining points give a slope of -3.83 eV^{-1} with a correlation factor of 0.91. By using the elastic data of Spetzler, Sammis and O'Connell [282] the $cB\Omega$-model gives from $(T^{-1} - F)/k$ the value 7.36 eV^{-1} which should be compared with the experimental value 8.83 (in a plot of natural logarithms).

Sodium iodide. Assuming that the Na and I ions diffuse in NaI with the same mechanisms we apply the same method to this material. The latest diffusion constants and enthalpies were given by Kostopoulos, Reddy and Beniere [293]. By extrapolating their values to 570 K we obtained a slope and an intercept of -22.42 eV^{-1} and $+6.29$ (in a plot of natural logarithms). On the basis of the elastic data mentioned in §9.6.5, the $cB\Omega$-model gives a slope of 15.4 eV^{-1}. The difference may be attributed to the usual experimental error in the quantity dB/dT and mainly to the fact that the values of h_{exp} for the two ions differ only by 12% so that the experimental errors in h_{exp} influence the slope considerably.

Silver bromide. The diffusion data for Cl^-, I^- and Na^+ have been published by Batra and Slifkin [305,368,71], while those for Ag^+ and Br^- are taken from fig. 4 of the paper of Friauf [65]. Rubidium has recently been measured by Cardegna and Laskar [308a] and K by Laskar and Cardegna

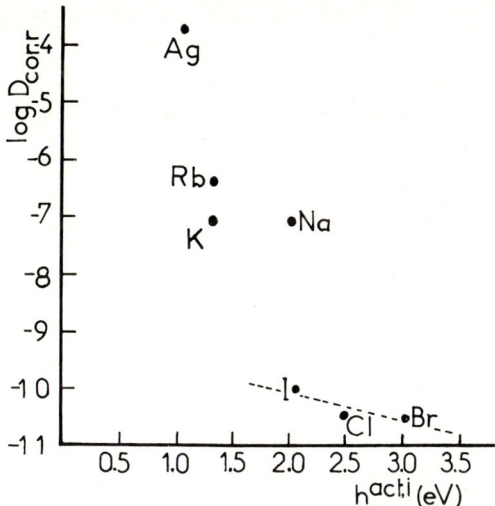

Fig. 11.12. Heterodiffusion in AgBr ($T = 648$ K). (References given in the text.)

[368a]. The results are shown in fig. 11.12 for $T = 648$ K. The dotted line is a fitting to the points for the anions; it gives a slope and an intercept of -0.507 eV^{-1} and 9.03 with a correlation factor of 0.9. The elastic data of Tannhauser, Brunner and Lawson [312] after correction to isothermal conditions give a slope of 0.95 eV^{-1} which must be compared with the experimental slope 1.17 eV^{-1} in the scale of natural logarithms. There is some indication that the diffusion for the three free anions follows the same mechanism. The cations have much larger diffusion coefficients. We observe that $D_{Ag} > D_{Na} > D_{anion}$. This fact has a natural explanation if we recall that the silver ions diffuse mainly with interstitials, sodium (and perhaps potassium and rubidium) ions through cation vacancies and anions through anion vacancies. The three diffusion coefficients scale roughly as follows

$$D_{Ag^+} \sim n_{int,cat} \exp\left(-\frac{g^{m,Ag^+}}{kT}\right),$$

$$D_{Na^+} \sim n_{vac,cat} \exp\left(-\frac{g^{m,Na^+}}{kT}\right),$$

$$D_{anion} \sim n_{vac,an} \exp\left(-\frac{g^{m,an}}{kT}\right).$$

11.2.4 Linearity of the log D_0-values as function of enthalpies

A slightly different approach is possible in cases where the Arrhenius plots are practically linear both for the hetero- and the self-diffusion, so that h and s can be taken as temperature independent. For such cases eqs. (7.43) and (7.45) can be written

$$D^i = D_0^i \exp(-h^i/kT),$$

where the preexponential factor

$$D_0^i = f^i \alpha^2 \nu^i \exp(s^i/k)$$

is practically temperature independent. After replacing s^i from eq. (11.5) one obtains

$$D_0^i = f^i \alpha^2 \nu^i \exp\left(\frac{F}{k} h^i\right).$$

The quantity ν^i for various diffusants in a given matrix can be replaced with the help of eq. (11.1) so that by setting ν^m equal to the Debye frequency one finally gets for the quantity $D_{0,\text{corr}}^i \equiv D_0^i (m^i/m^m)^{1/2}$ the expression:

$$D_{0,\text{corr}}^i = f^i \alpha^2 \nu_D \exp\left(\frac{F}{k} h^i\right). \tag{11.9}$$

If f^i is taken the same as for the case of heterodiffusion and self-diffusion, we see that for a given matrix material a plot of $\ln D_{0,\text{corr}}^i$ as function of h^i will give a straight line with a slope F/k.

Such a linearity has been noticed by Decker, Ross, Evenson and Vanfleet [359] for elements diffusing in Pb. A similar compilation of Lesage and Huntz [369] for the bcc-metals Mo, Nb, Ta and W showed only an indistinct correlation.

As an example we give in fig. 11.13 the plot of

$$RT \ln\left(\frac{D_0^i}{D_0^{\text{ref}}} \sqrt{\frac{m^i}{m^{\text{ref}}}}\right) \qquad \text{versus } h^i$$

for various elements diffusing in lead and which according to eq. (11.9) should have a slope equal to FT, where $T = 600$ K. Note that D_0^{ref} and m^{ref} refer to the bulk material (i.e. to Pb). The continuous line in the figure comes from a least squares fitting of the experimental points to a straight line whereas the dotted line corresponds to the prediction of the $cB\Omega$-model. The "experimental" line has a slope equal to 0.4005 whereas that of $cB\Omega$ is 0.337, i.e. only 16% lower; the latter slope is calculated from the value $F = 5.66 \times 10^{-4}$ K^{-1} (mentioned in § 9.2.4) and hence $FT = 0.337$. The contents of the figure and all the data used above have been kindly

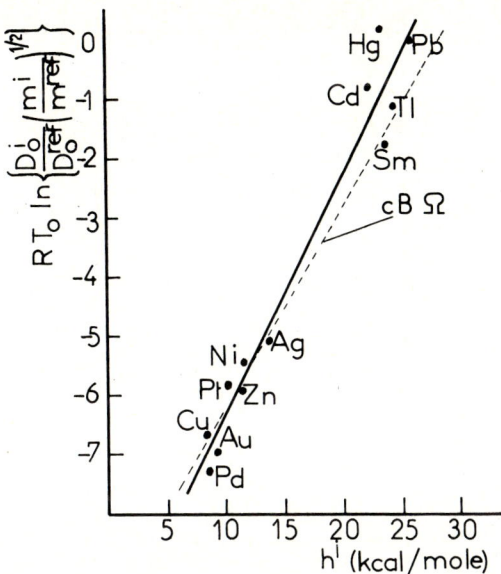

Fig. 11.13. Application of the D_0-method to the heterodiffusion in lead, according to Vanfleet (private communication); full line: best fit; dashed line: $cB\Omega$-model. ($T = T_0 = 600$ K.)

forwarded to the authors by H.B. Vanfleet; for the theoretical line plotted in the figure, Vanfleet has inserted into eq. (9.3) the values ($T = 600$ K) $\kappa = 2.99 \times 10^{-3}$ kbar^{-1}, $(\partial \kappa / \partial T)_P = 2.85 \times 10^{-6}$ kbar^{-1} K^{-1}, $\beta = 1.08 \times 10^{-4}$ K^{-1} and $V = 18.796$ cm^3/mole. In view of the fact that the various elements depicted in fig. 11.13 do not diffuse in lead with the same mechanism, the agreement between the full and the dashed line might indicate that the corresponding f^i-values do not vary by a large factor (see eq. 11.9).

12 MIXED ALKALI AND SILVER HALIDES

The basic problem in mixed halides is the determination of various properties (volume, elastic constants, etc.) as a function of composition. Of special interest is the composition x_m at which conductivity attains a maximum value; the latter can be well above the corresponding values of the end members. An explanation also of the cause for which this maximum occurs is desirable. The $cB\Omega$-model can give such an explanation and further predict with confidence the value of x_m in terms of well known properties of the end members. In order to reach this goal the following steps have to be taken: (i) Determine the mean atomic volume Ω of the mixed system from the mean atomic volumes Ω_1 and Ω_2 of the end members 1 and 2. (ii) Devise a method according to which the bulk modulus of the mixed system can be predicted as a function of composition. The composition x_m can then be determined. The above procedure allows also the possibility of determining the "excess conductivity" (or diffusivity) with respect to the corresponding values of the two end members.

12.1 Variation of the mean atomic (or molecular) volume with composition

Let v_1 be the volume per "molecule" of the pure component (1). For reasons of simplicity – without losing generality as will become clear below – we assume that v_1 is smaller than the volume v_2 per "molecule" of the pure component (2). Let now V_1 and V_2 denote the molar volumes of these two end members, i.e. $V_1 = Nv_1$ and $V_2 = Nv_2$. We define a "defect volume" v^d as the increase of the volume V_1 if one "molecule" of type (1) is replaced by one "molecule" of type (2). Due to the fact that the hard spheres model is a good approximation for alkali halides v^d should be equal to the decrease of the volume V_2 if one "molecule" of type (2) is replaced by a molecule of type (1). The validity of this relation will be checked below and will be found to be correct to a high degree of approximation.

The volume V_{N+n} of a crystal containing N molecules of type (1) and n molecules of type (2) is given by

$$V_{N+n} = \frac{m_1 + (n/N)m_2}{\rho} N, \qquad (12.1)$$

where m_1, m_2 are the masses of the "molecules" of the two pure components and ρ is the density of the mixed system.

In order to check whether v^d is independent of composition or not we make the following considerations; the addition of one molecule of type (2) to a crystal containing N molecules of type (1) will increase its volume by $v^d + v_1$. If therefore v^d is independent of composition, the volume V_{N+n} should be equal to (applying N = Avogadro's number):

$$V_{N+n} = V_1 + n(v^d + v_1) \qquad (12.2)$$

or

$$V_{N+n} = V_1 + \frac{n}{N}(Nv^d + V_1). \qquad (12.3)$$

In order to test the constancy of v^d we calculate the volumes V_{N+n} from eq. (12.1), as the density values are easily available, and plot them as a function of n/N. Such a check has been carried out in the system (KCl–KBr) where the density values are known from the measurements of Slagle and McKinstry [370,371]. The resulting values of V_{N+n} are given in table 12.1 and plotted in fig. 12.1. The figure shows that V_{N+n} versus n/N is actually a straight line. A least squares fitting to the straight line shows that its slope is $Nv^d + V_1 = 43.22$ cm³ whereas the intercept is $V_1 = 37.65$ cm³. The intercept is practically equal to the experimental value of the molecular volume of KCl, $V_1 = 37.58$ cm³, given by Slagle and McKinstry or to the value 37.51 cm³ given by Smith and Cain [276]. The difference between the slope and the molecular volume of KCl gives for Nv^d either 5.61 cm³ if the V_1-value of Slagle and McKinstry is used or 5.75 cm³ from the V_1-value of Smith and Cain. The difference $N(v_2 - v_1)$ is approximately equal to 5.85 cm³ [276]. We therefore conclude that v^d can be considered independent of composition to a very good approximation and determined from the end members by

$$v^d = v_2 - v_1. \qquad (12.4)$$

Exactly the same value of v^d results if one plots V_{N+n} versus n/N by taken alternatively KBr as the pure material; in this case v^d becomes negative, but again its absolute value is equal to $v_2 - v_1$.

Such a behaviour is expected for a hard spheres model because the increase v^d of the volume V (fig. 12.2) when the "smaller" atom (1) is

Table 12.1
Various properties of KCl + x mole% KBr at RT.

x (mole% KBr)	n/N	ρ^a (g cm^{-3})	$V_{N+n}^{calc.}$ [eq. (12.1)] (cm^3)	$\kappa_{exper.}^{S}$ [a] (10^{-12} dyn cm^{-2})	$\kappa_{calc.}^{S}$ [eq. (12.13)] (10^{-12} dyn cm^{-2})	$\kappa_{exp}^{S} V_{N+n}$ (10^{-10} cm^5 dyn^{-1})
0	0	1.984	37.58	5.464 (5.495) [c]		2.053 (2.065)
7.7	0.0834	2.04 [b]	41.41	5.525	5.517 (5.547) [d]	2.288
16.8	0.2019	2.129	46.31	5.659	5.643 (5.674)	2.621
17.1	0.2063	2.126	46.62	5.656	5.662 (5.693)	2.637
38.2	0.6181	2.302	64.34	5.905	5.854 (5.886)	3.799
38.7	0.6313	2.300	65.08	5.903	5.844 (5.876)	3.842
57.8	1.370	2.453	96.84	6.099	6.041 (6.074)	5.906
59.8	1.488	2.473	101.74	6.135	6.071 (6.105)	6.242
79.5	3.878	2.613	205.16	6.284	6.241 (6.275)	12.89
80.0	4	2.613	210.72	6.309	6.236 (6.271)	13.294
100	–	2.744	–	6.482	–	–

[a] From ref. [370 or 371] except for $x = 7.7\%$.
[b] From ref. [372].
[c] From ref. [376].
[d] The values between parentheses have been calculated from eq. (12.13) by using the value $B_1^S = 182$ kbar given in ref. [376]. The compressibility values are the adiabatic ones.

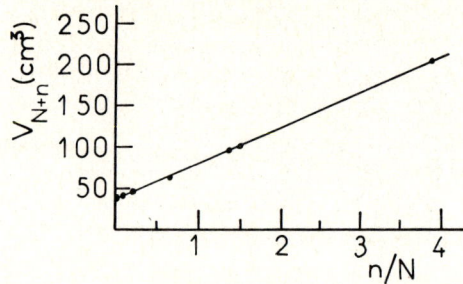

Fig. 12.1. Volume of a crystal containing N molecules KCl and n molecules KBr; the linearity indicates that the defect volume v^d is independent of composition (see eq. 12.3).

exchanged by a larger atom (2) is equal to the decrease which occurs when atom (2) is exchanged by atom (1).

Equation (12.3) gave the volume resulting from the addition of n molecules to a pure body consisting of N molecules. It can be set in a different form by considering the volume V in which n molecules have replaced n molecules of the pure body. The molar fraction x is connected to n/N by

$$\frac{n}{N} = \frac{x}{1-x}. \qquad (12.5)$$

Then with the help of eqs. (12.3) and (12.4) one gets

$$V = (1-x)V_1 + xV_2. \qquad (12.6)$$

The two quantities are connected by

$$V_{N+n} = \left(1 + \frac{n}{N}\right)V. \qquad (12.7)$$

Equation (12.6) is a relation analogous to Vegard's law [373], which suggests that the lattice parameter l of the mixed system is given by $l = (1-x)l_1 + xl_2$ in terms of the parameters l_1 and l_2. It is clear that this equation cannot be

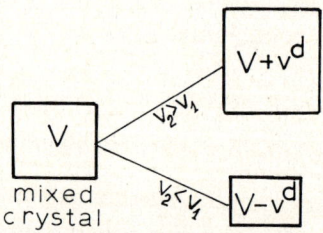

Fig. 12.2. Change of volume upon replacement of molecule 1 by molecule 2. (v^d: Absolute value of the defect volume.)

valid because it contradicts eq. (12.6) which was so successful in describing experimental data. It should be mentioned that the latter equation is compatible with the conclusion of Slagle and McKinstry [370] who found $l^{3.2} = (1-x)l_1^{3.2} + xl_2^{3.2}$ by using a fitting procedure.

We end this paragraph by noting that the above proposal of v^d being independendent of composition (and hence $v^d = v_2 - v_1$) is also valid in the case of the mixed crystal $K_{1-x}Na_xI$ for which the density data of Haget, Chanh, Garin and Bonpunt [374] are available.

12.2 Variation of the bulk modulus with composition

It is an experimental fact [371] that the bulk modulus B and its temperature derivative do not vary linearly with composition. The same holds for the compressibility κ. This is in contrast to theoretical treatments quoted in ref. [375] which predict a linear dependence of κ versus x. We will present below two methods for the determination of B at any desired composition x. The first, which is based on the assumption v^d = constant just verified, gives the requested answer for B from the bulk moduli of the end members. In the second, only the elastic data of one pure member are needed, but the $cB\Omega$-model has also to be employed.

First method: Differentiating eq. (12.6) with respect to pressure we get:

$$\kappa V = (1-x)\kappa_1 V_1 + x\kappa_2 V_2. \tag{12.8}$$

Inserting the value of V from eq. (12.6) we immediately get

$$B = B_1 \frac{1 + x[(V_2/V_1) - 1]}{1 + x[(B_1 V_2/B_2 V_1) - 1]}. \tag{12.9}$$

This equation permits the direct evaluation of B at any desired composition and temperature in terms of the elastic data of the two end members; it should be valid both for isothermal and adiabatic bulk moduli.

The validity of the method can be checked only for $KCl_{1-x}Br_x$ because only for this system are sufficient elastic data available.* In table 12.2 we have calculated the values of B^S for various compositions and temperatures. The values of V_1, V_2, B_1^S and B_2^S of the end components used in the calculation are those reported in ref. [371]. For the temperature variation of the volumes the values of β_1 and β_2 of Lindström [375] were used. In the same table the experimental values of B^S measured by Slagle and Mc-Kinstry are inserted for the sake of comparison. An inspection of the table

---

* Very recently the validity of the method has been checked for other alkali-halide mixed crystals, giving excellent results (M. Lazaridou and J. Grammatikakis, J. Phys. Chem. Sol. 1985, in press).

Table 12.2
Bulk modulus (kbar) of $KCl_{1-x}Br_x$, calculated from the data of the end members, according to eq. (12.9).

x (%)		25°C	100°C	250°C	300°C	400°C	dB^S/dT (10^{-2} kbar grad^{-1})
0	$B^S_{exp.}$	182.6	177.3	164.9	160.2	150.0	−8.6
16.8	$B^S_{exp.}$	176.8	171.4	159.2	154.7	145.1	−8.4
	$B^S_{calc.}$	176.7	169.9	157.2	152.7	143.2	−8.8
38.2	$B^S_{exp.}$	170.2	164.7	152.8	148.5	139.6	−8.1
	$B^S_{calc.}$	169.3	163.5	151.1	146.9	138.5	−8.2
59.8	$B^S_{exp.}$	164.5	158.9	147.2	143.1	134.7	−7.9
	$B^S_{calc.}$	163.0	157.6	145.5	141.3	132.6	−8.1
79.5	$B^S_{exp.}$	159.8	154.2	142.7	138.8	130.8	−7.7
	$B^S_{calc.}$	159.1	153.3	141.3	137.2	128.6	−8.1
100	$B^S_{exp.}$	155.5	149.8	138.5	134.7	127.1	−7.6

shows that the maximum difference between the calculated and the experimental values does not exceed 1.5% although B^S varies totally by 44%. In the last column of the table we give the dB^S/dT-values as they result from a least squares fitting to a straight line of the calculated B^S-values. We again see that the calculated dB^S/dT-values, as expected, show a satisfactory agreement with those obtained from experiments.

Second method. Each molecule added to a homogeneously mixed crystal can be considered as a defect. A compressibility κ^d of the defect volume can be defined in an analogous way to the compressibility of a vacancy by

$$\kappa^d = -\frac{1}{v^d}\left.\frac{\partial v^d}{\partial P}\right|_T.$$

Whenever κ^d is approximately constant throughout all the composition range we can calculate its value with the help of the $cB\Omega$-model, which connects κ^d with the elastic quantities of the bulk material (see §A.11, note 23). Differentiating eq. (12.3) with respect to pressure we get:

$$\kappa V_{N+n} = \kappa_1 V_1 + \frac{n}{N}\left(\kappa^d N v^d + \kappa_1 V_1\right). \tag{12.10}$$

In view of the already verified constancy of v^d throughout the composition range, if κ^d is also independent of composition, eq. (12.10) demands that a plot of κV versus n/N should be a straight line. A check of such a linearity of alkali halides is only possible in the case of $KCl_{1-x}Br_x$. Silver halides will be discussed separately. Using the experimental data given in table 12.1, the experimental values of $\kappa^S V_{n+N}$ versus n/N have been plotted in fig. 12.3. The linear fitting gives an excellent correlation factor which in view of the constancy of v^d justifies the assumption that also κ^d is independent of composition. Note that the quantities (κ, κ_1) in eq. (12.10) are both either isothermal or adiabatic; obviously the quantity κ^d in these two cases is slightly different. In fig. 12.3 the κ-values are adiabatic so that the κ^d-value determined below corresponds to this case. The slope of the graph is $\kappa^d N v^d + \kappa_1^S V_1 = 2.802 \times 10^{-10}$ cm^5 dyn^{-1}. Using the known quantities of the pure component (1) $\kappa_1^S = 5.464$ or 5.495×10^{-12} cm^2 dyn^{-1} and $V_1 = 37.66$ cm^3 one gets after fitting $\kappa^d = 13.2\,(\pm 1\%) \times 10^{-12}$ cm^2 dyn^{-1}. In the present example experimental values of κ^S were available and were used in order to show that κ^d is constant. Let us examine now whether the $cB\Omega$-model leads, by means of eq. (8.31), to reliable values of κ^d and therefrom to the determination of the value of κ for any desired composition with the help of eq. (12.10).

According to eq. (8.31) the compressibility κ^d of a defect in any solid can be calculated from the elastic properties of the bulk solid. The fact that κ^d does not depend on composition simplifies the question as one can choose

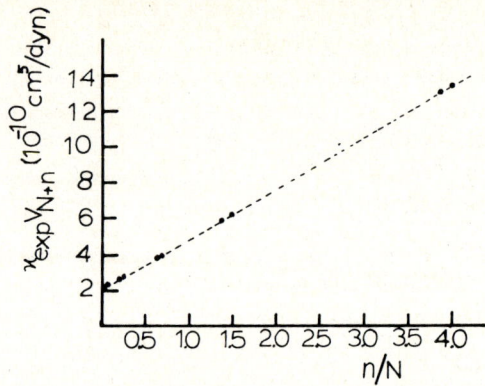

Fig. 12.3. The quantity $\kappa_{\text{exp}}^S V_{N+n}$ for $KCl+(n/N)$ KBr; in view of the constancy of v^d the linearity indicates that the defect compressibility κ^d is almost independent of composition.

for the calculation the elastic data of one of the end materials, e.g. the component (1). As measurements of the quantity $d^2 B_1/dP^2$ are frequently not available one can determine it accurately, as Smith and coworkers [376-379] have shown, from the modified Born model [652]:

$$B\frac{d^2 B}{dP^2} = -\tfrac{4}{9}(n^B + 3), \tag{12.11}$$

where n^B is the usual Born exponent. Equation (8.31) now reads

$$\kappa^d = \kappa_1 + \frac{4}{9}\frac{(n^B + 3)}{(dB_1/dP)-1}\kappa_1 \tag{12.12}$$

and hence eq. (12.10) becomes

$$\kappa = \kappa_1 \frac{V_1}{V_{N+n}}\left(1+\frac{n}{N}\right) + \kappa_1 \frac{n}{N}\frac{Nv^d}{V_{n+N}}\left\{1+\frac{4}{9}\frac{(n^B+3)}{(dB_1/dP)-1}\right\}. \tag{12.13}$$

This equation permits the direct evaluation of the compressibility κ of the mixed system for any desired composition in terms of the elastic data of the end member (1). The correctness of eq. (12.13) can be checked for $KCl_{1-x}Br_x$. We have chosen pure KCl as member (1). Using the values [376] $\kappa_1^S = 5.495 \times 10^{-12}$ cm^2/dyn, $dB_1^S/dP = 5.35$ and $n_1^B = 9.86$, eq. (12.12) leads to $\kappa^d = 12.72 \times 10^{-12}$, from which the compressibility for each composition studied by Slagle and McKinstry can be calculated. The results of this application are given in table 12.1. For each concentration two values of κ^S have been obtained. For the first one the value κ_1^S of Slagle and McKinstry [371] was used and for the second (between parentheses) that of Roberts and Smith [376]. Clearly the latter values agree within 0.6% with the experimental.

§12.2] Variation of the bulk modulus with composition 333

Table 12.3
Isothermal elastic data of the mixed crystal $AgCl_xBr_{1-x}$.

c (% AgCl)	n/N	ρ (g/cm^3)	V_{N+n} (cm^3)	B (10^{10} erg/cm^3)	κV_{N+n} (10^{-12} cm^5 dyn^{-1})
0	0	6.476	28.996	37.8	76.709
19.5	0.242	6.309	35.266	37.9	93.050
39.1	0.642	6.143	45.546	38.0	119.859
56.6	1.304	5.985	62.602	38.3	163.451
78.7	3.695	5.774	124.239	39.6	313.735
100	–	5.570	–	41.7	–

For the hard spheres model the calculated values of κ^d should not change (see also §A.11, note 23) if one takes as starting material the other end component, i.e. KBr. This statement can be understood by studying fig. 12.2: under increased pressure the volume on the left side will be $V(P)$ and on the right side $V(P) \pm |v^d(P)|$. For instance, if one plots $\kappa^S V_{N+n}$ versus n/N – but now taking as starting material KBr instead of KCl – the slope leads to $\kappa^d = 13.4 \times 10^{-12}$ cm^2/dyn. This value compares favourably with that (13.2×10^{-12}) found above from the corresponding plot of fig. 12.3 and with that (12.72×10^{-12}) determined from the $cB\Omega$-relation (8.31) with the help of the Born model.

Attention is drawn to cases like $AgBr_{1-x}Cl_x$ for which the Born model does not provide an adequate [380] description. In these cases one should rely upon eqs. (12.3), (12.4) and (12.10) but not on eq. (12.13). Let us apply this procedure to the mixed system (AgBr–AgCl); in this application we shall intentionally take as starting material AgBr ($V_1 = 28.996$ cm^3/mole) and by considering that for pure AgCl the volume is $V_2 = 25.731$ cm^3/mole one gets $Nv^d = V_2 - V_1 = -3.265$ cm^3. As expected from table 12.3 this value is comparable to -3.223 cm^3 obtained from the slope of the straight line V_{N+n} versus n/N (see eq. 12.3). Further using the κ-values of Cain [381] given in the same table for various concentrations one actually finds that κV_{N+n} versus n/N is a straight line the slope of which is $\kappa^d(Nv^d) + \kappa_1(AgBr)V_1(AgBr) = 63.99 \times 10^{-12}$ cm^5 dyn^{-1}. By inserting the v^d-value one finds $\kappa^d = 3.947 \times 10^{-12}$ cm^2/dyn. Note that the κ^d-value is appreciably *higher* than the compressibility of AgBr ($\kappa_1 = 2.645 \times 10^{-12}$ cm^2/dyn) and AgCl ($\kappa_2 = 2.398 \times 10^{-12}$ cm^2/dyn). If one alternatively selects AgCl as starting material one obtains from the slope of the plots of V_{N+n} versus n/N and κV_{N+n} versus n/N practically the same value for $|v^d|$ but a slightly higher value for κ^d (i.e. 4.68×10^{-12} instead of 3.95×10^{-12} cm^2/dyn); this is not unexpected because the hard spheres model is not equally successful for silver halides as it is for alkali halides.

Pressure variation of the bulk modulus in mixed alkali halides. By differentiating eq. (12.9) with respect to pressure one finds

$$\frac{dB}{dP} = \frac{1 + x\left(\frac{V_2}{V_1} - 1\right)}{1 + x\left(\frac{V_2 B_1}{V_1 B_2} - 1\right)} \frac{dB_1}{dP} - x \frac{V_2 B_1}{V_1 B_2}$$

$$\times \frac{(1-x)\left(1 - \frac{B_2}{B_1}\right)\left(1 - \frac{B_1}{B_2}\right) + \left(1 - x + x\frac{V_2}{V_1}\right)\left(\frac{dB_1}{dP} - \frac{B_1}{B_2}\frac{dB_2}{dP}\right)}{\left(1 - x + \frac{V_2 B_1}{V_1 B_2}\right)^2}$$

(12.14)

It is clear that dB/dP does not vary linearly with the concentration. An application of this formula to the *adiabatic* elastic data of the mixed system (KCl + KBr) gives the results included in fig. 12.4. In this application we have used the values $V_2/V_1 = 1.15416$, $B_1^S/B_2^S = 1.18259$, $dB_1^S/dP = 5.35$ and $dB_2^S/dP = 5.39$ [376,276,372]. Cain [372] has found that at $x = 7.7\%$ KBr the quantity dB^S/dP is around 5.38, in agreement with the calculated value. An important remark arising from the inspection of the figure is that the function dB^S/dP versus x attains the maximum value of 5.42 at

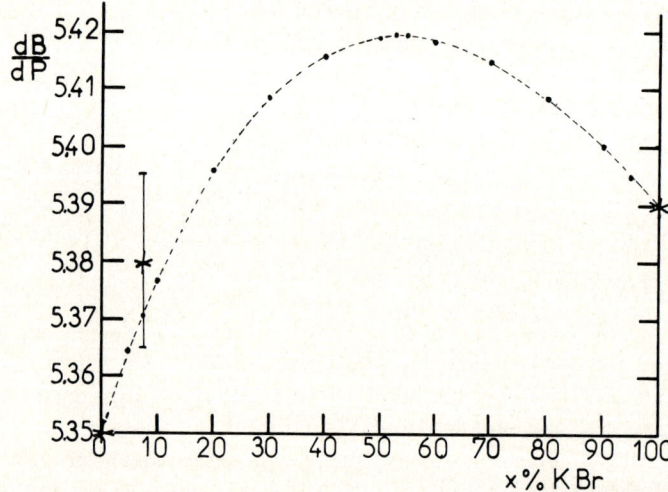

Fig. 12.4. The quantity dB^S/dP for $KCl_{1-x}Br_x$; solid dots: calculated from eq. (12.14); asterisks: experimental values (see text); the dashed line has been drawn as a guide to the eye.

approximatively $x = 53\%$; the value of 5.42 is well above the corresponding values of the end members so that this difference could be detected experimentally.

12.3 Composition for maximum conductivity and diffusivity from the $cB\Omega$-model

A combination of eqs. (12.4), (12.5) and (12.10) gives

$$\frac{B}{B_1} = \frac{1 + x\varepsilon}{1 + x\mu^d\varepsilon}, \tag{12.15}$$

where $\mu^d \equiv \kappa^d/\kappa_1$ and $\varepsilon \equiv (V_2/V_1) - 1$; the latter will be assumed to be positive. We note that B varies in a hyperbolical way with the composition x. By multiplying eq. (12.6) with (12.15) and replacing V/V_1 by Ω/Ω_1 – where Ω and Ω_1 denote the mean volume per atom of the mixed crystal and the "pure" crystal (1) respectively – one gets

$$\frac{B\Omega}{B_1\Omega_1} = \frac{(1 + x\varepsilon)^2}{1 + x\mu^d\varepsilon}. \tag{12.16}$$

In the case of a single diffusion (or conduction) mechanism we know that σT or D is proportional to $\exp(-g^{act}/kT)$. Therefore, if g^{act}, $g^{act,1}$ and σ, σ_1 (or D, D_1) denote the activation Gibbs energy and the conductivity (or diffusion coefficient) of the mixed crystal and of the pure component (1), one can write

$$\frac{\sigma \text{ (or } D\text{)}}{\sigma_1 \text{ (or } D_1\text{)}} = \exp\left(-\frac{g^{act} - g^{act,1}}{kT}\right). \tag{12.17}$$

In eq. (12.17) we have disregarded a factor due to the difference between the preexponential factors arising from the variation of the lattice constant and the Debye frequency with the concentration. This is allowed because the influence of their variation is very small in comparison to that arising from the exponential functions; one can write $g^{act} = cB\Omega$ and $g^{act,1} = c_1 B_1 \Omega_1$ and hence eq. (12.17) can be written

$$\frac{\sigma \text{ (or } D\text{)}}{\sigma_1 \text{ (or } D_1\text{)}} = \exp\left\{-\frac{c_1 B_1 \Omega_1}{kT}\left(\frac{cB\Omega}{c_1 B_1 \Omega_1} - 1\right)\right\}. \tag{12.18}$$

In connection with eq. (12.16), eq. (12.18) indicates that the function $[c(1 + x\varepsilon)^2]/[c_1(1 + x\mu^d\varepsilon)] = f(x)$ reflects basically the variation of the conductivity (or diffusion coefficient) with the composition x. Assuming that c

Table 12.4
Properties of mixed alkali halides: prediction of the concentration x_m of maximum conductivity or diffusivity from the $c B\Omega$-model.[a]

End members[c] (1) (2)	n^B	dB_1/dP	$\mu^d \equiv \kappa^d/\kappa_1$	$\varepsilon \equiv (V_2/V_1)-1$	$x_{m,calc}$ [eq. (12.19)] (%)	$x_{m,exper}$ (%)	Method of measurement
NaCl–NaBr	8.73	5.38	2.190	0.191	46	45–47	Conductivity (ref. [382], 179–393°C)
NaBr–NaCl	9.21	5.44	2.222		52		
KI–RbI	10.39	5.797[b]	2.241	0.122	88	85–90	Diffusion (ref. [383], 545–595°C)
RbI–KI	11.26	6.087[b]	2.246		90		
NaI–KI	9.79	5.58	2.241	0.302	36		
		5.597[b]	2.237		35	45 ± 5	Diffusion (ref. [374], 480°C)
KI–NaI	10.39	5.56	2.305		44		
		5.797[b]	2.241		35		
KCl–RbCl	9.86	5.46	2.281	0.145	85	80–85	Diffusion (ref. [384], 625–687°C)
RbCl–KCl	10.57	5.62	2.305		90		

[a] The values of n^B and V_2/V_1 are from Smith and Cain [276] whereas those for dB/dP are from Roberts and Smith [376].
[b] Calculated from the formula $dB/dP = (n^B + 7)/3$ given in refs. [377] and [378].
[c] The underlined species have the higher molar volume.

is independent of x the function $f(x)$ reaches a minimum value (remember that by definition ε is positive), when the (relative) molar concentration of the component (2) takes the value

$$x_m = \frac{\mu^d - 2}{\varepsilon \mu^d}. \tag{12.19}$$

Inserting the values of ε and μ^d, the molar concentration of crystal (2) (which was arbitrarily selected to have the higher molar volume) at which $f(x)$ takes its minimum volume can be determined; this is the composition at which the conductivity (or diffusivity) reaches its *maximum value*. According to eq. (12.12) the calculation of μ^d only needs the elastic data n^B, B_1 and dB_1/dP of component (1) whereas ε is determined from the molar volumes of the end members.

We now proceed to the comparison of eq. (12.19) with experiments for those mixed alkali halides for which the conductivity or diffusivity has been measured [374,382–384]. Table 12.4 contains all necessary data. The calculated values of x_m match in a satisfactory way the experimental ones, if one considers the experimental error and the fact that all the elastic data [276] used refer to RT so that the differences between the calculated x_m-values and the experimental ones might be smaller than those obtained from the table. The system KBr–KI is described in §A.7.

The value of x_m resulting from eq. (12.19) is practically invariant regardless which species is taken as (1) and which as (2). This can immediately be seen from column 4 where we give the value of μ^d for both choices. We remember that irrespective of the case, the value of ε is given by convention as the ratio of the component having the higher molar volume over that having the smaller one. In table 12.4, for the sake of clarity, we have underlined the species that have the higher molar volume.

Equation (12.18), in principle, permits the calculation of the fractional increase of σ (or D) at the concentration x_m, provided that the values of c_1 and c are accurately known. At present the scatter of the existing elastic data at high temperatures does not permit such a calculation.

The occurrence of a maximum conductivity or diffusivity can also be qualitatively explained by considering that the dielectric constant reaches a maximum [385] at a certain composition, which can be accurately predicted [386] from properties of the end members. To a very rough approximation the free energy of formation per Schottky defect is inversely proportional to the dielectric constant and therefore the conductivity or diffusivity reaches its maximum value at the same concentration. This reasoning, however, requires the further assumption that the migration energy is either constant or varies in a similar way as the formation energy.

In principle eq. (12.19) predicts that x_m slightly shifts with temperature, the reason being that not only the ratio V_2/V_1 varies with temperature but also the quantity κ^d/κ_1 should be temperature dependent. However, up to date the measurements of x_m have a large experimental uncertainty so that a strict verification of this prediction cannot be made.

Conditions for the occurrence of a maximum in the conductivity or diffusivity of mixed crystals. We have shown above that the composition at which the conductivity or the diffusivity of the mixed system attains its maximum value is given by eq. (12.19). Of course, the resulting value of x_m must lie between zero and unity, i.e.

$$0 \leqslant \frac{\mu^d - 2}{\varepsilon \mu^d} \leqslant 1. \tag{12.20}$$

As ε is positive by definition we have $(\mu^d - 2)/\mu \geqslant 0$ and $(\mu^d - 2 - \varepsilon \mu^d)\varepsilon \mu^d \leqslant 0$. The first inequality gives $\mu^d \leqslant 0$ or $\mu^d \geqslant 2$, whereas the second leads to $(1 - 2/\mu^d) \leqslant \varepsilon$. We therefore conclude that *mixed systems with physical properties that do not satisfy these inequalities should not show a maximum conductivity or diffusivity regardless of their composition.*

Let us now examine the validity of the above conditions in mixed alkali and silver halides. An inspection of table 12.4 shows that for some mixed alkali halides the above conditions are actually fulfilled; this is in agreement with the experimental data because for all these systems a maximum in the conductivity (or diffusivity) has been detected at a concentration labelled with $x_{m,\text{exp}}$.

We now turn to mixed silver halides; the system (AgCl–AgBr) has been studied in detail by Cain and Slifkin [387]. For convenience we shall use the RT elastic and expansivity data. The values of $V_1 = 25.73$ cm^3/mole and $V_2 = 28.996$ cm^3/mole give $\varepsilon = 0.127$. As previously explained the plot of κV_{N+n} versus n/N has led to the conclusion that κ^d is equal to 3.95×10^{-12} cm^2/dyn (or 4.68×10^{-12} cm^2/dyn) so that by comparing it to the compressibility of AgCl ($\kappa_1 = 2.398 \times 10^{-12}$ cm^2/dyn) we find that the ratio $\mu^d \equiv \kappa^d/\kappa_1$ is smaller than 2. We therefore see that although the inequality $1 - 2/\mu^d \leqslant \varepsilon$ is obeyed, the other inequality ($\mu^d \geqslant 2$) is violated, resulting in a *monotonic* variation in the conductivity (or diffusivity) when going from one pure component to the other at any fixed temperature. This result is in excellent agreement with the recent experimental results of Cain and Slifkin [387] who found that at any fixed temperature in the intrinsic region there is a *monotonic* decrease in the conductivity on going from pure AgBr to pure AgCl. Another important point should be noticed. The concentration x_m at which the conductivity (or diffusivity) attains its maximum value will *not* necessarily coincide with the concentration at which the activation enthalpy

h^{act} attains its minimum value. This is obvious if one considers that according to eq. (8.16) h^{act} varies with composition depending on the way the quantities on the right side change with composition. On the other hand σT (or D) maximizes at the composition at which the quantity $c^{act}B\Omega$ of the exponent in eq. (7.52) attains its minimum value. The two compositions do not have to coincide as $B_0^{SL}\Omega_0$ and $B\Omega$ do not have to show extremal values at the same composition. As a striking example we again refer to the mixed system $AgBr_{1-x}Cl_x$ in which, although the conductivity does not show a maximum value, as already mentioned, the activation enthalpy seems to have a minimum at about 40%. Unfortunately a quantitative justification of this value cannot be given because c^{act} is not known as a function of composition. However, within the frame of the $cB\Omega$-model one can explain the experimental fact [387] that $h^{act}_{AgCl}/h^{act}_{AgBr} \simeq 1.16$. In the straight part of the intrinsic region of the conductivity curve the activation enthalpy h^{act} is approximately given by $h^{act} \simeq \frac{1}{2}h^f + h^m$, where h^f is the formation enthalpy per cation Frenkel defect and h^m the migration enthalpy for the cation interstitial motion. The values of c for formation, migration and activation can be obtained from eq. (8.19) by introducing the corresponding values v^f, v^m and v^{act} which are connected by $v^{act} = v^m + v^f/2$. From eq. (8.19) we have

$$c^i = \frac{v^i}{\Omega}\left(\frac{dB}{dP} - 1\right)^{-1}, \qquad (12.21)$$

where i refers to a formation, activation or migration process. For AgBr we have (§10.4.3) $v^f = 14 \pm 1.5$ cm^3/mole, $v^m = 3.6$ cm^3/mole and hence $v^{act} = 10.6$ cm^3/mole. Setting $\Omega = 23.36$ Å^3 and $dB/dP = 6.96$, eq. (12.21) gives $c^f = 0.162$, $c^m = 0.0417$ and hence $c^{act} = 0.1227$ (see table 12.5). In the case of AgCl we have $v^f = 16.7$ cm^3/mole, $v^m = 3.25$ cm^3/mole, $v^{act} = 10.6$ cm^3/mole and $dB/dP = 6.681$, from which we get $c^f = 0.228$, $c^m = 0.0445$ and $c^{act} = 0.1585$. By now using the relation $h^i = c^i B_0^{SL}\Omega_0$ ($i = m, f, act$) [see eq. (8.16)] we obtain the values inserted in table 12.5. If one considers that v^i has an experimental uncertainty of $\sim 10\%$ the values of c^i (and hence those of h^i) are uncertain to the same extent. In view of this uncertainty the calculated values of h^f and h^m are in reasonable agreement with the experimental ones. These calculated values lead to $h^{act}(AgCl)/h^{act}(AgBr) = 1.21 \pm 10\%$, which is in favourable agreement with the experimental value 1.16 reported by Cain and Slifkin.

A combination of eqs. (8.16) and (12.21) gives

$$h^{act} = \frac{v^{act}/\Omega}{\frac{dB}{dP} - 1} B_0^{SL}\Omega_0. \qquad (12.22)$$

The values of h^{act} are known [387] for $AgCl_xBr_{1-x}$ for various values x.

Table 12.5
Experimental and calculated defect parameters for pure silver halides.

	Ω [a] (Å^3)	B_0^{SL} [b] (10^{10} dyn/cm^2)	c^f	c^m	c^{act}	$h_{calc.}^f$ [eq. (12.22)] (eV)	$h_{exper.}^f$ [c] (eV)	h_{calc}^m [eq. (12.22)] (eV)	h_{exper}^m [c] (eV)
AgBr	23.36	51.10	0.162	0.0417	0.1227	1.21±10%	1.1	0.31	0.28
AgCl	20.81	53.59	0.228	0.0445	0.1585	1.59±10%	1.45	0.31	0.31

[a] Cain [381].
[b] By using the values of B and dB/dT given in ref. [381] at RT.
[c] Experimental values; for references, see tables 10.11 and 10.12.

Thermal expansivity

Table 12.6
Activation parameters for cation Frenkel defects in $AgCl_xBr_{1-x}$.[a]

x (% AgCl)	Ω_0[d] (Å^3)	B_0^{SL} [c] (10^{10} dyn/cm^2)	c^{act}	dB/dP [b]	v^{act} (Å^3)
0	23.36	51.10	0.1227	6.961	17.60 [e]
19.5	22.86	51.18	0.1239	6.895	(16.70)
39.1	22.36	51.43	0.1254	6.866	(16.45)
56.6	21.91	51.38	0.1301	6.770	(16.45)
78.7	21.35	52.17	0.1374	6.743	(16.85)
100	20.81	53.59	0.1585	6.681	17.60 [e]

[a] The values between parentheses are predicted from eq. (12.22).
[b] Cain [381].
[c] By using the values of B and dB/dT given in ref. [381] at RT.
[d] By linear extrapolation of the lattice parameters given for 195 K and 300 K by Loje and Schuele [309].
[e] Experimental values for v^{act} can be found from the values of v^f and v^m reported in tables 10.11 and 10.12.

Therefore eq. (12.22) gives v^{act} at each composition. The results are given in table 12.6. A direct experimental confirmation of these values is highly desirable.

12.4 Thermal expansivity

12.4.1 Mixed alkali halides

Differentiating eq. (12.3) with respect to temperature ($P = $ constant) we get

$$\beta V_{N+n} = \beta_1 V_1 + \frac{n}{N}(\beta^d N v^d + \beta_1 V_1), \tag{12.23}$$

where

$$\beta^d \equiv \frac{1}{v^d} \frac{dv^d}{dT}\bigg|_P \tag{12.24}$$

denotes the expansivity of the defect volume. Introducing eqs. (12.5) and (12.7), eq. (12.23) can alternatively be written as

$$\beta = \beta_1 \frac{1 + x \dfrac{\beta^d}{\beta^1} \dfrac{Nv^d}{V_1}}{1 + xNv^d/V_1}. \tag{12.25}$$

As mentioned above (see eq. 12.4), the value of v^d is reliably known from the relation

$$Nv^d = V_2 - V_1;$$

therefore eq. (12.25) permits the evaluation of the expansion coefficient β of the mixed crystal provided that β^d is known. We further assume that a molecule added to a homogeneous mixed crystal (1) can be considered as a defect for which eq. (8.26)

$$\beta^d = \beta + \frac{\dfrac{d}{dT}\dfrac{dB_1}{dP}}{\dfrac{dB_1}{dP} - 1} \tag{12.26}$$

can be applied; this basically means that β^d is independent of concentration. Because of the latter assumption the quantities β, dB/dP, etc., occurring in eq. (8.26) can be chosen as those of one of the end components, e.g. (1). Therefore a combination of eqs. (12.25) and (12.26) gives

$$\beta = \beta_1 \frac{1 + x\dfrac{Nv^d}{V_1}\left\{1 + \dfrac{\dfrac{d}{dT}\dfrac{dB_1}{dP}}{\beta_1\left(\dfrac{dB_1}{dP} - 1\right)}\right\}}{1 + x\dfrac{Nv^d}{V_1}}. \tag{12.27}$$

We proceed to the application of eq. (12.27) to the mixed system KCl + $x\%$ KBr. In the literature direct experimental data for $d(dB/dP)/dT$ do not exist for KCl. However, this quantity can be estimated from the pressure derivative of the adiabatic bulk modulus B^S. Roberts and Smith [376] indicate that

$$\frac{dB_1}{dP} - \frac{dB_1^S}{dP} \simeq 2T\beta_1\gamma$$

where γ is the Grüneisen constant of KCl. Disregarding the small temperature dependencies of dB^S/dP and γ we get

$$\frac{d}{dT}\frac{dB_1}{dP} \simeq 2\beta_1\gamma + 2T\gamma\frac{d\beta_1}{dT}.$$

Inserting the data quoted in refs. [289] and [376] we get for the second term on the right-hand side of eq. (12.26) the value $0.9\beta_1$. Using the above value and considering that $Nv^d/V_1 = 0.1512$ (see §12.1) and $dB_1^S/dP \simeq 5$, eq. (12.27) gives the values of β/β_1 for various concentrations. The results are plotted in the form of a line in fig. 12.5. In the same figure we have inserted as an asterisk the experimental value of Salimäki [388] for $x = 25\%$ KBr; it agrees with the calculated value within the experimental error.

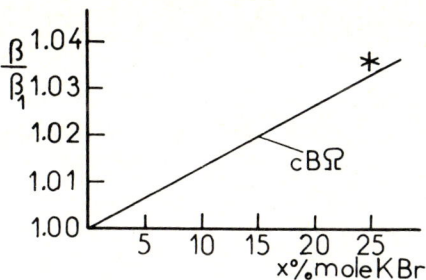

Fig. 12.5. Thermal expansivity of KCl–KBr; asterisk: experimental point from ref. [388]; line: calculated with $cB\Omega$ model (from eq. 12.27).

In alkali halides with NaCl-structure the derivative dB/dP increases with temperature [282] so that the quantity included in the brackets on the right side of eq. (12.27) is greater than unity; therefore, if β_1 has been selected to represent the expansion coefficient of the material having the smaller molar volume, eq. (12.27) indicates that β increases hyperbolically with x. This is in agreement with the general trend of the experimental data [389,390] which shows that the thermal expansion coefficient β of a binary solid solution of NaCl-type alkali halides makes a deviation from the linear concentration dependence.

Equation (12.27) predicts that under certain conditions negative deviations in the linear initial part of the hyperbolic concentration dependence of β could appear. Fullam [391] has observed such a negative deviation in $CsCl + x\%$ KCl; unfortunately, no experimental results for the temperature dependence of dB/dP in CsCl are known so that the validity of eq. (12.27) in this case cannot be checked.

Calculation of the ratio β/κ. Dividing the expression of β and κ given by eqs. (12.27) and (12.13) we get

$$\frac{\beta}{\kappa} = \frac{\beta_1}{\kappa_1} \frac{1 + x\dfrac{Nv^d}{V_1}\left\{1 + \dfrac{\dfrac{d}{dT}\dfrac{dB_1}{dP}}{\beta_1\left(\dfrac{dB_1}{dP} - 1\right)}\right\}}{1 + x\dfrac{Nv^d}{V_1}\left\{1 + \dfrac{4}{9}(n^B + 3)\left(\dfrac{dB_1}{dP} - 1\right)^{-1}\right\}}.$$

This equation shows that β/κ depends on properties of the pure component (1). Its sole connection to the end component (2) is the quantity v^d. This justifies Lindström's prediction [390a] that β/κ is governed by the properties of the pure components except for the lattice parameter.

12.4.2 Ionic solids doped with aliovalent impurities

Up to now we have studied mixed systems of the form $A^+B^- - C^+D^-$ which do not contain extrinsic vacancies. We now turn to the discussion of the properties of an ionic crystal of the form A^+B^- doped with small amounts of $M^{++}B_2^-$, which is the most usual case met in experiments dealing with point defects.

If a "molecule" of type $M^{++}B_2^-$ is added to an A^+B^--crystal, the M^{++}-ion replaces an A^+-ion which then occupies a site on the surface. The two B^--ions can also be considered as settling on the surface. The M^{++}-ion is doubly charged, so that the charge compensation demands the transport of a further A^+-ion to the surface with the simultaneous production of a cation vacancy; in other words, the introduction of an "MB_2 molecule" produces a cation vacancy and further obliges the two A-ions and the two B-ions to settle on the surface. The creation of the cation vacancy changes the volume of the crystal by an amount equal to the formation volume of the vacancy as defined in §3.2. On the other hand, the M-ion *plus* an A-ion settling on the surface changes the volume of the crystal by v^M which is generally different from v^f. Furthermore, the settling of two B-ions on the surface changes the volume by $2\Omega^0$, where Ω^0 is the mean value per atom of the crystal AB. Therefore the *addition* of the molecule "MB_2" to the crystal AB produces a total increase of its volume by (v^f is *solely* the formation volume per vacancy)

$$v^M + v^f + 2\Omega^0;$$

of course, in the above discussion we have assumed that the M-ions are not close to the cation vacancy. This assumption disregards the usual "association effect" according to which "M^{++} plus the neighbouring vacancy" produces a dipole [653]; the latter effect is well-known from the reorientation of the dipole upon the application of an external electric field. The addition therefore of n molecules MB_2 to the host material AB will change the volume according to

$$V_{N+n} = V^0 + n(v^M + v^f + 2\Omega^0), \tag{12.28}$$

where $V^0 \equiv 2N\Omega^0$ is the volume of the pure crystal AB (N molecules), and V_{N+n} is the resulting total volume. Assuming that $n \ll N$, the volume V has to be a linear function of n. Differentiating eq. (12.28) with respect to temperature and introducing the definitions of the expansion coefficients β, β^0, β^M and β^f of the doped crystal, the pure crystal, the volume v^M and the volume v^f we get

$$\beta V_{N+n} = \beta^0 V^0 + n(\beta^M v^M + \beta^f v^f + \beta^0 2\Omega^0),$$

§12.4] Thermal expansivity

which – considering that $n/N \ll 1$ – can also be written as

$$\beta - \beta^0 \simeq \frac{n}{N} \frac{1}{2\Omega^0} \{(\beta^M - \beta^0)v^M + (\beta^f - \beta^0)v^f\}. \tag{12.29}$$

In a similar way by differentiating eq. (12.28) with respect to pressure and introducing the definitions of the "compressibilities" κ, κ^0, κ^M and κ^f, of the doped crystal, the pure crystal, the volume v^M and the volume v^f we get (when $n/N \ll 1$):

$$\kappa - \kappa^0 \simeq \frac{n}{N} \frac{1}{2\Omega^0} \{(\kappa^M - \kappa^0)v^M + (\kappa^f - \kappa^0)v^f\}. \tag{12.30}$$

In principle the quantities v^f, κ^f and β^f can be estimated from the variation of the ionic conductivity (or diffusivity) of the pure material at various temperatures and pressures. More precisely, in the case of a single operating mechanism the knowledge of the initial slopes ($P \simeq 0$) of the isothermal plots $\ln \sigma T$ (or $\ln D$) versus P of the pure material leads to values of v^f (at $P = 0$) for various temperatures from which the determination of β^f is obvious [654]. Furthermore, the curvature of the above isothermal plots at high pressures provides the value of κ^f at each temperature (see §7.3.4). When the above quantities have been determined the values of v^M, κ^M and β^M can be obtained by combining to solubility experiments in which MB_2 is dissolved in AB. More precisely, the quantities v^M and κ^M can in principle be deduced from the measurements of (isothermal) solubility at various pressures; then from the temperature variation of the (isobaric) solubility the quantity β^M can be determined with a procedure analogous to that followed for the determination of β^f.

The above treatment of the experimental data, although tedious, leads to quantities that can be directly compared to microscopic calculations because the latter treat separately the influence of an isolated M^{++}-ion and the corresponding influence of the presence of a vacancy.

Analysis of the expansivity data of $A^+B^- + x\% \ M^{++}B_2^-$ and comparison to the $c B\Omega$-model. We present as an example the analysis of the expansivity data of AgBr doped with small amounts of $CdBr_2$ [391a]. Zieten [392] has measured the change of the volume expansion coefficient β^0 of AgBr (N molecules) when various small amounts (n molecules) of $CdBr_2$ have been added for concentrations up to $y = n/N = 20.2 \times 10^{-3}$. (Note that $n/N \simeq (n/N)/[(n/N) + 1]$). Extracting the various values of β from fig. 4 of Zieten, which refers to 100°C, we obtain the results plotted in fig. 12.6 in the form of $\beta - \beta^0$ versus y. The extraction of the values from Zieten's paper has introduced some uncertainties; however, clearly $\beta - \beta^0$ is a strongly increasing (and also linear) function of y. In view of eq. (12.29) we immediately conclude that (at least one of) the quantities β^f, β^M have to be

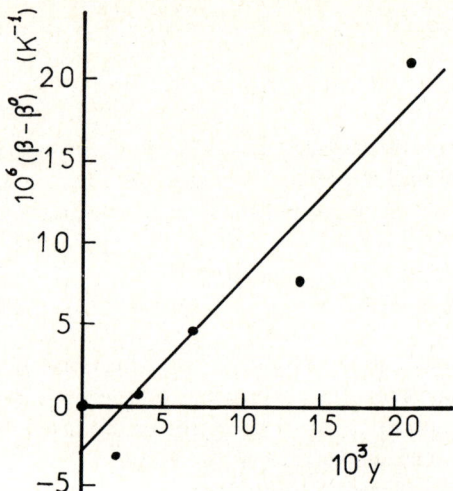

Fig. 12.6. Contribution of the "Cd^{2+} + vacancy" defects to the thermal expansion coefficient of AgBr; points: ref. [392]; line: best linear fit. (Courtesy Phys. Stat. Sol.)

quite different from the bulk coefficient β^0. This, of course, indicates that anharmonicity plays an extremely important role when the combined defect "M^{++} + cation vacancy" is introduced into AgBr. The linear fitting of fig. 12.6 gives for the slope the value

$$\frac{1}{2\Omega^0}\{(\beta^M - \beta^0)v^M + (\beta^f - \beta^0)v^f\} = 12 \times 10^{-4} \text{ K}^{-1}.$$

In order to further simplify the above result we recall that the ionic radius of Cd is appreciably ($\sim 100\%$) greater than that of Ag so that we expect a strong outward relaxation of the host lattice around Cd^{++}. This behaviour may be similar to the situation around the cation vacancy because, as experimentally confirmed by various techniques, in ionic crystals outward relaxations occur around the vacancy; this is also consistent with the experimental results of Kurnick [393] who found that in AgBr the formation volume per Schottky defect (i.e. cation plus anion vacancy) lies between 1.1 and 1.4 in units of the molecular volume $2\Omega^0$. In other words, the result of Kurnick indicates that around each vacancy we have a positive relaxation of the order of 0.1 to $0.4\Omega^0$; this relaxation volume is comparable to the expected relaxation volume around each Cd^{++}. In the light of the above discussion we are not going too far by assuming that

$$(\beta^M - \beta^0)v^M \simeq (\beta^f - \beta^0)v^f \tag{12.31}$$

and hence the slope becomes:

$$\frac{v^f}{\Omega^0}(\beta^f - \beta^0) \simeq 12 \times 10^{-4} \text{ K}^{-1}. \tag{12.32}$$

Due to the approximation (12.31) the left member of eq. (12.32) may be in error by at most a factor of two; but here we are interested only in the order of magnitude of β^f. According to Kurnick's experiments [393] $v^f/2\Omega^0$ is of the order of unity; as $\beta \simeq 10^{-4}$ K^{-1} we can definitely conclude that β^f is greater than β^0 by one order of magnitude. The above result is interesting in two respects: (1) It indicates that local anharmonicity is a pertinent factor in the transport properties of AgBr; (2) it is in fundamental agreement with the diffusion experiments of Slifkin and co-workers where a strong temperature dependence of h^f was observed in AgBr. Accordingly large values of β^f/β^0 are consistent from eq. (6.45) with large values of $(\partial h^f/\partial T)_P$. We now proceed to the comparison of the experimental result $(\beta^f/\beta^0 \simeq 10)$ with the value predicted from the $cB\Omega$-model. Loje and Schuelle [394] give for dB/dP at 300 K the value 7.49. For 195 K they give $dB^S/dP = 6.26 \pm 3\%$; by adding a correction $T\beta\gamma$ as indicated by Roberts and Smith [376] we obtain, for 195 K, $dB/dP = 6.6$. From these two values one obtains the temperature derivative of dB/dP and by introducing it into eq. (8.26) we find $\beta^f \simeq 14 \times 10^{-4}$ K^{-1}. If we consider the experimental errors quoted by Loje and Schuelle, it should lie between 7×10^{-4} and 20×10^{-4} K^{-1}, i.e. $7 < \beta^f/\beta^0 < 20$. This result is in quite good agreement with the value of β^f/β^0 obtained from the experimental data. Similarly to the way indicated above for the expansivity case, one could analyse the compressibility data of a system $AB + x\%$ MB$_2$; unfortunately there are no direct elastic data for the system AgBr $+ x\%$ CdBr$_2$.

Concluding this section some general remarks could be made. We have seen that from a general thermodynamic viewpoint there is a close interconnection between the conductivity or diffusivity of the pure crystal A^+B^- and the expansivity (or compressibility) data of the same crystal doped with small amounts of aliovalent impurities. Let us take for instance a case where the analysis of the expansivity data of the crystal "A^+B^- + small amounts of $M^{++}B_2^-$ (or $A_2^+C^{--}$)" shows that the expansion coefficient of the ensuing defect is appreciably larger than that of the host material; this fact immediately leads to the conclusion that it is inconsistent to proceed to an analysis of the conductivity or diffusion data of the pure material A^+B^- with temperature independent parameters (see eq. 6.45, where $c_P^0 = (dh^f/dT)_P$).

13 | EXPLANATION OF VARIOUS EMPIRICAL LAWS

In ch. 8, relations between various defect parameters have been derived from the $cB\Omega$-model and in chs. 9 to 12 their correctness has been investigated. In this chapter we proceed to the explanation of various empirical laws [395–403,655,656] which connect the defect parameters to macroscopic quantities as the melting point. Classical empirical laws as those of Van Liempt [396], Zener [399], Nachtrieb [397,398], Keyes [401], Lawson [400,402], etc., find a unified justification within the frame of the $cB\Omega$-model; furthermore, some other well-known empirical facts are also proved in this chapter. However, one should note in advance that these laws are known to be of approximative nature and will result from $cB\Omega$-formulae only after some reasonable approximation. For some laws more accurate forms can be achieved. The connection of the ratios v/g or $v/h_{\exp}$ to the quantity $d\ln T_M/dP$ (and therefrom to the latent heat of fusion) has already been discussed in §10.1 and 10.5.

13.1 Connection of the enthalpy to the melting point

A very interesting fact in the study of point defects in solids is that a number of defect parameters scale with some melting property of the bulk material [395]. It has repeatedly been stated that such a connection was anticipated because a disordering occurs in the region of the solid around the defect. However, this correlation of defect properties with melting quantities can be explained without introducing the concept of "local melting". For this reason we calculate $s/h_{\exp}$ by connecting eqs. (3.16), (8.1), (8.2) and (8.15). One gets

$$\frac{s}{h_{\exp}} = \frac{-c\left.\dfrac{d(B\Omega)}{dT}\right|_P}{cB_0^{SL}\Omega^0}, \qquad (13.1)$$

which is equivalent to eq. (9.4). For materials for which $B\Omega$ is a linear function of T up to the melting point, the derivative $d(B\Omega)/dT$ can be approximated by $\{B_M \Omega_M - (B_0 \Omega_0)^{SL}\}/T_M$ where the subscript M denotes the values at the melting point. By considering also that $\Omega_0^{SL} \simeq \Omega_0$, eq. (13.1) can be written as

$$h_{exp} = \frac{s}{A} T_M, \qquad (13.2)$$

where

$$A \equiv 1 - \frac{B_M}{B_0^{SL}} \frac{\Omega_M}{\Omega_0}. \qquad (13.3)$$

Equation (13.2) is a general relation of the $cB\Omega$-model and can be applied to formation, migration and activation processes. A different interpretation of the proportionality of h_{exp} with T_M will be given later. For a given category of solids the ratios B_M/B_0^{SL} and Ω_M/Ω_0 are, to a good approximation, constants, and therefore the quantity A becomes a characteristic property of each category.

Equation (13.2) has been recognized for a number of years as an important empirical fact. The way it was just developed shows that it should give good results only [659] for materials that have an almost linear connection between $B\Omega$ and T up to the melting point, i.e. temperature independent values of the enthalpy and the entropy. Its accuracy will now be demonstrated for a variety of materials and processes.

13.1.1 Self-diffusion in metals

Dienes [403] collected a number of empirical laws that connect the preexponential factor D_0 of self-diffusing atoms with h^{act}/T_M in various forms. An extensive analysis of experimental data for a number of metals showed a linear relation between $\log[D_0/(\nu_D a^2)]$ and h^{act}/T_M, where a is the interatomic distance and ν_D is the Debye frequency. In other words, Dienes suggested that s^{act} is roughly proportional to h^{act}/T_M in agreement with eq. (13.2).

In the case of metals we have $B_0^{SL}/B_M \simeq 1.25$ to 1.35 and $\Omega_M/\Omega_0 = b_M = 1.08$ which gives $A = 0.14$ to 0.2. This result not only, explains the rule noticed by Dienes but it also gives a direct physical interpretation for the slope, which was hitherto lacking. Considering that in metals s^{act} is around $3k$, eq. (13.2) gives $s^{act}/A = 1.3$ to 1.9×10^{-3} eV/K $= 30$ to 43.8 kcal mole^{-1} K^{-1}. The mean values proposed by Lazarus [395] and Flynn [404] were 32 to ~ 37 kcal mole^{-1} K^{-1}.

13.1.2 Vacancy formation in metals

The value of A for metals lies, as already mentioned, between 0.14 and 0.2. The formation entropy for most metals lies between 2 and 2.5 k. Equation (13.2) therefore gives

$$s^f/A = (1.4 \sim 0.97) \times 10^{-3} \text{ eV/K}$$

and hence

$$h^f(\text{eV}) = (1.4 \sim 0.97) \times 10^{-3} T_M.$$

This rule has already been mentioned by Lazarus [395], Flynn [404], Maier, Peo, Saile, Schaefer and Seeger [405] and others [406] with the coefficients around 10^{-3} eV/K and was considered to be important. Similar considerations apply to rare gas solids [404].

13.1.3 Formation and migration of defects in alkali halides

In alkali halides the ratio B_0/B_M is appreciably larger than for metals (i.e. around 1.6), whereas Ω_M/Ω_0 lies between 1.1 and 1.15; this gives $A \simeq 0.3$. Considering further that s^f lies between 7.5 and 10 k, one obtains from eq. (13.2) $h^f(\text{eV}) \simeq 2.1$ to $2.9 \times 10^{-3} T_M$. The migration entropy s^m for cation vacancies in alkali halides lies between 2 and 3 k. Therefore eq. (13.2) becomes

$$h^m(eV) = \frac{s^m}{A} T_M = (0.57 \sim 0.9) \times 10^{-3} T_M.$$

Both the above results explain the Barr–Lidiard relations [407], a theoretical basis of which was lacking.

13.1.4 Organic compounds

Analysing the isobaric specific heat data at high temperatures Baughman and Turnbull [408] have determined the vacancy formation enthalpy and entropy for a number of organic compounds. Although, as discussed in ch. 7 the usual way of analysing C_P-experimental data through linear extrapolation overestimates the value of s^f, one can accept from their experimental data that at least the general trend of s is proportional to h^f/T_M. This trend indicates that A must have an approximately constant value for all materials of this category. In table 13.1 the quantity $s^f T_M/h^f$ is given; recalling eq.

Table 13.1
Defect formation parameters in organic compounds.

Compound	h (eV)	s (k-units)	sT_M/h (dimensionless)
2,2,3-trimethylbutane	0.201	4.5	0.48
cis-1,2-dimethylcyclopentane	0.222	6.3	0.54
cyclooctane	0.303	6.5	0.53
1,1-dimethylcyclohexane	0.31	8.8	0.59
benzene	0.542	15.5	0.69

(13.2) we see that the values of this quantity lie within the region expected for the quantity A, i.e. smaller than unity and around one half.

13.1.5 An alternative explanation for the proportionality of formation enthalpy and the melting temperature

The proportionality of the experimental formation enthalpy $h_{\exp}^f$ to the melting point can also be explained, within the frame of the $cB\Omega$-model as follows. Inserting into the relation $h_{\exp}^f = c^f B_0^{SL} \Omega_0$ (see eq. 8.16) the value of c^f from eq. (8.19) we find

$$h_{\exp}^f = \frac{v^f}{\Omega}\left(\frac{dB}{dP} - 1\right)^{-1} B_0^{SL} \Omega_0 \tag{13.4}$$

or

$$h_{\exp}^f = \frac{v^f}{\Omega}\left(\frac{dB}{dP} - 1\right)^{-1} \frac{B_0^{SL}}{B_M} \frac{\Omega_0}{\Omega_M} B_M \Omega_M. \tag{13.5}$$

Let us consider first the case of monoatomic crystals. A simplified calculation [409] can show

$$B_M \Omega_M \simeq \text{const} \cdot M\Theta_D^2 \Omega_M^{2/3}, \tag{13.6}$$

where the constant depends on crystal structure. However, from Lindemann's law [658] we know

$$M\Theta_D^2 \Omega_M^{2/3} \simeq \text{const}' T_M \tag{13.7}$$

where again "const'" has approximately the same value for each category of solids. Recalling also that each of the quantities v^f/Ω, dB/dP, B_0^{SL}/B_M and Ω_0/Ω_M have approximatively a common value for each category of solids the combination of eqs. (13.5), (13.6) and (13.7) directly leads to the conclusion that h^f is proportional to T_M with a proportionality constant depending on the category of materials studied.

In the case of alkali halides the proportionality between h^f_{exp} and T_M results when one inserts eq. (13.11) into eq. (13.5).

13.1.6 Relation between formation parameters and the Debye temperature in alkali halides

In the following we shall need a connection between the bulk modulus of a crystal and the corresponding Debye temperature Θ_∞ at high temperatures. The value of Θ_∞ is connected to the mean square frequency $\langle \omega^2 \rangle$ of the normal modes by the relation [410,411]

$$\Theta_\infty^2 = \frac{5}{3} \frac{\hbar^2}{k^2} \langle \omega^2 \rangle. \tag{13.8}$$

The quantity $\langle \omega^2 \rangle$ can be calculated from the elastic constants c_{ij} under the simplifications mentioned below.

Taking into account only the Coulomb forces and the short-range repulsive forces between nearest neighbours the value of $\langle \omega^2 \rangle$ is given by [410,411]

$$\langle \omega^2 \rangle = \frac{l}{\mu}(c_{11} + 2c_{12}), \tag{13.9}$$

where μ is the reduced mass and l the nn-distance. Due to eq. (13.8) eq. (13.9) gives

$$B\Omega = \frac{k^2}{5\hbar^2} \mu \Theta_\infty^2 \Omega^{2/3} \tag{13.10}$$

because $\Omega = l^3$. According to the Lindemann relation [658] the term $\mu \Theta_\infty^2 \Omega^{2/3}$ at the melting point is proportional to T_M and therefore one gets the relation

$$B_M \Omega_M \simeq \text{const} \cdot T_M. \tag{13.11}$$

The experimental fact that the ratios B_M/B_0, dB/dP, Ω_0/Ω_M and v^f/Ω have roughly the same values for all alkali halides with NaCl-structure has already been mentioned. Introducing eq. (13.10) into eq. (13.5) one gets the approximate Sastry–Mulimani [412] relation

$$h_{exp} \simeq \text{const} \cdot \mu \Theta_\infty^2 \Omega_M^{2/3}.$$

For metals a similar correlation has been proposed [413–415].

13.2 Explanation of the empirical laws connecting activation entropy and enthalpy to the activation volume

According to the relation

$$\left.\frac{dB}{dT}\right|_P = \left.\frac{dB}{dT}\right|_V - \beta B \left.\frac{dB}{dP}\right|_T$$

eq. (9.1) is equivalent to

$$s^{act} = c^{act}\Omega\left\{\beta B(\eta - 1) - \left.\frac{dB}{dT}\right|_V\right\},$$

where $\eta \equiv (dB/dP)_T$; by dividing with v^{act} [see eq. (8.19)] one gets

$$\frac{s^{act}}{v^{act}} = \frac{\beta B(\eta - 1) - \left.\frac{dB}{dT}\right|_V}{\eta - 1}. \tag{13.12}$$

This expression is what we believe to be the general form of the law

$$s^{act}/v^{act} \simeq \beta B, \tag{13.13}$$

which resulted from proposals of Zener, Lawson and Keyes for self-diffusion activation processes in metals. Equation (13.13) immediately results from eq. (13.12) by introducing Swenson's empirical rule [415a] for metals which can be written as (see also ref. [660] for alkali halides)

$$\left|\left.\frac{\partial B}{\partial T}\right|_V\right| \ll \beta B(\eta - 1). \tag{13.14}$$

A formula of the form of eq. (13.13) is valid for all three types of processes and finds a thermodynamical proof from eq. (3.126) for cases where $|s^*| \ll s$ (see below). Furthermore, by dividing eqs. (8.19) and (8.16) we get

$$v^{act}/h^{act} = b\left(\frac{dB}{dP} - 1\right)/B_0^{SL}, \tag{13.15}$$

which is what we believe to be the accurate form of Keyes' empirical formula

$$v^{act}/h^{act} \simeq 4/B \tag{13.16}$$

for (self-diffusion) activation in metals. The empirical factor 4 resulted from adjustments to a number of experimental results. Obviously eq. (13.16) results from eq. (13.15) of the $cB\Omega$-model by using the approximation $bB \approx B_0^{SL}$ and by considering that the average value of dB/dP in metals is close to 5.

As already mentioned, the existing empirical laws can be explained on the basis of the $cB\Omega$-model. It should be stressed, however, that some empirical relations can also be explained by simple thermodynamics, bearing in mind that there are two families of defect parameters which result from the comparison of the real crystal to the "isobaric" or "isochoric" perfect crystal (see §3.1). As an example we give below the explanation of the empirical law reported by Zener, Lawson and Keyes (see eq. 13.13). We have seen that the formation entropies s^* and s^f are connected through the relation

$$s^* = s^f - v^f \beta B. \tag{13.17}$$

As shown in §3.5 [see eq. (3.156)] s^* is connected to the change of frequencies upon the production of the defect under the conditions of constant temperature and volume; it is therefore logical to assume that in some cases

$$|s^*| \ll s^f \tag{13.18}$$

because s^f is a measure of the change of frequencies when the volume of the crystal changes from V^0 to $V^0 + v$ or from $V - v$ to V, the reason being that the frequencies depend mainly on volume and exhibit only a slight explicit temperature dependence. A combination of eqs. (13.17) and (13.18) gives

$$s^f \simeq v^f \beta B. \tag{13.19}$$

A similar procedure for migration entropies leads to

$$s^m \simeq v^m \beta B. \tag{13.20}$$

The combination of eqs. (13.19) and (13.20) immediately leads to eq. (13.13).

13.3 Connection of defect parameters to the Grüneisen constant

13.3.1 Connection of the defect entropy to the Grüneisen constant; correlation between the defect enthalpy and the expansion coefficient

Using eq. (8.16), eq. (8.8) can be written as

$$s = -h_{\exp} \beta \frac{B}{B_0^{SL}} b \left(1 + \frac{1}{\beta B} \frac{\partial B}{\partial T} \bigg|_P \right). \tag{13.21}$$

In the case of metals Swenson's relation (13.14) leads to

$$-\frac{\partial B}{\partial P} \bigg|_T \simeq \frac{1}{\beta B} \frac{\partial B}{\partial T} \bigg|_P. \tag{13.22}$$

A combination of eqs. (13.21) and (13.22) gives

$$s \simeq h_{\exp}\beta \frac{B}{B_0^{SL}} b \left(\left.\frac{\partial B}{\partial P}\right|_T - 1 \right). \tag{13.23}$$

Considering Slater's [328] formulation of the mean Grüneisen constant

$$\gamma = -\frac{1}{6} + \frac{1}{2} \left.\frac{\partial B}{\partial P}\right|_T, \tag{13.24}$$

eq. (13.23) gives [416]

$$s \simeq 2 h_{\exp}\beta \frac{B}{B_0^{SL}} b(\gamma - \tfrac{1}{3}). \tag{13.25}$$

In the case of metals at high temperatures we have $Bb \simeq B_0^{SL}$ within 20%. As γ is much larger than $\tfrac{1}{3}$, eq. (13.25) simplifies to

$$s \simeq 2 h_{\exp}\beta\gamma. \tag{13.26}$$

This equation connects the defect entropy for vacancies in metals to the mean Grüneisen constant at high temperatures. A number of remarks can now be made on eq. (13.26). It is an experimental fact that the term $b_M \equiv \exp(\int_0^{T_M}\beta dT)$ at the melting point T_M is practically the same for all metals (around 1.08). Therefore for practically all metals we can write

$$\bar{\beta}T_M \simeq \text{constant} \simeq 0.08, \tag{13.27}$$

where $\bar{\beta}$ is the mean volume expansion coefficient. Combining eqs. (13.26) and (13.27) we get

$$h_{\exp} \approx \frac{1}{2 \times 0.08} \frac{s}{\gamma} T_M \tag{13.28}$$

Bearing in mind that s and γ do not vary much from metal to metal we conclude that in a rough approximation

$$h_{\exp} \simeq \text{constant}/\bar{\beta} \tag{13.29}$$

or

$$h_{\exp} \simeq \text{constant} \times T_M. \tag{13.30}$$

Equations (13.29) and (13.30) express two long known empirical laws. Obviously the "constants" in these relations are different for different processes. Equation (13.30) has alternatively been derived in §13.1; Equation (13.29) has been noticed by Flynn [35] as an empirical fact.

13.3.2 Relation between the migration volume and the Grüneisen constant

Ionic crystals. The definition of the isothermal bulk modulus implies that

$$\left.\frac{\partial B}{\partial \Omega}\right|_T = \left.\frac{\partial B}{\partial P}\right|_T \left.\frac{\partial P}{\partial \Omega}\right|_T.$$

With the help of eq. (2.25) this gives

$$\left.\frac{\partial B}{\partial \Omega}\right|_T = -\frac{B}{\Omega} \left.\frac{\partial B}{\partial P}\right|_T. \quad (13.31)$$

On the other hand the transverse optical Grüneisen constant is

$$\gamma_{TO} = -\frac{\Omega}{\omega_{TO}} \left.\frac{\partial \omega_{TO}}{\partial \Omega}\right|_T. \quad (13.32)$$

For ionic crystals with solely central forces, the frequency ω_{TO} of the transversal optical mode of zero wave number was expressed by Ludwig to a first approximation [411,417–420] by

$$\omega_{TO} \simeq \text{constant} \times (Bl/\mu)^{1/2} \quad (13.33)$$

where l is the nearest-neighbour distance and μ is the reduced mass. Introducing eq. (13.33) into (13.32) and with the help of eq. (13.31) one gets

$$\gamma_{TO} \simeq -\frac{1}{6} + \frac{1}{2} \left.\frac{\partial B}{\partial P}\right|_T. \quad (13.34)$$

This formula for γ_{TO} has been checked [419] and found to agree with experimental values within 10% for a large number of ionic solids.

A connection of γ_{TO} with the migration volume results from the introduction of eq. (13.34) into eq. (8.21) of the $cB\Omega$-model; one obtains

$$v^m \simeq \frac{h^m_{\text{exp}}}{B_0^{SL}\Omega_0} 2(\gamma_{TO} - \tfrac{1}{3})\Omega. \quad (13.35)$$

A connection of v^m to the γ-value of a transverse mode is physically acceptable, because the migration of an ion in ionic crystals is expected to take place through the TO modes rather than through the LO modes.

Monoatomic solids. In this case Slater [328] has indicated that within the Debye approximation the mean Grüneisen constant γ is given by eq. (13.24). Combining this relation with eq. (8.21) one gets [416]

$$v^m = \frac{h^m_{\text{exp}}}{B_0^{SL}\Omega_0} 2(\gamma - \tfrac{1}{3})\Omega. \quad (13.36)$$

Equations (13.35) and (13.36) are similar in form but contain different Grüneisen constants. Both of them are found to be much larger than $1/3$;

the value 1/3 usually constitutes only 10 to 15 percent of the Grüneisen constant. At high temperatures Ω exceeds Ω_0 at most by 15 percent. Therefore to a good approximation eqs. (13.35) and (13.36) can be written as

$$v^m \simeq 2\frac{h^m_{\exp}}{B_0^{SL}}\gamma_{TO} \quad \text{for ionic solids,} \tag{13.37}$$

and

$$v^m \simeq 2\frac{h^m_{\exp}}{B_0^{SL}}\gamma. \quad \text{for monoatomic solids.} \tag{13.38}$$

Formula (13.37) has already been proposed by Flynn. In his theory γ is the Grüneisen constant γ_i for the mode (or modes) responsible for the migration process, which however for the time being cannot be determined. Obviously eq. (13.37) of the $cB\Omega$-model has the advantage of having the same form but containing the quantity γ_{TO}, which is experimentally available.

13.3.3 Relation between the migration volumes and the corresponding enthalpies for a given solid

The application of eq. (8.21) of the $cB\Omega$-model to the migration volumes $v^{m,+}$ and $v^{m,-}$ of a cation or an anion vacancy in a given ionic material leads to

$$v^{m,+}/v^{m,-} \simeq h^{m,+}_{\exp}/h^{m,-}_{\exp}. \tag{13.39}$$

The same relation holds for the more general case where two different defects are migrating in the same lattice. As an example we present SrF_2. Andeen et al. [421] have found that the migration volume $v^{m,I}$ for type I dipoles in SrF_2 doped with lanthanum is 3.26 cm^3/mole and Andeen et al. [422] have stated $v^{m,II}$ for type II dipoles in erbium-doped SrF_2 to be 4.73 cm^3/mole. The corresponding enthalpies are [423] $h^{m,I} = 0.478$ eV and $h^{m,II} = 0.701$ eV. We see that $h^{m,II}/h^{m,I} = 1.466$ and $v^{m,II}/v^{m,I} = 1.45$, in excellent agreement with eq. (13.39). Other verifications of eq. (13.39) can be found in refs. [642,643,663].

13.4 Compatibility with the Jost-model of the dielectric continuum

The Jost model, extensively discussed by Flynn [35], leads to formulae for parameters of the formation of vacancies in ionic crystals. This is an interesting fact as the $cB\Omega$-model also gives correct results, although it emerges from completely different considerations. In the following we shall discuss *only* the qualitative agreement of the two models. A more accurate discussion of the Jost-model is given in ch. 14.

13.4.1 Formation entropy

Flynn [35] applied the Jost-model to the calculation of the formation energy of a point defect in an ionic crystal. Within the frame of this approximation the formation entropy is given by:

$$s \simeq \frac{h}{\varepsilon} \left.\frac{\partial \varepsilon}{\partial T}\right|_P, \tag{13.40}$$

where ε is the static dielectric constant. In the case of alkali halides the term $\varepsilon^{-1}(\partial \varepsilon/\partial T)_P$ is around $3\text{--}4 \times 10^{-4}$ K^{-1} whereas h is around 2.5 eV. The insertion of these values into eq. (13.40) gives for s a value between 7 and 10 k which agrees with the experimental values.

The ionic polarisability α in an alkali halide crystal A^+B^- with NaCl-structure is approximately given by [419,420]

$$\alpha = \text{constant} \times e^{*2}(Bl)^{-1} \tag{13.41}$$

where e^* is the so-called Szigeti charge, and the constant is practically independent of temperature and pressure (see §A.11, note 20).

With increasing temperature the value of B decrease whereas l increases so that the temperature variation of B and l changes in different directions. Taking into account that the percentage of the temperature variation of B is greater than that of l, eq. (13.41) reveals that α (and hence ε) increases with T. These considerations show that the Jost model can be interpreted as giving a connection between the formation entropy and the temperature decrease of the bulk modulus; the temperature increase of l works in the opposite direction but is of minor importance. We note that both effects are purely anharmonic bulk qualities. According to this model the formation entropy of a Schottky defect is due, at least in the major part, to the temperature dependence of the dielectric constant.

According to eq. (8.17) the previous conclusions are in qualitative agreement with those of the $cB\Omega$-model:

$$s = -\frac{hb}{B_0^{SL}}\left(\beta B + \left.\frac{\partial B}{\partial T}\right|_P\right).$$

In this equation the entropy is mainly governed by the term $\partial B/\partial T$, while βB has an opposite sign but is of minor importance. We again note that both are anharmonic qualities. The $cB\Omega$-model is therefore compatible with the Jost model as the derivatives $\partial \varepsilon / \partial T$ and $\partial B / \partial T$ are interconnected [420].

13.4.2 Formation volume

A similarity in the formal expressions of the Jost model and the $cB\Omega$-model can also be noticed in the formulae for the formation volume. Flynn's considerations led to the relation

$$\delta V = -\frac{h}{\varepsilon} \left.\frac{\partial \varepsilon}{\partial P}\right|_T, \quad (13.42)$$

where δV is the relaxation volume ($v^f - 2\Omega$) when a Schottky defect is created in alkali halides. Experiments show that in ionic crystals the dielectric constant ε decreases upon compression; furthermore in alkali halides we have [419]:

$$\frac{1}{\varepsilon} \left.\frac{\partial \varepsilon}{\partial P}\right|_T \approx -10^{-11} \, \text{cm}^2/\text{dyn},$$

and therefore eq. (13.42) indicates that the relaxation volume is positive.

On the other hand the $cB\Omega$-model indicates that the large positive value of v^f (and therefore the large positive value of δV for alkali halides) is mainly due to the pressure variation of the bulk modulus [see eq. (8.19)]. In order to find a contact between the two models we have to show that $\partial \varepsilon / \partial P$ and $\partial B / \partial P$ are interconnected. Taking as a first approximation a volume-independent value [420] of e^*, eq. (13.41) finally gives

$$\left.\frac{\partial \alpha}{\partial P}\right|_T \approx -\alpha \frac{1}{B} \left(-\frac{1}{3} + \left.\frac{\partial B}{\partial P}\right|_T \right). \quad (13.43)$$

The value of $\partial B / \partial P$ for alkali halides is around 5, so that eq. (13.43) gives a negative value for $\partial \alpha / \partial P$. This means that α (and hence ε) decreases on compression mainly because of the large pressure variation of the bulk modulus. Equation (13.42) of the Jost model can therefore be interpreted as a connection between the relaxation volume and the pressure variation of the bulk modulus.

The latter conclusion is in qualitative agreement with the $cB\Omega$-model which indicates that in alkali halides the large positive value of v^f (and hence of δV) is mainly due to the anharmonicity of the solid, i.e. to the value of $\partial B / \partial P$.

A last important remark should be added: in the case of an harmonic approximation ($\partial B / \partial P = 0$, $\partial \varepsilon / \partial P > 0$ [419]) both models lead to negative relaxation volumes.

13.5 Correlation of activation parameters resulting from dielectric loss or ITC-experiments

As described in §7.4.2 the experimental results of dielectric loss or ionic thermocurrent (ITC) techniques are analysed with the help of equation:

$$\tau = \tau_0 \exp\left(\frac{E}{kT}\right) \tag{13.44}$$

Combining eq. (13.44) with eq. (7.67) and assuming $\nu \simeq \nu_{TO}$ we find:

$$s^{m,b} = k \ln \frac{\tau_0^{-1}}{\lambda \nu_{TO}}, \tag{13.45}$$

where the symbols m, b indicate that these effects are due exclusively to the migration (m) of a bound defect (b). In the case of linear Arrhenius plots the energy E is by definition the experimental value of h_{exp}, which is connected to h_0 by eq. (8.15). Introducing the experimental value $h_{exp}(=E)$ into eq. (8.17) one gets

$$\frac{s^{m,b}}{E} = -\frac{b}{B_0^{SL}}\left(\beta B + \left.\frac{\partial B}{\partial T}\right|_P\right)$$

or due to eq. (13.45)

$$\frac{k \ln(\lambda \tau_0 \nu_{TO})^{-1}}{E} = -\frac{b}{B_0^{SL}}\left(\beta B + \left.\frac{\partial B}{\partial T}\right|_P\right). \tag{13.46}$$

When considering various aliovalent dopants (i) in a given (ionic) host crystal, the reorientation processes of the resulting dipoles are described with the help of parameters τ_0^i, E^i, which according to the $cB\Omega$-model have to be interconnected through eq. (13.46). In table 13.2 we give the experimental quantity $k \ln(\tau_0^i \lambda \nu_{TO})^{-1}/E^i$ for a number of trivalent dopants in CaF_2. The values of τ_0 and E are given by Andeen, Link and Fontanella [425]. The quantity ν_{LO} has been measured by Lowndess [426], by Bosomworth [427] and by Denham, Field, Morse and Wilkinson [428]; then by using the approximate relation $\nu_{LO} \simeq 2\nu_{TO}$ we find ν_{TO}. The elastic and expansivity data are from Wong and Schuele [429]. We see that within the experimental error this quantity is equal to the "bulk property"

$$-\frac{b}{B_0^{SL}}\left(\beta B + \left.\frac{\partial B}{\partial T}\right|_P\right)$$

as demanded by eq. (13.46). The same holds for SrF_2 and BaF_2 [424].

Table 13.2

Comparison of $-k\ln(2\nu_{TO}\tau_0)/E$ for R_1-relaxation in CaF_2 with the bulk quantity $-(b/B_0^{SL})[\beta B+(\partial B/\partial T)_P]$ of the $cB\Omega$-model.[a]

dopant	$-k\ln(2\tau_0\nu_{TO})/E$			$\dfrac{-b}{B_0^{SL}}\left(\beta B+\left.\dfrac{\partial B}{\partial T}\right\vert_P\right)$
	$(10^{-4}\,K^{-1})$			$(10^{-4}\,K^{-1})$
	ref. [426]	ref. [427]	ref. [428]	ref. [429]
Tb	2.77	2.84	2.82	~ 3.08
Er	2.76	2.82	2.80	~ 3.08
Dy	2.80	2.87	2.82	~ 3.08
Ho	2.91	2.97	2.95	~ 3.08
Gd	2.94	3.00	2.98	~ 3.08
Eu	3.06	3.12	3.10	~ 3.08
Nd	3.17	3.24	3.22	~ 3.08
Ce	3.20	3.26	3.24	~ 3.08
Pr	3.35	3.41	3.39	~ 3.08

[a] For details see ref. [424]; $\lambda=2$.

Equation (13.46) explains the well known empirical fact that the plot of $\ln\tau_0^i$ versus E_i for various dopants in the same host material is a straight line, which furthermore gives the direct physical meaning of its slope.

14 THEORETICAL BASIS OF THE $cB\Omega$-MODEL

14.1 Introduction

The connection of the defect parameters of migration, formation and (self-diffusion) activation processes to bulk properties was raised at least 3 decades ago. Zener [430] first proposed that the migration entropy for interstitial motion depends on the temperature derivative of the shear modulus μ of the bulk material. His basic idea was that, when the atom passes through the saddle point, shear stresses become important, thus leading to a connection of s^m with $d\mu/dT$ [431]. Twenty-two years later Flynn developed the dynamical theory of migration in monoatomic solids. The spirit of Zener's idea found a justification by Flynn in the Debye approximation under the condition that μ has to be replaced by another effective shear modulus $\tilde{C}$ given by [432]:

$$\frac{15}{2\tilde{C}} = \frac{3}{C_{11}} + \frac{2}{C_{11} - C_{12}} + \frac{1}{C_{44}}.$$

Within the frame of the dynamical theory, the migration volume v^m is connected to the migration Gibbs energy g^m through the relation

$$\frac{v^m}{g^m} = -\frac{1}{B} + \frac{1}{\tilde{C}}\frac{d\tilde{C}}{dP}.$$

This equation considerably differs from that predicted by the $cB\Omega$-model. However, in deriving the above equation two assumptions were inherently made: (1) that the motion entropy depends only on the volume and (2) that the volume dependence of the elastic constants at high temperatures gives a reliable measure of the volume dependence of the phonon frequencies. As Flynn himself pointed out neither assumption is very accurate so that the results can maintain only a semiquantitative validity. He also stressed that Zener's model, even in its new form, must fail in the case of ionic crystals.

A striking example of the disparity is the case of PbF_2. The elastic data under pressure have recently been reported by Rimai and Sladek [358]; their data for $T = 296$ K lead to $B = 630$ kbar, $\tilde{C} = 625.5$ kbar, $dB/dP = 7.13$ and $d\tilde{C}/dP = 2.56$. The values of B and dB/dP correspond [662] to the adiabatic elastic constants but the difference with the isothermal ones is very small. According to Flynn's formula these values give $v^m/g^m = 2.5 \times$

10^{-12} cm^3 erg^{-1}. From Samara's measurements one gets for v^m/g^m the value 9.4×10^{-12} cm^3 erg^{-1} for fluorine vacancy motion and 8.3×10^{-12} cm^3 erg^{-1} for the migration of fluorine interstitials (with a plausible error of $\pm 15\%$, assuming the migration entropy to be 2 to 3 k (see table 10.13). One sees that the Flynn–Zener model gives for the ratio v^m/g^m a value that differs from the experimental one by a factor of 3 to 4. On the other hand the $cB\Omega$-model according to eq. (8.22) gives $v^m/g^m \simeq 9.7 \times 10^{-12}$ cm^3 erg^{-1}, which agrees with the experimental values within the experimental error (see §10.4.4 and the footnote of table 10.13).

All the above discussions were restricted to migration processes; according to Flynn [35] a connection between bulk qualities and formation parameters appears to have no justification.

Keyes [433] has earlier extended Zener's ideas to defect volumes. He proposed a connection of the activation volume to the pressure derivative of the shear modulus. An application of his formula [434] finds deviations up to 50% for elemental crystals and fails for ionic crystals [435,644,663].

Similarly the empirical formula $s^{act}/v^{act} = \beta B$ of Lawson, Zener and Keyes (see §13.2), although giving values of the correct order of magnitude, is untenable from a theoretical standpoint, because it can be obtained only when $s^{*,act} \simeq 0$. In the case of the $cB\Omega$-model the situation is different.

In chs. 9 and 10 it was shown that the application of the $cB\Omega$-model works equally well for formation and activation processes and this for a large variety of defects in various categories of solids. Such an agreement cannot be fortuitous, the more so, as the model does not contain any factor that has to be empirically adjusted. Therefore it is a logical conclusion that this agreement must somehow have a thermodynamical origin. An explanation is attempted in the present chapter.

In §14.2 we give some preliminary aspects that will be needed. More precisely we discuss the range of values of the quantity $B(\mathrm{d}^2 B/\mathrm{d} P^2)[(\mathrm{d} B/\mathrm{d} P)_T - 1]^{-1}$, which seems to play an essential role. Furthermore, we discuss the experimental and theoretical evidence about the values of the quantity κ^{act}/κ or κ^f/κ and about the pressure variation of $[(\mathrm{d} B/\mathrm{d} P)_T - 1]^{-1} \times (v/\Omega)$. In §14.3 we first derive the thermodynamical identity

$$\left.\frac{\partial g}{\partial (B\Omega)}\right|_T = \frac{v}{\Omega}\left(\frac{\mathrm{d} B}{\mathrm{d} P} - 1\right)^{-1},$$

which is valid for formation, migration and (self-diffusion) activation processes. Then, after presenting strong theoretical and experimental arguments that the right-hand side of this equation (which we will label with c) is practically *pressure independent*, we arrive at the conclusion that

$$g = Bv\left(\frac{\mathrm{d} B}{\mathrm{d} P} - 1\right)^{-1},$$

or equivalently

$$g = cB\Omega,$$

where

$$c \equiv \frac{v}{\Omega}\left(\frac{dB}{dP} - 1\right)^{-1}.$$

From a purely physical point of view it will be shown that the above formulae are derived as an immediate consequence of the following requirement: "the compressibility of the defect volume cannot exceed the bulk compressibility by more than a few times". From a microscopic point of view this physical requirement seems to hold better in the case of a vacancy formation process than in the case of an activation (formation plus migration) process. This is important, because Flynn raised doubts as to the validity of an equation of the form of eq. (8.1), especially in the case of a formation process. In §14.4 we will show that c can be considered also as *temperature independent*. In the quasi-harmonic approximation this statement is exact. In real solids, however, c may depend slightly on temperature; for this reason the bounds of the temperature variation of c will be determined in terms of well-known macroscopic properties of the solid, so that the possible errors introduced into the calculation of g can be evaluated for each case. Quantitative assessments lead to the conclusion that in most cases within the present experimental accuracy this approximation is sufficient.

14.2 Preliminary aspects

14.2.1 Evidence for a limiting value of the compressibility of defect volumes

In a great number of fcc (Ag, Cu, Au, Pb, etc.), bcc (Na, β–Ti), tetragonal (Sn, In) and hexagonal (Cd, Zn) metals self-diffusion has been studied for pressures up to 10 kbar except in Pb, where the study has been extended up to 40 kbar. With the exception of Na (see below) the isothermal plots of $\ln D$ versus P have not shown a noticeable curvature. From a thermodynamical point of view the absence of a curvature demands (see §7.3.4) the term $\exp\int_0^P \kappa^{act} dP$ (where κ^{act} denotes the compressibility of the activation, i.e. formation plus migration, volume) to be appreciably smaller than 1.05 or 1.1. This is the limiting value if one accepts the usual maximum experimental error of 5 to 10% in the determination of activation volumes. One is immediately led to the conclusion that a curvature cannot be observed in experiments for which $\bar{\kappa}^{act}\Delta P < 0.05$, where $\bar{\kappa}^{act}$ is the mean defect compressibility in the pressure range studied. The above limiting condition predicts that pressure experiments extending up to 10 kbar cannot show a detectable curvature for materials for which $\kappa^{act} < 5 \times 10^{-3}$ kbar^{-1}. For all metals mentioned above (except Na) the high-temperature isothermal bulk compressibility κ lies roughly within the range of 1×10^{-3} to 2.5×10^{-3} kbar^{-1}. One can therefore conclude that for all the above materials for which no curvature has been detected with pressures up to 10 kbar, the ratio κ^{act}/κ is appreciably smaller than 5. Of course we discuss only positive values of κ^{act}; if these were negative the plot of $\ln D$ versus P should show a downward curvature, which has been never observed. Theoretical considerations below also show that κ^{act} is positive.

Let us now consider highly compressible materials with κ of the order of 10^{-2} kbar^{-1} or larger; the limiting value 0.05 of $\bar{\kappa}^{act}\Delta P$ is already reached at pressures below 10 kbar. For the detection of a curvature for such materials κ^{act}/κ does not have to reach up to 5. We state for example the case of Na for which the curvature can be explained (§7.3.4) with a ratio κ^{act}/κ lying between 2 and 3 [37].

Until now the considerations dealt with the compressibility κ^{act} of the (self-diffusion) activation volume. Experiments leading directly to the formation volume studied the effect of pressure on the electrical resistance in quenched metals. Such experiments [436–440,664] have been reported only for gold and aluminium [436] up to 6 kbar and platinum [440] up to 9 kbar. In all these experiments the plot of "resistivity increment after quenching" versus pressure does not show any detectable curvature, similar to the self-diffusion experiments. In view of this fact it is now reasonable to assume that the limiting values for the ratio κ^{act}/κ also hold for the ratio κ^f/κ, where κ^f is the compressibility of the formation volume.

We now turn our attention to conductivity or diffusion experiments under pressure on sodium- and potassium-halides; they have [322,323] curved plots of $\ln D$ (or $\ln \sigma T$) versus P already for pressures of 8 kbar and their bulk compressibilities, in the temperature region where the experiments have been carried out, lie between 5×10^{-3} and 10×10^{-3} kbar^{-1}. As they have been studied both in the extrinsic and the intrinsic range, their curvature is consistent [441] with a value of κ^{act}/κ (or κ^f/κ or κ^m/κ) around 4 or 5 (see §10.4.2). This upper limit of κ^{act}/κ seems to be valid also for CsCl, for which no noticeable curvature has been reported up to 8 kbar [442]. Recent conductivity experiments [356] for PbF$_2$ up to 6 kbar also have not detected any curvature.

Summarising, we state that all pressure experiments known up to date, either showing a curvature or not, are consistent with a value of κ^d/κ smaller than or at most equal to 5, where the superscript d stands for any kind of defect.

The existence of a limiting value for κ^d/κ is further substantiated by theory. Its evaluation is based on the thermodynamical formula (3.57) which gives the change $\kappa - \kappa^0$ of the compressibility of a crystal due to the vacancies it contains; κ^f is the compressibility of the formation volume of the vacancy. It can be written as:

$$\frac{\kappa - \kappa^0}{\kappa^0} = \frac{n}{N} \frac{v^f}{\Omega} \left(\frac{\kappa^f}{\kappa^0} - 1 \right). \tag{14.1}$$

This equation holds for any type of defect provided that they do not interact and that v^f stands for the (total) formation volume of each defect. The quantities $(\kappa - \kappa^0)/\kappa^0$ and v^f/Ω are obtained from microscopic calculations so that one can determine κ^f/κ^0. This can be done in two completely equivalent methods which differ only by the starting point of the calculation. The first method [443–445] requires the calculation of those Green's functions which relate the displacements in the "lattice with defects" to those of the ideal lattice. The other general method (energy-method) has been proposed by Ludwig [446] and is an extension of the Born–Huang method for nonprimitive lattices. Various calculations have been carried out following both methods. Ludwig's method has been applied to bcc and fcc crystals by Pistorius [447], who concluded that "the elastic constants are *reduced* by vacancies and that the relative change is about two or three times larger than the atomic concentration of vacancies". In view of eq. (14.1) and considering that v^f/Ω is of the order of unity, we conclude that for vacancies the ratio κ^f/κ^0 is around 3 or 4. Exactly the same conclusion can be drawn from the Green's function calculations by the Jülich [444,445] group for vacancies in Cu.

The theoretical results with Green's functions for isolated interstitials are slightly different. Dederichs, Lehmann and Scholz [444] give $(\kappa - \kappa^0)/[\kappa^0(n/N)] \simeq 7.9$ and $v/\Omega = 2.41$ for dumb-bell interstitials in Cu. Inserting these values into eq. (14.1) we get $\kappa^f/\kappa^0 \simeq 4.3$. On the other hand the experiment of Holder, Granato and Rehn [448] gives for isolated interstitials in Cu $(\kappa - \kappa^0)/[\kappa^0(n/N)] = 0 \pm 1$ and $v/\Omega \simeq 1.5$. When inserted into eq. (14.1) these values give for κ^f/κ^0 the value of 1 ± 0.7, which has to be combined with the value of one vacancy in order to get the total effect per Frenkel defect. Note also that the experimental results of Robrock and Schilling on Al [449] are in agreement with those mentioned above for Cu. Summarising, one sees that the theory gives an increase of the compressibility (i.e. $\kappa^f > \kappa^0$) upon the introduction of a defect. The theoretical values are larger than the direct experimental ones. This small disagreement is attributed [445] to the strong repulsive part of the used Morse potential.

We shall now turn our attention to the compressibility of the migration volume. Direct microscopic calculations for the quantity κ^m do not exist in the literature. A rough idea of its magnitude may be drawn from the following speculations: the volume v^m can be written as

$$v^m = V_S - V,$$

where V_S is the volume of the crystal when the moving ion is on the saddle point (S) and V the equilibrium volume of the crystal when the ion lies at its normal lattice site. From a physical point of view the occurrence of the saddle point configuration (i.e. moving ion at S and thus creation of a vacancy at its previous lattice site) might be analogous to the addition of a Frenkel pair. This correlation, however, may well be questionable for various reasons because the saddle point configuration is a dynamical process occurring during a time-interval of the order of the period of the lattice vibrations, whereas the addition of a Frenkel pair leads to a stable configuration of long duration. Furthermore, the symmetries of the relaxations in these two cases are different and in the case of a Frenkel defect the interstitial lies far from the vacancy. The influence of Frenkel pairs on the compressibility of fcc metals has been studied by Dederichs and coworkers [444,445]. Microscopic calculations showed that Frenkel pairs affect the bulk modulus to an amount only 2 to 3 times greater than single vacancies. In this sense we can draw the rough conclusion that κ^m/κ^0 should be of the same order of magnitude as κ^f/κ^0; according to Ludwig [450] it might exceed it at most by a factor 3 to 4 (see also ref. [665]).

Finally, we will determine a maximum value for the compressibility κ^{act} of the activation volume. For single vacancies in monoatomic solids we have

$$v^{act} = v^f + v^m \tag{14.2}$$

and by differentiating with respect to pressure this becomes

$$\kappa^{act} v^{act} = \kappa^f v^f + \kappa^m v^m$$

or

$$\frac{\kappa^{act}}{\kappa^0} = \frac{\kappa^f}{\kappa^0} \frac{v^f}{v^{act}} + \frac{\kappa^m}{\kappa^0} \frac{v^m}{v^{act}}. \tag{14.3}$$

It is known that v^m is usually smaller than v^f and usually the ratio v^{act}/v^m is around 3. Let us assume for the moment that κ^m exceeds κ^0 by a large amount, e.g. $\kappa^m/\kappa^0 \approx 10$; recalling now that theoretical considerations and experimental evidence have shown κ^f/κ^0 to be around 3, equation (14.3) reveals that the ratio κ^{act}/κ^0 is close to 5. Similar results are found for Schottky defects in alkali halides, considering that eq. (14.2) changes to $v^{act} = v^m + v^f/2$, from which we get $v^m/v^f \approx 1/6$ [322,323].

The above theoretical considerations have shown that even if the compressibility of the migration volume exceeds that of the bulk material by one

§14.2] *Preliminary aspects* 369

order of magnitude the ratio $\kappa^{\mathrm{act}}/\kappa^0$ continues to have values at most around 5; this upper limit is consistent with the conclusions that were drawn from the study of experimental data. Differentiating the relation [35] $v^{\mathrm{m}}/g^{\mathrm{m}} = 2\gamma_i/B$ with respect to pressure we find:

$$\frac{\kappa^{\mathrm{m}}}{\kappa} = \frac{\mathrm{d}\ln\gamma_i}{\mathrm{d}\ln V}\bigg|_T + \frac{\mathrm{d}B}{\mathrm{d}P}\bigg|_T - \gamma_i.$$

This shows that a large value of $\kappa^{\mathrm{m}}/\kappa$ demands a large volume variation of the Grüneisen constant γ_i of the driving mode i; it is only a speculation that $\mathrm{d}\ln\gamma_i/\mathrm{d}\ln V$ should be larger for interstitial migration.

14.2.2 *The limiting values of the quantity* $-B(\mathrm{d}^2B/\mathrm{d}P^2)$ $\times [(\mathrm{d}B/\mathrm{d}P)_T - 1]^{-1}$

The pressure variation of the quantity $\Omega[(\mathrm{d}B/\mathrm{d}P) - 1]$ is important in the $cB\Omega$-model; it will be studied with the help of the quantity A defined by

$$A \equiv B\frac{\mathrm{d}^2B}{\mathrm{d}P^2}\left(\frac{\mathrm{d}B}{\mathrm{d}P}\bigg|_T - 1\right)^{-1}. \tag{14.4}$$

Evaluating each of its factors, narrow bounds of A can be determined. The physical meaning of $(\mathrm{d}B/\mathrm{d}P) - 1$ has already been discussed in ch. 2. It was initially thought that the Murnaghan equation of state [27], according to which $\mathrm{d}B/\mathrm{d}P$ is pressure independent, could well describe the real behaviour of solids. Its idea lies in the prediction of the quasi-harmonic approximation that the quantity $\mathrm{d}B/\mathrm{d}P$ must be pressure and temperature independent [1]. However, it has long been experimentally verified and theoretically justified that this quantity decreases slightly on compression. Generally speaking one can say [254] that in all categories of solids $\mathrm{d}^2B/\mathrm{d}P^2$ is negative and its absolute value is of the order of the quantity $B^{-1}(\partial B/\partial P)_T$. More precisely, for each category of solids the following experimental and theoretical evidence exists.

Rare gas solids (RGS) Using the usual Lennard–Jones potential one finds $\mathrm{d}B/\mathrm{d}P$ to be 8 and $B(\mathrm{d}^2B/\mathrm{d}P^2)$ equal to -15; the theory therefore suggests that in RGS the quantity A is close to -2. Birch [254] has analysed the elastic data from the piston displacement experiments of Anderson and Swenson [259] for Ar, Kr and Xe and those of Anderson, Fugate and Swenson [253] for Ne. He found $\mathrm{d}B/\mathrm{d}P$ to lie between 7.5 and 10 and $-B\mathrm{d}^2B/\mathrm{d}P^2$ to vary from 7.5 to 17 except for Ar at 77 K and Xe at 159 K, where the values are 33.3 and 38.0, respectively. One can therefore safely state that the dimensionless quantity A lies roughly between -1 and -4.2.

Metals. The quantity dB/dP varies from 3.5 up to 8; the lower values pertain to alkali metals. Furthermore, measurements [200] show that for many of them the quantity $-B(d^2B/dP^2)$ is roughly equal to dB/dP; therefore one can state that in metals the quantity $-A$ is somewhat higher than unity. This experimental result is not without theoretical support [451,452].

Alkali halides. According to the Born-model as modified by Smith and coworkers [376–379] the quantities B and $B(d^2B/dP^2)$ are given by

$$\frac{dB}{dP} = \frac{n^B + 7}{3} \quad \text{and} \quad B\frac{d^2B}{dP^2} = -\tfrac{4}{9}(n^B + 3),$$

where n^B is the Born exponent, which is close to 10. Therefore, the theory suggests that $-A$ is around unity [652]. According to Smith [377] the above values describe rather well the experimental data of alkali halides. However, it should be mentioned that the experiments of Spetzler et al. [282] on NaCl lead to higher values of $-A$, i.e. around 3 or 4. On the other hand Barsch and Shull's experiments [292] on NaI and KI verify Smith's statement as they found $-A$ in NaI and KI to be 1.63 and 1.44, respectively.

Silver halides. Loje and Schuele [309] have studied AgCl and AgBr under pressure and found a significant nonlinear term in the function B versus P. This has to be included in computing the equation of state in preference of the ordinary Murnaghan equation. Their results show that in AgBr the quantity $-A$ is around 1.67, a value similar to that found in alkali halides. AgI seems to be of special interest because dB/dP is close to unity at room temperature and becomes negative at higher temperatures (see table 10.11). The quantity d^2B/dP^2 in AgI is still unknown.

Summarising the above remarks we can say that both experiment and theory suggest that for various categories of solids the quantity $-A$ always remains between narrow bounds (1–4.2). It is of interest here to repeat the following remark made by Birch [254]: "It is a striking characteristic of inorganic crystalline solids that, despite a variation of B by several thousandfold, the dimensionless quantities dB/dP and $B(d^2B/dP^2)$ remain within *narrow bounds* regardless of composition or crystal structure."

14.2.3 *The pressure variation of the quantity $(v/\Omega)[(dB/dP) - 1]^{-1}$*

We shall label with c the expression mentioned in the heading:

$$c(P, T) \equiv \frac{v}{\Omega}\left(\left.\frac{dB}{dP}\right|_T - 1\right)^{-1}. \tag{14.5}$$

This is an important quantity in the present theory because, as will be shown later, it can be considered to a certain extent as being independent of pressure and temperature. Its variation with pressure is best studied from the value of its isothermal compressibility κ^c defined by

$$\kappa^c \equiv -\frac{1}{c} \frac{dc}{dP}\bigg|_T. \tag{14.6}$$

Inserting the above definition of c one gets after some manipulation

$$\frac{\kappa^c}{\kappa} = \frac{\kappa^d}{\kappa} - 1 + A, \tag{14.7}$$

where the defect compressibility κ^d stands for κ^{act}, κ^f or κ^m, depending on the mechanism to which c refers in eq. (14.5). We will study the order of magnitude of the quantity κ^c/κ. This study is now straightforward because in the previous sections 14.2.1 and 14.2.2 the limiting values of κ^d/κ and A have been determined. More precisely, in §14.2.2 both theoretical and experimental results prove that in various categories of solids the quantity $-A$ exceeds unity but does not take values greater than 4.2; in most cases it is around 2. In §14.2.1 all experimental evidence showed that κ^{act}/κ must be smaller than 5, only in alkali halides is it close to 5. We have also stated from restricted experimental evidence that the same limit should hold for the ratio κ^f/κ. The latter statement is strongly strengthened by the detailed theoretical calculations which suggest that κ^f/κ lies between 3 and 4.

Combining the above considerations we conclude that the absolute value of the right side of eq. (14.7) lies between zero and unity and therefore $|\kappa^c/\kappa| \lesssim 1$. It is likely that positive values of the ratio κ^c/κ correspond to vacancies whereas for interstitials slightly negative values should hold; this is so if one recalls the theoretical results of Dederichs and coworkers [444,445]. As we shall see in §14.3 the well-founded conclusion of this paragraph that $|\kappa^c/\kappa|$ is at most of the order of unity is of paramount importance for the proof of the $cB\Omega$-model.

14.3 Proof of the $cB\Omega$-model with a pressure-independent c

14.3.1 Introduction

In this paragraph we shall show that the defect Gibbs energy g and the defect volume v are interconnected through the relation (see ref. [666] for a clarification)

$$g = \frac{Bv}{\frac{dB}{dP}\bigg|_T - 1} \quad \text{or equivalently} \quad g(P, T) = c(T)B\Omega,$$

where $c(T)$ is a practically *pressure-independent* constant defined as

$$c(T) \equiv \frac{v}{\Omega}\left(\left.\frac{dB}{dP}\right|_T - 1\right)^{-1}.$$

In order to do so, we shall go through the following steps: In §14.3.2 the thermodynamical identity

$$\left.\frac{dg}{d(B\Omega)}\right|_T = \frac{v}{\Omega}\left(\left.\frac{dB}{dP}\right|_T - 1\right)^{-1}$$

will be proved. The quantity on the right side, which has been defined as c, is to a large extent (see §14.2.3) pressure independent. Therefore, in §14.3.4 we shall write

$$dg(P, T) = c(T)d(B\Omega), \qquad T = \text{constant},$$

and upon integration get

$$g(P, T) = c(T)B\Omega + \Lambda(T)$$

or equivalently,

$$g = \frac{Bv}{\left.\frac{dB}{dP}\right|_T - 1} + \Lambda.$$

This relation can also be obtained in an alternative procedure, shown in §14.3.5. The existence of the integration constant Λ cannot be ruled out from general thermodynamics; however, this is possible by means of microscopic considerations. Actually, in §14.3.6 we show that for two characteristic cases (i.e. rare gas solids and alkali halides) the formation Gibbs energy at $T = 0$ K is directly proportional to $B\Omega$ without any additive constant (i.e. $\Lambda = 0$). This will also be shown in §14.3.7 by proper use of the classical macroscopic Jost-model in ionic crystals. Finally, in §14.3.8 we shall indicate that the dynamical theory of migration might also lead to the statement that the Gibbs migration energy is proportional to $B\Omega$, again without any additive constant.

14.3.2 Proof of the thermodynamical identity $[dg/d(B\Omega)]_T = (v/\Omega)[(dB/dP) - 1]^{-1}$

The definition of defect volumes implies that

$$dg = \frac{v}{\Omega}\Omega\, dP, \qquad T = \text{constant}. \tag{14.8}$$

Both symbols, g and v, correspond to a formation or a migration (and therefore to an activation) process. The definition of the isothermal bulk modulus can be given in the form:

$$B = -\Omega \left.\frac{dP}{d\Omega}\right|_T \tag{14.9}$$

so that eq. (14.8) can be written as

$$dg = -\frac{v}{\Omega} B \, d\Omega, \quad T = \text{constant}. \tag{14.10}$$

By combining the relation

$$\left.\frac{dB}{dP}\right|_T = \left.\frac{dB}{d\Omega}\right|_T \left.\frac{d\Omega}{dP}\right|_T$$

with eq. (14.9) one immediately finds

$$B \, d\Omega = -\left(\frac{dB}{dP} - 1\right)^{-1} d(B\Omega) \tag{14.11}$$

and due to eq. (14.10)

$$dg = \frac{v}{\Omega}\left(\frac{dB}{dP} - 1\right)^{-1} d(B\Omega), \quad T = \text{constant}. \tag{14.12}$$

By recalling the definition of c

$$c = \frac{v}{\Omega}\left(\frac{dB}{dP} - 1\right)^{-1}, \tag{14.13}$$

eq. (14.12) reads:

$$dg = c \, d(B\Omega), \tag{14.14}$$

which is just the relation given in the heading.

14.3.3 Possible error of g by considering c as pressure independent

At this point the quantity c is assumed as pressure independent and errors resulting from this simplifying assumption are determined. Consider a real solid in the initial state (P_0, V_0, T) which is isothermally compressed to the final state (P, V, T) by increasing the pressure by $\Delta P = P - P_0$. We are interested to find the difference $g(P) - g(P_0)$ of the defect Gibbs energies between these two states.

Integrating eq. (14.14) between the initial (i) and the final (f) state we get:

$$g(P) - g(P_0) = \int_i^f c \, d(B\Omega). \tag{14.15}$$

If c depends on pressure the quantity $g(P) - g(P_0)$ appears in fig. 14.1 in

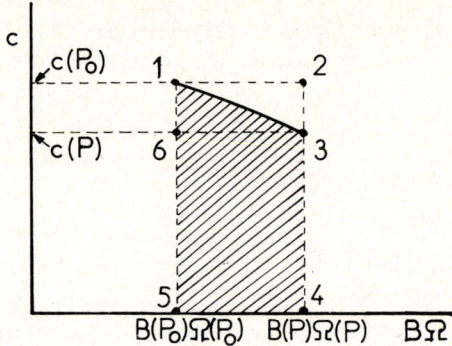

Fig. 14.1. The shaded area is $g(P) - g(P_0)$ for $\kappa^c < 0$; the area 1231 corresponds to the error introduced by the statement that c is pressure independent.

the form of the shaded area A_{13451} (taking $v > 0$, i.e. $g(P) > g(P_0)$). Let us determine also the simplified case when c does not change during the process, thus giving the area $A_{12451} = c(P_0)|B(P)\Omega(P) - B(P_0)\Omega(P_0)|$ and evaluate the error which is introduced by this assumption. To a good approximation the error is

$$A_{1231} \simeq \tfrac{1}{2}\{c(P_0) - c(P)\}\{B(P)\Omega(P) - B(P_0)\Omega(P_0)\}.$$

The relative error of $g(P) - g(P_0)$ is approximatively

$$E_{\text{rel}} = A_{1231}/A_{12451} \simeq \frac{1}{2}\left\{1 - \frac{c(P)}{c(P_0)}\right\}.$$

Considering the definition of κ^c [see eq. (14.6)] one gets

$$c(P) = c(P_0)\exp\int_{P_0}^{P} -\kappa^c \, dP$$

and hence we have

$$E_{\text{rel}} \simeq \tfrac{1}{2}\left\{1 - \exp\int_{P_0}^{P} -\kappa^c \, dP\right\}. \tag{14.16}$$

This is a general thermodynamical result and it is valid regardless the value and the algebraic sign of κ^c. We now restrict ourselves to a pressure range for which

$$\left|\int_{P_0}^{P} -\kappa^c \, dP\right| < 0.1 \tag{14.17}$$

and hence

$$\exp\int_{P_0}^{P} \kappa^c \, dP \simeq 1 + \int_{P_0}^{P} \kappa^c \, dP + \text{small terms}. \tag{14.18}$$

We have already shown that $|\kappa^c|$ lies between 0 and κ (see §14.2.3). Accepting that $|\kappa^c| = \kappa$ one gets the maximum possible relative error

$$|E_{\text{rel}}^{\max}| \simeq \tfrac{1}{2} \int_{P_0}^{P} \kappa \, \mathrm{d}P$$

or approximately,

$$|E_{\text{rel}}^{\max}| \simeq \tfrac{1}{2} \bar{\kappa} \Delta P, \tag{14.19}$$

where $\bar{\kappa}$ is the mean value in the pressure range (P_0, P).

This is a quite interesting result. It shows that when we are working, as usual, in a pressure range appreciably smaller than the bulk modulus ($\kappa \Delta P \ll 1$) the assumption that c is pressure independent, i.e. $c(P) = c(P_0)$, does not introduce appreciable errors.

As will be shown later [see eq. (14.36)] the order of magnitude of $g(P) - g(P_0)$ is

$$g(P) - g(P_0) \simeq g(P_0)\left(\frac{\mathrm{d}B}{\mathrm{d}P} - 1\right)\frac{\Delta P}{B}. \tag{14.20}$$

If $g(P)$ is to be predicted from an accurate value of $g(P_0)$, it will be of interest to know the maximum error, $E^{\max}$, in the predicted value. By definition $E_{\text{rel}}^{\max}$ is $E^{\max}/\{g(P) - g(P_0)\}$; the insertion of the appropriate expressions gives

$$|E^{\max}| \simeq \tfrac{1}{2} g(P_0)\left(\frac{\Delta P}{B}\right)^2\left(\frac{\mathrm{d}B}{\mathrm{d}P} - 1\right) \tag{14.21}$$

or in terms of $g(P)$

$$|E^{\max}| \simeq \tfrac{1}{2} g(P)\left(\frac{\Delta P}{B}\right)^2\left(\frac{\mathrm{d}B}{\mathrm{d}P} - 1\right)\left\{1 + \left(\frac{\mathrm{d}B}{\mathrm{d}P} - 1\right)\frac{\Delta P}{B}\right\}^{-1}. \tag{14.22}$$

In table 14.1 we quote the rough values of the error as resulting from eq. (14.19) and (14.22) together with the typical experimental data used for their calculation. The error starts being of influence when $\Delta P/B$ is comparable to 0.1. Such experimental cases are rare and pertain to highly compressible

Table 14.1
Typical maximum errors in the prediction of g at high pressures (by assuming c as pressure independent)

$B(0)$ (kbar)	$\mathrm{d}B/\mathrm{d}P$	ΔP (kbar)	$\Delta P/B(0)$	$E^{\max}/g$ [a]
1000 (e.g. Au, Cu)	5	10	10×10^{-3}	0.019%
400 (e.g. Pb)	5	10	25×10^{-3}	0.11%
60 (e.g. Na)	4	7	120×10^{-3}	1.5%

[a] Maximum error of g from eq. (14.22).

materials like alkali metals or RGS. Equation (14.21), however, allows, even in these extreme cases, the estimation of the uncertainty introduced by the assumption that c is pressure independent. In these cases the expansion given in eq. (14.18) has to be somewhat improved by some small additional terms.

Summarising the considerations up to the present point we can state that for all materials and, at least, for pressure ranges ΔP for which $\Delta P/B < 0.1$, the quantity $g(P)$ can be predicted according to eq. (14.15) from $g(P_0)$ according to the equation

$$g(P) - g(P_0) = c\{B(P)\Omega(P) - B(P_0)\Omega(P_0)\} \qquad (14.23)$$

in which c is considered to be a constant. Remember that the errors given by eqs. (14.19) and (14.21) or (14.22) are the maximum possible and occur when $|\kappa^c| = \kappa$.

14.3.4 Proof of the equations $g = [(dB/dP) - 1]^{-1}Bv$ and $g(P, T) = c(T)B\Omega$

In view of the results of §14.3.3 the function g at a given pressure P_1 and at any temperature T can be calculated from

$$g = \left.\frac{dg}{d(B\Omega)}\right|_{T, P=P_1} B\Omega + \Lambda \qquad (14.24)$$

(see also §14.3.5) or by using the identity (14.12)

$$g = \left(\frac{dB}{dP} - 1\right)^{-1} Bv + \Lambda, \qquad (14.25)$$

where Λ is a constant which does not depend on $B\Omega$ so that for a given pressure P_1 it may depend only on temperature.

The plausible existence of the quantity Λ slightly complicates the situation. However, microscopic considerations developed in §§14.3.6 and 14.3.8 indicate that Λ must be zero in formation and in migration processes. There is another strong argument based on the result of measurements. Because of the fact that all the other quantities in eq. (14.25) have been directly measured one can immediately judge about the plausible values of Λ. In the tables of ch. 10 we have calculated the quantity $Bv[(dB/dP) - 1]^{-1}$ and compared the result with the experimentally known quantity g; this has been carried out for a large variety of solids, i.e. fcc metals, bcc metals, tetragonal metals, hexagonal metals, rare gas solids, alkali halides with NaCl-structure, various salts with CsCl-structure, etc. Considering the experimental errors (up to 10%, mainly because of v) an overall agreement

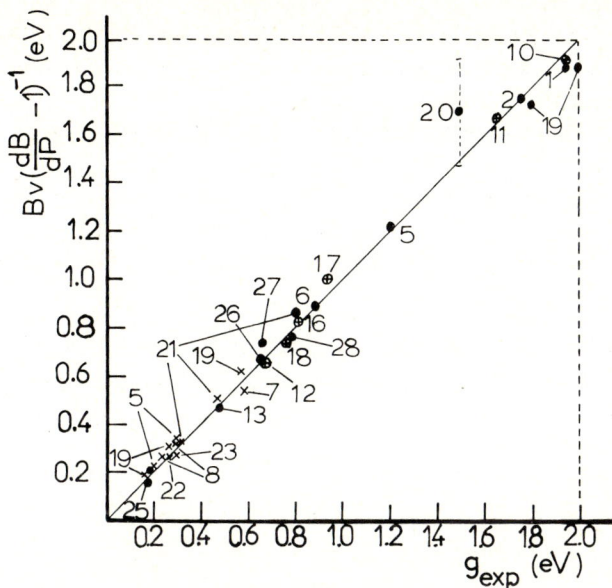

Fig. 14.2. Test of the correctness of eq. (14.26); ●: formation; ×: migration; ⊕: activation; the numbers 1–18 refer to table 10.14; 19: CsCl; 20: NaBr; 21: TlBr; 22: Na; 23: K; 25: Ne; 26: Ar; 27: Kr; 28: Xe; the coordinates of the rare gas solids are multiplied with 10.

between the values g and $Bv[(dB/dP) - 1]^{-1}$ emerges and hence we are justified in setting Λ equal to zero. As eq. (14.25) must hold irrespective of the temperature at which the measurements are made we can collect the totality of the results in fig. 14.2. All points should lie on a line under 45° which is the condition for the validity of eq. (14.25). We note a good fit in spite of the fact that the ordinates vary within two orders of magnitude. The deviations from the line are random. For NaBr the deviation is large but still lies within the error bars. Of special interest is the point for B4-AgI for which both v and dB/dP are negative, thus giving a positive value for $Bv[(dB/dP) - 1]^{-1}$, which coincides with the experimental g-value. The overall agreement shown in this figure is convincing, considering that the comparison has been made between quantities that are extracted from different experimental procedures: g is obtained from isobaric heating experiments while $Bv[(dB/dP) - 1]^{-1}$ contains quantities resulting from isothermal pressure experiments. Therefore eq. (14.25) can be written as

$$g = \left(\frac{dB}{dP} - 1\right)^{-1} Bv \tag{14.26}$$

or, considering the definition of c – see eq. (14.13) –

$$g = cB\Omega, \tag{14.27}$$

where, in principle, all the quantities depend on (P, T) or on (V, T). Considering the results of the previous paragraph (§14.3.3) that c, at least for pressure ranges $\Delta P/B < 0.1$, can be considered pressure independent, the previous relation reads [666]

$$g(P, T) = c(T) B\Omega \tag{14.28}$$

or [669]

$$g(V, T) = c(T) B\Omega. \tag{14.29}$$

We emphasize that eq. (14.26) interconnects two defect properties (i.e. g and v) with bulk properties. Already the validity of such an equation itself seems to be important; for instance, at $T = 0$ K it provides an extremely useful constraint for microscopic calculations in which the energy of formation h_0^f is simultaneously computed with v^f. It reveals that such a microscopic calculation can be considered to be reliable only when the computed values of h_0^f and v^f obey this connection.

14.3.5 Alternative derivation of the relation $g(P, T) = c(T)B\Omega$

A Taylor expansion of g around the initial conditions (P_0, V_0, T) gives (for $T =$ constant)

$$g(P) = g(P_0) + \left.\frac{\mathrm{d}g}{\mathrm{d}(B\Omega)}\right|_{T, P=P_0} \{B\Omega - B(P_0)\Omega(P_0)\}$$

$$+ \frac{1}{2} \left.\frac{\mathrm{d}^2 g}{\mathrm{d}(B\Omega)^2}\right|_{T, P=P_0} \{B\Omega - B(P_0)\Omega(P_0)\}^2 + \cdots. \tag{14.30}$$

We now calculate the derivative $[\mathrm{d}^2 g/\mathrm{d}(B\Omega)^2]_T$ with the help of the identity written in the heading of §14.3.2 and of eq. (14.6):

$$\left.\frac{\mathrm{d}^2 g}{\mathrm{d}(B\Omega)^2}\right|_T = \frac{\mathrm{d}}{\mathrm{d}(B\Omega)}\left\{\left.\frac{\mathrm{d}g}{\mathrm{d}(B\Omega)}\right|_T\right\} = \left.\frac{\mathrm{d}c}{\mathrm{d}(B\Omega)}\right|_T$$

$$= \left.\frac{\mathrm{d}c}{\mathrm{d}P}\right|_T \left.\frac{\mathrm{d}P}{\mathrm{d}(B\Omega)}\right|_T = -\kappa^c c \left.\frac{\mathrm{d}P}{\mathrm{d}(B\Omega)}\right|_T. \tag{14.31}$$

but we know that

$$\left.\frac{\mathrm{d}(B\Omega)}{\mathrm{d}P}\right|_T = \Omega \left.\frac{\mathrm{d}B}{\mathrm{d}P}\right|_T + B \left.\frac{\mathrm{d}\Omega}{\mathrm{d}P}\right|_T = \Omega\left\{\left.\frac{\mathrm{d}B}{\mathrm{d}P}\right|_T - 1\right\}$$

and hence eq. (14.31) becomes

§14.3] Proof of the $cB\Omega$-model

$$\left.\frac{d^2g}{d(B\Omega)^2}\right|_T = -\frac{\kappa^c c}{\Omega}\left(\frac{dB}{dP}-1\right)^{-1}.$$

In view of this relation, eq. (14.30) is rewritten as

$$g(P) = g(P_0) + c(P_0)\{B\Omega - B(P_0)\Omega(P_0)\}$$
$$-\frac{1}{2}\frac{\kappa^c(P_0)c(P_0)}{\Omega(P_0)}\left\{\left.\frac{dB}{dP}\right|_{T,P=P_0}-1\right\}^{-1}$$
$$\times \{B\Omega - B(P_0)\Omega(P_0)\}^2 + \cdots. \quad (14.32)$$

The relative importance of the second term of this series in comparison to the first one can be judged from their ratio

$$\frac{\text{2nd term}}{\text{1st term}} = -\frac{1}{2}\frac{\kappa^c(P_0)}{\Omega(P_0)}\left\{\left.\frac{dB}{dP}\right|_{T,P=P_0}-1\right\}^{-1}\{B\Omega - B(P_0)\Omega(P_0)\}$$

$$= -\frac{1}{2}\frac{\kappa^c(P_0)}{\kappa(P_0)}\left\{\left.\frac{dB}{dP}\right|_{T,P=P_0}-1\right\}^{-1}\left\{\frac{B\Omega}{B(P_0)\Omega(P_0)}-1\right\}. \quad (14.33)$$

In order to make an estimation we are not going far wrong by using the Murnaghan equation of state according to which dB/dP does not vary with the pressure; the condition $dB/dP =$ constant finally gives

$$B/B(P_0) = \{\Omega(P_0)/\Omega\}^{dB/dP} \quad (14.33a)$$

and hence (with $\Delta\Omega < 0$):

$$\frac{B\Omega}{B(P_0)\Omega(P_0)} = \{\Omega(P_0)/\Omega\}^{[(dB/dP)-1]} = \{1-\Delta\Omega/\Omega\}^{[(dB/dP)-1]}$$
$$\simeq 1 - [(dB/dP)-1](\Delta\Omega/\Omega). \quad (14.34)$$

Due to eq. (14.34) eq. (14.33) gives

$$\frac{\text{2nd term}}{\text{1st term}} \simeq \frac{1}{2}\frac{\kappa^c(P_0)}{\kappa(P_0)}\frac{\Delta\Omega}{\Omega} \simeq -\tfrac{1}{2}\kappa^c(P_0)\Delta P.$$

In view of the fact that the ratio $|\kappa^c|/\kappa$ is of the order of unity (see §14.2.3) and furthermore that $|\Delta\Omega/\Omega| \ll 1$ we conclude that the second term of the expansion given in eq. (14.30) can be deleted and hence we get

$$g(P) - g(P_0) = c(P_0)\{B\Omega - B(P_0)\Omega(P_0)\} \quad (14.35)$$

which is just eq. (14.23). The integration of this equation immediately leads to eq. (14.28) apart from the constant Λ discussed in the previous paragraph.

Sometimes it is useful to know the order of magnitude of the difference $g(P) - g(P_0)$; inserting eq. (14.34) into eq. (14.35) we get

$$g(P) - g(P_0) \simeq -g(P_0)\left(\left.\frac{dB}{dP}\right|_T - 1\right)\frac{\Delta\Omega}{\Omega}. \tag{14.36}$$

14.3.6 Microscopic comments related to the $cB\Omega$-model

In this paragraph we shall show, from a microscopic point of view, that the energy of the defect formation at $T = 0$ K is actually proportional to $B\Omega$ without the existence of any additive constant. This will be carried out first for vacancy formation in rare gas solids by considering interactions up to 2nd neighbours. The method will be extended to alkali halides for the formation of a Schottky defect.

Basic relations. For reasons of brevity we shall label with V_z the volume per atom in rare gas solids (or the volume per "molecule" in alkali halides); the symbol z_i stands for the number of neighbours ($i = 1$ first neighbours, $i = 2$ second neighbours, etc.) whereas l_i denotes the distances between the atoms and a is the lattice constant. The symbol ϕ stands for the potential energy. For other details we refer to Ludwig [411].

The external pressure P and the bulk modulus are connected to the derivatives of the cohesive energy E_z through the relations

$$-3P = 3\frac{dE_z}{dV_z} = \frac{a}{V_z}\frac{dE_z}{da}, \quad B = \frac{a^2}{9V_z}\frac{d^2E_z}{da^2} - \frac{2a}{9V_z}\frac{dE_z}{da} \tag{14.37}$$

and hence, due to eq. (14.37) we find

$$B = \frac{a^2}{9V_z}\frac{d^2E_z}{da^2} + \tfrac{2}{3}P. \tag{14.38}$$

The cohesive energy E_z is found by $E_z = \tfrac{1}{2}\sum_i z_i\varphi(l_i)$ and hence:

$$\frac{dE_z}{da} = \frac{1}{2a}\sum_i z_i l_i \varphi'(l_i), \tag{14.39}$$

$$\frac{d^2E_z}{da^2} = \frac{1}{2a^2}\sum_i z_i l_i^2 \varphi''(l_i), \tag{14.40}$$

where the prime denotes derivatives with respect to distance.

Rare gas solids. * We adopt the usual potential ($R = l_i$):

$$\varphi(R) = 4\varepsilon\left\{\left(\frac{\sigma}{R}\right)^{12} - \left(\frac{\sigma}{R}\right)^{6}\right\}$$

and hence

$$\varphi'(R) = 4\varepsilon\left\{-12\frac{\sigma^{12}}{R^{13}} + 6\frac{\sigma^{6}}{R^{7}}\right\},$$

$$\varphi''(R) = 4\varepsilon\left\{156\frac{\sigma^{12}}{R^{14}} - 42\frac{\sigma^{6}}{R^{8}}\right\}.$$

Considering interactions up to 2nd neighbours we have the following sum (we denote with l the nn distance and hence $l_1 = l$, $l_2 = l\sqrt{2}$, $z_1 = 12$, $z_2 = 6$):

$$\sum_{i=1}^{2} l_i \varphi'(l_i) z_i = 24\varepsilon(z_1 + z_2/8)\left(\frac{\sigma}{l}\right)^{6}\left\{1 - \frac{z_1 + z_2/64}{z_1 + z_2/8}\left(\frac{\sigma}{l}\right)^{6}\right\}.$$

In a similar way we get

$$\sum_{i=1}^{2} z_i l_i^2 \varphi''(l_i) = 24\varepsilon\left(\frac{\sigma}{l}\right)^{6}\left\{26(z_1 + z_2/64)\left(\frac{\sigma}{l}\right)^{6} - 7(z_1 + z_2/8)\right\}.$$

The quantity

$$y \equiv \frac{z_1 + z_2/64}{z_1 + z_2/8} = \frac{129}{136} = 0.949$$

comes as a correction if one includes interactions not only up to 1st neighbours but also up to 2nd neighbours. Therefore the previous sums can be rewritten as

$$\sum_{i=1}^{2} l_i z_i \varphi'(l_i) = 24\varepsilon(z_1 + z_2/8)\left(\frac{\sigma}{l}\right)^{6}\left\{1 - 2y\left(\frac{\sigma}{l}\right)^{6}\right\}, \tag{14.41}$$

$$\sum_{i=1}^{2} z_i l_i^2 \varphi''(l_i) = 24\varepsilon(z_1 + z_2/8)\left(\frac{\sigma}{l}\right)^{6}\left\{26y\left(\frac{\sigma}{l}\right)^{6} - 7\right\}. \tag{14.42}$$

The combination of eqs. (14.37), (14.39) and (14.41) finally gives

$$\left(\frac{\sigma}{l}\right)^{6} = \frac{1}{2y} + \frac{V_z}{4\varepsilon(z_1 + z_2/8)} P, \tag{14.43}$$

whereas the combination of eqs. (14.38), (14.40) and (14.42) gives

$$B = \frac{4\varepsilon}{yV_z}(z_1 + z_2/8) + 7P. \tag{14.44}$$

---

* This proof and the next one have been kindly forwarded to the authors by Prof. W. Ludwig.

The above procedure is general and can be used for any solid described by a "two-parameter" potential (e.g. ε and σ); these parameters can then always be written as a function of B and V_z. Equation (14.44) gives

$$\varepsilon = \frac{y}{4(z_1 + z_2/8)} BV_z - \frac{7y}{4(z_1 + z_2/8)} PV_z$$

and hence

$$\varepsilon = \frac{0.893}{4z_1} BV_z - \frac{7 \times 0.893}{4z_1} PV_z. \tag{14.45}$$

This relation holds for every rare gas solid. Let us now proceed to the calculation of the formation energy h_0^f in the case of argon at $T = 0$ K. Using the Maradudin and the Flinn Green functions Ludwig [1] finds $h_0^f = 6.69\varepsilon$ and states that the numerical constant can only be slightly different for other rare gas solids. Inserting the value of ε from eq. (14.45) we get

$$h_0^f = 0.124 BV_z - 0.871 PV_z. \tag{14.46}$$

Considering that B is of the order of a few kilobars the second term is negligible ($P/B \ll 1$) so that eq. (14.46) gives (for $V_z \equiv \Omega$)

$$h_0^f = 0.124 B\Omega.$$

This result shows that in rare gas solids at $T = 0$ the energy g^f is directly proportional to $B\Omega$ and therefore we are justified in setting into eq. (14.24) the constant Λ equal to zero.

Alkali halides. Considering Born–Mayer repulsive interactions only between nearest neighbours we can immediately write

$$E_z = -\frac{Ze^2}{l} + zA \exp(-l/\rho), \tag{14.47}$$

$$\frac{dE_z}{dl} = -\frac{Ze^2}{l^2} - \frac{zA}{\rho} \exp(-l/\rho), \tag{14.48}$$

$$\frac{d^2 E_z}{dl^2} = -\frac{2Ze^2}{l^3} + \frac{zA}{\rho^2} \exp(-l/\rho), \tag{14.49}$$

where Z is the usual Madelung constant, l is the nn-distance, z is the number of nn neighbours and A and ρ the usual parameters of the Born–Mayer potential. Because of eq. (14.48) the equilibrium condition [i.e. eq. (14.37)] gives

$$-3P = \frac{l}{V_z} \frac{dE_z}{dl} \Rightarrow \frac{l}{\rho} zA \exp(-l/\rho) = \frac{Ze^2}{l} + 3PV_z, \tag{14.50}$$

whereas due to eqs. (14.50) and (14.49) eq. (14.38) also gives

$$BV_z - \frac{l}{\rho}PV_z = \frac{Ze^2}{l}\frac{l}{9\rho}\left\{1 - \frac{2\rho}{l}\right\}. \tag{14.51}$$

As is known, the system of eqs. (14.50) and (14.51) permits the determination of the parameters ρ and A as a function of B and V_z (in alkali halides with NaCl-structure we have $Z = 1.7476$, $V_z = 2l^3 = 2\Omega$).

Consider now, for the sake of simplification, the case of small pressures ($P/B \ll 1$). Then the cohesive energy given by eq. (14.47) becomes due to eq. (14.50)

$$E_z = -\frac{Ze^2}{l}\left(1 - \frac{\rho}{l}\right). \tag{14.52}$$

The quantity e^2/l is given as a function of B, from eq. (14.51):

$$Z\frac{e^2}{l} = 18\left(\frac{l}{\rho} - 2\right)^{-1} B\Omega \tag{14.53}$$

and hence eq. (14.52) reads

$$E_z = -18\left(1 - \frac{\rho}{l}\right)\left(\frac{l}{\rho} - 2\right)^{-1} B\Omega. \tag{14.54}$$

Due to the Born–Mayer repulsive interaction the potential energy φ_{+-} can also be written in terms of $B\Omega$ by introducing eq. (14.53) into eq. (14.50):

$$z\varphi_{+-} \equiv zA \exp\left(-\frac{l}{\rho}\right) = \frac{18\rho}{l}\left(\frac{l}{\rho} - 2\right)^{-1} B\Omega. \tag{14.55}$$

Equations (14.53) to (14.55) express the Madelung energy, the Born–Mayer repulsive contribution and the cohesive energy in terms of $B\Omega$. With the help of the above procedure the corresponding expressions are easy to derive by including also the Born–Mayer interactions ($\varphi_{++}, \varphi_{--}$) of second neighbours.

We now wish to find the expression of the Schottky formation energy $h_0^f \equiv g^f$ at $T = 0$ K in alkali halides. All the recent microscopic calculations are refinements of the classical Mott–Littleton [453] method. In this frame the energy h_0^f results finally in the form (for NaCl-structure):

$$h_0^f = 3.4952\frac{e^2}{l} - 12\left[\varphi_{+-}(l) + \varphi_{++}(\sqrt{2}\,l) + \varphi_{--}(\sqrt{2}\,l)\right]$$

$$-\frac{e^2}{l}\frac{1}{4\pi}\left(1 - \frac{1}{\varepsilon}\right) \times 16.5323. \tag{14.56}$$

The coefficients are taken from the recent work of Tharmalingam [454]. *
The symbol l again denotes the nn (equilibrium) distance.

Inserting the values of e^2/l and φ_{+-} given by eqs. (14.53) and (14.55) into eq. (14.56) and disregarding – to a rough approximation – the terms φ_{++} and φ_{--} one immediately gets the desired result:

$$h_0^f = f\left(\frac{l}{\rho}, \frac{1}{\varepsilon}\right) B\Omega. \tag{14.57}$$

Of course, as noticed above, a more accurate procedure is to write eqs. (14.47) and (14.55) again, including "second neighbour Born–Mayer repulsive terms", and then inserting them into eq. (14.56); this is omitted here because it is simply numerical work not changing the basic conclusion. Equation (14.57) clearly shows that the Gibbs energy g^f at $T=0$ is proportional to $B\Omega$.

14.3.7 Justification of the $cB\Omega$-formula from the Jost-model

In this paragraph we shall deal with the macroscopic model of Jost. We have done so in order to show that a proper use of Jost's model actually leads to the conclusion that g scales with $B\Omega$. In the following we shall employ the ordinary suggestion of Jost's model which might have to be extended because of the following reasons [35]:

(1) The cohesive energy employed must be corrected for vibrational energy (recall that the Jost-model is a static one).

(2) The Jost-model treats cells of the lattice as adjacent regions of a uniformly polarizable continuum which is polarized by electric fields diverging from an empty cell whose ion is removed thereby, forming a vacancy which is effectively charged. The polarization appearing in the lattice is presumed spherical and extending over a radius R and hence the relaxation energy released by this polarization is given by (if the charge-misfit is equal to one) $(e^2/2R)(1-\varepsilon^{-1})$.

However, the assumption of continuum dielectric behaviour (which is anyhow correct for slowly varying fields far from the defect) may fail to describe the polarization in the discrete lattice structure close to the vacancy cells; it is then better to ascribe a modified dielectric constant different from ε to the volume occupied by the shell (or shells) of ions closest to a vacancy.

(3) In the case of Schottky pair formation we have two (poorly defined) radii R_+ and R_- identifying anion $(-)$ and cation $(+)$ vacancy radii;

---

* A first check showed that the numerical values of Tharmalingam should be slightly different; however, it does not influence our conclusions here.

however, in this case we do not go too far by writing

$$\tfrac{1}{2}(R_+^{-1} + R_-^{-1}) = \overline{R}^{-1}$$

with $\overline{R}$ the atomic cell radius of the lattice.

As already mentioned, we shall employ in spite of the above remarks the ordinary suggestion of Jost without further corrections because, as will become clear, we are interested only in qualitative results.

Within this frame the formation Gibbs energy in ionic crystals is usually written as [35]

$$g = |E_z| - \frac{e^2}{2R}\left(1 - \frac{1}{\varepsilon}\right)$$

where R is the effective mean "radius" of the cavities (which are of the order of half the interatomic distance l); writing $R \equiv \xi l$ the previous equation reads

$$g = |E_z| - \frac{e^2}{l}\frac{1}{2\xi}\left(1 - \frac{1}{\varepsilon}\right)$$

or by inserting the values of E_z and e^2/l from eqs. (14.52) and (14.53) we get

$$g = 18\left\{\left(1 - \frac{\rho}{l}\right) - \frac{1}{2\xi Z}\left(1 - \frac{1}{\varepsilon}\right)\right\}\left(\frac{l}{\rho} - 2\right)^{-1} B\Omega. \tag{14.58}$$

This expression shows that g is directly proportional to $B\Omega$ (for its coefficient see below) without any additive constant. Therefore in ionic crystals the deletion of the constant Λ in the expression $g = cB\Omega + \Lambda$ is justified. There are also many useful points emerging from eq. (14.58).

(1) The coefficient in front of $B\Omega$ is practically pressure and temperature independent. (a) The term ρ/l varies only slightly with temperature and pressure but this variation reflects only an insignificant change to the whole coefficient; this is so because with the application of pressure (or with decreasing temperature) both terms, $(1 - \rho/l)$ and $[(l/\rho) - 2]$, show a slight increase so that their ratio remains practically constant. (b) The static dielectric constant decreases on compression (by 1% per kbar) and increases with temperature. Therefore, upon application of pressure (or with decreasing temperature) both terms $(1 - 1/\varepsilon)$ and $[(l/\rho) - 2]$ show a slight decrease so that their ratio remains practically constant. *

(2) Let us now proceed to a numerical example in the characteristic case of NaCl. Inserting the (low temperature) values $l = 2.8$ Å and $\rho = 0.32$ Å

---

* A more consistent discussion should also take into account the temperature and pressure variation of ξ which is a measure of the relaxation volume.

$Z = 1.7476$, $\varepsilon = 5.87$, $B = 270$ kbar and $g(\simeq h_0) \simeq 2.4$ eV, we estimate that the appropriate value of ξ is 0.369; this value shows that R actually lies close to the half of the nn distance. Inserting these values into (14.58) we get

$$g \simeq h_0 = \left(0.294 + 2.068 \frac{1}{\varepsilon}\right) B\Omega. \tag{14.59}$$

Note that the two terms between parentheses are comparable. Let us now attempt, in spite of the shortcomings mentioned, the calculation for temperatures close to the melting point by roughly assuming the same values of ρ and ξ to be appropriate for high temperatures as well. Then by increasing the value of l by 3% one finds the first number to decrease only by 0.8%, whereas the other (i.e. 2.068) decreases by $\sim 4\%$. In view of the fact that the dielectric constant increases by $\sim 17\%$ up to the melting point we conclude that the term between parentheses decreases by a total amount close to 10% * whereas the term $B\Omega$ – for the same temperature range – is reduced by $\sim 40\%$. Similar results are drawn from the study of the influence of pressure on the various terms of eq. (14.59). Recalling that ε decreases by $\sim 1\%$ per kbar whereas B increases by $\sim 2.3\%$ per kbar one immediately concludes that the pressure variation of the quantity $B\Omega$ is by a factor 4 to 5 greater than that of the whole term between parentheses.

The general conclusion is now clear: the temperature and the pressure variation of g – as they result from an alternative use of the Jost-model – is governed mainly by the corresponding variation of the quantity $B\Omega$ and *not* by that of the dielectric constant. This conclusion is important because numerous publications in the last three decades have stated that

$$g = \text{const} \times \frac{1}{\varepsilon}.$$

It is true that in some cases this expression has led to a good qualitative (but not quantitative) description of the experimental data; this is not fortuitous and is due to the fact that the quantity $1/\varepsilon$ varies to a certain degree in a similar fashion with $B\Omega$, as one can see from the study of the classical Lorentz–Lorenz equation when the ionic polarizability is expressed in terms of the elastic constants within the frame of the central forces approximation [419,420]

$$\frac{\varepsilon - 1}{\varepsilon + 2} = \frac{A}{\Omega} + \frac{A'}{B\Omega l}$$

where A and A' are almost constants. The above aspects thus show that g is actually proportional to $B\Omega$ and not to $1/\varepsilon$.

---

* This amount may be smaller if one uses the appropriate values of ρ and ξ for high temperatures.

14.3.8 Justification of the $cB\Omega$-model for the migration process in the framework of dynamical theory

The dynamical theory suggests that the quantities v^m and g^m are connected through the relation [35]:

$$v^m/g^m = 2\gamma_i/B \tag{14.60}$$

where γ_i is thought to be the Grüneisen constant of the mode that is mainly responsible for the migration process. A more detailed investigation [35] shows that γ_i is a "kind of average" referring to the frequency spectrum. We are probably not going far by assuming, in monoatomic crystals, that γ_i is approximatelly equal to the mean Grüneisen constant $\bar{\gamma}_D$ in the Debye approximation. In this approximation Slater [328] has found that

$$\bar{\gamma}_D = -\frac{1}{6} + \frac{1}{2}\left.\frac{dB}{dP}\right|_T$$

whereas later this relation has been corrected to [329,330]

$$\bar{\gamma}_D = -\frac{1}{2} + \frac{1}{2}\left.\frac{dB}{dP}\right|_T. \tag{14.61}$$

Combining eqs. (14.60) and (14.61) one gets

$$g^m = \frac{Bv^m}{\left.\frac{dB}{dP}\right|_T - 1} \tag{14.62}$$

without an additive constant. This equation is analogous to eq. (14.26) or (14.28) obtained for a formation process.

Introducing into eq. (14.62) the definition of the migration volume, i.e. $v^m = (dg^m/dP)_T$, one finally gets with the help of eq. (14.11) $dg/g = d(B\Omega)/B\Omega$ (with $T = $ constant) and upon integration [666]

$$g^m = c^m(T)B\Omega \tag{14.63}$$

where $c^m(T) \equiv (v^m/\Omega)/[(dB/dP)_T - 1]$ is a constant which *has to be* pressure independent. Similarly eq. (14.63) can also be proved for ionic crystals if one uses – instead of eq. (14.61) – the relation [see eq. (13.34)]

$$\gamma_{TO} = \frac{1}{2}\left.\frac{dB}{dP}\right|_T - \frac{1}{6} \simeq \frac{1}{2}\left.\frac{dB}{dP}\right|_T - \frac{1}{2}$$

and further approximates γ_i with γ_{TO}. The latter approximation is justified from the fact that in ionic crystals the long-wavelength TO-modes seem to be mainly responsible for the migration process.

14.4 On the temperature dependence of c

The former paragraph showed that only small, if any, errors result when c is considered as pressure (or volume) independent. There still remains the question about its temperature dependence. The answer is treated in this paragraph.

In cases of linear or slightly (upward) curved Arrhenius plots (with one operating mechanism) it is sure that the temperature dependence of c^{act} is very small. This is expected from [667] eq. (7.49):

$$\frac{\mathrm{d}h^{\text{act}}}{\mathrm{d}T}\bigg|_P = \frac{Tv^{\text{act}}\beta^{0\,2}}{\kappa^0}\left(2\frac{\beta^{\text{act}}}{\beta^0} - \frac{\kappa^{\text{act}}}{\kappa^0}\right) + \text{a small term.}$$

By following the considerations of Nowick and Dienes [76] or those of Gilder and Lazarus [37] we can say that for almost linear Arrhenius plots the quantity between parentheses must be of the order of unity. In view of the fact that we have already shown that κ^{act} exceeds κ^0 only a few times (see §14.2.3) one immediately sees that β^{act} may not exceed β^0 by more than a few times. Therefore, although both v^{act} and Ω increase with temperature their ratio increases only slightly. As $\mathrm{d}B/\mathrm{d}P$ increases also slightly with temperature, one can argue that the quantity $c^{\text{act}} = (v/\Omega)[(\mathrm{d}B/\mathrm{d}P) - 1]^{-1}$ is practically temperature independent.

The above aspect should be considered a speculation only; undertaking a number of steps, a more detailed justification will follow. In §14.4.1 we shall show c from general aspects to be *exactly* temperature independent in the quasi-harmonic approximation. In real solids, however, it may depend slightly on temperature. In view of this possibility we shall proceed in §14.4.2 to the determination of the bounds of this possible temperature dependence; they are found from the requirement that the "constant-volume defect entropy" is expected to be zero or negative and furthermore, that its absolute value must be smaller or equal to that of the constant-pressure defect entropy. Then, in §14.4.3, we shall discuss quantitatively the extent to which this possible temperature dependence of c influences the prediction of the value of the entropy when calculated from the $cB\Omega$-formula. We will find that the bounds of the possible temperature dependence of c are narrow enough to achieve always a reliable determination of the values of the entropy if one considers the current experimental accuracy. Finally, in §14.4.4 an alternative manner of discussing the possible temperature dependence of c will be given.

14.4.1 The temperature dependence of c

In §14.3 it was proved that c can be considered as pressure independent, provided that it is examined in the region $\Delta P/B < 0.1$. Here we will study the extent to which c can be considered to be temperature independent as well.

We first show that the pressure developed upon heating a solid under constant volume from absolute zero (initial state i) up to the melting point (final state f) is appreciably smaller than its bulk modulus. This is evident from the thermodynamical relation (2.27)

$$\Delta P = \int_i^f dP = \int_i^f \left.\frac{dP}{dT}\right|_V dT$$
$$= \int_{T=0,V}^{T=T_M,V} \beta(V,T) B(V,T) dT \simeq \bar{B}(V) \int_{T=0,V}^{T=T_M,V} \beta \, dT,$$

if one further considers that, even at ambient constant pressure, the term $\int_{T=0}^{T=T_m} \beta dT$ is of the order of 10^{-1} and that the bulk modulus varies very little when the volume of the solid remains constant, even upon increasing temperature. We therefore conclude that $\Delta P/\bar{B} \lesssim 0.1$.

Recalling the definition of s^*,

$$s^* \equiv -\left.\frac{\partial g}{\partial T}\right|_V = -\left.\frac{\partial g}{\partial (B\Omega)}\right|_V \left.\frac{\partial (B\Omega)}{\partial T}\right|_V$$

one sees that for isochoric (V = constant and hence Ω = constant) heating of a solid the following thermodynamical identity holds:

$$\left.\frac{dg}{d(B\Omega)}\right|_V = -\frac{s^*}{\Omega \left.\frac{dB}{dT}\right|_V}. \tag{14.64}$$

This identity holds for all processes: formation, migration, and self-diffusion activation.

In a quasi-harmonic approximation (QA) the quantity s^* is given by eq. (3.156):

$$s^* = -k \sum_i \ln \frac{\omega_i'(V)}{\omega_i(V)};$$

it is temperature independent, when $T > T_E$ (T_E is the Einstein temperature), because the frequencies do not change when the "macroscopic volume" remains constant [668]. Furthermore, in QA the quantity $(dB/dT)_V$ is also a constant. Therefore within the frame of QA the right-hand side of eq. (14.64) is *exactly* temperature independent if the volume is kept constant.

In real solids however, the frequencies show a very small explicit temperature dependence, so that both the quantities s^* and $(dB/dT)_V$ may show a slight temperature dependence. We are therefore led to the conclusion that, to a good approximation, their ratio might be temperature independent; under such an approximation we can write

$$-\frac{s^*}{\Omega \left.\dfrac{dB}{dT}\right|_V} \equiv c_1 = \text{constant}. \tag{14.65}$$

Of course the quantity c_1 may depend on the volume. Therefore eq. (14.64) can be written as

$$\left.\frac{dg}{d(B\Omega)}\right|_V = c_1(V) = \text{temperature independent}. \tag{14.66}$$

But we have already shown that within the constraint of eq. (14.17) the (formation or activation) Gibbs energy g takes the form

$$g(V, T) = c(T) B\Omega. \tag{14.67}$$

Obviously, the combination of eqs. (14.66) and (14.67) leads to the conclusion that in the formula $g = cB\Omega$ "c" is pressure (volume) and temperature independent ($c = c_1$).

Summarising, we remember that the justification of the temperature independence of the quantity c for formation and activation processes is based on the validity of the assumption (14.65). This assumption is *absolutely correct* in QA and there are also good arguments that it also holds in a real solid. In the latter case, however, the interesting question remains whether the small *explicit* temperature variations of s^* and $(dB/dT)_V$ (both of which are exclusively due to the slight *explicit* temperature variation of the frequencies) are in the same direction, so that their ratio remains exactly temperature independent. In view of this open question it is worthwhile to determine the bounds of this possible temperature dependence of c.

14.4.2 Bounds for the possible temperature variation of c

The temperature variation of c can best be discussed with its "thermal expansion coefficient" defined by

$$\beta^c \equiv \frac{1}{c} \left.\frac{dc}{dT}\right|_P. \tag{14.68}$$

Introducing the definition of c we have

$$\frac{\beta^c}{\beta} = \frac{\beta^d}{\beta} - 1 - \frac{1}{\beta} \frac{\dfrac{d}{dT}\dfrac{dB}{dP}}{\left.\dfrac{dB}{dP}\right|_T - 1}, \qquad (14.69)$$

where β^d denotes the expansion coefficient of the volumes v^f, v^{act} or v^m, depending on the mechanism to which c refers.

By differentiating eq. (14.67) with respect to temperature we get

$$-s^* = \left.\frac{\partial g}{\partial T}\right|_V = c(T)\Omega \left.\frac{\partial B}{\partial T}\right|_V + B\Omega \left.\frac{dc}{dT}\right|_V,$$

or by considering the definition of β^c [see eq. (14.68)]

$$s^* = -c\Omega \left.\frac{\partial B}{\partial T}\right|_V - \beta^c cB\Omega \qquad (14.70)$$

or

$$-\frac{s^*}{g} = \frac{1}{B}\left.\frac{\partial B}{\partial T}\right|_V + \beta^c. \qquad (14.71)$$

The lower limit of β^c comes from the knowledge that in usual cases s^* is zero or negative:

$$\beta^c \geq -\frac{1}{B}\left.\frac{dB}{dT}\right|_V,$$

or by considering eq. (2.34) we have

$$\beta^c/\beta \geq \delta - \left.\frac{dB}{dP}\right|_T, \qquad (14.72)$$

where δ is the usual Anderson–Grüneisen constant defined in eq. (8.10). The upper limit can be found by differentiating eq. (14.67) with respect to temperature:

$$s = -\left.\frac{dg}{dT}\right|_P = -c\Omega\left(\beta B + \left.\frac{dB}{dT}\right|_P\right) - \beta^c cB\Omega. \qquad (14.73)$$

If ξ denotes the ratio of the absolute values of the two entropies s and s^*, i.e.

$$\xi \equiv |s|/|s^*|, \qquad (14.74)$$

and one further considers that in usual cases $s > 0$ and $s^* < 0$, the introduction of their values from eqs. (14.70) and (14.73) into eq. (14.74) finally gives

$$\frac{\beta^c}{\beta} = \delta - 1 - \xi(1+\xi)^{-1}\left(\left.\frac{dB}{dP}\right|_T - 1\right).$$

Setting the limiting value $\xi = 1$ (see §3.4) we get the upper limit of β^c/β:

$$\frac{\beta^c}{\beta} \lesssim \delta - \frac{1}{2}\left(\left.\frac{dB}{dP}\right|_T + 1\right). \tag{14.75}$$

When elastic and expansivity data are available the quantities δ and dB/dP are directly known and therefore one can find the limits of β^c from eq. (14.72) and (14.75); a rough estimation can be obtained by using the relation of Chang [455]: $\delta \simeq (dB/dP) - 1$ (or $\delta \simeq 2\gamma$) which, however, has been criticised by Barron [456], mainly because it disregards the difference between the adiabatic (δ_S) and isothermal (δ) Anderson–Grüneisen parameter.

An important point should be noted. We have found the possible temperature dependence of c (i.e. β^c) by demanding that in the usual cases the value of s^*, in contrast to s, is zero or negative and that $|s^*| \lesssim |s|$. (Of course eq. (14.72) and (14.75) should change in cases when s and v are negative). These requirements pertain to the normal cases of formation processes ($s^f > 0$, $v^f > 0$) but it is not absolutely sure that they are valid for migration processes as well. This is so because, although it is rather sure from the relation $s^m - s^{m*} = v^m \beta B$ that $|s^{m*}|$ is appreciably smaller than s^m (when $\beta v^m > 0$), it is very difficult to theoretically predict with confidence the sign of s^{m*}. However, a first inspection of the known relation $s^m - s^{m*} = 3\gamma k v^m/\Omega$ shows that in the case of metals ($\gamma \simeq 2.5$, $v^m/\Omega \simeq 0.15$) the difference $s^m - s^{m*}$ should be around $1\ k$ whereas in alkali halides ($\gamma \simeq 1.8$, $v^m/\Omega \simeq 0.5$) we get $s^m - s^{m*} \simeq 2.7\ k$. In view of the fact that s^m in metals is $1-2\ k$ and in alkali halides $2-3\ k$ we actually conclude that s^{m*} is likely to be either practically zero or slightly negative.

14.4.3 Bounds of the formation entropy

Equation (14.73) calculates the entropy without disregarding any temperature variation of c; the exact determination of s requires the knowledge of β^c, which is usually unknown. However, the bounds of β^c are already known from eqs. (14.72) and (14.75) so that both the upper and lower limit of s can safely be found. Due to eq. (14.67), eq. (14.73) can equivalently be written as

$$-\frac{s}{g} = \beta + \frac{1}{B}\left.\frac{dB}{dT}\right|_P + \beta^c \tag{14.76}$$

or by considering the bounds given by eqs. (14.72) and (14.75)

$$\tfrac{1}{2}\beta\left(\left.\frac{dB}{dP}\right|_T - 1\right) \lesssim \frac{s}{g} \lesssim \beta\left(\left.\frac{dB}{dP}\right|_T - 1\right). \tag{14.77}$$

The latter inequality is useful in the computational procedure for the analysis of experimental data in which the quantities s and h are treated as independent parameters. Inequality (14.77) is a *constraint* between them because the quantities β and $(dB/dP)_T$ are well-known. For instance in a "Simmons–Balluffi" experiment the experimental errors in the determination of n/N (and hence of g^f) do not allow an unambiguous determination of a single pair of parameters (s, h); in such cases various plausible pairs (but leading, practically, to the same values of g) are reported. However, the knowledge of the bounds of s/g may help towards a better selection of a single (s, h)-pair.

Another use of eq. (14.77) may lie in the analysis of isobaric specific heat data: for reasons explained in §7.2 the usual ways of analysing C_P-data, although leading to a good estimate of the formation enthalpy h^f, give unacceptably high values of s^f; the knowledge of the bounds of s^f may lead to an estimation of the anharmonic contribution to C_P-values in the high temperature region when these bounds are combined to a reasonable h^f-value. The latter is of interest from a microscopic point of view.

We finally report an application of eq. (14.77) in the analysis of conductivity and diffusion data in alkali halides. This analysis is highly complicated due to the great number of parameters involved and therefore the knowledge of a constraint between s^f and h^f is of primary importance. We take two characteristic examples.

Jacobs and Pantelis' [271] analysis of conductivity data of KCl give $s^f = 4.2\,k$ and $h^f = 2.3$ eV for the formation process of a Schottky defect. These parameters lead to $s^f/g^f \simeq 1.7 \times 10^{-4}$ K^{-1} for 300 K. However, by using the values $\beta = 1.11 \times 10^{-4}$ K^{-1} and $(dB/dP)_T = 5.46$ one gets $\frac{1}{2}\beta[(dB/dP) - 1] \simeq 2.5 \times 10^{-4}$ K^{-1} which shows that the left inequality in eq. (14.77) is violated. The same violation occurs if one examines the parameters $h = 2.02$ eV and $s = 4.3\,k$ found by Brown and Jacobs [457] in their analysis of KBr by adopting Debye–Hückel corrections. In this case one finds, for 300 K, $s^f/g^f = 1.94 \times 10^{-4}$ K^{-1}, whereas by using $dB/dP = 5.47$ and $\beta = 1.16 \times 10^{-4}$ K^{-1} the value of $\frac{1}{2}\beta[(dB/dP) - 1]$ is equal to 2.59×10^{-4} K^{-1}.

At this point it is useful to recapitulate the procedure which led to eq. (14.77): starting from the general relation $g = c(T)B\Omega$ the right-hand side of eq. (14.77) comes from the requirement that s^* has to be negative or zero whereas the left-hand side results from the fact that $|s^*| \leq s$. However, if one does not wish to rely on a $cB\Omega$-expression one could use – instead of eq. (14.77) – the relation:

$$\tfrac{1}{2}v\beta B \leq s \leq v\beta B, \qquad (14.78)$$

which clearly is a *thermodynamical* result provided that one combines eq. (3.126) with the relations $s^* \leq 0$ and $|s^*| \leq s$.

Consequences of the calculation of entropy by setting c as temperature independent. Allowing c to depend on temperature we have seen that s is given by eq. (14.76) which also reads: $s = g\beta(\delta - 1) - g\beta^c$. Let us now estimate the relative importance of the last term $g\beta^c$, which contains the temperature dependence of c, with respect to the other term; their ratio r reads $r = (\delta - 1)^{-1}\beta^c/\beta$.

Recalling the bounds of β^c/β given by eqs. (14.72) and (14.75) and setting $\delta \sim (\mathrm{d}B/\mathrm{d}P)_T$ one finds that $-50\% < r \lesssim 0\%$. The latter results show that the statement that c is temperature independent – i.e. the use of the relation

$$s = g\beta(\delta - 1), \tag{14.79}$$

which is just eq. (8.11) of ch. 8 – might overestimate the entropy by an amount which is in any case smaller than 50%. However, in view of the uncertainty of the current experimental entropies (which is usually 50 to 100%), we are justified to rely on eq. (14.79) at least as a first approximation.

14.4.4 *Alternative discussion of the possible temperature dependence of c*

We now discuss a possible variation of c with temperature along the same lines as in §14.3.5 for a pressure-dependent c. Writing

$$c_1 \equiv -\frac{s^*}{\Omega \left.\dfrac{\mathrm{d}B}{\mathrm{d}T}\right|_V} \tag{14.80}$$

and expanding $g(V, T)$ in a Taylor series around the initial (i) conditions (V, T_i) we have (see ref. [669] for a clarification)

$$g(V, T_i) = g(V, T_i) + \left.\frac{\mathrm{d}g}{\mathrm{d}(B\Omega)}\right|_{V, T = T_i} (B\Omega - B_i\Omega)$$

$$+ \frac{1}{2} \left.\frac{\mathrm{d}^2 g}{\mathrm{d}(B\Omega)^2}\right|_{V, T = T_i} (B\Omega - B_i\Omega)^2 + \cdots. \tag{14.81}$$

Using now the identity (14.64) we immediately find that the ratio of the term of second order over that of the first order is:

$$\Gamma \equiv \frac{\text{2nd}}{\text{1st}} \approx \tfrac{1}{2}\beta^{c_1}(T - T_i) \tag{14.82}$$

where β^{c_1} is the temperature expansion coefficient of c_1 defined by

$$\beta^{c_1} = \frac{1}{c_1}\frac{\mathrm{d}c_1}{\mathrm{d}T};$$

inserting eq. (14.80) we get

$$\beta^{c_1} \equiv \frac{1}{s^*} \left.\frac{ds^*}{dT}\right|_V - \left.\frac{d^2 B}{dT^2}\right|_V \bigg/ \left.\frac{dB}{dT}\right|_V. \qquad (14.83)$$

In the case of QA (see §14.4.1) the quantities s^* and $(dB/dT)_V$ do not depend on T so that $\beta^{c_1} = 0$ and hence $c_1 \neq f(T)$. In cases of real solids, however, inserting eq. (14.83) and recalling that

$$c_V^* = T \left.\frac{ds^*}{dT}\right|_V$$

we have (see eq. 14.82)

$$\Gamma = \tfrac{1}{2}(T - T_i) \left\{ \frac{1}{s^*} \frac{c_V^*}{T} - \left.\frac{d^2 B}{dT^2}\right|_V \bigg/ \left.\frac{dB}{dT}\right|_V \right\}. \qquad (14.84)$$

If this result is sufficiently small eq. (14.81) suggests that $g(V, T)$ must be linear with $B\Omega$ with a proportionality factor c_1 independent of temperature, i.e. $c_1 \neq c_1(T)$. Up to this point we have used only general thermodynamical aspects but a similar problem remains as in §14.3.5: "to determine within which temperature intervals $\Delta T \equiv T - T_i$ we are allowed to use a temperature independent c_1 in order to predict $g(V, T)$ from $g(V, T_i)$ without significant errors". This, of course, depends on the desired accuracy of the value $g(V, T)$ which is to be predicted. Considering the experimental errors we do not require an accuracy of the "predicted $g(V, T)$-value" better than 1%; this is so because B already has an experimental error of $\sim 1\%$. Now the message is clearly: once we have determined an interval ΔT for which the quantity Γ is smaller than 1% thermodynamics alone allow the statement that $g(V, T)$ is linear in respect to $B\Omega$ with a coefficient c_1 independent of temperature. The error introduced by writing the above equation is known from eq. (14.84). The determination of the range ΔT then faces the same difficulties as that of ΔP in a previous paragraph. We either have to rely upon microscopic calculations or to trust our experience on the trend of the experimental data. Both theoretical calculations and experimental data suggest that the term $(d^2 B/dT^2)_V/(dB/dT)_V$ in eq. (14.84) is negligible. We therefore turn to the remaining term, i.e.

$$\frac{1}{2s^*} \frac{c_V^*}{T} (T - T_i).$$

In the case of formation and self-diffusion activation (i.e. migration plus formation) processes the quantity s^* (see §3.4) is negative and its absolute value comparable to but smaller than s and hence it is of the order of k. The absolute value of c_V^* is usually of the order of a few tenths of k (see §6.4.3 for a clarification). We therefore conclude that c_V^*/s^* is of the order of 10^{-1} and, even if we are working in a large temperature range, i.e.: up to T_M, the term Γ [(see eq. (14.84)] is of the order of at most 5%. Of course we can always select a shorter temperature interval and in this way reduce the error to the desired extent. From a purely mathematical point of view we have seen that the Taylor series rapidly converges so that we can always select a temperature interval ΔT in which g is linear in $B\Omega$, with a temperature independent coefficient c_1 within a given accuracy of g.

We therefore have arrived at a similar conclusion as in the previous paragraph: the suggestion that

"g is proportional to $B\Omega$ when V is constant" (14.85)

is easily accessible from thermodynamical considerations, and the only remaining point is the temperature region ΔT for which we are sure that the proportionality factor can be considered temperature-independent: this, of course, depends on the desired accuracy, e.g. $a\%$, in the prediction of the $g(V, T)$-value from $g(V, T_i)$. Such a value of ΔT can always be chosen so that $\Gamma < a\%$; we then can write [669]

$$g(V, T) = c_1(V) B\Omega \qquad (14.86)$$

or by introducing the definition of c_1

$$g = -\left(\left.\frac{dB}{dT}\right|_V\right)^{-1} Bs^*. \qquad (14.87)$$

The relations (14.29) and (14.86) are compatible with

$$g = cB\Omega,$$

where now the coefficient c does not depend on V and T.

14.5 Survey of the proof of the $cB\Omega$-formula

The route for the justification of the $cB\Omega$-formula faced two different problems.

1st problem: whether we are justified in writing that the function $g(P, T)$ or [669] $g(V, T)$ is directly proportional * to $B\Omega$ where the proportionality factor is pressure (or volume) independent but may depend on temperature. This problem was discussed in §14.3.

2nd problem: whether we can consider the function $g(V, T)$ to be directly proportional to $B\Omega$ where now the proportionality factor is temperature independent. This problem was discussed in §14.4.

---

* In the following we use the term "proportional" instead of "linear" because as we have seen the integration constant is likely to be zero in all cases.

In the present section – after combining (in §§14.5.1 and 14.5.2) summaries of the above two problems with some comments – we shall proceed in §14.5.3 to a discussion of the physical connection between the conclusions drawn in §§14.3 and 14.4.

14.5.1 Summary and comments on the proof of the relation $g(V, T) = c(T)B\Omega$

In §14.3 it was definitely proved that *for* $T = constant$ $g(P, T)$ is proportional to $B\Omega$ with the proportionality factor $c \equiv [(dB/dP) - 1]^{-1} \times (v/\Omega)$. This is an immediate consequence of the thermodynamical identity $(dg/d(B\Omega))_T = c$ and of the fact that expanding g with respect to $B\Omega$ the ratio of the 2nd order term to the first one is $-\frac{1}{2}\kappa^c \Delta P$ (when $\exp\int_{P_0}^{P}\kappa^c dP < 1.1$) where κ^c is the "compressibility" of the quantity c. In other words, pure thermodynamics demand that: "In a pressure range (P_1, P_2) for which $\int_{P_1}^{P_2}\kappa^c dP < 0.1$ we can always set $g(P, T)$ proportional to $B\Omega$ with a pressure independent proportionality factor $c(P_1)$; the relative error of the predicted g-value at a pressure P *is known* and is given by $-\frac{1}{2}\kappa^c(P - P_1)$".

The only point that remains to be discussed is the determination of the value of κ^c; this is desirable in order to know within which pressure range we are allowed to use the statement that c is pressure independent when we are satisfied to determine g with a known accuracy. Simple physical expectations suggest that $|\kappa^c|$ must be at most of the order of κ, because κ^c – by definition the measure of the reduced pressure variation of the quantity $(v/\Omega)[(dB/dP) - 1]^{-1}$ – takes values that depend explicitly on κ and because the second pressure derivatives of the quantities g and B are expected to be quite small. However, due to the fact that thermodynamics alone cannot give any definite value for κ^c, we have to revert either to reliable microscopic calculations or to our experience from some well-known experimental facts. We have intentionally followed both ways in order to assure our conclusions.

According to eq. (14.7), κ^c/κ is a sum of three terms. The term $-A$ is defined by eq. (14.4) and varies within narrow bounds (1 to 4). This has been not only theoretically predicted but also experimentally established in §14.2.2 for various categories of solids. There remains therefore the estimation of the term $(\kappa^d/\kappa) - 1$. As κ^d refers to formation, migration and self-diffusion activation we have to distinguish between three cases:

(1) In the case of formation processes we have the detailed microscopic calculations (verified by experiments) of Ludwig [446], Pistorius [447], and Dederichs et al. [444,445], which by independent methods showed that $(\kappa^f/\kappa) - 1$ exceeds unity only by a few times. Therefore the quantity $|\kappa^c/\kappa| \equiv |(\kappa^f/\kappa) - 1 + A|$ must be of the order of unity.

(2) In the case of an activation process there are no direct microscopic calculations for κ^{act} but only estimates which suggest that $(\kappa^{act}/\kappa) - 1$ must be comparable to $(\kappa^f/\kappa) - 1$. This is also obvious from the fact (see §14.2.1) that even if κ^m exceeds κ by one order of magnitude the value of $(\kappa^{act}/\kappa) - 1$ has to be around 4. In order to ascertain this we have also examined the experimental diffusion or conductivity isothermal plots under pressure and have again found that $(\kappa^{act}/\kappa) - 1$ cannot exceed 4. We have therefore concluded that the quantity $|\kappa^c/\kappa| = |(\kappa^{act}/\kappa) - 1 + A|$ must be again of the order of unity.

(3) In the case of a migration process there is a complete lack of microscopic calculations for the value of κ^m. This is exactly the reason why we have presented in §14.3.8 a justification of the expression $g^m = c^m B\Omega$ in a quite different manner, i.e. by means of the dynamical theory of diffusion. Although it cannot be precluded, it seems unphysical that κ^m might exceed κ by more than one order of magnitude. We have therefore concluded that although no sure upper limit for κ^m/κ exists, according to eq. (14.7) the quantity κ^c/κ should probably lie between 5 and 7. This upper limit is also consistent with the value of κ^m/κ required for the explanation of the curvature observed in the isothermal plots of $\ln \sigma$ versus P in the extrinsic conductivity range of some alkali halides (§14.2.2 and §10.4.2).

The above remarks, developed from the cases (1) to (3), lead to the conclusion that the proportionality of g^f or g^{act} to $B\Omega$ – with a proportionality factor independent of pressure – should pertain to pressure ranges ΔP for which $\int_{P_1}^{P_2} \kappa^c dP < 0.1$ or $\overline{\kappa^c} \Delta P < 0.1$. Taking $|\kappa^c/\kappa|$ of the order of unity we get $\kappa \Delta P < 0.1$ or $\Delta P/B < 0.1$. The error introduced in this way into the predicted values of g is given by eq. (14.22).

In the case of a migration process the range ΔP where the proportionality between g^m and B/Ω is valid may be smaller, at most, by a factor of 5 to 8. The pressure range given by the above inequality covers the usual experimental pressure ranges, e.g. for a material with $B = 1000$ kbar we can consider c a constant up to 100 kbars with a relative error in the pressure variation of g^{act} or g^f smaller than $\frac{1}{2} \Delta P/B$, i.e. $< 5\%$.

As became clear from the above recapitulations the validity of the proportionality between g and $B\Omega$ is beyond any doubt. We again stress that the only "point" remaining for discussion is the pressure range ΔP for which we can consider c as pressure independent without introducing significant (but anyhow known) errors. However, on this "point" there is not much room for discussion; this is so because thermodynamics simply demand that in order to have relative errors of the quantity $g(P) - g(P_0)$ smaller than e.g. $\frac{1}{2}\kappa^c \Delta P$, we have to work in a pressure range ΔP given by: $|\int_{P_1}^{P_2} \kappa^c dP| < 0.1$ i.e. $|\overline{\kappa}^c \Delta P| < 0.1$. This is equivalent to $|(\kappa^f - \kappa + A\kappa)\Delta P|$ < 0.1 or $|(\kappa^f/\kappa) - 1 + A|(\Delta P/B) < 0.1$. Therefore, denying the proportionality between e.g. g^f and $B\Omega$ is equivalent to stating that there is *no* reasonable pressure range for which we can write the latter inequality. Of course, such a statement is far from true, because in view of the *known* values of A and B for each material it must be inversely proved that κ^f exceeds κ by order(s) of magnitude. The invalidity of such a statement is clarified by the following example for copper ($B \simeq 10^3$ kbar): For a typical experimental pressure range of $\Delta P = 5$ kbar the last inequality is violated *only* when κ^f/κ appreciably exceeds 20, which means that upon the creation of 1% of defects (e.g. by irradiation when $v \simeq \Omega$) the bulk modulus should change by an amount appreciably larger than 20%, which, of course, contradicts the experimental facts [448].

It is useful to describe the above conclusions in the following way: "We have shown that for T = constant the volume dependence of g can well be described on the basis of the proportionality between g and $B\Omega$, the proportionality constant of which can be considered to be volume independent in a pressure (volume) region prescribed by the above mentioned inequality".

The extension of the above conclusions to other kinds of "defects" (e.g. Na^+ in KCl) is straightforward. In that case the quantities g and v have the following meaning: g is the variation of the thermal Gibbs energy upon the introduction of this "defect" which creates a variation v of the volume of the matrix material; due to the fact that v is again given by $-(dg/dP)_T$ the above procedure for vacancies can equally be extended to other defects. The only difference is that now the values of $(\kappa^d/\kappa) - 1$ may result in a different region, but this affects only the extent of the pressure range ΔP up to which the new value of c can be considered to be pressure independent. The other considerations remain practically the same.

14.5.2 Discussion and comments on the proof of the relation $g(V, T) = c(V)B\Omega$

In §14.4. we have seen that the validity of the relation given in the heading is incontestable in the case of the quasi-harmonic approximation (QA) because of the occurrence of the thermodynamical identity

$$\left.\frac{dg}{d(B\Omega)}\right|_V = -\frac{s^*}{\Omega \left.\frac{dB}{dT}\right|_V} \quad (\equiv c_1),$$

where both the quantities s^* and $(dB/dT)_V$ have to be temperature independent within the frame of QA (for V = constant). We have also presented good arguments that the second member of this identity is almost temperature independent in real solids as well. In order, however, to discuss all possible cases we have allowed for an explicit temperature dependence of c_1.* In this general case we were able to determine the bounds of the temperature variation of c_1 from the physical requirement that the constant-volume entropy s^* has to be negative (or zero) and that its absolute value is smaller than that of the constant-pressure entropy s. The bounds of this possible temperature dependence of c_1 are given – in terms of its "expansion coefficient" β^c – by

$$\delta - \frac{dB}{dP} < \beta^{c_1}/\beta < \delta - \tfrac{1}{2}\left(\frac{dB}{dP} + 1\right).$$

For a given solid the possible temperature variation of c_1 can therefore be determined as a function of well-known macroscopic properties. This is useful for practical applications; one can always work with the (quasi-harmonic) assumption that c_1 is temperature independent while the region of the possible errors introduced by this assumption can be be estimated.

§14.4 shows that the quantity $(dg/d(B\Omega))_V$ can be considered temperature independent either in the whole temperature range above the Einstein temperature (within the QA) or in a restricted interval ΔT depending on the desired accuracy which is known anyhow from the inequality given above. Consequently, we can always write

$$g(V, T) = c_1(V) B\Omega + K(V).$$

14.5.3 Interconnection of the conclusions drawn in §§14.3 and 14.4

We recall the results:

$$g(V, T) = c(T) B\Omega + \Lambda(T) \quad \text{(from §14.3)},$$
$$g(V, T) = c_1(V) B\Omega + K(V) \quad \text{(from §14.4 and §14.5.2)},$$

where c and c_1 have been defined by eqs. (14.13) and (14.65) and the integration constants Λ and K may depend only on T and V, respectively. The last two relations are consistent with [669]:

$$g(V, T) = \text{const} \times B\Omega + \text{const}',$$

where now the constants *have to be* independent of temperature and volume. It is therefore sufficient to determine their value at a *single state*, e.g. at absolute zero. In §14.3.6 we have seen with the help of microscopic calculations that at least in two general, quite different cases (rare gas solids and alkali halides) the constant "const'" is actually zero and this is likely to

---

* The quantity c_1 is exactly equal to c, see §§14.4.1 or 14.5.3.

be so in all cases. We have therefore concluded that the results of the considerations developed in §§14.3 and 14.4 are consistent with the relation

$$g(V, T) = \text{const} \times B\Omega,$$

where "const" ($= c = c_1$) has to be *temperature* and *pressure* (*volume*) *independent*.

§14.5.4. A more direct procedure for the foundation of the $cB\Omega$-model for the formation process.

We start by writing (recall the identity 14.12):

$$g^f(P, T) \text{ or } f^*(V, T) = c^f B\Omega, \text{ where } c^f \equiv \frac{(v^f/\Omega)}{(\mathrm{d}B/\mathrm{d}P)_T - 1}. \quad (14.88)$$

In order to prove now the validity of the $cB\Omega$-model we have to show that c^f does *not* depend on volume and temperature:

(a) *non-dependence on volume*; concerning the numerator of eq. (14.88) microscopic concepts show (see §14.2.1) that κ^f cannot exceed κ by a large factor. Concerning the denominator of eq. (14.88) simple microscopic considerations indicate that the quantity "$-A$" (see §14.2.2) is at most equal to 4.2. Combining these two conclusions we immediately find (see §§14.3.3 or 14.3.5) that their ratio, i.e. the quantity c^f, can well be considered as volume independent (at least for macroscopic volume variations ΔV for which $\Delta P/B \leq 0.1$).

(b) *non-dependence on temperature (V = constant)*; within the frame of QA one easily shows [1] that the denominator of (14.88) is temperature independent [660]. The same holds for the numerator of eq. (14.88) because in QA the volume v^f (when $V = $ constant and hence $\Omega = $ constant) must not show any *explicit* temperature dependence (we remark that in QA the quantity v^f depends *only implicitly* on T, i.e., through the variation of the volume). Obviously in the above procedure we have disregarded the (anyhow) very small fractional variation of the quantities B and Ω resulting from the thermal creation of vacancies.

APPENDIX

A.1 Introduction

Since the completion of the main part of the text of this book several papers related to its contents have appeared in the literature. In this Appendix the most important of these papers are described in brief. The numbers between parentheses in the headings of the following paragraphs refer to the corresponding section numbers of the main text.

A.2 Linear and curved Simmons–Balluffi plots (§7.3)

The anharmonic behaviour of solids leads to vacancy formation parameters that may change with temperature, often causing plots of ln n versus $1/T$ to show a curvature. According to the $cB\Omega$-model a linearization is achieved if the abcissa $1/T$ is replaced by $B\Omega/kT$ [593a]. Such a linearization has recently been successfully applied on CD_4 by Prokhvatilov and Isakina [593].

A.3 Stimulated current emission in the earth (§7.4)

In §7.4 the theory of the thermally stimulated current emission has been extensively discussed and a large number of experimental cases presented. Lately, experiments have been extended to rocks by Dologlou-Revelioti [594]. She investigated rocks of widely differing petrographic composition and found an emission of thermally stimulated current in the region around room temperature. (See §A.11, note 12).

Current emission from rocks has been studied by Sobolev and coworkers. In a recent publication [595] on a large granite sample they report a current at a critical stress around 50% of the fracture stress. In this experiment the rate b of the pressure increase was around $4 \times 10^7 \text{dyn cm}^{-2} \text{ h}^{-1}$ and the duration of the current emission equal to a few hours. Accepting the latter as a relaxation time τ, the application of eq. (7.72) indicates that $J(P)$ maximizes when

$$\frac{bv^m}{kT} = \frac{1}{\tau}.$$

This leads to a migration volume around 10^{-21} cm^3, which is quite large in comparison to the mean volume per atom and probably indicates an orientation process of large complexes.

As mentioned, the application of the theory of pressure-induced electric currents to geophysics has recently led to important results in the field of earthquake prediction. Attention is drawn to the point that thermally, or pressure, stimulated *polarization* currents can be observed *even in the absence* of any external electric field provided that the material contains piezoelectric inclusions not uniformly distributed (e.g. rocks containing piezoelectric materials like quartz). Such currents have been actually detected in a number of samples from Greek rocks. It is therefore evident that the physical model that explains the emission of precursor electric signals (see §7.4.2) does *not* require the existence of any *external* electric field. The polarising electric field depicted in fig. 7.5 comes from the gradual variation of stresses during the preseismic period in the focal region (of course this field is not necessarily constant in time; see also §A.11, note 13). Precursor variations of the electric field in the earth have been monitored in Greece since March 1981. They can be separated into two types. The first [123a] consists of a strong pulse of a few hundred mV/m with a duration of a few ms and a lead time between 1/2 and 4 min. Due to its minor importance we are disregarding it. The second type is a considerably weaker variation (1–100 mV/km) with a duration between 1 and 90 min. and a lead time usually between $6\frac{1}{2}$ and 115 hours [596]. In a few cases it can reach 1 week. Both the duration and the lead time are independent of the magnitude and of the seismic region at which the earthquake originates (see also §A.11, note 14).

We will briefly describe certain empirical rules that have been established [596,597]. The most important feature that allows the exploitation of these signals for earthquake prediction is the fact that earthquakes of the same magnitude occurring in a *given seismic region* and measured at a *given site* give the same variation of the field (maximum value ΔE of the detected deviation of the electric field). Exploiting the above rule the following connection was established between magnitude M and the quantity ΔE:

$$\log \Delta E = 0.37M + \text{constant}, \tag{A.1}$$

which means that an increase of M by one unit increases the value of ΔE by a factor of 2.3; an explanation for this connection is given in §A.11, note 11.

When comparing values of ΔE measured at a *given station* for earthquakes of the same magnitude but occurring at *various epicentral distances* r, it was found that ΔE decreases with increasing r according to a $1/r$-law (for $r > 50$ km). As the field is produced by an electric current flowing in the earth, this law indicates that the current density attenuates according to $1/r$, a dependence that theoretically would not immediately be expected. It might be due to the current density lines dispersing practically in two dimensions.

In order to take care of the diminuation of the quantity ΔE for increasing epicentral distance, eq. (A.1) has to be written as

$$\log(r\Delta E) = 0.37M + \text{constant}'.$$

A plot of $\log(r\Delta E)$ versus M is valid for all detectable signals at a *given station*.

Variations of the electric field propagate with the velocity of the order of the velocity of light, so that the signal arrives simultaneously at the stations where it is detected. This is the case whenever the magnitude and epicentral distance warrant a detectable signal; the values of ΔE, however, simultaneously recorded at these stations do not scale with $1/r$. This is due to the fact that the current density should be multiplied with an "effective resistivity" ρ in order to supply the value of ΔE. The factor ρ depends on the local conditions of the earth, but it is not clear how it depends on the inhomogeneities of the various layers. As its value is found to depend on the direction in which ΔE is being measured, it has to be studied for *each* direction (i), e.g. NS and EW. According to the above considerations the quantity that is representative of the current density for each direction (i) is

$$J_i = \Delta E_i / \rho_i$$

In order to be able to compare the current density at various stations we have to revert to relative values (ρ_{rel}) by dividing each value with the ρ-value of a *isotropic* base station for which we set $\rho_{\text{base,NS}} = \rho_{\text{base,EW}} = 1$. This has to be done empirically for each station; variations of ρ_{rel} up to a factor of 30 have been found. One thus obtains a working formula for the "signal strength" of the current density:

$$J_{\text{rel}} = \left\{ \left.\frac{\Delta E}{\rho_{\text{rel}}}\right|_{\text{NS}}^2 + \left.\frac{\Delta E}{\rho_{\text{rel}}}\right|_{\text{EW}}^2 \right\}^{1/2}.$$

An empirical plot of $\log(J_{\text{rel}} r)$ versus M can now be drawn, valid for all stations and all seismic areas; it is also a straight line with a slope of 0.37.

The use of the quantity J_{rel} has been found to be an efficient method for predicting the epicenter of an earthquake [597] by making measurements in a network of stations. As the signal strength scales with $1/r$ the epicenter can be determined, usually, with an error less than 120 km, which is at present the mean distance in Greece between the stations. The magnitude can now be found from the plot connecting $\log(J_{\text{rel}} r)$ to M. This is possible because by predicting the epicenter the epicentral distance of each station is automatically determined.

An unexpected phenomenon, described as *"directivity"* was discovered (its plausible physical origin is discussed in §A.11, note 10). The electric signal of a given earthquake does not appear at all stations for which the expected strength would warrant a detection. Nevertheless, all *detected* signals show the $1/r$-behaviour, as already described, so that the directivity does not introduce errors in the determination of the epicenter. The directivity effect appears at a station only in respect to certain seismic areas. As an example a seismic site A might give a well observable signal at station B, while an earthquake originating at B might never give a signal at site A. As a result of this effect each station, *irrespective* of the earthquake magnitude and epicentral distance, records precursor electric signals *only* from specific seismic regions that are empirically determined. These seismic regions constitute the so-called *"effective region map"* of each station. Obviously an earthquake that is going to happen at a seismic region (i) produces simultaneous electric signals at two or more stations *only* when this region (i) belongs to the earlier constructed *"effective region maps"* of these stations. At this point we mention a further interesting observation; the *"polarity effect"*. Signals from a given seismic area and detected at a given station are always deviations in an unique sense, e.g. always an increase or always a decrease of the measured field component(s). Although *directivity* reduces the number of stations which contribute to the determination of the epicenter, it turns out (together with *polarity*) to be a useful tool because it gives hints in identifying the seismic area from which the earthquake is expected. The reliability of predictions lies around 80% at the present moment. A denser network of stations will, undoubtedly, increase the efficiency. An extensive description of the physical properties of the pre-seismic signals has just been published by Varotsos and Alexopoulos [596,597].

A.4 Self-diffusion (§9.3)

Self-diffusion was extensively treated in ch. 9 for a large number of solids. The diffusion measurements for sodium ranged from 194.4 to 370 K, while the data on the bulk modulus had to be based on a combination of experimental results, because no single series of measurements covered the whole temperature region (see §9.1). As shown, the curved self-diffusion curve can be explained on hand of the above data with the assumption of a single mechanism.

This result can now be substantially strengthened by using the latest measurements of Anderson and Swenson [598]. As shown by Varotsos, Alexopoulos and Lazaridou [599] the curved diffusivity plot can be explained in an unified way for the whole region with a single mechanism and with a reasonable accuracy, although the diffusion coefficient changes by 5 orders of magnitude.

A similar study has been made for the curved self-diffusion plot of lithium. In the past, measurements of the elastic constants were available only up to 295 K. Very recently, improved isothermal elastic data for temperatures up to 350 K were published by Anderson and Swenson [600]. Eftaxias, Grammatikakis and Varotsos [601] have shown that the new data reproduce the self-diffusion curve very accurately within a range of 8 orders of magnitude. (See also §A.11, note 17).

A number of heterodiffusion studies (§11.2) has also been published during the recent years. The diffusivity of Co in Cu and Pt in Cu has been reported by Döhl et al. [604] and by Neuman et al. [605]. The diffusion coefficient of oxygen in Cu was studied by Albert, Kirchheim and Dietz [606]. Finally Hood et al. [607] measured the diffusivity of Co in aluminum single crystals.

A.5 Discussion on the values of s/h and v/g (§§9.1 and 10.1)

The constancy of the value of s/h for various defect mechanisms of a given bulk material was proven in ch. 9 for many cases. The plot of s versus

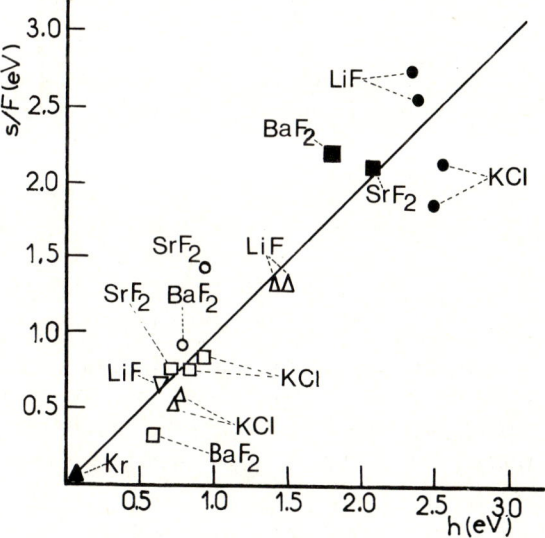

Fig. A.1. Demonstration of eq. (9.4) The line is the locus for $s/F = h$. The plotted points are taken from table A.1. The symbols are explained in the table.

Table A.1
Connection between s and h (h in eV, F in 10^{-4} K^{-1}).[a]

	Kr $F_{100}=32$ ref. [617]		LiF $F_{295}=3$ ref. [618]		KCl $F_{298}=3.66$ ref. [619]		SrF$_2$ $F_{400}=2.6$ ref. [620]		BaF$_2$ $F_{298}=3.07$ ref. [621]	
	s/k	h	s/k	h	s/k	h	s/k	h	s/k	h
Schottky formation (●)			8.9	2.37	8.00	2.49				
			9.6	2.34	8.99	2.54				
vacancy formation (▲)	2.0	0.077								
anion Frenkel formation (■)							6.35	2.05	7.85	1.81
anion interstitial migration (○)							4.34	0.93	3.26	0.79
free cation vacancy migration (△)			2.3	0.75	2.2	0.72				
			2.3	0.71	2.4	0.73				
free anion vacancy migration (□)					3.6	0.92	2.12	0.70	1.12	0.59
					3.2	0.85				
bound cation vacancy migration (▽)			2.3	0.66						

[a] The subscript to F indicates the temperature for which it has been calculated.

h for each material should give more or less a straight line with a slope

$$F \equiv -\frac{\beta B + \left.\dfrac{dB}{dT}\right|_P}{B - T\beta B - T\left.\dfrac{dB}{dT}\right|_P},$$

demanded by the $cB\Omega$-model (eqs. 9.3 or 9.4). The results for the totality of s/F-values should lie therefore on a common line for all materials when plotted against h. This is shown in fig. A.1 for LiF and for some newer results not included in the main text. The coordinates of the plotted points are given in table A.1, together with their corresponding references; the latter mention either the publication or the source from which they were obtained.

Another important result of the $cB\Omega$-model is eq. (10.1) which can be written in the form $v/L = g$, where L is the bulk quantity

$$L \equiv \left(\frac{dB}{dP} - 1\right) B^{-1}.$$

In ch. 10 a number of figures depicted its validity for a variety of materials and defect mechanisms. Here we give data from our latest publications. The plot of fig. A.2 depicts v/L versus g for the values of table A.2 When g is not available we use h instead of g in view of their relatively small difference. Emphasis was given to ionic solids, mainly to fluorides, for which reviews by Schoonman [602] and Samara [603] have appeared.

Table A.2
Demonstration of eq. (10.1)

		L (10^{-3} kbar^{-1})	v (cm^3/mol)	h (eV)	g (eV)	Ref.
1 Kr	vacancy formation	521 (110 K)	28.44		0.058	[617]
2 Ar	vacancy formation	366 (60 K)	23.63		0.068	[617]
3 LiF	cation vacancy migration	6.5	3.2–4		0.69	[618]
4 LiF	Schottky formation	6.5	12.6–13.2		2.14	[618]
5 LiBr	Schottky formation	19.9	29.5			[624]
6 PbCl$_2$	anion vacancy migration	19	4.7	1.7		[624]
7 PbCl$_2$	activation (conduction intrinsic region)			0.32		
8 PbBr$_2$	anion vacancy migration	19	19.8	1.12		[624]
9 PbBr$_2$	activation (conduction intrinsic region)	19	4.4	0.3		[624]
10 CaF$_2$	free anion migration	19	21.6	1.0		[624]
11 CaF$_2$	bound F-vacancy migration	4.4	2.1–3.2	0.52		[603]
12 SrF$_2$	bound interstitial migration	4.4	1.72	0.502		[625]
13 SrF$_2$	bound F-vacancy migration	5.52	4.7	0.7		[626]
14 Mg$_2$SiO$_4$	activation (conduction intrinsic region)	5.52	3.0	0.59		[626]
15 Mg$_2$SiO$_4$	dislocation migration	3.13	18	5.4		[624]
16 Mn-Zn-ferrite	disaccomodation activation	3.13	11	3.1		[624]
		2.3	1.2	0.59		[627]

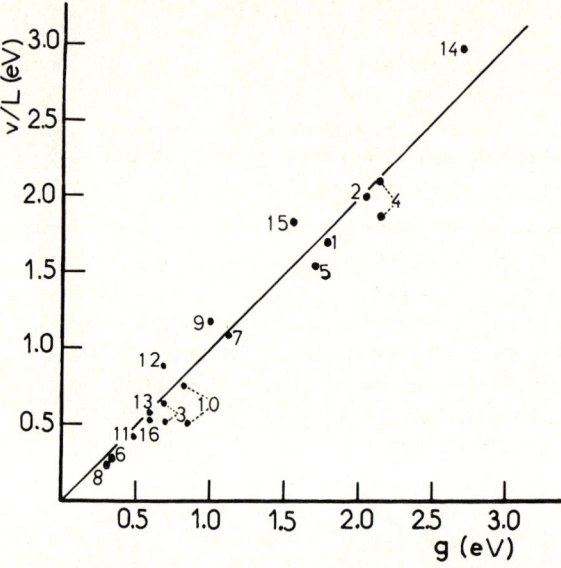

Fig. A.2. Demonstration of eq. (10.1). The line is the locus for $v/L = g$. The plotted points are taken from table A.2. The coordinates of points 14 and 15 have been divided by two.

A.6 Recent developments in the detection of preseismic electric signals (§§7.4.2 and A.3)

Since October 1982 the electric field of the earth (electrotelluric field) is continuously monitored at sixteen sites ("stations") dispersed over Continental Greece. The measurements of each station are automatically transmitted to a central station (Athens) through telephone lines. This telemetric network allows the real-time observation of the preseismic electric signals described in §§7.4.2 and in A.3; therefore the epicenter and the magnitude (M) of an impending earthquake can be determined well before its occurrence. The predicted parameters (i.e. epicentral coordinates and magnitude) of each case are *officially documented* by expediting a telegram *before* the earthquake occurrence. Examples of documented predictions – for the period until October 1983 – can be found in ref. [597]. The above method has been extensively discussed in a specially organised international meeting [628] held at Athens (November 1984); a basic conclusion was that "this method provides a potentially very powerful tool for earthquake prediction". Some of the physical aspects and documented predictions discussed in this meeting have been recently reviewed by S. Uyeda (Earthquake Research

Institute of Japan) [629]. In the same conference D. Lazarus and L. Slifkin presented two physical models that can explain, in a satisfactory way, the creation of a transient electric signal before the earthquake occurrence.

Variations of the electrotelluric field are usually detected by measuring the potential difference of pairs of (non-polarised or metal) electrodes placed at distances of the order of 30 m to 300 m. These measurements should be carried out at regions very far from sources of electric noise (e.g. industries). At the above meeting Varotsos and Alexopoulos presented a new experimental technique that can significantly increase the "signal over noise" ratio; this technique consists of measuring the potential difference (and therefrom the electric field) between points of the surface of the earth that have distances between them of the order of some kilometers. The latter is experimentally achieved by measuring the potential difference between two telephone lines (wires) *grounded* at two sites far apart, e.g. at two suburbs of a city. This technique is currently applied to Greece by employing measurements of this type at all sites at which telemetric stations have been installed (their sites are depicted in ref. [596]). At a given station a transient variation of the electric field can be considered with confidence as a true precursor when it appears simultaneously at the dipoles with small length (30 to 300 m) and with large length (some km) and, of course, when it is not due to magneto-telluric variations. This technique definitely excludes any possibility of misinterpreting a local noise as a precursor [670] and leads to an increase of the efficiency of the issued predictions. As an example we give the results of the first four months of 1985: within the area $N^{42.5}_{35}$ E^{25}_{20} only three earthquakes(EQ), with (surface) magnitude (M_S) equal to 5 units or larger, occurred on March 12, April 21 and April 30 with epicenters at (39.3° N, 24.0° E), (35.6, 22.0), (39.3, 22.9) and with M_S equal to 5.0, 5.5 and 5.7 units respectively. These EQ were predicted well in advance ($2\frac{1}{2}$, 5 and 5 days before each EQ, respectively) with the official telegrams No. 333, 338 and 340, respectively; the deviations between the actual and the predicted parameters were: 50, 80 and 50 km in the epicentral location and 0.4, 0.3 and 0.7 units in the magnitude, respectively. For a statistical evaluation we emphasize that, during the above period of four months, *no* other telegram announcing EQ with $M_S > 5$ was issued except No. 330; the latter predicted a $M_S = 5.4$-event and was actually followed by a much weaker EQ; the latter occurred on February 21 with $M_S = 4.7$ at an epicenter: (40.0° N, 24.4° E), i.e. 100 km from the predicted point.

The above three larger events were (relatively) *isolated* in time and space. By following the procedure indicated by Varotsos and Alexopoulos [623] one easily obtains that the probability of predicting the parameters (time, epicenter and magnitude) of each of the above events by chance is of the order of 10^{-3}.

A.7 Mixed ionic crystals (§§12.2 and 12.3)

In §12.2 a method was described which predicts the bulk moduli and the conductivity of mixed ionic crystals from the values of the pure constituents. Shanker and Jain [608] have recently calculated the elastic moduli B and dB/dP of a number of mixed alkali halides. Their results for KCl–KBr coincide with those of refs. [609] and [610]. In another publication the aforementioned authors extended their calculations to elastic constants of higher order [611].

The correctness of eq. (12.19) concerning the conductivity has been checked by Lazaridou and Varotsos [612] based on the conductivity measurements of Shahi and Wagner [613] of KBr–KI mixed crystals which suggest a maximum of conductivity at 64 at% KI while eq. (12.19) predicted the maximum at 62 at%.

A recent discussion on the analysis of the conductivity of mixed silver halides has been given by Cain [614]. The parameters for various defect mechanisms have been determined for AgBr–AgCl from one end of the composition range to the other. A proportionality between B and the defect energy was observed, as expected from the $cB\Omega$-model. Measurements of the ionic conductivity of the system $AgI_{1-x}Cl_x$, similar to those published by Cain and Slifkin [387], have been recently presented in ref. [671].

A theoretical calculation has been made by Singh, Neb and Sanyal [615] for the bulk modulus of mixtures of rare gas solids. The trend of the results is similar to the behaviour described in ch. 12 for mixed alkali halides. Calculations of the variation of B with composition have also been recently published for the following mixed systems: KCl–KBr [672], KBr–KI [672], CaF_2–SrF_2 [673,674]. The results are in agreement with the aspects presented in ch. 12.

A.8 Calculation of defect parameters from first principles (§§9.8 and 14.3.6)

Recently a theory was proposed by Bhatia and March [616] which calculates the formation energy for vacancies in rare gas solids. It is based on the assumption of suitable forces between the atoms and in a special case leads to results similar to those of the $cB\Omega$-model.

Until 1981 there were the following two basic disagreements between microscopic calculations of defect parameters in alkali and silver halides and experimental data:

(1) The calculations showed that the formation energy of a Schottky defect in alkali halides or a cation Frenkel defect in silver halides decreases significantly when the temperature increases [90,96,97,314,647]. This calcu-

lated formation energy was thought to be [631] the internal energy u^f (see §3.4) and then by approximating $u^f \approx h^f$ it was compared [325,90,97] with the experimental value of the formation enthalpy (see also §A.11, note 4). The temperature *decrease* of the calculated quantity was considered to be the explanation of the upward curvature observed in conductivity and diffusion plots [96,97,631]. However, thermodynamics demands (see §§3.6.1 and 9.6.2 and ref. [352]) that this curvature can be explained *only* when the defect enthalpy *increases* with increasing temperature (see also §A.11, note 4).

(2) Since 1971, the microscopic calculations [325–327] showed that the relaxation volume Δv of a Schottky defect in alkali halides is negative, thus leading to a formation volume v^f appreciably smaller than the molecular volume. This result however is in contrast to accurate experimental data [322,323] which definitely showed that Δv is undoubtedly positive (see §10.4.2).

During the recent years the above two points have been clarified as follows:

Point 1. In ref. [630] it was stressed that the hitherto published calculated quantity is *not* u^f or h^f but the *isochoric* internal energy u^*. Therefore it was indicated that one should add, at each temperature, to u^* the term $T\beta B v^f$ in order to get the desired quantity h^f (see eq. 3.128). Since then a lot of microscopic publications have been published in ionic crystals which considered the above indication and actually found values of h^f not only comparable to the experimental data but also increasing with temperature.

Point 2. Until 1980 the microscopic calculations have been carried out [325,326] by the so-called "lattice statics technique" by employing an expression for v^f which demands a strict *harmonic* model for the lattice surrounding the defect region, i.e. it depends upon the assumption that the harmonic force constants are *independent* of crystal volume. In refs. [36,630] however it was clarified that the reason for which the calculations led to negative values of Δv was the use of the *harmonic* approximation which does not provide a reliable description of the real behaviour of solids. It was therefore suggested that the calculations should be repeated either in the frame of quasi-harmonic approximation or in another frame that could consider the anharmonic behaviour of solids; such a re-calculation should also lead [630] to values of v^f that increase significantly with temperature so that $\beta^f \gg \beta$. The microscopic calculations were actually repeated [632,633] considering the above indications, i.e. in the frame of quasi-harmonic approximation, and it was found that Δv is positive and that v^f increases with temperature.

A final remark should be added concerning the microscopic calculations of the formation entropy of a Frenkel or a Schottky defect in ionic crystals;

until 1980 these did not take into account the concept of the distinction between "isobaric" and "isochoric" formation entropy which plays a prominent role in such type of calculations (see §A.11, note 5).

A.9 Determination of the concentration of vacancies from a constant volume X-ray study at various temperatures. Thermal vacancies in solid ^{3}He (§§7.1.1 and 10.3)

Equation (7.6b) can be equivalently written as:

$$x = 3\left(\frac{\Delta L}{L^0} - \frac{\Delta \alpha^X}{\alpha^0}\right),$$

where L denotes the macroscopic length, α^X the X-ray lattice parameter and the superscript 0 refers to values in the absence of thermal vacancies, i.e. to values of the isobaric perfect crystal. When the experiment is carried out at *constant* macroscopic volume conditions, i.e. $\Delta L = 0$, the above equation becomes:

$$x = -3\frac{\Delta \alpha^X}{\alpha^0},$$

which means that the concentration x of vacancies can be obtained solely from X-ray data by measuring the X-ray lattice parameter at various temperatures but at constant macroscopic volume conditions. Such an experiment has just been described by Heald, Baer and Simmons [634] in solid ^{3}He. It was found that bcc ^{3}He contains about 0.5% thermal vacancies at melting, essentially independent of molar volume. A solid hcp ^{3}He was also studied and was found to have 0.1% thermal vacancies at melting. By comparing these results to older NMR and ultrasonic experiments one can conclude that in the bcc phase vacancies move by a tunneling process, while in the hcp phase their motion is thermally activated. In the bcc phase the vacancy formation volume v^f was found to vary from 0.4 to 0.3 Ω depending on the macroscopic volume under which the measurements have been carried out. It was finally concluded that in bcc ^{3}He a *localized* vacancy picture is inadequate.

A.10 Self-diffusion probed by spin–lattice relaxation (§§7.3.6 and 9.1)

Self-diffusion in solid Li was recently studied [635] via the spin–lattice relaxation of polarised radioactive ^{8}Li nuclei using their asymmetric β-decay radiation. The diffusion-induced spin–lattice relaxation rate was measured as a function of temperature from the melting point T_M down to about

$\frac{1}{2}T_M$. The self-diffusion coefficient obtained in such a way varies over almost seven decades and a clear deviation from Arrhenius behaviour is observed. This curvature in the ln D versus $1/T$ plot can be immediately explained, in the frame of a single mechanism, as follows (see §9.1): by using the recent isothermal elastic data of Anderson and Swenson [600] one calculates $B\Omega/kT$ for each temperature; then by plotting ln D versus $B\Omega/kT$ a *straight* line results the slope of which directly leads to c^{act}. This value of c^{act} allows the determination of the activation enthalpy and entropy for each temperature by means of eqs. (9.1) and (9.2).

A11. Notes added after the completion of the main text

General Thermodynamics

Note 1: physical meaning of the quantities H, U, F and G. The study of eqs. (2.11) to (2.14) leads to the following remarks concerning the meaning of the quantities H, U, F and G: (1) The change of the enthalpy is just equal to the heat added to the system in an isobaric process. (2) The quantity F is the "effective potential energy" at constant temperature whereas U is the "effective potential energy" at constant entropy; in other words, F is a measure of the energy that the system *can develop* (i.e., it tells how much work can be performed by the system) in an isothermal process. Equation (2.11) reveals that for a reversible isothermal change dF is just equal to the work dW ($= -PdV$) performed by the environment on the system. (3) Concerning G, one sees from eq. (2.8) that in an isobaric process dG exceeds dF by the work pdV ($= -dW$) exerted by the system on the environment, i.e., d$G = dF + pdV$; this term pdV cancels the reversible work in dF and hence enables G to monitor the net entropy of the system plus environment.

It should be noted that eqs. (2.11) to (2.14) are all based upon taking the heat entry as given by TdS. Therefore, in equilibrium states the quantities F, G, H and U are *minimum values* of the expressions on the right-hand side for the particular variables and virtual changes contemplated.

Note 2: variation of dB/dT with pressure. When comparing the plots $B(0)$ versus T and $B(P)$ versus T, the experimental data indicate, in contrast to the case shown in fig. 9.2, that *usually* the absolute value of dB/dT ($P = $ const) *decreases* upon increasing pressure (see e.g. table 4 of ref. [282]). This can be alternatively understood from the fact that at higher temperatures the value of dB/dP ($T = $ const) increases and that hence $B(0)$ should fall faster with T than $B(P)$.

Note 3: symbolism of isochoric defect formation parameters. An extensive presentation of the formulae that connect the two families of parameters can be found in ref. [630]; the impact of these formulae on microscopic calculations in ionic crystals has been nicely discussed in a interesting review paper by L. Slifkin, European Scientific Notes **34** (1980) 67. A few years later the *same* formulae were published by others (see: C.R.A. Catlow, J. Corish, P.W.M. Jacobs and A.B. Lidiard, J. Phys. **C14** (1981) L121, and later papers by the same group) but with a slightly different notation; more precisely Catlow et al. preferred a subscript "V" instead of the asterisk already used by Varotsos and Alexopoulos in ref. [630] in order to discriminate the "isochoric" parameters. Because these formulae are now widely used in the literature and in order to avoid any confusion we keep throughout the present book the notation of ref. [630].

Microscopic calculations

Note 4: thermodynamical considerations and microscopic calculations in ionic crystals. In ref. [631] Jacobs repeated the aspect that the calculated values in refs. [96,97] correspond to the defect formation energy u^f which is equal to the formation enthalpy h^f at ordinary pressures (obviously by using the relation $h^f = u^f + Pv^f$ and considering that $Pv^f \ll u^f$). Furthermore, in ref. [631] he expressed the view that the calculations in refs. [96,97] can be considered to provide the proof that the extra decrease Δg^f of g^f at high temperatures (which explains the upward curvature of the conductivity plot) is due to a *decrease* in the formation energy u^f; as an example, Jacobs presented for AgCl the values u^f ($T = 50°C$) = 1.361 eV and u^f ($T = 450°C$) = 1.122 eV. A quite different view however was *simultaneously* presented by Varotsos and Alexopoulos [630], based on the distinction between "isochoric" and "isobaric" formation parameters (see also § A.8); using eq. (3.128), $u^* = h^f - Tv^f\beta B$ (or for small pressures $u^* \simeq u^f - Tv^f\beta B$), ref. [630] stated that the calculated values in refs. [96,97] represent u^*, not u^f. Furthermore, ref. [630] proved that the difference of u^f and u^* at high temperatures is not negligible (i.e., the term $Tv^f\beta B$ is roughly around 50% of u^f or u^*) so that the following procedure is necessary: "for each temperature the correction term $Tv^f\beta B$ should be added to the calculated quantity (i.e., to u^*) and then compared to the experimental value of h^f; the quantity u^* *decreases* with temperature in contrast to h^f that has to *increase* in order to lead to an upward curvature of the conductivity plot." The above procedure was actually followed in later calculations (e.g., by C.R.A. Catlow, J. Corish, P.W.M. Jacobs and A.B. Lidiard, J. Phys. **C14** (1981) L121, and in other publications of the same group) and confirmed the suggestions of ref. [630].

Note 5: microscopic calculations of the formation entropy in ionic crystals. Until 1979 only few microscopic calculations of the *formation entropy* of a Schottky or Frenkel defect in ionic crystals have been published, e.g., by T.M. Haridasan, J. Govindarajan, M.A.H. Nerenberg and P.W.M. Jacobs, Phys. Rev. **B20** (1979) 3481; their calculated values were directly compared to the experimental ones. In ref. [630], however, it was stressed that this comparison is not allowed because the calculations give s^*, in contrast to (isobaric) experiments which lead to s^f; furthermore, it was suggested [630] that one should first add the term $v^f \beta B$ to s^* and then compare the sum with the experimental s^f (see eq. 3.126). This suggestion was actually followed in later publications: J.H. Harding and A.M. Stoneham, Phil. Mag. **B43** (1981) 705; V.C. Sahni and P.W.M. Jacobs, Phil. Mag. A **46** (1982) 817; P.W.M. Jacobs, M.A.H. Nerenberg, J. Govindarajan and T.M. Haridasan, J. Phys. **C15** (1982) 4245; M.J. Gillan and P.W.M. Jacobs, Phys. Rev. **B28** (1983) 759.

Note 6: defect calculations in β-PbF_2. In a recent publication (P.W.M. Jacobs, V.C. Sahni and C.S. Vempati, Phil. Mag. A **49** (1984) 301) the formation volume of an anion Frenkel defect and the migration volumes for the anion vacancy and anion interstitial have been theoretically estimated. The consequences of the distinction between "isochoric" and "isobaric" defect parameters, as they were developed in ref. [630], have been considered in this calculation (see § A.8); however, the calculated values are appreciably lower (by a factor close to 2) than the experimental ones. The results show a fast increase of the defect volumes with temperature, thus confirming earlier suggestions [36,37].

Note 7: the Gilder–Lazarus relation in the frame of the quasi-harmonic approximation. If the usual assumption $(dv^{act}/dT)_V = 0$ of QA is adopted, then eq. (2.34a) gives $\beta^{act}/\beta = \kappa^{act}/\kappa$, which simplifies eq. (7.49) to $c_p \approx T\beta v^{act}\beta^{act}$, or

$$c_p \approx T\beta^{act}\gamma\tilde{c}_V v^{act}/\Omega \tag{7.49a}.$$

Piezostimulated currents and related geophysical aspects

Note 8: comments on the detection of piezostimulated currents. When dealing with piezostimulated currents (PSC) we remind that the solid contains dipoles, the relaxation time $\tau(P, T)$ of which has different values at various pressures. The orientation of a dipole changes through jumps of a *bound* ("b") defect (e.g., vacancy, interstitial, etc.) and the corresponding migration volume $v^{m,b}$ may be positive or negative (§ 7.4.2). As the attempt frequency ν varies only *slightly* on compression (see eq. 7.70) the pressure variation of $\tau(P, T)$ is *solely* regulated by the sign of $v^{m,b}$; more precisely, when $v^{m,b} > 0$ the value of τ increases on compression, whereas when $v^{m,b} < 0$ an increase of P reflects a *decrease* of τ.

In § 7.4.2 it was briefly noted that the piezostimulated currents should be classified into two categories: Depolarization (PSDC) or Polarization (PSPC). The PSPC are observed *during* the continuous action of an electric field, in contrast to PSDC, in which case the current is detected when the electric field has been switched off. The detection of the currents is therefore carried out either by gradually decreasing or increasing the pressure, depending on the sign of $v^{m,b}$. The *maximum* absolute value of the transient current is observed at a pressure P_{cr} at which eq. (7.75a) is obeyed. The measurements are carried out in the following way:

PSD currents; when $v^{m,b}$ is positive, PSDC are observed when the pressure decreases from a (higher) pressure P_h (see § 7.4) to a (lower) pressure P_l. The sample is initially polarized by applying an electric field at the lower pressure for a time appreciably larger than the relaxation time and then increasing the pressure up to P_h, after which the field is switched off. In the case of a negative $v^{m,b}$ one starts by applying the field at P_h, switching it off after the pressure has been reduced to P_l; then the depolarization current is observed during a gradual increase of pressure from $P = P_l$ to $P = P_h$.

PSP currents; when $v^{m,b} > 0$, PSPC are detected when decreasing the pressure from P_h to P_l (the field is applied at $P = P_h$ until $P = P_l$ is reached). On the other hand, when $v^{m,b} < 0$ the detection is carried out when increasing the pressure from P_l to P_h (applying the field from $P = P_l$ to $P = P_h$).

Note 9: detection of a piezoelectric effect in rocks. Baird and Kennan (G.A. Baird and P.S. Kennan, Tectonophysics **111** (1985) 147) have recently found a piezoelectric effect in tourmalinite rocks of the order of 25 to 100 times weaker than that obtained from (natural) tourmaline. An extensive study restricted to the observation of piezoelectric effect in rocks containing quartz has been published by Parkhomenko (E.I. Parkhomenko, in: Electrification Phenomena in Rocks, Plenum Press, New York, 1971); see also the study of some mylonites by Bishop (J.R. Bishop, Tectonophysics **77** (1981) 297).

Note 10: physical origin of the directivity effect. Concerning the physical origin of the "directivity effect" the following comments might be useful. In principle, the absence of a SES at a station – although it may have an appreciably smaller epicentral distance than the station that actually recorded this SES – could be due to any of the following three causes: (1) anisotropic emission of the current at its source, (2) resistivity anomalies in the path of the current, and (3) inhomogeneities of the earth in the vicinity of the station. Although the directivity effect may be a combined result of the above three causes, we are inclined to believe that the third one plays the prominent role in view of the following: Consider two stations A and B placed at equal distances r from a seismic area C. Assume further that the distance d between the stations is appreciably smaller (e.g. by one order of magnitude) than the epicentral distances r. It might happen that (*all sizeable*) earthquakes in C produce clear SES at station A but not at B. As the main part of the current paths CA and CB are practically the same, it is evident that the above systematic absence of SES at station B can be explained from the existence of inhomogeneities in its vicinity.

Note 11: possible explanation of the connection between earthquake magnitude and amplitude of the electric signal(SES). The experimental fact that the *maximum* deviation ΔE of the variation of the electric field scales with M may be understood in the following way: for simplification, assume an ellipsoidal stressed volume V (semi-axes: L, w, h) of which the major semi-axis L is parallel to the polarising field. Upon increasing stress a simultaneous achievement of the stimulating stress σ_{cr} (i.e., $P = P_{cr}$ in fig. 7.5) at all points of the stressed volume V should be excluded because then the time integral of the electric field variation should be proportional to the "effective" emitting surface and therefore this integral would scale with magnitude, a behaviour which is not observed. On the other hand, the "points" obeying the condition $\sigma = \sigma_{cr}$ should lie on a surface A that sweeps through the stressed volume with a velocity u (parallel to L). At each moment t the amplitude of the signal measured at a station is proportional to the "emitting" area A and hence the *maximum* amplitude ΔE will be observed when this effective surface A reaches its *maximum* value, i.e. $A_{max} = wh$. Within such a scheme the duration T of the signal reflects the time needed for the surface A to travel through the stressed volume (of course we assume $L/u \gg \tau$); in view of the fact that u does *not* have the same value for different seismic events we see that T does *not* scale with L and hence with M. On the other hand, ΔE is proportional to (wh) and therefore it has to scale with the magnitude (because larger M means larger V and hence larger wh) [596]).

Note 12: compatibility of laboratory experiments in rocks with field measurements. Laboratory experiments (see also E. Dologlou and P. Varotsos, Z. Geophysik, 1985, submitted for publication) showed that the emitted current from rock samples has a density of the order of 10^{-9} to 10^{-10} A/cm^2. Accepting the same value for the emitting surface $wh \approx 1$ km^2 of the seismic source (see § A.11, note 11) and studying a surrounding cylindrical surface with radius $r = 100$ km and a depth of a few km (practically a two dimensional case) we easily find that at a station with an epicentral distance of 100 km the signal should have an amplitude of the order of 1 mV per 100 m (when $\rho \approx 10^3$ Ω m of the layer under the station), which is compatible with the field observations.

Note 13: presence of a polarising electric field in the earth as an the explanation of the nature of the SES. The evidence for the creation of a polarising electric field is strengthened by the following observation: in cases of *large earthquakes, as already mentioned in ref. [596], before* the detection of the pre-Seismic Electric Signal (SES) a gradual transient variation (GVEF) of the (telluric) electric field is generated, which appears as a gradual variation of the background of our measurements. The GVEF starts usually one to two weeks before the earthquake and reaches (at the time of the maximum deviation) an amplitude roughly one order of magnitude larger than the SES. When GVEF is detected it *always* obeys the following two rules: (1) it appears at a small number of stations, i.e. *only* at the stations at which SES are subsequently observed (recall that SES is recorded at a restricted number of stations due to the directivity effect), and (2) the polarity of GVEF was found to be the *same* as that of the SES subsequently observed at the same station.

Note 14: comments on the lead time Δt of SES. Earthquakes with relatively small values (e.g. 7h) of the lead time $\Delta t = (P_{fr} - P_{cr})/b$ (see fig. 7.5 and consider eq. 7.75a) should correspond to relatively large values of b, i.e. to a fast accumulation of stress (and might be of large stress–drop); on the other hand, earthquakes with larger Δt, e.g. 2 to 4 days, should correspond to lower values of b (and might have a smaller stress drop). The study of a limited number of cases indicates that main shocks after a period of quiescence have relatively high Δt-values, i.e. some days (up to 1 week).

Diffusion

Note 15: macroscopic and microscopic diffusion coefficients. The *microscopic* (or particle) diffusion coefficient d for a particular kind of defect, say a vacancy, is given by eq. (7.36) and is related to the mobility μ through the microscopic Einstein relation: $d/\mu = kT/e$. The experimentally measured quantities, the *macroscopic* diffusion coefficient D and conductivity σ, are connected to the above quantities by the relations (x denotes the concentration, see § 7.1): $D = xd$ and $\sigma = eCx\mu$. The combination of the above three equations leads immediately to the macroscopic Nernst–Einstein relation, i.e., to eq. (7.36a). Obviously, Cx denotes the number of defects per unit volume.

Note 16: diffusion of hydrogen in metals. For hydrogen isotopes diffusing in metals deviations from the predictions of classical rate theory ($D_1/D_2 = (m_1/m_2)^{1/2}$) have been observed. In the low-temperature region quantum-mechanical investigations can quantitatively account for the observed temperature dependence of the diffusion coefficient and for its isotope effect and relate it to a competition between over-barrier and tunnelling transitions of the H-isotopes; see e.g. B. Hohler and H. Kronmueller, Phil. Mag. A, **43** (1981) 1189; a very interesting review paper by A. Seeger has appeared in Hydrogen in Metals I, eds. G. Alefeld and J. Voelkl (Springer, Berlin, 1978).

Note 17: calculation of the self-diffusion coefficient of lithium by using recent isothermal elastic data. From columns 3 and 5 of table III of ref. [600] we have $B = 120.8, 118.9, 115.6, 109.4$ kbars and $\Omega = 21.329, 21.438, 21.617, 21.792$ Å^3 for the following temperatures: $T = 195, 238, 294$ and 350 K, respectively. According to the $cB\Omega$-model the *uncorrelated* self-diffusion coefficient is given by the formula (see eq. 9.7) $D = a^2 \nu \exp(-c^{\text{act}} B\Omega/kT)$; using the values $a = 3.5$ Å, $\nu = \nu_D = 7 \times 10^{12}$ s^{-1} and $D\,(T = 195\text{ K}) = 1.34 \times 10^{-15}$ cm^2/s the application of the formula, at the lowest temperature, i.e. $T = 195$ K, gives: $c^{\text{act}} = 0.308$. Inserting the latter value of c^{act} into the same formula we get for $T = 238, 294$ and 350 K the following results: $D = 3.58^{+0.8}_{-0.7} \times 10^{-13}$, $4.98^{+0.9}_{-0.8} \times 10^{-11}$ and $2.13^{+0.3}_{-0.3} \times 10^{-9}$ cm^2/s; these calculated values are in reasonable agreement with the experimental values (ref. [204]): 3.76×10^{-13}, 5.69×10^{-11} and 2.18×10^{-9} cm^2/s, respectively. At the melting point ($T_M = 453.4$ K) a linear extrapolation of the above elastic data gives $B = 98.1$ kbar, $\Omega = 22.115$ Å^3 and hence the calculated D-value is $1.96^{+0.3}_{-0.2} \times 10^{-7}$ cm^2/s, which is only around 25% smaller than the experimental one: 2.49×10^{-7} cm^2/s.

Dielectric losses and thermally stimulated currents

Note 18: comments on the analysis of dielectric loss data at various pressures and temperatures. Considering that the second term γ_{TO}/B of the right-hand side of eq. (7.63b) is usually of minor importance in comparison to the first one we see that the curvature of the plot $\ln \tau$ versus P leads to the value of $\kappa^{\text{m,b}}$ by following a procedure similar to that developed in § 7.3.4. Experimental data therefore lead to the ratios $\beta^{\text{m,b}}/\beta$ and $\kappa^{\text{m,b}}/\kappa$; If these ratios were found to be different from each other, doubt would be cast upon those QA calculations that assume that v^m does not depend explicitly on T, i.e., that take $(dv^m/dT)_V = 0$. The latter condition, when applied to eq. (2.34a) (see p. 21, taking $Q = v^m$), gives: $\beta^{\text{m,b}}/\beta = \kappa^{\text{m,b}}/\kappa$. In other words, experimental data showing $\beta^{\text{m,b}}/\beta \neq \kappa^{\text{m,b}}/\kappa$ directly prove that the above QA calculations are not self-consistent. Recent results in fluorides indeed give such indication, having a negative $\beta^{\text{m,b}}/\beta$ whereas $\kappa^{\text{m,b}}/\kappa$ is positive.

Note 19: compressibility of the migration volume from dielectric loss data. Dielectric relaxation measurements at various pressures in some fluorides indicate that the compressibility $\kappa^{m,b}$ of the volume that corresponds to a bound ("b") defect motion exceeds the bulk compressibility by one order of magnitude; it seems that $\kappa^{m,b}$ for bound vacancies is larger than for bound interstitials, see e.g. J. Fontanella, M. Wintersgill, A.V. Chadwick, S. Saghafian and C.G. Andeen, J. Phys. **C14** (1981) 2451. The phenomenon that the ratio κ^m/κ is larger than unity, i.e. that v^m *decreases with pressure faster than the bulk volume*, has been suggested in ref. [36] and also by P. Varotsos and K. Alexopoulos, Phys. Stat. Sol. (a) **47** (1978) K133.

Note 20: comments on the Szigeti charge e^*. If "f" denotes the short-range force constant and ϵ, ϵ_∞ are the low-frequency and high-frequency dielectric constants, we note that the so-called "dynamical charge" e^* in the usual Szigeti relation:

$$\epsilon - \epsilon_\infty = \frac{4\pi}{9V} \frac{(\epsilon_\infty + 2)^2}{\mu \omega_{TO}^2} (e^*)^2$$

is not exactly equal to the "static" effective charge in the relation $\alpha = (e^*)^2/f$. The force constant f is expressed through the other Szigeti relation:

$$f = (\epsilon_\infty + 2)^{-1}(\epsilon + 2)\mu\omega_{TO}^2.$$

It is therefore obvious that, as mentioned, eqs. (13.33) and (13.41) can *only* be considered as *rough* estimates.

Note 21: comments on the dielectric loss spectrum. The imaginary part ϵ_2 of the dielectric constant consists of two terms: ϵ_2' and ϵ_2''. The first term corresponds to the so called DC conductivity (e.g., motion of *free* vacancies) which is independent of frequency; the other term, ϵ_2'', corresponds to the reorientation of dipoles (e.g., motion of *bound* vacancies). For each temperature, ϵ_2'' depends on the frequency ω of the external electric field and *maximizes* at a frequency ω_m for which: $\omega_m(T)\tau \simeq 1$, where τ obeys eq. (7.64). For the term ϵ_2' we recall that, for T = constant, we have: $\sigma_{DC} \propto \epsilon_2'\omega$ = constant.

Note 22: peak cleaning procedure and some comments on ionic thermocurrents. The "peak cleaning procedure" consists of a selective polarisation and depolarization scheme [112,113]. Consider a TSDC experiment on a crystal having two types of dipoles, giving rise to two current maxima at temperatures T_{1m} and T_{2m} so that $T_{1m} < T_{2m}$. The crystal is polarized at a temperature T_p by applying the field for a time appreciably larger than the relaxation times $\tau_1(T_p)$ and $\tau_2(T_p)$ of both types of dipoles. By cooling the crystal down to T_L and switching off the field (fig. 7.3) the crystal is now heated just beyond T_{1m} so that only the type (1) dipoles decay; it is then cooled down again and reheated to detect the peak at T_{2m}.

Alternatively, one can detect solely the peak of type (1) dipoles by the following procedure: we select a proper polarisation temperature T_p at which $\tau_1(T_p) \ll \tau_2(T_p)$ and apply the external field for a time t appreciably smaller than $\tau_2(T_p)$ but larger than $\tau_1(T_p)$ so that the polarisation of the crystal will result in the saturation of type (1) dipoles only. Therefore, on a subsequent thermal stimulation only the current peak at T_{1m} will appear.

An extensive study of the influence of a temperature dependent equilibrium polarization on TSDC or TSPC is presented in ref. [122]. Among others the following interesting remarks were made: (1) a TSDC must always have a larger magnitude than the corresponding TSPC peak as a result of an increase in polarization during the cooling procedure; (2) for the latter reason the magnitude and area of a TSDC peak depends on the cooling rate.

Mixed crystals

Note 23: comments on the compressibility and expansivity of a mixed crystal. Even when a "true mixed crystal" exists over the entire composition range i.e., $0 \leq x \leq 1$, the serious point arises whether the elastic data of either of the two end members leads to the same κ^d-value when eq. (8.31) is used. This, of course, does not always hold. Therefore one should discriminate that, even when eq. (8.31) fails to give correct κ^d-values, eq. (12.10) can still be used as follows: once the density and the compressibility are experimentally known for a single composition ($n \neq 0$), the quantities v^d and κ^d can be determined from eqs. (12.3) and (12.10); then the compressibility κ is accessible for any other composition by means of eq. (12.10). Note that the procedure just mentioned does not presume the validity of eq. (12.4) or the existence of a true mixed crystal over the whole composition range, i.e., $0 \leq x \leq 1$; it only requires constant v^d and κ^d values over a restricted composition range which should include the single measurement and the predicted point.

Remarks analogous to those above for κ^d are also applicable to β^d, using β^d instead of κ^d, β instead of κ, eq. (8.26) instead of (8.31) and (12.25) instead of (12.10).

REFERENCES

[1] W. Ludwig, Springer Tracts in Modern Physics, Vol. 43 (Springer, Berlin 1967).
[1a] P. Varotsos and K. Alexopoulos, Phys. Rev. **B24** (1981) 904.
[2] Ya.Y. Kraftmakher, High Temp-High Pressure **5** (1973) 433, and references therein.
[3] T.B. Douglas and A.W. Harman, J. Res. Natl. Bur. Stand. **78A** (1974) 4, 515; C_P-measurements of KCl.
[4] A.J. Leadbetter, D.M.T. Newsham and G.R. Settatree, J. Phys. **C2** (1969) 393.
[5] A.J. Leadbetter and G.R. Settatree, J. Phys. **C2** (1969) 385; C_P-measurements of NaCl, KCl and KBr.
[6] Yu.P. Korenkov and R.G. Mints, High Temp. **12** (1974) 989; C_P of Ta and W at high T.
[7] I.G. Kozhevnikov, High Temp. **12** (1974) 785; C_P of Nb.
[8] Y.B. Fedorov, High Temp. **13** (1975) 608; C_P of W, Mo and Nb near their melting points.
[9] F. Bremer and J. Nölting, Ber. Bunsenges. Phys. Chem. **77** (1973) 398; C_P of mixed AgBr–CuBr.
[10] R.Q. Fugate and C.A. Swenson, J. Low Temp. Phys. **10** (1973) 317; C_V-measurement of Ne up to 93 K, the melting point for $P = 2.5$ kbar.
[11] H.J. Huebscher, Cryogenics **6** (1966) 168; C_V-measurement of Ar up to 95 K which is the melting point for $P = 0.6$ kbar.
[12] F. Haensler, K. Gamper and B. Serin, J. Low Temp. Phys. **3** (1970) 23; C_V-measurement of Ar up to 120 K which is the melting point for $P = 1.6$ kbar.
[13] K. Gamper, J. Low Temp. Phys. **6** (1972) 35; C_V-measurement of Xe up to 223 K which is the melting point for $P = 1.7$ kbar.
[14] P. Korpiun and E. Luescher, in: Rare Gas Solids, Vol. II, eds. M.L. Klein and J.A. Venables (Academic Press, New York, 1977) p. 746–762.
[15] J.P. Bastide and C. Loriers-Susse, High Temp.-High Pressures **10** (1978) 427; measurement of C_P for cerium at 300 K for pressures up to 20 kbars. Attention is drawn to the point that for this material and for pressures up to 6 kbars C_P *increases* with pressure. The sign of $(\partial C_P/\partial P)_T$ can be predicted from the thermodynamic relation $(\partial C_P/\partial P)_T = -TV\beta^2\{1 + \beta^{-2}(\partial \beta/\partial T)_P\}$, where V is the volume to which C_P refers.
[16] F.F. Voronov, V.A. Goncharova and O.V. Stal'gorova, Sov. Phys. JETP **49** (1979) 687; the compressibility of the γ-phase of Ce increases on compression.
[17] M.H. Manghnani, K. Katahara and E.S. Fisher, Phys. Rev. **B9** (1974) 1421; measurement of dB/dP for Re at 25°C.
[18] J.L. Tallon, J. Phys. Chem. Solids **41** (1980) 837; the paper contains useful aspects for the Equation of State for solids.
[19] J.P. Romain, A. Migault and J. Jacquesson, J. Phys. Chem. Solids **37** (1976) 1159; a useful compilation of dB/dP for various metals.
[20] H.L. Albert, E.S. Fisher, K.W. Katahara and M.H. Manghnani, J. Phys. **F9** (1979) L209.
[21] Z.P. Chang and G.R. Barsch, J. Phys. Chem. Solids **32** (1971) 27; a useful discussion for the Murnaghan Equation of State in RbCl, RbBr and RbI.
[22] D. Lazarus, Phys. Rev. **76** (1949) 545; this paper provides the pioneering experimental work for the measurement of the elastic constants under pressure in ionic crystals.
[23] M.R. Vukcevich, Phys. Stat. Sol. (b) **54** (1972) 219.

[24] R.C. Hollinger and G.R. Barsch, J. Phys. Chem. Solids **37** (1976) 845; theoretical calculation of higher order elastic constants of alkali halides.
[25] D.S. Puri and M.P. Verma, Phys. Rev. **B15** (1977) 2337; study of the many body effects on dB/dP in alkali halides with an extensive compilation of data.
[26] J. Frankel, M.A. Hussain and R.D. Scanlon, J. Phys. Chem. Solids **40** (1979) 67; Equation of State by methods of continuum mechanics.
[27] F.D. Murnaghan, in: Finite Deformations of an Elastic Solid (Dover, New York 1967).
[28] S. Mori and Y. Hiki, J. Phys. Soc. Jpn. **45** (1978) 1449.
[29] O.M. Krasil'nikov, Sov. Phys. Solid State **19** (1977) 764.
[30] R.D. Peters, M.A. Breazeale and V.K. Pare, Phys. Rev. **B1** (1970) 3245; these measurements show that dB/dP in Cu increases by 10% from 77 to 300 K.
[31] P. Van 't Klooster, N.J. Trappeniers and S.N. Biswas, Physica **B97** (1979) 65; these measurements on Cu show that dB/dP is 5.28 for $T = 79$ K and 5.33 for $T = 298$ K.
[32] C. Kittel, Introduction to Solid State Physics 4th Ed. (Wiley, New York, 1971) p. 142.
[33] J.P. Ganne and J. Von Stebut, Phys. Rev. Lett. **43** (1979) 634.
[34] J. Teltow, private communication.
[35] C.P. Flynn, in: Point Defects and Diffusion (Clarendon Press, Oxford, 1972).
[36] P. Varotsos, W. Ludwig and K. Alexopoulos, Phys. Rev. **B18** (1978) 2683.
[37] H.M. Gilder and D. Lazarus, Phys. Rev. **B11** (1975) 4916.
[38] Ref. [35]; ch. 7; for an excellent review of the dynamics of atomic migration see also Phys. Rev. **171** (1968) 682.
[39] C.A. Wert and C. Zener, Phys. Rev. **76** (1949) 1169.
[40] G.H. Vineyard, Physics Chem. Solids **3** (1957) 121.
[41] S.A. Rice, Phys. Rev. **112** (1958) 804.
[42] N.B. Slater, in: The Theory of Unimolecular Reactions (Cornell, 1959).
[43] H.R. Glyde, Rev. Mod. Phys. **39** (1967) 373; J. Phys. Chem. Solids **28** (1967) 2061.
[44] M. Kac, Am. J. Math. **65** (1943) 609.
[45] M.D. Feit, Phys. Rev. **B3** (1971) 1223.
[46] J.A. Van Vechten, Phys. Rev. **B12** (1975) 1247.
[47] R.A. Felice, J. Trivisonno and D.E. Schuele, Phys. Rev. **B16** (1977) 5173.
[48] Rare Gas Solids, Vol. II, eds. M.L. Klein and J.A. Venables (Academic Press, New York, 1977) p. 776, 802, 1215.
[49] W.E. Schoknecht and R.O. Simmons, in: Thermal Vacancies and Thermal Expansion, eds. M.G. Graham and H.E. Hagey, Symp. Thermal Expansion, Oct. 27–29, 1971 (American Institute of Physics, New York, 1972) pp. 169–182.
[50] Rare Gas Solids, Vol. II, ed. M.L. Klein and J.A. Venables (Academic Press, New York, 1977) ch. 12, P. Korpiun and E. Luescher; ch. 19, A.V. Chadwick and H.R. Glyde.
[51] P. Korpiun and H.J. Coufal, Phys. Stat. Sol. (a) **6** (1971) 187.
[51a] R.O. Simmons and R.W. Balluffi, Phys. Rev. **119** (1960) 600.
[52] W.E. Schoknecht, Ph. D. Thesis (University of Illinois, Urbana, 1971, unpublished).
[53] A.T. Macrander, Phys. Rev. **B21** (1980) 2549.
[54] C.R. Brooks, J. Phys. Chem. Solids **29** (1968) 1377; specific heat measurements of Cu.
[55] C.R. Brooks and R.E. Bingham, J. Phys. Chem. Solids **29** (1968) 1553; measurement of C_P of Al.
[56] C.C. Yeh and C.R. Brooks, High Temp. Sci. **5** (1973) 403; C_P of Pt.
[57] C.R. Brooks, Phys. Stat. Sol. (b) **89** (1978) K123; reanalysis of the data of ref. [56].
[58] W. Kramer and J. Nölting, Acta Metall. **20** (1972) 1353; one of the authors (P.V.) greatly appreciates a very useful discussion with Prof. J. Nölting on this point.
[59] Y. Adda and J. Philibert, in: La Diffusion dans les Solides (Presses Univ. de France, 1966).

[60] J. Bardeen and C. Herring, in: Atom Movements (A.S.M. Cleveland, 1951) p. 87; also in: Imperfections in Nearly Perfect Crystals, ed. W. Shockley (Wiley, New York, 1952) p. 26.
[61] A.D. Le Claire in: Physical Chemistry vol. 10, eds. H. Eyring, D. Henderson and W. Jost (Academic Press, New York, 1970) ch. 6.
[62] H. Mehrer, J. Nucl. Mat. Sci. **69–70** (1978) 38.
[63] J.G. Mullen, Phys. Rev. **124** (1961) 1723.
[64] E. Steiner, H. Mehrer and A. Seeger, Phys. Stat. Sol. (b) **75** (1976) 361.
[65] R.J. Friauf, J. Physique **38** (1977) 1077.
[66] N.L. Peterson, Isotope Effects in Diffusion, in: Diffusion in Solids, Recent Developments, eds. A.S. Nowick and J.J. Burton (Academic Press, New York, 1975) ch. 3.
[67] A. Seeger and H. Mehrer, in: Vacancies and Interstitials in Metals, eds. A. Seeger, D. Schumacher, W. Schilling and J. Diehl (North-Holland, Amsterdam, 1970) p. 1.
[68] W.J. Fredericks, Diffusion in Alkali Halides in: Diffusion in Solids, Recent Developments, eds. A.S. Nowick and J.J. Burton (Academic Press, New York, 1975) ch. 8.
[69] M. Beniere, M. Chemla and F. Beniere, J. Phys. Chem. Solids **37** (1976) 525.
[70] A.P. Batra and L.M. Slifkin, J. Phys. **C9** (1976) 947.
[71] A.P. Batra and L.M. Slifkin, J. Phys. **C11** (1978) L317.
[72] H. Mehrer and A. Seeger, Phys. Stat. Sol. **35** (1969) 313; **39** (1970) 647.
[73] N.L. Peterson, Comments Solid State Phys. **8** (1978) 107.
[74] P. Audit and H.M. Gilder, Phys. Rev. **B18** (1978) 4151.
[75] J.N. Mundy, Phys. Rev. **B3** (1971) 2431.
[76] A.S. Nowick and G.J. Dienes, Phys. Stat. Sol. **24** (1967) 461.
[77] B.N.N. Achar, Phys. Rev. **B2** (1970) 3848.
[78] R.E. Pawell, Rev. Sci. Instrum. **35** (1964) 1066.
[79] R.E. Pawel and T.S. Lundy, J. Phys. Chem. Solids **26** (1965) 937; J. Electrochem. Soc. **115** (1968) 233.
[80] D. Gupta and R.T.C. Tsui, Appl. Phys. Lett. **17** (1970) 294.
[81] K. Maier and W. Schuele, Euratom Report EUR 5234 d (1974).
[82] K. Maier, C. Bassani and W. Schuele, Phys. Lett. **A44** (1973) 539.
[83] K. Maier, H. Mehrer, E. Lessmann and W. Schuele, Phys. Stat. Sol. (b) **78** (1976) 689.
[84] J.H. Crawford and L.M. Slifkin, in: Defect in Solids, Vol. I (Plenum Press, New York, 1972).
[85] J.H. Crawford and L.M. Slifkin, Ann. Rev. Material Sci. **1** (1971) 139.
[86] L.M. Slifkin, Introductory talk in: Conference on Defects in Insulating Crystals, Catlinburg (1977).
[87] J. Corish, P.W.M. Jacobs and S. Radhakrishna, Point Defects in Ionic Crystals in: Surface and Defect Properties of Solids, Vol. 6, eds. M.W. Roberts and J.M. Thomas (The Chemical Society, London, 1977) p. 218.
[88] W. Bollmann, Phys. Stat. Sol. (a) **61** (1980) 395.
[89] L.A. Acuna and P.W.M. Jacobs, J. Phys. Chem. Solids **41** (1980) 595, and references therein.
[90] C.R.A. Catlow, J. Corish, K.M. Diller, P.W.M. Jacobs and M.J. Norgett, J. Physique Colloq. **37** (1976) C7-253, and references therein.
[91] N.F. Mott and R.W. Gurney, in: Electronic Processes in Ionic Crystals 2nd Ed. (Oxford University Press, London) ch. 11.
[92] P. Varotsos and K. Alexopoulos, J. Phys. Chem. Solids **38**, 997 (1977); J. Physique Lettres **38** (1977) L329; Phys. Stat. Sol. (b) **92** (1979) 633.
[93] P. Varotsos, J. Physique **37** (1976) C7-327.
[94] P. Varotsos and K. Alexopoulos, J. Phys. Chem. Solids **39** (1978) 759.

[95] L.A. Acuna and P.W.M. Jacobs, J. Physique **41** (1980) C6-72.
[96] C.R.A. Catlow, J. Corish, K.M. Diller, P.W.M. Jacobs and M.J. Norgett, J. Phys. C **12** (1979) 451.
[97] C.R.A. Catlow, J. Corish and P.W.M. Jacobs, J. Phys. **C12** (1979) 3433; **13** (1980) 1977.
[98] A.P. Batra and L.M. Slifkin, Phys. Rev. **B12** (1975) 3473.
[99] A.S. Nowick and W.R. Heller, Adv. Phys. **14** (1965) 101.
[100] A.S. Nowick, Adv. Phys. **16** (1967) 1.
[101] P. Varotsos and D. Miliotis, J. Phys. Chem. Solids **35** (1974) 927.
[102] R. Leib, Thesis (University of North Carolina, Greensboro, 1976).
[103] A.P. Batra, J.P. Hernandez and L.M. Slifkin, Phys. Rev. **B22** (1980) 734, and references therein.
[104] P. Varotsos and K. Alexopoulos, Philos. Mag. **A42** (1980) 13.
[105] C. Andeen, I.M. Hayden and J. Fontanella, Phys. Rev. **B21** (1980) 794.
[106] J. Fontanella, M.C. Wintersgill and C. Andeen, Phys. Stat. Sol. (b) **97** (1980) 303.
[107] C. Bucci and R. Fieschi, Phys. Rev. Lett. **12** (1964) 16.
[108] C. Bucci, R. Fieschi and G. Guidi, Phys. Rev. **148** (1966) 816.
[109] C. Bucci, Phys. Rev. **164** (1967) 1200.
[110] L. Nunes de Oliveira and G.F.L. Ferreira, Nuovo Cimento **B23** (1974) 385.
[111] P. Müller and J. Teltow, Phys. Stat. Sol. **12** (1972) 471.
[111a] D. Shaw and H.R. Moghaddam, Phys. Stat. Sol. (a) **65** (1981) 167.
[112] S.W.S. Mckeever and D.M. Hughes, J. Phys. **D8** (1975) 1520.
[113] S. Radhakrishna and S. Harridos, Cryst. Lattice Defects **7** (1978) 191.
[113a] E. Laredo, M. Puma, N. Suarez and D. Figueroa, Phys. Rev. **B23** (1981) 3009.
[114] R.D. Shelley and G.R. Miller, J. Solid State Chem. **1** (1970) 218.
[115] W. Van Weperen, B.P.M. Lenting, E.J. Bijvank and H.W. den Hartog, Phys. Rev. **B16** (1977) 2953; see also J. Physique **41** (1980) C6-275.
[115a] D. Triantis, Dissertation (University of Athens, Athens, 1983), in Greek.
[116] J.P. Stott and J.H. Crawford Jr., Phys. Rev. Lett. **26** (1971) 384.
[117] E.L. Kitts, Jr., M. Ikeya and J.H. Crawford Jr., Phys. Rev. **B8** (1973) 5840.
[118] E.L. Kitts Jr. and J.H. Crawford Jr., Phys. Rev. **B9** (1974) 5264.
[119] G.E. Matthews Jr. and J.H. Crawford Jr., Phys. Rev. **B15** (1977) 55, and references therein.
[119a] P.W.M. Jacobs and S.H. Ong, J. Phys. Chem. Solids **41** (1980) 431; **41** (1980) 437, and references therein.
[119b] A.B. Aalbers and H.W. den Hartog, Phys. Rev. **19** (1979) 2163.
[119c] H.W. den Hartog and J.C. Langevoort, Phys. Rev. **24** (1981) 3547.
[120] F. Cusso and F. Jaque, Solid State Commun. **29** (1979) 283; study of NaCl doped with Mn^{2+} from LN to 600 K.
[120a] F. Cusso and F. Jaque, Cryst. Latt. Defects **9** (1981) 127; KI doped with Ca^{2+}, Sr^{2+}, Pb^{2+} and Ba^{2+}.
[120b] F. Cusso and F. Jaque, Solid State Commun. **35** (1980) 965; J. of Phys. **C15** (1982) 2875.
[120c] J. Hernandez, H. Murrieta, F. Jaque and J. Rubio, Solid State Commun. **39** (1981) 1061; study of eight alkali halides doped with Eu^{2+}.
[120d] P. Varotsos, D. Kostopoulos and S. Mourikis, Phys. Stat. Sol. (a) **57** (1980) 331; study of NaF doped with Ca^{2+}.
[120e] D. Kostopoulos, P. Varotsos and S. Mourikis, J. of Phys. **C13** (1980) 3003.
[120f] D. Triantis and D. Kostopoulos, J. of Phys. **C15** (1982) 4037; study of NaI doped with Ca^{2+} as in ref. [120e].
[120g] D. Kostopoulos, S. Mourikis and P. Varotsos, J. Physique **42** (1981) 1481; measurements of KCl with divalent anions of SO_4^{2-}.
[120h] A. Kessler and K. Ebert, Z. Phys. **B34** (1979) 1; study of NH_4Cl doped with NH_3 or

with $NiCl_2$.
- [120i] M. Jacquet and M. Bathier, J. Physique Colloq. C6, **41** (1980) 447.
- [120k] H.J. Krokoszinski and K. Baerner, J. Chem. Phys. **72** (1980) 2616.
- [120ℓ] P. Varotsos, D. Kostopoulos, S. Mourikis and S. Kouremenou, Solid State Commun. **21** (1977) 831.
- [120m] M. Zielinski, T. Swiderski and M. Kryszewski, Polymer **19** (1978) 883.
- [120n] D. Kostopoulos, S. Kouremenou, P. Varotsos and S. Mourikis, Phys. Stat. Sol. (a) **70** (1982) K19.
- [120o] E. Dologlou-Reveliotti, private communication.
- [120p] J. Vanderschueren and A. Linkens, J. of Electrostatics **3** (1977) 155.
- [120q] D. Figueroa, E. Laredo and M. Puma, Solid State Commun. **25** (1978) 509; K. Suzuki, J. Phys. Soc. Jpn. **16** (1961) 67.
- [121] P.R. Moran and D.E. Fields, J. Appl. Phys. **45** (1974) 3266.
- [121a] G. Pfister and A. Ankowitz, J. Appl. Phys. **45** (1974) 1001.
- [121b] S.W. McKeever and E. Lilley, J. of Phys. **C14** (1981) 3547.
- [121c] N. Kristianpoller and Y. Kirsch, J. Phys. **C12** (1979) 1073.
- [122] J. Vanderschueren, A. Linkens, J. Gasiot, J.P. Fillard and P. Parot, J. Appl. Phys. **51** (1980) 4967.
- [122a] C.G. Andeen, J. Fontanella and M.C. Wintersgill, J. Phys. **C13** (1980) 3449.
- [122b] M. Siu Li, M. deSouza and S.E. Kapphan, Phys. Stat. Sol. (b) **112** (1982) 685.
- [123] Bui Ai, P. Destruel, Hoang The Gima and R. Loussier, Phys. Rev. Lett. **34** (1975) 84.
- [123a] P. Varotsos, K. Alexopoulos and K. Nomicos, Pract. Athens Acad. **56** (1981) 277; **56** (1981) 417; **57** (1982) 341.
- [123b] P. Varotsos, K. Alexopoulos and K. Nomicos, Phys. Stat. Sol. (b) **111** (1982) 581.
- [123c] P. Varotsos, K. Alexopoulos, K. Nomicos, G. Papaioannou, M. Varotsou and E. Dologlou-Revelioti, Pract. Athens Acad. **56** (1981) 434.
- [124] C.H.P. Lupis, Acta Metall. **25** (1977) 751.
- [125] R.E. Howard and A.B. Lidiard, Rep. Progr. Phys. **27** (1964) 161.
- [126] C. Wagner, Acta Metall. **19** (1971) 843.
- [127] A.D. Franklin, in: Point Defects in Solids, eds. J.H. Crawford and L.M. Slifkin (Plenum Press, New York 1972) ch. 1.
- [128] A. Mainwood and A.M. Stoneham, Phil. Mag. **37** (1978) 255; **37** (1978) 263.
- [129] R.B. McLellan, Acta Metall. **27** (1979) 793; Scr. Metall. **12** (1978) 345; J. Phys. Chem. Solids **39** (1978) 781; also in: Treatise in Materials Science, ed. H. Herman (Academic Press, New York, 1975).
- [130] F. Beniere, J. Phys. Lett. **37** (1976) L177.
- [131] R. Raab and H. Peisl, private communication.
- [132] C.M. Garza and D.A. Huckaby, J. Chem. Phys. **72** (1980) 4982.
- [132a] P. Varotsos and K. Alexopoulos, Phys. Rev. **B22** (1980) 3130.
- [133] K. Alexopoulos and P. Varotsos, J. Phys. **F8** (1978) 2227.
- [134] R. Guerard, H. Peisl and R. Zitzmann, Appl. Phys. **3** (1974) 37.
- [135] R.O. Simmons and R.W. Balluffi, Phys. Rev. **117** (1960) 52.
- [136] G. Bianchi, D. Mallejac, C. Janot and G. Champier, C.R. Hebd. Séances Acad. Sci. Ser. **B263** (1966) 1404.
- [137] J.L. Tallon and A. Wolfenden, J. Phys. Chem. Solids **40** (1979) 831.
- [138] A.J. Cornish and J. Burke, J. Scient. Instr. **42** (1955) 212.
- [139] A.S. Berger, S.T. Ockers, M.K. Chason and R.W. Siegel, J. Nucl. Mater. **69–70** (1978) 734.
- [140] J. Bass, Phil. Mag. **15** (1967) 717.
- [141] R.W. Siegel, J. Nucl. Mater. **69–70** (1978) 117.

[142] A. Seeger, J. Phys. **F3** (1973) 248; this is a review article with an extensive list of references for defect parameters in metals.
[143] P. Audit and H.M. Gilder, J. Nucl. Mater. **69–70** (1978) 641.
[144] H.M. Gilder and P. Audit, Phys. Rev. Lett. **38** (1977) 30.
[145] G.M. Hood and R.J. Schultz, J. Phys. **F10** (1980) 545.
[146] T.S. Lundy and J.F. Murdock, J. Appl. Phys. **33** (1962) 1671.
[147] M. Beyeler and Y. Adda, J. Physique **29** (1968) 345.
[148] Ch.Y. Sun, Ph.D.Thesis (University of Illinois, Urbana, 1971).
[149] C.P. Slichter and D.C. Ailion, Phys. Rev. **135A** (1964) 1099.
[150] R.C.G. Killean and E.J. Lisher, J. Phys. **C8** (1975) 3510; J. Phys. **F5** (1975) 1107.
[151] R.O. Simmons and R.W. Balluffi, Phys. Rev. **129** (1963) 1533.
[152] R.R. Bourassa and B. Lengeler, J. Phys. **F6** (1976) 1405.
[153] R. Scholz and W. Schuele, Phys. Lett. **64A** (1977) 340.
[154] A.S. Berger, S.T. Ockers and R.W. Siegel, J. Phys. **F9** (1979) 1023.
[155] D. Bartdorff, G. Neumann and P. Reimers, Phil. Mag. **A38** (1978) 157.
[156] Ref. [32], p. 219.
[157] Y.A. Chang and L. Himmel, J. Appl. Phys. **37** (1966) 3568.
[158] S. Nanao, K. Kuribayashi, S. Tanigawa and M. Doyama, J. Phys. **F3** (1973) L225; **F7** (1977) 1403.
[159] W. Triftshäuser and J.D. McGervey, Appl. Phys. **6** (1975) 177.
[160] M. Doyama, K. Kuribayashi, S. Nanao and S. Tanigawa, Appl. Phys. **4** (1974) 153.
[161] N.Q. Lam, S.J. Rothman and L.J. Nowicki, Phys. Stat. Sol. (a) **23** (1974) K35.
[162] R.O. Simmons and R.W. Balluffi, Phys. Rev. **119** (1960) 600.
[163] J.D. McGervey and W. Triftshauser, Phys. Lett. **A44** (1973) 53.
[164] J.L. Campbell, C.W. Schulte and J.A. Jackman, J. Phys. **F7** (1977) 1985.
[165] Y.P. Varshni, Phys. Rev. **B2** (1970) 3952.
[166] N.Q. Lam, S.J. Rothman, H. Mehrer and L.J. Nowicki, Phys. Stat. Sol. (b) **57** (1973) 225.
[167] F. Seitz, in: Modern Theory of Solids (McGraw–Hill, New York, 1940) p. 110.
[168] R. Feder and A.S. Nowick, Phys. Rev. **109** (1958) 1959.
[169] R. Feder and A.S. Nowick, Phil. Mag. **15** (1967) 805.
[170] N.H. Nachtrieb and G.S. Handler, J. Chem. Phys. **23** (1955) 1659.
[171] H.B. Vanfleet, Phys. Rev. **B21** (1980) 4340.
[172] A.D. LeClaire, in: Metals Reference Book, Vol. II, 4th Ed. C.J. Smithells (London, 1967) p. 637–653.
[173] H.B. Vanfleet, private communication.
[174] H. Mehrer and A. Seeger, Cryst. Lattice Defects **3** (1972) 1.
[175] J.B. Hudson and R.E. Hoffman, Trans. Met. Soc. AIME **221** (1961) 761.
[176] C.L. Vold, M.E. Glicksman, E.W. Kammer and L.C. Cardinal, J. Phys. Chem. Solids **38** (1977) 157.
[177] R.A. Miller and D.E. Schuele, J. Phys. Chem. Solids **30** (1969) 589.
[178] T.E. Pochapsky, Acta Metall. **1** (1953) 747.
[178a] P. Varotsos, Phys. Stat. Sol. (a) **55** (1979) K139.
[178b] L. De Schepper, J. Cornelis, G. Knuyt, J. Nihoul and L. Stals, Phys. Stat. Sol. (a) **61** (1980) 341.
[178c] K. Alexopoulos and P. Varotsos, Phys. Rev. **B24** (1981) 3606.
[179] G.A. Sullivan and J.W. Weymouth, Phys. Rev. **A136** (1964) 1141.
[180] R. Feder and H.P. Charbnau, Phys. Rev. **149** (1966) 464.
[181] M. Ritter, G. Fritsch and E. Luescher, J. Appl. Phys. **41** (1970) 5071.
[182] W. Adlhart, G. Fritsch, A. Heidemann and E. Luescher, Phys. Lett. **A47** (1974) 91.

[183] W. Adlhart, G. Fritsch and E. Luescher, J. Phys. Chem. Solids **36** (1975) 1405.
[184] J.N. Mundy, L.W. Barr and F.A. Smith, Phil. Mag. **15** (1967) 411.
[185] N.H. Nachtrieb, J.A. Weil, E.Catalano and A.W. Lawson, J. Chem. Phys. **20** (1952) 1189.
[186] G. Göltz, A. Heidemann, H. Mehrer, A. Seeger and D. Wolf, Phil. Mag. **A41** (1980) 723.
[187] M. Ait-Salem, Jülich-report no. 1322 (1976).
[188] G. Brünger, O. Kanert and W. Wolf, Phys. Rev. **B22** (1980) 4247.
[189] M.D. Feit, Phil. Mag. **25** (1972) 769.
[190] A. Da Fano and G. Jacucci, Phys. Rev. Lett. **39** (1977) 950.
[191] P. Varotsos and K. Alexopoulos, Phys. Rev. **B21** (1980) 3379.
[192] W. Franklin, J. Chem. Phys. **57** (1972) 2659.
[193] C. Herzig and H. Eckseler, Z. Metallkd. **70** (1979) 215.
[194] D.L. Martin, Phys. Rev. **154** (1967) 571.
[195] R.H. Martinson, Phys. Rev. **178** (1969) 902.
[196] G. Fritsch, F. Geipel and A. Prasetyo, J. Phys. Chem. Solids **34** (1973) 1961.
[197] R.I. Beecroft and C.A. Swenson, J. Phys. Chem. Solids **18** (1961) 329; see also C.E. Monfort and C.A. Swenson, J. Phys. Chem. Solids **26** (1965) 219.
[198] G. Fritsch, M. Nohman, P. Korpiun and E. Luescher, Phys. Stat. Sol. (a) **19** (1973) 555.
[199] M.E. Diederich and J. Trivisonno, J. Phys. Chem. Solids **27** (1966) 637.
[200] R. Grover, I.C. Getting and G.C. Kennedy, Phys. Rev. **B7** (1973) 567.
[201] P.S. Ho and A.L. Ruoff, J. Phys. Chem. Solids **29** (1968) 2101.
[202] R. Feder, Phys. Rev. **B2** (1970) 828.
[203] R. Messer, Ph. D. Thesis (Stuttgart, 1976).
[204] R. Messer and F. Noack, Appl. Phys. **6** (1975) 79.
[205] D.C. Ailion and C.P. Slichter, Phys. Rev. **137A** (1965) 235.
[206] R.A. Hultsch and R.G. Barnes, Phys. Rev. **125** (1962) 1832.
[207] R.G. Barnes, R.D. Engardt and R.A. Hultsch, Phys. Rev. Lett. **2** (1959) 202.
[208] J.M. Titman and B.M. Moores, J. Phys. **F2** (1972) 592.
[209] A. Lodding, J.N. Mundy and A. Ott, Phys. Stat. Sol. **38** (1970) 559.
[210] J.P. Day and A.L. Ruoff, Phys. Stat. Sol. (a) **25** (1974) 205.
[211] C.A. Swenson, J. Phys. Chem. Solids **27** (1966) 33.
[211a] M. Pokorny, J. Phys. **F12** (1982) 39.
[211b] G. Stetter, W. Adlhart, G. Fritsch, E. Steichele and E. Luescher, J. Phys. **F8** (1978) 2075.
[212] D.K.C. MacDonald, J. Chem. Phys. **21** (1953) 177.
[213] Ya.A. Kraftmakher and P.G. Strelkov, in: Vacancies and Interstitials in Metals, eds. A. Seeger, P. Schümacher, W. Schilling and J. Diehl (North-Holland, Amsterdam, 1970) p. 59.
[214] J.N. Mundy, T.E. Miller and R.J. Porte, Phys. Rev. **B3** (1971) 2445.
[215] D.R. Schouten and C.A. Swenson, Phys. Rev. **B6** (1974) 2175.
[216] G. Fritsch and H. Bube, Phys. Stat. Sol. (a) **30** (1975) 571.
[217] K. Maier, M. Peo, B. Saile, H.E. Schaefer and A. Seeger, Phil. Mag. **A40** (1979) 701.
[218] K.D. Rasch, R.W. Siegel and H. Schultz, Phil. Mag. **A41** (1980) 91.
[219] R.G. Gupta and R.W. Siegel, J. Phys. **F10** (1980) L7.
[220] J.N. Mundy, S.J. Rothman, N.Q. Lam, H.A. Hoff and L.J. Nowicki, Phys. Rev. **B18** (1978) 6566.
[221] W. Danneberg, Metall (Berlin) **15** (1961) 977.
[222] A.A. Korolev, L.V. Pavlinov and M.J. Gavrilyuk, Fiz. Met. Metalloved. **33** (1972) 295.
[223] R. Lowrie and A.M. Gonas, J. Appl. Phys. **38** (1967) 4505.
[224] G. Morizur, A. Radenac and J.C. Cretenet, High Temp.-High Pressures **8** (1976) 113.

[225] Metals Handbook: Properties and Selection of metals, Vol. I, 8th Ed., ed. Taylon-Lyndon (American Society for Metals, Cleveland, 1961).
[226] D. Ablitzer, Phil. Mag. **35** (1977) 1239, and references therein.
[227] R.E. Einziger, J.N. Mundy and H.A. Hoff, Phys. Rev. **B17** (1978) 440.
[228] Y. Talmor, E. Walker and S. Steinemann, Solid State Commun. **23** (1977) 649.
[229] V.A. Petukhov, V.Ya. Chekhovskoi and A.G. Mozgovoi, Teplofiz. Vys. Temp. **16** (1978) 421.
[230] R. Balzer and H. Sigvaldason, J. Phys. **F9** (1979) 171.
[231] M. Current and H.M. Gilder, Phys. Rev. **B16** (1977) 2386.
[232] H.M. Gilder and G.N. Wallmark, Phys. Rev. **182** (1969) 771; measurements of bulk thermal expansion of Zn.
[232a] P.D. Pathak and R.J. Desai, Phys. Stat. Sol. (a) **62** (1980) 625.
[233] J.P. Simon, P. Vostry, J. Hillairet and P. Vajda, Phys. Stat. Sol. (b) **64** (1974) 277.
[234] N.L. Peterson and S.J. Rothman, Phys. Rev. **163** (1967) 645.
[235] G.A. Shirn, E.S. Wajda and H.B. Huntington, Acta Metall. **1** (1953) 513.
[236] B.J. Buescher, H.M. Gilder and N. Shea, Phys. Rev. **B7** (1973) 2261.
[237] H.M. Ledbetter, Elastic Properties of Zinc, a Compilation and a Review, in: J. Phys. Chem. Ref. Data **6 (4)** (1977) 1181.
[238] L.C. Chhabildas and H.M. Gilder, Phys. Rev. **B5** (1972) 2135.
[239] S.N. Vaidya and G.C. Kennedy, J. Phys. Chem. Solids **31** (1970) 2329.
[240] K.D. Swartz and C. Elbaum, Phys. Rev. **B1** (1970) 1512.
[241] R. Balzer and H. Sigvaldason, Phys. Stat. Sol. (b) **92** (1979) 143.
[242] W. Desorbo, J. Phys. Chem. Solids **15** (1960) 7.
[242a] S. Dedoussis, S. Charalambous and M. Chardalas, Phys. Lett. **A62** (1977) 359.
[242b] I.K. Mackenzie and P.C. Lichtenberger, Appl. Phys. **9** (1976) 331.
[243] M.I. Current, Ph.D.Thesis (Rensselaer Polytechnic Institute, Troy, NY, 1974).
[244] E.W. Kammer, L.C. Cardinal, C.L. Vold and M.E. Glicksman, J. Phys. Chem. Solids **33** (1972) 1891.
[245] We shall assume that $D_{\perp c}$ can be expressed through a single Gibbs activation energy.
[246] C. Coston and N.H. Nachtrieb, J. Phys. Chem. **68** (1964) 2219.
[247] W.K. Warburton and D. Turnbull, Fast Diffusion in Metals, in: Diffusion in Solids, Recent Developments, eds. A.S. Nowick and J.J. Burton, (Academic Press, New York, 1975).
[248] B.F. Dyson, J. Appl. Phys. **37** (1966) 2375.
[249] B.F. Dyson, T.R. Anthony and D. Turnbull, J. Appl. Phys. **38** (1967) 3408.
[250] B.E. Sirovich and R.E. Norberg, Quoted by A.V. Chadwick and H.R. Glyde, in: Rare Gas Solids, Vol. II, eds. M.L. Klein and J.A. Venables (Academic Press, New York, 1977).
[251] R. Henry and R.E. Norberg, Phys. Rev. **B6** (1972) 1645.
[252] P. Korpiun and E. Luescher, in: Rare Gas Solids, Vol. II, eds. M.L. Klein and J.A. Venables (Academic Press, New York, 1977) p. 798 and the data plotted in fig. 27.
[253] M.S. Anderson, R.Q. Fugate and C.A. Swenson, J. Low Temp. Phys. **10** (1973) 345.
[254] F. Birch, J. Phys. Chem. Solids **38** (1977) 175.
[255] D.N. Batchelder, D.L. Losee and R.O. Simmons, Phys. Rev. **162** (1967) 767.
[256] A. Bernè, G. Boato and M. De Paz, Nuovo Cimento **46** (1966) 182.
[257] E.H.C. Parker, H.R. Glyde and B.L. Smith, Phys. Rev. **176** (1968) 1107.
[258] L.A. Schwalbe, Phys. Rev. **B14** (1976) 1722.
[259] M.S. Anderson and C.A. Swenson, J. Phys. Chem. Solids **36** (1975) 145.
[260] P. Korpiun, G. Kampfer and E. Luescher, quoted in: Rare Gas Solids, Vol. II, eds. M.L. Klein and J.A. Venables (Academic Press, New York, 1977) p. 809.

[261] D.G. Cowgill and R.E. Norberg, Phys. Rev. **B13** (1976) 2773.
[262] A.V. Chadwick and J.A. Morrison, Phys. Rev. **B1** (1970) 2748
[263] D.L. Losee and R.O. Simmons, Phys. Rev. **172** (1968) 934.
[264] W.M. Yen and R.E. Norberg, Phys. Rev. 131 (1963) 269.
[265] Rare Gas Solids, Vol. II, eds. M.L. Klein and J.A. Venables (Academic Press, New York, 1977) p. 782.
[266] A.V. Chadwick and H.R. Glyde, in ref. [265], ch. 19.
[267] R.A. Miller and C.S. Smith, J. Phys. Chem. Solids **25** (1964) 1279.
[268] P. Varotsos, D. Kostopoulos and S. Mourikis, Phys. Stat. Sol. (a) **57** (1980) 331.
[269] M. Lagu and B. Dayal, J. Phys. **C8** (1975) 961; the expansivity data have been obtained from fig. 4 of this reference.
[270] C.F. Bauer and D.H. Whitmore, Phys. Stat. Sol. **37** (1970) 585.
[271] P.W.M. Jacobs and P. Pantelis, Phys. Rev. **4** (1971) 3757.
[272] A.B. Lidiard, Phys. Rev. **94** (1954) 29.
[273] H. Yoshizawa and K. Hirakawa, J. Phys. Soc. Jpn. **43** (1977) 793.
[274] L.E.A. Jones, Phys. Earth Planet. Inter. **13** (1976) 105.
[275] S. Hart, J. Phys. **D10** (1977) L261; Jones (ref. [274]) and Hart observe definitely that B falls with T in a quadratic form; unfortunately Jones' measurements extend only up to 700 K whereas the coefficients of the quadratic fits given by Hart seem to have an error.
[276] C.S. Smith and L.S. Cain, J. Phys. Chem. Solids **36** (1975) 205.
[277] R.A. Bartels and D.E. Schuele, J. Phys. Chem. Solids **26** (1965) 537.
[278] G. Leibfried and W. Ludwig, Solid State Phys. **12** (1961) 276.
[279] K.K. Srivastava and H.D. Merchant, J. Phys. Chem. solids **34** (1973) 2069.
[280] S. Chandra and J. Rolfe, Can. J. Phys. **48** (1970) 412.
[281] See the pioneering experiments of Batra and Slifkin in silver halides in refs. [98] and [305].
[282] H. Spetzler, C.G. Sammis and R.J. O'Connell, J. Phys. Chem. Solids **33** (1972) 1727; these data are considered as reliable because, at low temperatures, they coincide with the excellent compilation reported in ref. [276] for RT.
[283] C. Nadler and J. Rossel, Phys. Stat. Sol. (a) **18** (1973) 711.
[284] A.R. Alnatt, P. Pantelis and S.J. Sime, J. Phys. **C4** (1971) 1778.
[285] J.S. Cook and J.S. Dryden, J. Phys. **C12** (1979) 4207.
[286] F. Beniere and R. Rokbani, J. Phys. Chem. Solids **36** (1975) 1151.
[287] H. Mashida and W.J. Fredericks, J. Physique **37** (1976) C7-385.
[288] J.S. Cook and J.S. Dryden, J. Phys. **C14** (1981) 1133; the point (1.24 eV, 5.43 k) lies on the line predicted in fig. 9.34 from $cB\Omega$-theory.
[289] G.R. Barsch and Z.P. Chang, Phys. Stat. sol. **19** (1967) 139.
[290] J.E. Rapp and H.D. Merchant, J. Appl. Phys. **44** (1973) 3919.
[291] I.N. Cyrbu, V.L. Ul'yanow and A.A. Botaki, Fiz. Tverd. Tela **15** (1973) 3389.
[292] G.R. Barsch and H.E. Shull, Phys. Stat. Sol. (b) **43** (1971) 637.
[293] D. Kostopoulos, K.V. Reddy and F. Beniere, J. Physique **41** (1980) C6-252, and references therein.
[294] L.E. Wylde, Thesis (University of Kent, 1972).
[295] Y.N. Pershits and V.L. Veisman, Sov. Phys. Solid State **12** (1971) 2566.
[296] D. Triantis and D. Kostopoulos, see ref. [120f].
[297] D. Kostopoulos, F. Beniere and K.V. Reddy, J. Phys. Chem. Solids **40** (1979) 357; **41** (1980) 727.
[298] S.C. Chandra and J. Rolfe, Can. J. Phys. **48** (1970) 397; analysis of the conductivity data of pure KI, $KI + Sr^{2+}$, $KI + CO_3^{2-}$.
[299] J. Teltow, Ann. Phys. (Germany) **5** (1949) 63.

[300] J. Ebert and J. Teltow, Ann. Phys. (Germany) **15** (1955) 268.
[301] P. Müller, Phys. Stat. Sol. **12** (1965) 775; **21** (1967) 693.
[302] J. Corish and P.W.M. Jacobs, J. Phys. Chem. Solids **33** (1972) 1799.
[303] J. Corish and P.W.M. Jacobs, Phys. Stat. Sol. (b) **67** (1975) 263.
[304] A. Batra and L. Slifkin, J. Physique **37** (1976) C7-396.
[305] A.P. Batra and L.M. Slifkin, J. Phys. Chem. solids **38** (1977) 687; diffusion of Na-tracer in AgBr.
[306] Ref. [98]; diffusion of Na-tracer in AgCl; this experiment is considered as the most intriguing and useful experiment up to date in silver halides. The same holds for ref. [305].
[307] J.K. Aboagye and R.J. Friauf, Phys. Rev. **B11** (1975) 1654.
[308] A.M. Raaen, I. Svare and T.A. Fjeldly, Phys. Rev. **B21** (1980) 4895.
[308a] P.A. Cardegna and A.L. Laskar, Phys. Rev. **B24** (1981) 530.
[309] K.F. Loje and D.E. Schuele, J. Phys. Chem. Solids **31** (1970) 2051.
[310] P.G. Strelkow, Phys. Z. Sowj. **12** (1937) 77.
[311] B.P. Lawn, Acta Crystallogr. **16** (1963) 1163.
[312] D.S. Tannhäuser, L.J. Bruner and A.W. Lawson, Phys. Rev. **102** (1956) 1276.
[313] W. Zierau and C. Falter, private communication.
[314] P.W.M. Jacobs, J. Corish, B.A. Devlin and C.R.A. Catlow, in: Fast Ion Transport in Solids, eds. P. Vashishta, J.N. Mundy and G.K. Shenoy, Conf. Lake Geneva, Wisconsin, May 1979, (North-Holland, New York, 1979) p. 589.
[315] H. Kanzaki, Phys. Rev. **81** (1951) 884.
[316] K. Kobayashi, Phys. Rev. **85** (1952) 150.
[316a] F. Bremer and J. Nölting, Ber. Bunsenges. **80** (1976) 12.
[317] D. Tannhäuser, J. Phys. Chem. Solids **5** (1958) 224.
[318] P. Varotsos and K. Alexopoulos, Phys. Stat. Sol. (b) **92** (1979) 633.
[319] R.D. Fouchaux and R.O. Simmons, Phys. Rev. **A136** (1964) 1664.
[320] R.D. Fouchaux, J. Phys. Chem. Solids **31** (1970) 1113.
[321] A.E. Abey and C.T. Tomizuka, J. Phys. Chem. Solids **27** (1966) 1149.
[322] D.N. Yoon and D. Lazarus, Phys. Rev. **B5** (1972) 4935.
[323] G. Martin, D. Lazarus and J.L. Mitchell, Phys. Rev. **B8** (1973) 1726.
[324] S. Lansiart and M. Beyeler, J. Phys. Chem. Solids **36** (1975) 703.
[325] I.D. Faux and A.B. Lidiard, Z. Naturf. **a26** (1971) 62.
[326] M.J. Norgett, AERE Harwell, Rep. **12** (1974) 7650.
[327] C.R.A. Catlow, I.D. Faux and M.J. Norgett, J. Phys. **C9** (1976) 419.
[328] J.C. Slater, in: Introduction to Chemical Physics (McGraw-Hill, New York, 1939).
[329] J.S. Dugdale and D.K.C. Macdonald, Phys. Rev. **89** (1953) 832.
[330] Jai Shanker, A.P. Gupta and O.P. Sharma, Phil. Mag. **B37** (1978) 329.
[331] J.A. Cornet, J. Phys. Chem. Solids **32** (1971) 1489.
[332] A. Languille, D. Calais and M. Fromont, J. Phys. Chem. Solids **35** (1974) 1373.
[333] M.W. Finnis and M. Sachdev, J. Phys. F **6** (1976) 965.
[334] M.W. Finnis and R.M. Nieminen, J. Phys. **F7** (1977) 1999.
[335] G. Jacucci and R. Taylor, J. Phys. **F9** (1979) 1489.
[336] G. Allan and M. Lannoo, J. Phys. Chem. Solids **37** (1976) 699.
[337] P.A. Smith and C.A. Smith, J. Phys. Chem. Solids **26** (1965) 279.
[338] C.R. Kohler and A.L. Ruoff, J. Appl. Phys. **36** (1965) 2444.
[339] R.N. Jeffery, Phys. Rev. **B3** (1971) 4044.
[340] R.F. Peart, Phys. Stat. Sol. **20** (1967) 545.
[341] D. Lazarus, from a private communication to R.N. Jeffery.
[342] E.S. Fisher and D. Dever, in: Science, Technology and Applications of Titanium, eds. R.I. Jaffee and N.E. Promisel (Pergamon, Oxford, 1970) p. 373.

[343] E.S. Fisher, M.H. Manghnani and K.W. Katahara, in: Proc. 4th Int. Conf. on High Pressure, Kyoto 1974, p. 393; the value given corresponds to the ω-phase.
[344] N.E. Walsöe de Reca and C.M. Libanati, Acta Metall. **16** (1968) 1297.
[345] M. Fromont, J. Phys. Chem. Solids **36** (1975) 1397.
[346] M.P. Dariel, D. Dayan and A. Languille, Phys. Rev. **B4** (1971) 4348.
[347] A. Languille, M.P. Dariel, D. Calais and B. Coqblin, Mem. Sci. Rev. Metall. **LXX** (1973), no. 3.
[348] M. Dupuy, C.R. Acad. Sci. Paris **263** (1966) 35; M. Dupuy and D. Calais, Trans. Met. Soc. AIME **242** (1968) 1679.
[349] Ref. [266], p. 1222.
[350] P. Varotsos, Phys. Stat. Sol. (a) **26** (1974) 311.
[351] H. Spalt, H. Lohstöter and H. Peisl, Phys. Stat. Sol. (b) **56** (1973) 469.
[352] P. Varotsos and K. Alexopoulos, J. Phys. **C12** (1979) L761; J. Physique **41** (1980) C6-526.
[353] W. Biermann, Z. Phys. Chem. (Frankfurt) **25** (1960) 90; **25** (1960) 253.
[354] P. Varotsos and K. Alexopoulos, Phys. Rev. **B21** (1980) 4898.
[355] D.C. Allen and D. Lazarus, Phys. Rev. **B17** (1978) 1913; the value $\beta^{act} \simeq 6.5 \times 10^{-3}$ K^{-1} can be easily extracted from fig. 11 of this reference.
[356] G.A. Samara, J. Phys. Chem. Solids **40** (1979) 509.
[357] J. Oberschmidt and D. Lazarus, Phys. Rev. **B21** (1980) 2952; **21** (1980) 5813.
[358] D.S. Rimai and R.J. Sladek, Phys. Rev. **B21** (1980) 843.
[359] D.L. Decker, R.A. Ross, W.E. Evenson and H.B. Vanfleet, Phys. Rev. **B15** (1977) 507.
[360] R.E. Einziger and J.N. Mundy, Phys. Rev. **B17** (1978) 449.
[360a] See ref. [178c].
[361] A.E. Lord Jr. and D.N. Beshers, Acta Metall. **14** (1966) 1659.
[362] D.J. Dever, J. Appl. Phys. **43** (1972) 3293.
[362a] S. Mrowec, Defects and Diffusion in Solids (North-Holland, Amsterdam, 1980).
[363] D.B. Butrymowicz, J.R. Manning and M.E. Read, INCRA Monograph V, The Metallurgy of Copper (National Bureau of Standards, Washington, July 1977).
[364] V.A. Gorbachev, S.M. Klotsman, Ya A. Rabovskii, V.K. Talinskii and A.N. Timofeev, Ag2, Cd, In1: Fiz. Met. Metalloved. **34** (1972) 879, Rh: **34** (1972) 1104, Sn3, Sb3: **35** (1973) 889, Au3, Pb, Bi: **44** (1977) 214.
[365] J. Askill and G.B. Gibbs, Phys. Stat. Sol. **11** (1965) 557.
[366] G.M. Hood, J. Phys. F: Metal Physics **6** (1976) 19.
[367] M.S. Jackson and D. Lazarus, Phys. Rev. **B15** (1977) 4644; it contains a very useful discussion for the "anomalous" bcc metals.
[367a] J.I. Federer and J.N. Lundy, Trans. Metall. Soc. AIME **227** (1963) 592.
[368] A.P. Batra and L.M. Slifkin, J. Phys. Chem. Solids **30** (1969) 1315.
[368a] See A.L. Laskar and P.A. Cardegna, Radiat. Eff. **75** (1983) 27.
[369] B. Lesage and A.M. Huntz, J. Less Common Met. **52** (1977) 197.
[370] O.D. Slagle and H.A. McKinstry, Acta Crystallogr. **21** (1966) 1013.
[371] O.D. Slagle and H.A. McKinstry, J. Appl. Phys. **38** (1967) 446.
[372] L.S. Cain, J. Phys. Chem. Solids **37** (1976) 1178.
[373] L. Vegard, Z. Phys. **5** (1921) 17.
[374] Y. Haget, N.B. Chanh, P. Garin and L. Bonpunt, Acta Metall. **23** (1975) 723.
[375] R. Lindström, Ann. Acad. Sci. Fenn. Ser. **AVI**, 399 (1972).
[376] R.W. Roberts and C.S. Smith, J. Phys. Chem. Solids **31** (1970) 619.
[377] C.S. Smith, in: Note on the Murnaghan Equation of State, Atomic Energy Commesion Tech. Rep. No. 3 (Contract No. AT40-I-3802) October 1968.
[378] C.S. Smith and R.W. Roberts, in: Generalized Born model, Atomic Energy Commission, Tech. Report No. 1 (contract No. AT40-I-3802) 1968.

[379] C.S. Smith and K.O. McLean, J. Phys. Chem. Solids **34** (1973) 1143.
[380] W.G. Kleppmann and H. Bilz, Commun. Phys. **1** (1976) 105.
[381] L.S. Cain, J. Phys. Chem. Solids **38** (1977) 73.
[382] Bhima Sankaram and K.G. Bansigir, Cryst. Lattice Defects **7** (1978) 209.
[383] L. Bonput and Y. Haget, J. Chimie Physique **71** (1974) 537.
[384] M. Kantola and R. Lindström, Ann. Acad. Sci. Fenn. **AVI** (1967) 222.
[385] K. Kamiyoshi and Y. Nigara, Phys. Stat. Sol. (a) **6** (1971) 223.
[386] P. Varotsos, Phys. Stat. Sol. (b) **100** (1980) K 133.
[387] L.S. Cain and L.M. Slifkin, J. Phys. Chem. Solids **41** (1980) 173.
[388] K.E. Salimäki, Ann. Acad. Sci. Fenn. Ser. **AVI**, 56 (1960).
[389] R. Lindström, T. Blömberg and H. Yli-Pietila, Ann. Acad. Sci. Fenn. Ser. **AVI**, 396 (1972).
[390] J. Hietala, Ann. Acad. Sci. Fenn. Ser. **AVI**, 121 to 123 (1963).
[390a] R. Lindström, Ann. Acad. Sci. Fenn. Ser. VI (1963) 121.
[391] H.T. Fullam, Mat. Res. Bull. **7** (1972) 289.
[391a] P. Varotsos and K. Alexopoulos, Phys. Stat. Sol. (b) **102** (1980) K63.
[392] W. Zieten, Z. Physik **145** (1956) 125.
[393] S.W. Kurnick, J. Chem. Phys. **20** (1952) 218.
[394] K.F. Loje and D.E. Schuelle, J. Phys. Chem. Solids **31** (1970) 2051.
[395] D. Lazarus, in: Solid State Physics, vol. 10, eds. F. Seitz and D. Turnbull (Academic Press, New York, 1960). p. 71.
[396] J. Van Liempt, Z. Physik **96** (1935) 534; he proposed that $h^{act} = 32\, T_M$ kcal/mole.
[397] N.H. Nachtrieb, Proc. 3rd Int. Conf. on Reactivity of Solids, Madrid, 1956.
[398] N.H. Nachtrieb and G.S. Handler, Acta Metall. **2** (1954) 797; they propose $h^{act} = 16.5\, L$ kcal/mole, where L is the latent heat of fusion.
[399] C. Zener, in: Imperfections in Nearly Perfect Crystals ed. W. Shockley (Wiley, New York, 1952) p. 289; J. Appl. Phys. **22** (1951) 372.
[400] A.W. Lawson, J. Phys. Chem. Solids **3** (1957) 250; J. Appl. Phys. (Suppl.) **33** (1962) 466.
[401] R.W. Keyes, J. Chem. Phys. **29** (1958) 467.
[402] A.W. Lawson, S.A. Rice, R.D. Corneliussen and N.H. Nachtrieb, J. Chem. Phys. **32** (1960) 447.
[403] G.J. Dienes, J. Appl. Phys. **21** (1950) 1189.
[404] Ref. [35], p. 784, 785, 786.
[405] K. Maier, M. Peo, B. Saile, H.E. Schaefer and A. Seeger, Proc. 5th Int. Conf. Positron Annihilation (Japan, 1979) 9A-IV-4, 205; they suggest that in bcc transition metals h^f (eV) $= 0.98 \times 10^{-3}\, T_M$.
[406] K. Maier, G. Rein, B. Saile, P. Valenta and H.E. Schaefer, Proc. 5th Int. Conf. Positron Annihilation (Japan 1979) 9A-I-4, 101; they suggest $h^f = (8.1 \pm 0.3)\, kT_M$ for Al, Fe, Cd, Au, In, Zn, Pb and Ag.
[407] L.W. Barr and A.B. Lidiard, in: Physical Chemistry, Vol. X, ed. W. Jost (Academic Press, New York, 1970) p. 151; they suggest for alkali halides h^f (eV) $= 2.14 \times 10^{-3}\, T_M$. However, newer experimental data show higher values of h^f so that their coefficient 2.14×10^{-3} must be increased.
[408] R.H. Baughman and D. Turnbull; specific heat measurements for organic molecular crystals (quoted in ref. [35], p. 538).
[409] F. Reif, in: Statistical Physics, Berkeley Physics Course, Vol. 5 (McGraw–Hill, New York 1965) p. 254, 255; see also J.M. Ziman, in: Principles of the Theory of Solids (Cambridge University Press, 1971) pp. 62, 63.
[410] G. Leibfried, in: Gittertheorie der mechanischen und thermischen Eigenschaften der Kristalle, Handbuch der Physik, Vol. VII/1 (Springer Berlin, 1955).

[411] W. Ludwig, in: Festkörper-Physik I (Akademische Verlagsgesellschaft, Frankfurt am Main, 1970).
[412] P.V. Sastry and B.G. Mulimani, Phil. Mag. **20** (1969) 859.
[413] S.I. Ben-Abraham, A. Rabinovitch and J. Pelleg, Phys. Stat. Sol. (b) **84** (1977) 435; the interaction between a vacancy and the lattice vibrations in metals is represented as a collision between a sine-Gordon soliton and a breather; they than conclude that the energies of formation and migration are related to Θ_D.
[414] N.H. March, Phys. Lett. **20** (1966) 231; he employed screening theory to show that in metals h^f is connected to Θ_D; the last correlation has been empirically found in metals by K. Mukherjee, Phil. Mag. **12** (1965) 915.
[415] B.N. Oshcherin, Phys. Stat. Sol. (a) **29** (1975) K69, and references therein; he suggests that h^{act} is explicitly proportional to Θ_D^2 for compounds with Spinel structure.
[415a] C.A. Swenson, J. Phys. Chem. Solids **29** (1968) 1337.
[416] When one uses eq. (10.2) – instead of eq. (13.24) – the more accurate expression $s \simeq 2h_{exp}\beta\gamma B/B_0^{SL}b$ results; with the same arguments the term $1/3$ in eq. (13.36) can be deleted, wherefrom eq. (13.38) is directly derived.
[417] C. Postmus, J.R. Ferraro and S.S. Mitra, Phys. Rev. **174** (1968) 983.
[418] J.R. Ferraro, S.S. Mitra and A. Quattrochi, J. Appl. Phys. **42** (1971) 3677.
[419] P. Varotsos, Phys. Stat. Sol. (b) **90** (1978) 339.
[420] P. Varotsos, J. Appl. Phys. **51** (1980) 4553; the Szigeti charge varies only slightly on volume so that $(l/e^*)(de^*/dl) \approx -0.5$.
[421] C. Andeen, J. Fontanella and M.C. Wintersgill, J. Phys. **C13** (1980) 3449; data for the migration volume for type I dipoles in SrF_2 doped with lanthanum.
[422] Ref. [105]; data for v^m-values for type II dipoles in erbium-doped SrF_2.
[423] J. Fontanella, D.L. Jones and C. Andeen, Phys. Rev. **B18** (1978) 4454.
[424] P. Varotsos and K. Alexopoulos, J. Phys. Chem. Solids **41** (1980) 443.
[425] C. Andeen, D. Link and J. Fontanella, Phys. Rev. **B16** (1977) 3762.
[426] R.P. Lowndess, J. Phys. **C2** (1969) 1595; **4** (1971) 3083.
[427] D.R. Bosomworth, Phys. Rev. **157** (1967) 709.
[428] P. Denham, G.R. Field, P.L.R. Morse and G.R. Wilkinson, Proc. Roy. Soc. London **A317** (1970) 55.
[429] C. Wong and D.E. Schuelle, J. Phys. Chem. Solids **29** (1968) 1309.
[430] C. Zener, J. Appl. Phys. **22** (1951) 372; he suggests that $s^{act} = \lambda h^{act} d(\mu/\mu_0)/dT$, where μ_0 is the shear modulus at $T = 0$ K and λ an empirical numerical constant of the order of unity.
[431] R.B. McLellan, Scr. Metall. **10** (1976) 1107; Trans. AIME **245** (1969) 379; he suggests that $s^f = 8\pi r^3\{2\mu\beta\delta - \delta^2(d\mu/dT)_P\}$ and $h^f = 8\pi\mu r^3\delta^2$, where r is the Goldschmidt radius and δ the linear inward relaxation of the vacancy.
[432] Ref. [35], p. 332.
[433] R.W. Keyes, in: Solids Under Pressure eds. W. Paul and D.M. Warschauer (McGraw–Hill, New York, 1963); he suggests $v^{act} = g^{act}\{\mu^{-1}(d\mu/dP)_T - B^{-1}\}$.
[434] Ref. [35], p. 97, see table 3.4 therein.
[435] J. Holder, Thesis (University of Illinois, Urbana, 1968); a useful discussion of the connection of g^{act} with shear modulus.
[436] R.M. Emrick and P.B. McArdle, Phys. Rev. **188** (1969) 1156; effect of pressure on the electrical resistance quenched into gold and aluminum.
[437] R.P. Huebener and C.G. Homan, Phys. Rev. **129** (1963) 1162; measurement of v^f for Au.
[438] H.H. Grimes, J. Phys. Chem. Solids **26** (1965) 509; v^f-measurement for Au.
[439] R.M. Emrick, Phys. Rev. B **22** (1980) 3563; measurement of v^f in Au.

[440] M. Charles, J. Hillairet, M. Beyeler and J. Delaplace, Phys. Rev. **B11** (1975) 4808; effect of hydrostatic pressure up to 9 kbar on the electrical resistance quenched in platinum.
[441] P. Varotsos and K. Alexopoulos, Phys. Stat. Sol. (a) **47** (1978) K133.
[442] G.A. Samara, Phys. Rev. Lett. **44** (1980) 670.
[443] G. Leibfried and N. Breuer, in: Point Defects in Metals I, Springer Tracts in Modern Physics, Vol. 81 (Springer Verlag, Berlin, 1978).
[444] P.H. Dederichs, C. Lehmann and A. Scholz, Zs. Phys B **20** (1975) 155; see also R. Zeller and P.H. Dederichs, Zs. Phys. B **25** (1976) 139.
[445] P.H. Dederichs, C. Lehmann, H.R. Schober, A. Scholz and R. Zeller, J. Nucl. Mater. **69–70** (1978) 176; a useful review of the influence of the defects on the elastic constants is given.
[446] W. Ludwig, in: Calculation of the Properties of Vacancies and Interstitials, Proc Conf. May 1–5, 1966 (United States Dept. of Commerce, National Bureau of Standards) Miscellaneous Publications **287** (1966) 151.
[447] M. Pistorius, Z. Angew. Phys. **29** (1970) 145.
[448] J. Holder, A.V. Granato and L.E. Rehn, Phys. Rev. **B10** (1974) 363; **10** (1974) 349.
[449] K.H. Robrock and W. Schilling, J. Phys. **F6** (1976) 303.
[450] W. Ludwig, private communication.
[451] S.P. Singh, Phys. Rev. **B9** (1974) 3313.
[452] K.J. Dunn, Phys. Rev. **B12** (1975) 3497.
[453] N.F. Mott and M.J. Littleton, Trans. Faraday Soc. **34** (1938) 485.
[454] K. Tharmalingam, Phil. Mag. **A37** (1978) 201.
[455] Y.A. Chang, J. Phys. Chem. Solids **28** (1967) 697.
[456] T.H.K. Barron, J. Phys. **C12** (1979) L155.
[457] N. Brown and P.W.M. Jacobs, J. Physique **34** (1973) C9-437.
[458] S. Shih and J.P. Stark, J. Appl. Phys. **49** (1978) 648.
[459] G. Krautheim, A. Neidhardt and U. Reinhold, Phys. Stat. Sol. (a) **49** (1978) K125.
[460] G. Krautheim, A. Neidhardt, U. Reinhold and A. Zehe, Phys. Lett. **72A** (1979) 181.
[461] G. Krautheim, A. Neidhardt and U. Reinhold, Krist. and Techn. **13** (1979) 1335.
[462] K. Hoshino, Y. Iijima and K-I. Hirano, Metall. Trans. **A8** (1977) 469.
[463] H. Wenzl, V. Sorojic and B. Bischof, Jülich Reports 1384 (1977)
[464] S.K. Sen, M.B. Dutt and A.K. Barua, Phys. Stat. Sol. (a) **45** (1978) 657.
[465] J.L. Seran, Acta Metall. **24** (1976) 627.
[466] M.B. Dutt, S.K. Sen and A.K. Barua, Phys. Stat. Sol. (a) **56** (1979) 149.
[467] S.K. Sen, M.B. Dutt and A.K. Barua, Phys. Stat. Sol. (a) **32** (1975) 345.
[468] M.B. Dutt and S.K. Sen, Jpn. J. Appl. Phys. **18** (1979) 1025.
[469] C.T. Tomizuka and E. Sonder, Phys. Rev. **103** (1956) 1182.
[470] S.J. Rothman, N.L. Peterson and J.T. Robinson, Phys. Stat. Sol. **39** (1970) 635.
[471] J. Bihr, H. Mehrer and K. Maier, Phys. Stat. Sol. (a) **50** (1978) 171.
[472] P. Gas and J. Bernardini, Scr. Metall. **12** (1978) 367.
[473] T. Hehenkam and R. Wubbenhorst, Z. Metallk. **66** (1975) 275.
[474] C.T. Tomizuka and L. Slifkin, Phys. Rev. **96** (1954) 610.
[475] J. Bernardini, A. Combe-Brun and J. Cabane, Scr. Metall. **4** (1970) 985;
[476] J. Bernardini and J. Cabane, Act. Metall. **21** (1973) 1561;
[477] S. Bharati, S. Badrinarayan and A.P.B. Sinha, Phys. Stat. Sol. (a) **43** (1977) 653;
[478] R.E. Hoffman, Acta metall. **6** (1958) 95.
[479] C.B. Pierce and D. Lazarus, Phys. Rev. **114** (1959) 686.
[480] F.E. Jaumot Jr. and A. Sawatzky, J. Appl. Phys. **27** (1956) 1186.
[481] E. Sonder, L. Slifkin and C.T. Tomizuka, Phys. Rev. **93** (1954) 970.
[482] A. Sawatzky and F.E. Jaumot, Phys. Rev. **100** (1955) 1627.

[483] K. Hirano and S. Fujikawa, J. Nucl. Mat. **69–70** (1978) 564.
[484] W.B. Alexander and L.M. Slifkin, Phys. Rev. **B1** (1970) 3274.
[485] N.L. Peterson and S.J. Rothman, Phys. Rev. **B1** (1970) 3264.
[486] F. Sawayanagi and R.R. Hasiguti, J. Jpn. Inst. Met. **42** (1978) 1155.
[487] J.E. Hilliard, B.L. Averbach and M. Cohen, Acta Metall. **7** (1979) 86.
[488] B.F. Dyson, T. Antony and D. Turnbull J. Appl. Phys. **37** (1966) 2370.
[489] D.L. Decker, J.G. Melville and H.B. Vanfleet, Phys. Rev. **B20** (1979) 3036.
[490] W.K. Wartburton, Phys. Rev. **B11** (1975) 4945.
[491] H.B. Vanfleet, J.D. Jorgensen, J.D. Schmutz and D.L. Decker, Phys. Rev. **B15** (1977) 5545.
[492] K. Kusunoki and S. Nishikawa, Scr. Metall. **12** (1978) 615.
[493] W.K. Wartburton, Phys. Rev. **B7** (1973) 1330.
[494] C.W. Owens and D. Turnbull, J. Appl. Phys. **43** (1972) 3933.
[495] H.A. Resing and N.H. Nachtrieb, J. Phys. Chem. Solids **21** (1961) 40.
[496] A.G. Baker and H.M. Gilder, Bull. Am. Phys. Soc. **20** (1975) 442.
[497] D.L. Decker, Jonathan D. Weiss and H.B. Vanfleet, Phys. Rev. **B16** (1977) 2392.
[498] B.F. Donovan-Vojtovic, S.A Dods and M.S. Nasser, Phil. Mag. **34** (1976) 893.
[499] J. Pelleg, Phil. Mag. **33** (1976) 165.
[500] R.F. Peart, D. Graham and D.H. Tomlin, Acta Metall. **10** (1962) 519.
[501] J. Pelleg, J. Less Common Met. **17** (1970) 319.
[502] J. Pelleg. Phil. Mag. **19** (1969) 25.
[503] D. Ablitzer, Phil. Mag. **36** (1977) 391.
[504] F. Roux and A. Vignes, Rev. Phys. Appl. **5** (1970) 393.
[505] Int. Conf. on Diffusion in bcc Materials, Gatlinburg, 1964, p. 58.
[506] T.S. Lundy, F.R. Winslow, R.E. Pawel and G.J. McHargue, Trans. metall. Soc. AIME **233** (1965) 1533.
[507] J. Pelleg and G.M. Lindberg, Trans. Metall. Soc. AIME **245** (1969) 1654.
[508] R.P. Agarwala, Trans. Inst. Met. Jpn. **13** (1972) 425.
[509] K.V. Sathyaraj, D. Ablitzer and C. Demangeat, Phil. Mag. **A40** (1979) 541.
[510] J. Askill, Phys. Stat. Sol. **2** (1965) K167.
[511] F. Roux, Thesis (Nancy, 1972).
[512] R. P. Agarwala, S.P. Murarka and M.S. Anand, Acta Metall. **16** (1968) 61.
[513] D.S. Gornyy and R.M. Altovskiy, Fiz. Met. Metalloved **31** (1971) 781.
[514] A. Ott and A. Norden-Ott, Z. Naturf. **23A** (1968) 473.
[515] J.N. Mundy and W.D. McFall, Phys. Rev. **B7** (1973) 4363.
[516] A. Ott, Z. Naturf. **23A** (1968) 1683.
[517] A. Ott, A. Lodding and D. Lazarus Phys. Rev. **188** (1969) 1088.
[518] A. Ott, Z. Naturf. **25A** (1970) 1477.
[519] A. Ott, J. Appl. Phys. **40** (1969) 2395.
[520] A. Ott, Z. Naturf. **23A** (1968) 2126.
[521] A. Ott, J.N. Mundy, L. Loewenberg and A. Lodding, Z. Naturf. **23A** (1968) 771.
[522] J.N. Mundy, A. Ott and L. Loewenberg, Z. Naturf. **22A** (1964) 2113.
[523] J.N. Mundy, A. Ott, L. Loewenberg and A. Lodding Phys. Stat. Sol. **35** (1969) 359.
[524] J. Askill, Phys. Stat. Sol. (b) **43** (1971) K1.
[525] A.E. Pontau and D. Lazarus, Phys. Rev. **B19** (1979) 4027.
[526] J.F. Murdock, T.S. Lundy and E.E. Stansbury, Acta Metall. **12** (1964) 1033.
[527] J. Askill, Phys. Stat. Sol. **16** (1966) K63.
[528] S.J. Rothman, N.L. Peterson, A.L. Laskar and L.C. Robinson, J. Phys. Chem. Solids **33** (1972) 1061.
[529] O.R. Dobrovinskaya, V.I. Solunskii and A.G. Shakhova, Ukr. Fiz. Zh. **12** (1967) 868.

[529a] F. Beniere, M. Beniere and M. Chemla, J. Chimie Physique (Paris) **16** (1969) 898.
[530] M. Beniere, F. Beniere, C.R.A. Catlow, A.K. Shukla and C.N.R. Rao, J. Phys. Chem. Solids **38** (1977) 521.
[531] K. Furukawa, J. Takamura, N. Kuwana, R. Tahara and M. Abe, J. Phys. Soc. Jpn. **41** (1976) 1584.
[532] P. Tzanetakis, J. Hillairet and G. Revel, Phys. Stat. Sol. (b) **75** (1976) 433.
[533] B.T.A. Mckee, W. Triftshauser and A.T. Stewart, Phys. Rev. Lett. **28** (1972) 358.
[534] S.M. Kim, W.J.L. Buyers, P. Martel and G.M. Hood, J. Phys. **F4** (1974) 343.
[535] W. Triftshauser, Phys. Rev. **B12** (1975) 4634.
[536] M. Fluss, L.C. Smedzkjaer, M.K. Chason, D.G. Legnidi and R.W. Siegel, Phys. Rev. **B17** (1978) 3444.
[537] B.T.A. Mckee and N.K. Davé, unpublished.
[538] S.J. Rothman and N.L. Peterson, Phys. Stat. Sol. **35** (1969) 305.
[539] J.G.E.M. Backus, H. Bakker and H. Mehrer, Phys. Stat. Sol. (b) **64** (1974) 151.
[540] B.M. Cohen and W.K. Warburton, Phys. Rev. **B12** (1975) 5682.
[541] H.B. Vanfleet, Phys. Rev. **B21** (1980) 4337.
[542] F. Dworschak, Chr. Herzig and J.N. Mundy, J. Phys. **F10** (1980) 367.
[543] G.A. Alers and J.R. Neighbours, J. Phys. Chem. Solids **7** (1958) 58.
[544] J.D. Meakin and D. Klokholm, Trans. AIME **218** (1960) 463;
[545] W. Kramer and J. Nölting, Acta Metall. **20** (1972) 1353.
[546] M. Beniere, F. Beniere and M. Chemla, Solid State Commun. **13** (1973) 1339.
[547] A. Brun, P. Dansas and F. Beniere, J. Phys. Chem. Solids **35** (1974) 249.
[548] J. Rolfe, Can. J. Phys. **42** (1964) 2195.
[549] S. Chandra and J. Rolfe, Can. J. Phys. **49** (1971) 2098.
[550] L.W. Barr and D.K. Dawson, Atomic Energy Research Establishment Report No. R. 6234 (Harwell, U.K., 1969).
[551] D. Kostopoulos, Thesis (Rennes, 1977).
[552] N.H. Nachtrieb, H.A. Resing and S.A. Rice, J. Chem. Phys. **31** (1959) 135;
[553] O. Vollmer and R. Kolhaas, Z. Metallk. **59** (1968) 273; Z. Angew. Physik **25** (1968) 365.
[554] W.B. Daniels and C.S. Smith, Phys. Rev. **111** (1958) 713.
[555] C.T. Tomizuka, R.C. Lowell and A.W. Lawson, Bull. Am. Phys. Soc. II, Vol. **5** (1960) 181.
[556] C. Kittel, Introduction to Solid State Physics, 4th Ed. (Wiley, New York, 1971) p. 143.
[557] R. Chiarodo, J. Green, I.L. Spain and P. Bolsaitis, J. Phys. Chem. Solids **33** (1972) 1905.
[558] Y. Hiki and A.V. Granato, Phys. Rev. **144** (1966) 411.
[559] R.E. Schmunk and C.S. Smith, J. Phys. Chem. Solids **9** (1959) 100.
[560] R.D. Engardt and R.G. Barnes, Phys. Rev. **B3** (1971) 2391.
[561] C. Kittel, in: Introduction to Solid State Physics, 4th Ed. (Wiley, New York, 1971).
[562] M.W. Guinan and D.J. Steinberg, J. Phys. Chem. Solids **35** (1974) 1501.
[563] J.E. Dickman, R.N. Jeffery and D.R. Gustafson, Phys. Rev. **B16** (1977) 3334.
[564] A. Ott and A. Norden-Ott, J. Appl. Phys. **42** (1971) 3745.
[565] J.E. Dickey, Acta Metall. **7** (1959) 350.
[566] N.H. Nachtrieb and C. Coston, in: Physics of Solids at High Pressure, eds. C.T. Tomizuka and R.M. Emrick (Academic Press, New York, 1965) p. 336–348.
[567] F.H. Huang and H.B. Huntington, Phys. Rev. **B9** (1974) 479.
[568] M.S. Anderson and C.A. Swenson, J. Phys. Chem. Solids **36** (1975) 145.
[569] Landolt–Börnstein Tables, Group III, Vol. 2 (1969) p. 31.
[570] V.K. Gard, D.S. Puri and M.P. Verma, Phys. Stat. Sol. (b) **82** (1977) 481.
[571] A.R. Allnatt and P. Pantelis, Solid State Commun. **6** (1968) 309.
[572] C.P. Flynn, in: Point defects and Diffusion (Clarendon, Oxford, 1972) p. 595.

[573] J.H. Crawford and L.M. Slifkin, in: Point Defects in Solids (Plenum, New York, 1972) p. 135.
[574] N. Brown and P.W.M. Jacobs, J. Physique **34** (1973) C9-444.
[575] D. Lazarus, Proc. Int. Conf. on Crystal Defects, 1962, Conf. J. Phys. Soc. Jpn. 18, Suppl. II (1963) 224.
[576] G.H. Shaw, J. Phys. Chem. Solids **35** (1974) 911.
[577] G.H. Shaw, J. Phys. Chem. Solids **41** (1980) 155.
[578] M. Lallemand, Thesis (Université de Paris VI, Paris, 1972).
[579] K.O. McLean and C.S. Smith, J. Phys. Chem. Solids **33** (1972) 279.
[580] Quoted in P. Varotsos, J. Phys. Chem. Solids **39** (1978) 513; J. Physique **39** (1978) 1247.
[581] Quoted in P. Varotsos and K. Alexopoulos, Phys. Stat. Sol. (a) **47** (1978) K133.
[582] Quoted in R.W. Siegel, J. Nucl. Mater. **69–70** (1978) 117.
[583] Quoted in H.M. Gilder and D. Lazarus, Phys. Rev. **B11** (1975) 4916.
[584] P.S. Ho, J.P. Poivier and A.L. Ruoff, Phys. Stat. Sol. **35** (1969) 1017.
[585] B.T.A. Mckee, private communication to A. Seeger, J. Phys. **F3** (1973) 248.
[586] Value given in ref. [583] from a reanalysis of the data reported by J.M. Mundy, Phys. Rev. **B3** (1971) 2431.
[587] Quoted in P. Varotsos and W. Ludwig, J. Phys. **C11** (1978) L309.
[588] D.C. Allen and D. Lazarus, Phys. Rev. **B17** (1978) 1913.
[589] G.H. Shaw, J. Geophys. Res. **83** (1978) 3519.
[590] For details of the application of the $cB\Omega$-model to B4-AgI, see P. Varotsos and K. Alexopoulos, Phys. Rev. **B21** (1980) 4898.
[591] P.S. Ho, Phys. Rev. **B7** (1973) 3550.
[592] W.B. Daniels, Phys. Rev. **119** (1960) 1246.
[593] A.I. Prokhvatilov and A.P. Isakina, Phys. Stat. Sol. (a) **78** (1983) 147.
[593a] P. Varotsos and K. Alexopoulos, Phys. Stat. Sol. (b) **110** (1982) 9; review article concerning the recent aspects of the analysis of differential dilatometry and self-diffusion experiments.
[594] E. Dologlou-Revelioti, Dissertation, Athens 1984 (in Greek).
[595] G.A. Sobolev, A.A. Semercham, B.G. Salov, H.A. Spetzler, C.H. Sondergeld, V.N. Badanov, A.V. Kol'stov, V.F. Los, R.M. Nasimov, A.V. Ponomarev, I.R. Stakhovskii, V.A. Terentiev and I.M. Turetskii, Izv. Acad. Sci USSR, Phys. Solid Earth **18** (1983) 572.
[596] P. Varotsos and K. Alexopoulos, Tectonophysics **110** (1984) 73.
[597] P. Varotsos and K. Alexopoulos, Tectonophysics **110** (1984) 99.
[598] M.S. Anderson and C.A. Swenson, Phys. Rev. B **28** (1983) 5395.
[599] P. Varotsos, K. Alexopoulos and M. Lazaridou, Phys. Rev. **B31** (1985) 8263.
[600] M.S. Anderson and C.A. Swenson, Phys. Rev. **B31** (1985) 668.
[601] K. Eftaxias, J. Grammatikakis and P. Varotsos Phys. Rev. B (to appear Oct. 15, 1985).
[602] J. Schoonman, Solid State Ionics **5** (1981) 71.
[603] G.A. Samara, in: Solid State Physics 38, Eds. H. Ehrenreich, D. Turnbull and F. Seitz (Academic Press, New York, 1984).
[604] R. Döhl, M.-P. Macht and V. Naundorf, Phys. Stat. Sol. (a) **86** (1984) 603.
[605] G. Neuman, M. Pfundstein and P. Reimers, Phil. Mag. **A45** (1982) 499.
[606] E. Albert, R. Kirchheim and H. Dietz, Scr. Metall. **15** (1981) 673.
[607] G.M. Hood, R.J. Schultz and J. Armstrong, Phil. Mag. **A47** (1983) 775.
[608] J. Shanker and G.D. Jain, Phys. Stat. Sol. (b) **110** (1982) 257.
[609] P. Varotsos, Phys. Stat. Sol. (b) **99** (1980) K93.
[610] P. Varotsos, Phys. Stat. Sol. (b) **100** (1980) K47.
[611] J. Shanker and G.D. Jain, Phys. Rev. **B27** (1983) 2515.

[612] M. Lazaridou, J. Phys. Chem. Solids **46** (1985) 643.
[613] K. Shahi and J.B. Wagner Jr, J. Phys. Chem. Solids **44** (1983) 89.
[614] L.S. Cain, J. Phys. Chem. Solids **45** (1984) 907.
[615] R.K. Singh, D.K. Neb and S.P. Sanyal, J. Phys. **C16** (1983) 3409.
[616] A.B. Bhatia and N.H. March, J. Chem. Phys. **80** (1984) 2076.
[617] P. Varotsos and K. Alexopoulos, Phys. Rev. **B30** (1984) 7305.
[618] M. Lazaridou, C. Varotsos, K. Alexopoulos and P. Varotsos, J. Phys. **C18** (1985) 3891.
[619] M. Lazaridou, K. Alexopoulos and P. Varotsos, Phys. Stat. Sol. (b) **130** (1985) K105.
[620] P. Varotsos, K. Alexopoulos and M. Lazaridou, J. Phys. Chem. Solids **46** (1985) 1109.
[621] P. Varotsos, K. Alexopoulos and M. Lazaridou, Phys. Stat. Sol. (a) **88** (1985) K137.
[622] See ref. (617).
[623] P. Varotsos and K. Alexopoulos, Prak. Akad. Athinon **59** (1984) 51.
[624] K. Alexopoulos, M. Lazaridou and P. Varotsos, Phys. Rev. **B** (accepted for publication); the activation volumes in lead halides treated in this paper come from the data of J. Oberschmidt and D. Lazarus, Phys. Rev. **B21** (1980) 5813.
[625] M. Lazaridou, Phys. Stat. Sol. (b) **126** (1984) K123.
[626] P. Varotsos, K. Alexopoulos and M. Lazaridou, Phys. Stat. Sol. (b) **125** (1984) K109.
[627] M. Lazaridou, K. Alexopoulos and P. Varotsos, Jpn. J. Appl. Phys. **24** (1985) 781.
[628] The participants of this meeting came from the following countries: U.S.A. (D. Lazarus, L. Slifkin), Japan (S. Uyeda), China (Lü Dajiong), W. Germany (W. Ludwig, J. Zschau), U.K. (St. Crampin), Sweden (Ota Kulhànek, K. Meyer), Poland (R. Teisseyre) and Greece (K. Alexopoulos, P. Varotsos, K. Nomicos, E. Dologlou and M. Lazaridou).
[629] S. Uyeda, Kagaku (Japan) **55** (1985) 180; table 1 of this paper provides a list of all earthquakes with (surface) magnitude larger or equal to 4.8-units that occurred in Greece during 1984 along with the corresponding documented predictions.
[630] P. Varotsos and K. Alexopoulos, Canterbury Conference (September 1979) on "Lattice defects in Ionic crystals"; proceedings in J. Physique **41** (1980) C6-526; see also P. Varotsos and K. Alexopoulos, Phys. Stat. Sol. (a) **58** (1980) 639.
[631] P.W.M. Jacobs, J. Physique **41** (1980) C6-207.
[632] A.B. Lidiard, Phil. Mag. **A43** (1981) 301.
[633] M.J. Gillan, Phil. Mag. **A43** (1981) 291.
[634] S.M. Heald, D.R. Baer and R.O. Simmons, Phys. Rev. **B30** (1984) 2531.
[635] P. Heitjans, A. Körblein, H. Ackermann, D. Dubbers, F. Fujara and H-J. Stöckmann, J. Phys. **F15** (1985) 41.
[636] The verification of this correlation can also be found in a recent paper by M. Lazaridou and K. Eftaxias, Phys. Stat. sol. (a) **90** (1985) K147. The suggestion that "melting starts when the vacancy concentration in solid metals reaches a critical value of the order of 10^{-3}" has been supported by T. Gorecki in a series of publications, e.g. J. Physique **40** (1979) C5-63; Z. Metallkd. **70** (1979) 121; Acta Phys. Pol. **A56** (1979) 523; High Temp.-High Pressures **11** (1979) 683 and references therein.
[637] P. Varotsos and K. Alexopoulos, Phys. Stat. Sol. (a) **58** (1980) 639; this paper describes the consequences of the introduction of the "isochoric perfect crystal" in the treatment of the specific heat data.
[638] Experiments on fluorides (e.g. J. Fontanella, M.C. Wintersgill, A.V. Chadwick, R. Saghafian and C.G. Andeen, J. Phys. **C14** (1981) 2451) show that β^m of the volume that corresponds to a bound interstitial (or vacancy) motion is negative. In another class of ionic crystals, the alkali halides (see §10.4.2), experimental results on the cation vacancy motion are compatible with a positive value of β^m. We emphasize that a negative value of β^m does *not* imply a negative value of β^{act} because the latter – in the high temperature region of most materials – is mainly governed from β^f, which has *always* been experimentaly found positive.

[639] P. Varotsos, Phys. Rev. **B21** (1980) 874; see also ref. [424].
[640] P. Varotsos and K. Alexopoulos, J. Phys. Chem. Solids **42** (1981) 409; the migration volume for the so-called type-I dipoles in SrF_2 doped with lanthanum (at $T = 250$ K) is 3.26 cm^3/mole, whereas for type-II dipoles in erbium-doped SrF_2 this is 4.73 cm^3/mole. The migration enthalpies are 0.478 and 0.701 eV, respectively. We see therefore that the ratio of these two enthalpies is 1.466 whereas the ratio of the corresponding volumes is practically the same, i.e. 1.45. The same behaviour has been found for other cases as well (refs. [642] to [644]).
[641] B.A. Devlin and J. Corish, Phys. Stat. Sol. (a) **78** (1983) 53; they detected two overlapping dipole reorientation peaks at 92.6 and 99.5 K. The lower temperature peak ($h^{m,b} = 0.25$ eV, $\tau_0 = 1.84 \times 10^{-12}$ s) was attributed to nearest-neighbour cadmium vacancy dipoles.
[642] D.R. Figueroa, J. Fontanella, M.C. Wintersgill and C.G. Andeen, Phys. Rev. **B29** (1984) 5909; they compared their migration parameters for the bound (b) anion motion in CdF_2 doped with Na^+ with those determined in ref. [661] for the free ("f") vacancy motion. They found: $v^{m,b}/v^{m,f} = 0.85$ and $h^{m,b}/h^{m,f} = 0.83$, in excellent agreement with eq. (13.39).
[643] D.R. Figueroa, J. Fontanella, M.C. Wintersgill, A.V. Chadwick and C.G. Andeen, J. Phys. **C17** (1984) 4399; the values of v^{act}/h^{act} in the association region of the conductivity plot of PbF_2 doped with Li, Na, K and Rb are: 6.1, 6.58, 6.86 and 6.55 cm^3 mole^{-1} eV^{-1}, i.e. practically the same as predicted from eq. (13.39). For pure PbF_2 the corresponding value is 6.64 to 7.0, i.e. roughly equal to the previous ones.
[644] The activation volumes v^{act} in the association region of the conductivity plot for "undoped" CdF_2 or for CdF_2 doped with Li or Na have been recently measured (see ref. [642]); they are: 3.99, 4.08 and 3.94 cm^3/mole, respectively. Using: (1) $B = 10^{11}$ N/m^2, $dB/dP = 7$ (C. Varotsos, M. Lazaridou and P. Varotsos, Phys. Rev. **B32** (1985) 2634), (2) the activation enthalpy values 0.715, 0.715, 0.679 eV given in ref. [642], and (3) the rough approximation ($T = 270$ K) $g^{act} = h^{act}$, the $cB\Omega$-model (i.e. eq. 10.1) gives: $v^{act} = 4.1$, 4.1 and 3.9 cm^3/mole, respectively, in excellent agreement with the experimental values mentioned above. On the other hand the Keyes-model leads to v^{act}-values that are appreciably smaller (i.e. by a factor 2 to 3) than the experimental ones.
[645] When comparing the two plots (i.e. for $P = 0$ and $P > 0$) attention is drawn to the following two points: (1) contrary to that showing in fig. 9.3, in most cases the absolute value of the slope of ln D versus $1/T$ *increases* upon increasing the pressure (i.e. $v_h^{act} > 0$, see §3.6.3), and (2) for higher pressures the deviation [133] from linearity becomes *less*, i.e. δ ln $D(P) < \delta$ ln $D(0)$.
[646] By differentiating the relation $g = cB\Omega$ and considering eqs. (2.25), (3.31) and (10.6a), we find eq. (10.5) without using *any* approximation. Then eq. (10.1) immediately results by setting $\kappa^c = 0$.
[647] P.W.M. Jacobs, J. Corish and C.R.A. Catlow, J. Phys. **C13** (1980) 1977.
[648] B.E. Mellander and D. Lazarus, Phys. Rev. **B29** (1984) 2148; these measurements in AgCl and AgBr demonstrate that the activation volume increases considerably with increasing temperature and decreases with increasing pressure; for the case of AgCl a significant difference between the experimental value of v^f and the theoretical result of C.R.A. Catlow, J. Corish, P.W.M. Jacobs and A.B. Lidiard, J. Phys. **C14** (1981) L121 has been noticed. This difference, however, cannot be attributed to experimental uncertainties in view of the excellent quality of the data of Lazarus and coworkers since 1949.
[649] G.A. Samara, Phys. Rev. **B23** (1981) 575.

[650] A. Azimi, V.M. Carr, A.V. Chadwick, F.G. Kirkwood and R. Saghafian, J. Phys. Chem. Solids **45** (1984) 23.
[651] M.O. Manasreh and D.O. Pederson, Phys. Rev. **B30** (1984) 3482; adiabatic elastic data of PbF_2 from 300 to 850 K. The bulk modulus decreases nearly linearly with temperature until a dramatic decrease occurs near the Faraday temperature $T_C = 717$ K. According to the $cB\Omega$-model this behaviour reflects a strong upward curvature of the conductivity plot, a fact that is actually observed.
[652] One of the authors (P.A.V.) cordially acknowledges quite useful discussions with Prof. C.S. Smith on this point; a lot of Prof. Smith's unpublished work has been forwarded to (P.A.V.) and provided the basis for the derivation of a large part of chapter 12.
[653] When "association effect" is present the considerations followed in §12.4.2. should be slightly changed in order to account also for the temperature variation of the ratio of the "non-associated" and "associated" defects, which results in a further slight variation of the volume of the crystal with temperature.
[654] The low-temperature conductivity data of alkali halides with NaCl-structure obtained from measurements at various pressures lead to β^m, κ^m and v^m, whereas the data of high-temperatures give β^{act}, κ^{act} and v^{act}; the combination of these values gives β^f, κ^f and v^f. The latter quantities correspond to a Schottky defect (cation vacancy and anion vacancy) whereas the corresponding quantities β^f, κ^f, v^f in eqs. (12.29) and (12.30) correspond *only* to a cation vacancy. One can estimate, however, that in both cases the values of β^f and κ^f are roughly the same, and further that v^f (per cation vacancy) is of the order of one half of v^f (per Schottky defect).
[655] D. Lazarus, paper presented at the Indo–US Workshop on Diffusion in Solids (Bombay, India, January 1984); it presents a useful overview of the current state of the research on diffusion in metals.
[656] See ref. [88] which among other provides an excellent compilation of data concerning h^f for Schottky defects in alkali halides and anion Frenkel defects in CaF_2-type crystals; the following empirical relations are suggested: $h^f = 8L/N$ (Schottky defects in alkali halides), $h^f = 10.55 L/N$ (bcc metals), $h^f = 8.88 L/N$ (fcc metals), $h^f = 6.72 L/N$ (hcp metals), $s^f = (2$ to $3) L/NT_M$ for Schottky defects in alkali halides, $h^f = 1.37$ eV $+ 3.44 L/N$ for anion Frenkel defects in CaF_2-type crystals; L denotes the heat of fusion.
[657] In §13.1 we give a plausible justification of the proportionality between h_{exp} and T_M for various processes and types of solids on the basis of the $cB\Omega$-model. Once this point has been settled the consideration of the well-known fact that the entropy of fusion L/NT_M (where L denotes the latent heat of fusion) is practically constant for each type of solids immediately justifies the proportionality between h_{exp} and L.
[658] Lindemann (F.A. Lindemann, Phys. Z. **11** (1911) 609) proposed that crystals melt when the rms vibrational amplitude $\langle u^2 \rangle^{1/2}$ reaches a characteristic fraction $\delta \equiv \langle u^2 \rangle^{1/2}/R$ of the interatomic spacing R. There are today a lot of accurate microscopic calculations that find a common value of $\delta = 0.16$ immediately below the melting point in a variety of crystals. It should be mentioned that the stability of crystals has recently been studied with use of an accurate model including atomic vibrations and thermal creation of vacancies (see the interesting paper of L.K. Moleko and H.R. Glyde, Phys. Rev. **B30** (1984) 4215). This paper provides a useful review concerning the validity of the Lindemann suggestion in classical crystals and proves that the suggestion is *not* valid in the quantum limit, in which δ can take any value between 0.04 and 0.35. It is easily shown that the above Lindemann proposal coincides with eq. (13.7) in the case of a Debye solid.
[659] This point just explains the empirical fact noticed in ref. [655] that the "proportionality" between h and T_M is "destroyed" for materials with strongly curved Arrhenius plots.

[660] C.S. Smith and L.S. Cain, J. Phys. Chem. Solids **41** (1980) 199; it provides an excellent compilation of the currently accepted values of the isochoric temperature derivatives of B, C_{44} and $(C_{11}\text{-}C_{12})/2$ of the 16 Li, Na, K and Rb halides; they also found that $(dB/dT)_V$ is negative and has an absolute value appreciably less than that of $(dB/dT)_P$.
[661] J. Obserschmidt and D. Lazarus, Phys. Rev. **B21** (1980) 5823.
[662] There is *no* isothermal–adiabatic distinction for the shear constants; the two independent pure shear constants of a cubic crystal are: C_{44} and $C' = (C_{11} - C_{12})/2$.
[663] According to the Keyes-model the computation of the migration volume for the free anion vacancy in CdF$_2$ gives [603] 1.2 cm^3/mole, in disagreement with the accurate experimental value of 3.1 ± 0.2 cm^3/mole reported by Oberschmidt and Lazarus in ref. [661]. On the other hand, the $cB\Omega$-model leads to a value between 2.6 and 3.9 cm^3/mole (C. Varotsos, M. Lazaridou and P. Varotsos, Phys. Rev. **B32** (1985) 2634).
[664] R.M. Emrick, J. Phys. **F12** (1982) 1327; systematic errors have been investigated by using the same measuring apparatus for Pt-specimens in several quenching environments. It is argued that extrapolations of slow-quench data are as accurate as uncorrected fast-quench values. For $T = 800°C$ the value $v^f = 0.72 \pm 0.04\Omega$ is given.
[665] See §A.11, note 19.
[666] The only meaning of the notation $c(T)$ is that in this paragraph we justify only that "c" is *pressure (volume) independent*. The explicit temperature variation of "c", if any, is discussed separately.
[667] We should remind here that an equation similar to that obtained by Gilder and Lazarus for an activation process has been derived in §6.4.2 but *only for a formation process* (see eq. 6.44 or 6.45).
[668] Upon isochoric heating of a solid the vacancy concentration increases so that the (true) interatomic distance decreases; however, when considering that the latter decrease is quite small we can still assume that – within the frame of QA – the frequencies ω_i and ω'_i do not change with increasing temperature.
[669] Instead of using the notation $g(V, T)$ it is more consistent to deal with $f^*(V, T)$; however, considering that the reader is more familiar with the function g and in view of the equality $g(P, T) = f^*(V, T)$, in chapter 14 we sometimes use the notation $g(V, T)$.
[670] The simultaneous appearance of an SES at dipoles with small length (30 to 300 m) and at those with larger length (some km) is *not* an obligatory rule; there are cases in which an SES was observed at the large dipoles but *not* at the small and vice versa. This may be a combined result of the nature of the "directivity effect" (see §A.11, note 10) and of the fact that the ρ_{rel}-value of a large dipole may differ *appreciably* from that of its parallel small dipole.
[671] S. Ihara, Y. Warita and K. Suzuki, Phys. Stat. Sol. (a) **86** (1984) 729.
[672] B.R.K. Gupta, Phys. Stat. Sol. (b) **128** (1985) 99.
[673] C. Varotsos and M. Lazaridou, Phys. Stat. Sol. (b) **129** (1985) K95.
[674] R.K. Singh, N. Agnihotri, S.V. Suryanarayana and S.P. Sanyal, Phys. Stat. Sol. (b) **123** (1984) 453.

LIST OF MAIN SYMBOLS

a	Lattice constant
α	Ionic polarisability
β	Volume thermal expansion coefficient
$\beta^f, \beta^m, \beta^{act}, \beta^d$	Thermal expansion coefficient of the formation, migration, activation, or in general the defect volume
b	$\exp \int_0^T \beta \, dT$
B	Isothermal bulk modulus
B^S	Adiabatic bulk modulus
B_0^{SL}	The intercept of the linear part of the graph $B = f(T)$ with the vertical axis
C, C^0, C^{0*}	Molar specific heat
c	$\left(\dfrac{v}{\Omega}\right) \Big/ \left(\left.\dfrac{dB}{dP}\right|_T - 1\right)$ or $-s^* \Big/ \left(\Omega \dfrac{dB}{dT}\right)_V$
c_P, c_V	Specific heat per defect
$\tilde{c}_V$	Specific heat per atom
γ	Grüneisen constant
D^{SD}	Self-diffusion coefficient
D^T, D	Tracer self-diffusion coefficient
d	Microscopic diffusion coefficient
δ	isothermal Anderson–Grüneisen constant $\left(\delta \equiv -\dfrac{1}{\beta B}\left.\dfrac{\partial B}{\partial T}\right|_P\right)$
E	Activation energy, or isotope effect, or electric field
$\varepsilon, \varepsilon_\infty$	Static or high-frequency dielectric constant, or the ratio $V_2/V_1 - 1$ in mixed crystals
e^*	Szigeti charge
f	Correlation factor, or Helmholtz formation energy per defect
f^*	Constant volume Helmholtz energy per defect
F	The quantity $-\left(\beta B + \left.\dfrac{\partial B}{\partial T}\right|_P\right) \Big/ \left(B - T\beta B - T\left.\dfrac{\partial B}{\partial T}\right|_P\right)$
F, F^0, F^{0*}	Helmholtz energy
g, g^*	Defect Gibbs energy
G, G^0, G^{0*}	Gibbs energy
h, h^*	Defect enthalpy
H, H^0, H^{0*}	Enthalpy
Θ, Θ_D	Debye temperature
J	Current density
J_{rel}	Signal strength (relative) of a preseismic electric signal
k	Boltzmann constant
κ^S	Adiabatic compressibility
κ	Isothermal compressibility
$\kappa^f, \kappa^m, \kappa^{act}, \kappa^d$	Compressibility of the formation, migration, activation, or in general of the defect volume

List of main symbols

L	Latent heat of fusion, or the quantity: $B^{-1}\left(\dfrac{dB}{dP}-1\right)$, or macroscopic length
l	Nearest-neighbour distance
M	Mass of the atom
N	Avogadro's number
η	Number of defects (per mole)
ν	Attempt frequency
$\tilde{\nu}$	$\alpha^{-1}\sqrt{h^m/M}$
ν_{TO}	Frequency of the zone-centre transverse optic mode
ν_D	Debye frequency
p	Pressure
P	Pressure or probability
Π	polarization
R	Universal gas constant ($=Nk$)
ρ	Density, or resistivity
ρ_{rel}	Relative (effective) resistivity of a station
S, S^0, S^{0*}	Entropy
S_c	Configurational entropy
s, s^*	Thermal defect entropy
σ	Conductivity
U, U^0, U^{0*}	Internal energy
u, u^*	Defect internal energy
V, V^X	Volume, or volume measured by X-rays
v^f, v^m, v^{act}, v^d	Formation, migration, activation or in general defect volume.
w	Jump rate
x	Concentration
z	Number of nearest neighbours, or charge number
Z	Madelung constant
Ω	Mean volume per atom
NGS	Noble gas solids
PAS	Positron annihilation spectroscopy
QA	Quasi-harmonic approximation
RGS	Rare gas solids
SES	Pre-seismic electric signals

AUTHOR INDEX

Aalbers, A.B. 136
Abe, M., *see* Furukawa, K. 173
Abey, M. 266, 296, 299
Ablitzer, D. 225, 314
Ablitzer, D., *see* Sathyaray, K.V. 314
Aboagye, J.K. 260, 261, 262, 263, 264, 267, 296, 299, 305
Achar, B.N.N. 128, 198
Ackermann, H., *see* Heitjans, P. 414
Acuna, L.A. 129, 252, 259, 289
Adda, Y. 117
Adda, Y., *see* Beyeler, M. 174, 175, 273, 305
Adlhart, W. 195, 196, 197, 198, 199, 201, 204, 205
Adlhart, W., *see* Stetter, G. 214
Agarwala, R.P. 314, 319
Agnihotri, N., *see* Singh, R.K. 412
Ailion, D.C. 174, 208, 209
Ailion D.C. *see* Slichter, C.P. 174
Ait Salem, M. 196, 197, 204
Albert, E.L. 407
Albert, H.L. 19
Alers, G.A. 229
Alexander, W.B. 313
Alexopoulos, K. 167, 195, 308, 309
Alexopoulos, K., *see* Lazaridou, M. 408, 409
Alexopoulos, K., *see* Varotsos, P. 5, 60, 62, 86, 95, 129, 131, 137, 138, 140, 165, 198, 206, 249, 258, 264, 267, 266, 268, 287, 290, 292, 293, 297, 298, 305, 309, 345, 361, 362, 366, 368, 403, 404, 405, 406, 408, 409, 410, 411, 416, 417, 419, 420, 422
Allan, G. 275
Allen, D.C. 296, 297, 298, 305
Allnatt, A.R. 253, 254, 289
Altovskii, R.M., *see* Gornyy, D.S. 314
Anand, M.S., *see* Agarwala, R.P. 305, 314, 319
Andeen, C. 131, 137, 305, 358, 361
Andeen, C.G., *see* Figueroa, D.R. 137, 303, 358
Andeen, C., *see* Fontanella, J. 123, 131, 305, 358, 368, 422

Anderson, M.S. 243, 244, 245, 282, 369, 406, 407, 415, 421
Ankowitz, A., *see* Pfister, G. 136
Antony, T.R., *see* Dyson, B.F. 240, 313
Armstrong, J., *see* Hood, G.M. 407
Askill, J. 316, 317, 319
Audit, P. 123, 173, 198, 205, 275
Audit, P., *see* Gilder, H.M. 173
Averbach, B.L., *see* Hilliard, J.E. 313
Azimi, A. 302

Backus, J.G.E.M. 183, 184, 186, 272, 273
Badanov, V.N., *see* Sobolev, G.A. 403
Badrinarayan, S., *see* Bharati, S. 312
Baer, D.R., *see* Heald, S.M. 414
Baerner, K., *see* Krokoszinski, M.J. 136
Baird, G.A. 418
Baker, A.G. 313
Bakker, H., *see* Backus, J.G.E.M. 183, 184, 186, 272, 273
Balluffi, R.W., *see* Simmons, R.O. 88, 112, 169, 170, 172, 176, 177, 178, 180, 181, 184, 185, 186, 187, 190, 242, 243, 244, 393, 403
Balzer, R. 226, 227, 228, 229, 232, 233, 234, 235, 236, 237, 238
Bansigir, K.G., *see* Sankaram Bhima 336, 337
Bardeen, J. 117
Barnes, R.G. 208, 209
Barnes, R.G., *see* Engart, R.D. 273
Barnes, R.G., *see* Hultsch, R.A. 208, 209
Barr, L.W. 256, 351
Barr, L.W., *see* Mundy, J.N. 196, 197
Barron, T.H.K. 392
Barsch, G.R. 255, 257, 260, 342, 370
Barsch, G.R., *see* Chang, Z.P. 19
Barsch, G.R., *see* Hallinger, R.C. 19
Bartdorff, D. 177, 178, 179, 183, 273
Bartels, R.A. 251
Barua, A.K., *see* Dutt, M.B. 311, 312
Barua, A.K., *see* Sen, S.K. 311, 312
Bass, J. 171, 172, 173
Bassani, C., *see* Maier, K. 128, 177, 179, 183, 273

449

Author Index

Bastide, J.P. 18
Batchelder, D.N. 243
Bathier, M., *see* Jaquet, M. 136
Batra, A.P. 121, 122, 130, 131, 252, 260, 261, 262, 263, 264, 266, 267, 320
Bauer, C.F. 247, 248, 250
Baughman, R.H. 351
Beecroft, R.I. 199, 200, 204, 206
Ben-Abraham, S.I. 353
Beniere, F. 144, 253, 254, 255, 320
Beniere, F., *see* Beniere, M. 121, 122, 247, 248, 252, 253, 254, 255, 289, 291, 305, 320
Beniere, F., *see* Brun, A. 252
Beniere, F., *see* Kostopoulos, D. 257, 258, 259, 260, 320
Beniere, M. 121, 122, 247, 248, 252, 253, 254, 255, 289, 291, 305, 320
Beniere, M., *see* Beniere, F. 320
Berger, A.S. 171, 172, 173, 177, 178, 180, 181, 182
Bernardini, J. 312
Bernardini, J., *see* Gas, J.P. 312
Berne, A. 243
Beshers, D.N., *see* Lord, A.E. 308
Beyeler, M. 174, 175, 273, 305
Beyeler, M., *see* Charles, M. 366
Beyeler, M., *see* Lansiart, S. 267, 296, 299, 300, 305
Bharati, S. 312
Bhatia, A.B. 412
Bhima Sankaram 336, 337,
Bianchi, G. 169, 170, 172
Biermann, W. 289, 294
Bihr, J. 312
Bijvank, E.J., *see* Van Weperen, W. 136
Bilz, H., *see* Kleppmann, W.G. 333
Bingham, R.E., *see* Brooks, C.R. 115
Birch, F. 243, 244, 245, 280, 282, 369, 370
Bishof, B., *see* Wenzl, H. 311
Bishop, J.R. 418
Biswas, S.N., *see* van't Klooster 19, 157
Blomberg, T., *see* Lindström, R. 343
Boato, G., *see* Berne A. 243
Bollmann, W. 129, 349
Bolsaitis, P., *see* Chiarodo, R. 273
Bonput, L. 336, 337
Bonput, L., *see* Haget, Y. 329, 336, 337
Bosomworth, D.R. 361, 362
Botaki, A.A., *see* Cyrbu, I.N. 257

Bourassa, R.R. 176, 177, 178, 181
Breazeale, M.A., *see* Peters, R.D. 19, 157
Bremer, F. 14, 264
Breuer, M., *see* Leibfried, G. 367
Brooks, C.R. 5, 115, 180, 273
Brooks, C.R., *see* Yeh, C.C. 115
Brown, N. 256, 285, 289, 393
Brun, A. 252
Bruner, L.J., *see* Tannhäuser, D.S. 263, 321
Brünger, G. 197, 198
Bube, H., *see* Fritsch, G. 215, 216, 217
Bucci, C. 132, 133
Bui Ai 137
Burke, J., *see* Cornish, A.J. 171, 174
Büscher, B.J. 228, 231
Butrymowicz, D.B. 310, 311
Buyers, W.J.L., *see* Kim, S.M. 173

Cabane, J., *see* Bernardini, J. 312
Cain, L.S. 296, 305, 327, 333, 334, 338, 339, 340, 341, 412
Cain, L.S., *see* Smith, C.S. 251, 255, 260, 276, 326, 334, 336, 337, 354, 401
Calais, C.A., *see* Languille, A. 271, 279
Campbell, J.L. 185
Cardegna, P.A. 261, 262, 320
Cardegna, P.A., *see* Laskar, A.L. 320
Cardinal, L.C., *see* Kammer, E.W. 235, 236, 280
Cardinal, L.C., *see* Vold, C.L. 192, 194, 272, 312
Carr, V.M., *see* Azimi, A. 302
Catalano, E., *see* Nachtrieb, N.H. 196, 197, 204
Catlow, C.R.A. 44, 129, 130, 252, 254, 259, 264, 267, 268, 270, 287, 292, 297, 412, 413, 416
Catlow, C.R.A., *see* Beniere, M. 320
Catlow, C.R.A., *see* Jacobs, P.W.M. 264, 297, 298, 412
Chadwick, A.V. 109, 243, 245, 246, 283
Chadwick, A.V., *see* Azimi, A. 302
Chadwick, A.V., *see* Figueroa, D.R. 137, 358
Chadwick, A.V., *see* Fontanella, J. 123, 368, 422
Champier, G., *see* Bianchi, G. 169, 170, 172
Chandra, S. 252, 256, 260

Chang, Y.A.   178, 185, 186, 187, 273, 310, 392
Chang, Z.P.   19
Chang, Z.P., *see* Barsh, C.R.   255, 260, 342
Chanh, N.B., *see* Haget, Y.   329, 336, 337
Charalambous, S., *see* Dedoussis, S.   233
Charbnau, H.P., *see* Feder, R.   195, 196
Chardalas, M., *see* Dedoussis, S.   233
Charles, M.   366
Chason, M.K.   173
Chason, M.K., *see* Berger, A.S.   171, 172, 173
Chason, M.K., *see* Fluss, M.   173
Chemla, M., *see* Beniere, F.   320
Chemla, M., *see* Beniere, M.   121, 122, 247, 248, 252, 253, 254, 255, 289, 291, 305, 320
Chekhovskoi, V.Ya., *see* Petukhov, V.A.   225
Chhabildas, D.C.,   227, 228, 229
Chiarodo, R.   273
Cohen, B.M.   191
Cohen, M., *see* Hilliard, J.E.   313
Combe-Brun, A., *see* Bernardini, J.   312
Cook, J.S.   253, 254, 255
Coqblin, B., *see* Languille, A.   279
Corish, J.   129, 260, 261, 299
Corish, J., *see* Catlow, C.R.A.   44, 129, 130, 252, 254, 259, 264, 268, 267, 292, 297, 412, 413, 416
Corish, J., *see* Devlin, B.A.   136
Corish, J., *see* Jacobs, P.W.M.   264, 297, 298, 412
Cornelis, J., *see* De Schepper, L.   195
Corneliussen, R.D., *see* Lawson, A.W.   349
Cornet, J.A.   271, 279
Cornish, A.J.   171
Coston, C.   233, 236, 237, 238
Coston, C., *see* Nachtrieb, N.H.   280, 305
Coufal, H.J., *see* Korpium, P.   112, 244, 245, 282, 283
Cowgill, D.G.   243, 245
Crampin, S.T.   410
Crawford, J.H.   128, 132, 289
Crawford, J.H., *see* Kitts, E.   136
Crawford, J.H., *see* Matthews, G.E.   136
Crawford, J.H., *see* Stott, J.P.   136
Cretenet, J.C., *see* Morizur, G.   222
Current, M.   226, 227, 229, 233, 234, 235
Cusso, F.   136
Cyrbu, I.N.   257

Da Fano, A.   198
Daniels, W.B.   200, 273, 305
Danneberg, W.   219, 220
Dansas, P., *see* Brun, A.   252
Dariel, M.P.   278
Dariel, M.P., *see* Languille, A.   279
Dave, N.K., *see* Mckee, B.T.L.   173
Dawson, D.K., *see* Barr, L.M.   256
Day, J.P.   210
Dayal, B., *see* Lagu, M.   247
Dayan, D., *see* Dariel, M.P.   278
De Paz, M., *see* Berne, A.   243
deSouza, M., *see* Siu Li, M.   137
Decker, D.L.   307, 313, 322
Decker, D.L., *see* Vanfleet, H.B.   313
Dederichs, P.H.   367, 368, 371, 397
Dederichs, P.H., *see* Zeller, R.   367
Dedoussis, S.   233
Delaplace, J., *see* Charles, M.   366
Demangeat, C., *see* Sathyaray, K.V.   314
den Hartog, H.W.   136
den Hartog, H.W., *see* Aalbers, A.B.   136
den Hartog, H.W., *see* van Weperen, W.   136
Denham, P.   361, 362
de Schepper, L.   195
Desai, R.J., *see* Pathak, P.D.   226
Desorbo, W.   233
Destruel, P., *see* Bui Ai   137
Dever, D.   308
Dever, D., *see* Fisher, E.S.   277, 318
Devlin, B.A.   136
Devlin, B.A., *see* Jacobs, P.W.M. 264, 297, 412
Dickey, J.E.   280, 305
Dickman, J.E.   280, 305
Diederich, M.E.   200, 201
Dienes, G.J.   349, 350
Dienes, G.J., *see* Nowick, A.S.   125, 168, 388
Dietz, H., *see* Albert, E.L.   407
Diller, K.M., *see* Catlow, C.R.A.   129, 254, 259, 268, 412, 413, 416
Dobrovinskaya, O.R.   320
Dods, S.A., *see* Donovan-Vojtovic, B.F.   313
Döhl, R.   407
Dologlou-Revelioti, E.   136, 403, 410, 419
Dologlou-Revelioti, E., *see* Varotsos, P.   141
Donovan-Vojtovic, B.F.   313
Douglas, T.B.   14

Doyama, M. 182
Doyama, M., *see* Nanao, S. 182
Dryden, J.S., *see* Cook, J.S. 253, 254, 255
Dubbers, D., *see* Heitjans, P. 413
Dugdale, J.S. 270, 387
Dunn, K.J. 370
Dupuy, M. 279
Dutt, M.B. 311, 312
Dutt, M.B., *see* Sen, S.K. 311, 312
Dworschak, F. 191
Dyson, B.F. 238, 239, 240, 313

Ebert, J. 260
Ebert, J., *see* Kessler, A. 136
Eckseler, H., *see* Herzig, C. 199, 319
Eftaxias, K. 406
Eftaxias, K., *see* Lazaridou, M. 46, 160
Einziger, R.E. 225, 307, 308, 314
Elbaum, C., *see* Swartz, K.D. 232
Emrick, R.M. 366
Emrick, R.M., *see* Nachtrieb 305
Engart, R.D. 273
Engart, R.D., *see* Barnes, R.G. 208, 209
Evenson, W.E., *see* Decker, D.L. 307, 313, 322

Falter, C., *see* Zierau W. 264
Faux, I.D. 268, 270, 287, 297, 413
Faux, I.D., *see* Catlow, C.R.A. 268, 270, 287, 297, 413
Feder, R. 190, 191, 192, 194, 195, 196, 208, 209, 211, 212, 213, 214
Federer, J.I. 319
Fedorov, Y.B. 14
Feit, M.D. 81, 196, 198
Felice, R.A. 102, 210, 211
Ferraro, J.R. 357
Ferraro, J.R., *see* Postmus, C. 357
Ferreira, G.F.L., *see* Nunes de Oliveira, L. 133
Field, G.R., *see* Denham, P. 361, 362
Fields, D.E., *see* Moran, P.R. 136
Fieschi, R., *see* Bucci, C. 132, 133
Figueroa, D. 136, 137, 303, 358
Figueroa, D., *see* Laredo, E. 135
Fillard, J.P., *see* Vanderschueren, J. 136, 423
Finnis, M.W. 275
Fisher, E.S. 273, 277, 318
Fisher, E.S., *see* Albert, H.L. 19
Fisher, E.S., *see* Manghnani, M.H. 19

Fjeldly, T.A., *see* Raaen, A.M. 261
Fluss, M. 173
Flynn, C.P. 47, 71, 72, 74, 75, 80, 81, 82, 83, 84, 123, 144, 275, 289, 350, 351, 356, 358, 359, 360, 363, 364, 365, 369, 384, 387
Fontanella, J. 123, 131, 305, 358, 368, 422
Fontanella, J., *see* Andeen, C. 131, 137, 305, 358, 361
Fontanella, J., *see* Figueroa, D.R. 137, 303, 358
Fouchaux, R.D. 266, 267
Frankel, J. 19
Franklin, W. 144, 198
Fredericks, W.J. 121, 122
Fredericks, W.J., *see* Machida, H. 253, 254, 255, 257
Friauf, R.J. 118, 121, 122, 168, 261, 265
Friauf, R.J., *see* Aboagye, J.K. 260, 261, 262, 263, 267, 296, 299, 305
Fritsch, G. 199, 200, 201, 203, 205, 215, 216, 217
Fritsch, G., *see* Adlhart, W. 195, 196, 197, 198, 199, 201, 204, 205
Fritsch, G., *see* Ritter, M. 195, 196, 204
Fritsch, G., *see* Stetter, G. 214
Fromont, M. 278
Fromont, M., *see* Languille, A. 271, 279
Fugate, R.Q. 16
Fugate, R.Q., *see* Anderson, M.S. 243, 282, 349, 369
Fujara, F., *see* Heitjans, P. 413
Fujikawa, S., *see* Hirano, K. 313
Fullam, H.T. 343
Furukawa, K. 173

Gamper, K. 16
Gamper, K., *see* Haensler, F. 16
Ganne, J.P. 32
Gard, V.K. 285
Garin, P., *see* Haget, Y. 329, 336, 337
Garza, C.M. 161
Gas, P. 312
Gasiot, J., *see* Vanderschueren, J. 136, 423
Gavrilyuk, M.J., *see* Korolev, A.A. 219, 220
Geipel, F., *see* Fritsch, G. 199, 200
Getting, I.C., *see* Grover, R. 206, 207, 276, 370
Gibbs, G.B., *see* Askill, J. 316, 319

Gilder, H.M. 61, 99, 102, 12, 123, 125, 126, 127, 128, 158, 161, 167, 168, 173, 198, 205, 206, 208, 226, 291, 305, 366, 388, 417
Gilder, H.M., see Audit, P. 123, 173, 198, 205, 275
Gilder, H.M., see Baker, A.G. 313
Gilder, H.M., see Buescher, B.J. 228, 231
Gilder, H.M., see Chhabildas, L.C. 227, 228, 229
Gilder, H.M., see Current, M. 226, 227, 229
Gillan, M.J. 417
Glicksman, M.E., see Kammer, E.W. 235, 236, 280
Glicksman, M.E., see Vold, C.L. 192, 194, 272, 312
Glyde, H.R. 80
Glyde, H.R., see Chadwick, A.V. 109, 243, 246, 283
Glyde, H.R., see Moleko, L.K. 352, 353
Glyde, H.R., see Parker, E.H.C. 243
Göltz, G. 197, 198
Gonas, A.M., see Lowrie, R. 222
Goncharova, V.A., see Voronov, F.F. 19
Gorbachev, V.A. 310, 311
Gorecki, T. 46
Gornyy, D.S. 314
Govindarajan, J., see Haridasan, T.M. 417
Govindarajan, J., see Jacobs, P.W.M. 417
Graham, D., see Peart, R.F. 314
Grammatikakis, J., see Eftaxias, K. 406
Grammatikakis, J., see Lazaridou, M. 329
Granato, A.V., see Hiki, Y. 273, 305
Granato, A.V., see Holder, J. 367, 399
Green, J., see Chiarodo, R. 273
Grimes, H.H. 366
Grover, R. 206, 207, 276, 370
Guerard, R. 169, 170, 171, 172, 173
Guidi, G., see Bucci, C. 132
Guinan, M.W. 280, 305
Gupta, A.P., see Shanker Jai 270, 387
Gupta, B.R.K. 412
Gupta, D. 128
Gupta, R.G. 219
Gurney, R.W., see Mott, N.F. 129
Gustafson, D.R., see Dickman, J.E. 280, 305

Haensler, F. 16

Haget, Y. 329, 336, 337
Haget, Y., see Bonput, L. 336, 337
Handler, G.S., see Nachtrieb, N.H. 191, 194, 272, 349
Harding, J.H. 417
Haridasan, T.M. 417
Haridasan, T.M., see Jacobs, P.W.M. 417
Harman, A.W., see Douglas, T.B. 14
Harridos, S., see Radhakrishna, S. 134, 135, 422
Hart, S. 249
Hasiguti, R.R., see Sawayanagi, F. 313
Hayden, I.M., see Andeen, C. 131, 358
Heald, S.M. 414
Hehenkam, T. 312
Heidemann, A., see Adhlhart, W. 195, 196, 199, 201, 204
Heidemann, A., see Göltz, G. 197, 198
Heitjans, P. 414
Heller, W.R., see Nowick, A.S. 131
Henry, R. 242
Hernandez, J. 136
Hernandez, J., see Batra, A.P. 131
Herring, C., see Bardeen, J. 117
Herzig, C. 199, 319
Herzig, C., see Dworschak, F. 191
Hietala, J. 343
Hiki, Y. 273, 305
Hiki, Y., see Mori, S. 19
Hillairet, J., see Charles, M. 366
Hillairet, J., see Simon, J.P. 227
Hillairet, J., see Tzanetakis, P. 173
Hilliard, J.E. 313
Himmel, L., see Chang, Y.A. 178, 185, 186, 187, 273, 310
Hirakawa, K., see Yoshizawa, H. 248
Hirano, K. 313
Hirano, K., see Agarwala, R.P. 314
Hirano, K., see Hoshino, K. 311
Ho, P.S. 196, 200, 206, 207, 305
Hoang The Gima, see Bui Ai 137
Hoff, H.A., see Einziger, R.E. 225, 314
Hoff, H.A., see Mundy, J.N. 219, 220, 221, 222, 223, 224
Hoffman, R.E. 312
Hoffman, R.E., see Hudson, J.B. 192, 272
Hohler, B. 421
Holder, J. 364, 367, 399
Hollinger, R.C. 19
Homan, C.G., see Huebener, R.P. 366

Hood, G.M. 173, 316, 317, 407
Hood, G.M., see Kim, S.M. 173
Hoshino, K. 311
Howard, R.E. 144
Huang, F.H. 280, 305
Huckaby, D.A., see Garza, C.M. 161
Hudson, J.B. 192, 272
Huebener, R.P. 366
Huebscher, H.J. 16
Hughes, D.M., see McKeever, S.W.S. 422
Hultsch, R.A. 208, 209
Hultsch, R.A., see Barnes, R.G. 208, 209
Huntington, H.B., see Huang, F.H. 280, 305
Huntington, H.B., see Shirn, G.A. 228
Huntz, A.M., see Lesage, B. 322
Hussain, M.A., see Frankel, J. 19

Ihara, S. 412
Iijima, Y., see Hoshino, K. 311
Ikeya, M., see Kitts, E.L., Jr. 136
Ikonomopoulos, P. 142
Isakina, A.P., see Prokhvatilov, A.I. 403

Jackman, J.A., see Campbell, J.L. 185
Jackson, M.S. 317, 319
Jacobs, P.W.M. 130, 136, 248, 264, 297, 298, 393, 412, 413, 416, 417
Jacobs, P.W.M., see Acuna, L.A. 129, 130, 252, 259, 289
Jacobs, P.W.M., see Brown, N. 256, 289, 393
Jacobs, P.W.M., see Catlow, C.R.A. 44, 129, 130, 252, 254, 259, 264, 267, 268, 292, 297, 412, 413, 416
Jacobs, P.W.M., see Corish, J. 129, 260, 261, 299
Jacobs, P.W.M., see Gillan, M.J. 417
Jacobs, P.W.M., see Haridasan, T.M. 417
Jacobs, P.W.M., see Sahni, V.C. 417
Jacque, F., see Cusso, F. 136
Jacque, F., see Hernandez, J. 136
Jacquesson, J., see Romain, J.P. 19
Jacucci, G. 275
Jacucci, G., see Da Fanno 198
Jain, G.D., see Shanker, J. 412
Janot, C., see Bianchi, G. 169, 170, 172
Jaquet, M. 136
Jaumot, F.E. 312
Jaumot, F.E., see Sawatzky, A. 312
Jeffery, R.N. 277
Jeffery, R.N., see Dickman, J.E. 280, 305

Jones, D.L., see Fontanella, J. 358
Jones, L.E.A. 249
Jorgensen, J.D., see Vanfleet, H.B. 313
Jost, W. 359, 360, 372, 384, 385, 386

Kac, M. 80
Kamiyoshi, K. 337
Kammer, E.W. 235, 236, 280
Kammer, E.W., see Vold, C.L. 192, 194, 272, 312
Kampfer, G., see Korpium, P. 244
Kanert, O., see Brunger, G. 197, 198
Kantola, M. 336, 337
Kanzaki, H. 264
Kapphan, S.E., see Siu Li, M. 137
Katahara, K.W., see Albert, H.L. 19
Katahara, K.W., see Fischer, E.S. 273, 277
Katahara, K.W., see Manghnani, M.H. 19
Kennan, P.S., see Baird, G.A. 418
Kennedy, G.C., see Grover, R. 206, 207, 276, 370
Kennedy, G.C., see Vaidya, S.N. 232, 238, 280, 305
Kessler, A. 136
Keyes, R.W. 5, 55, 138, 303, 349, 354, 355, 364
Keyes, R.W., see Lawson, A.W. 364
Keyes, R.W., see Zener, C. 354, 355
Killean, R.C.G. 174, 192, 200, 222, 308
Kim, S.M. 173
Kirchheim, R., see Albert, E. 407
Kirkwood, F.G., see Azimi, A. 302
Kirsch, Y., see Kristianpoller, N. 136
Kittel, C. 20, 178, 235, 273, 280
Kitts, E.L., Jr. 136
Kleppman, W.G. 333
Klokholm, D., see Meakin, J.D. 233, 236, 237
Klotsman, S.M., see Gorbachev, V.A. 310, 311
Knuyt, G., see de Schepper, L. 195
Kobayashi, K. 264
Kohler, C.R. 276
Kolhas, R., see Volmer, O. 273
Kol'stov, A.V., see Sobolev, G.A. 403
Körblein, A., see Heitjans, P. 414
Korenkov, Yu.P. 14
Korolev, A.A. 219, 220
Korpiun, P. 16, 41, 109, 112, 243, 244, 245, 282, 283

Korpiun, P., see Fritsch, G. 200, 201, 203, 205
Kostopoulos, D. 136, 247, 257, 258, 259, 260, 320
Kostopoulos, D., see Triantis, D. 136, 257
Kostopoulos, D., see Varotsos, P. 136, 247
Kouremenou, S., see Kostopoulos, D. 136
Kouremenou, S., see Varotsos, P. 136
Kozhevnikov, I.G. 14
Kraftmakher, Ya.A. 14, 214, 215
Kramer, W. 116, 229, 236
Krasil'nikov, O.M. 19
Krautheim, G. 311
Kristianpoller, N. 136
Krokoszinski, H.J. 136
Kronmueller, H., see Hohler, B. 421
Kryszewski, M., see Zielinski, M. 136
Kulhanek, O. 410
Kuribayashi, K., see Doyama, N. 182
Kuribayashi, K., see Nanao, S. 182
Kurnick, S.W. 296, 300, 346, 347
Kusunoki, K. 313
Kuwana, N., see Furukawa, K. 173

Lagu, M. 247
Lallemand, M. 305
Lam, N.Q. 177, 179, 182, 183, 184, 186, 187, 273
Lam, N.Q., see Mundy, J.N. 219, 220, 221, 222, 223, 224
Langevoort, J.C., see den Hartog, H.W. 136
Languille, A. 271, 279
Languille, A., see Dariel, M.P. 278
Lannoo, M., see Allan, G. 275
Lansiart, S. 267, 296, 299, 300, 305
Laredo, E. 135
Laredo, E., see Figueroa, D. 136
Laskar, A.L. 320
Laskar, A.L., see Cardegna, P.A. 261, 262, 320
Laskar, A.L., see Rothman, S.J. 320
Lawn, B.P. 263
Lawson, A.W. 349, 354, 364
Lawson, A.W., see Nachtrieb, N.H. 196, 197, 204
Lawson, A.W., see Tannhäuser, D.S. 263, 321
Lawson, A.W., dee Tomizuka, C.T. 273, 305
Lawson, A.W., see Zener, C. 354, 355

Lazaridou, M. 46, 160, 329, 408, 409, 410, 412
Lazaridou, M., see Alexopoulos, K. 409, 410
Lazaridou, M., see Eftaxias, K. 46, 160
Lazaridou, M., see Varotsos, C. 303, 364, 412
Lazaridou, M., see Varotsos, P. 137, 406, 408, 409, 412
Lazarus, D. 19, 233, 277, 287, 294, 296, 349, 350, 351, 410, 411
Lazarus, D., see Allen, D.C. 296, 297, 298, 305
Lazarus, D., see Gilder, H.M. 61, 99, 102, 122, 123, 125, 126, 127, 128, 158, 161, 167, 168, 198, 205, 206, 208, 291, 305, 366, 388, 417
Lazarus, D., see Jackson, M.S. 317, 319
Lazarus, D., see Martin, G. 266, 287, 291, 292, 293, 366, 368, 413
Lazarus, D., see Mellander, B.E. 301
Lazarus, D., see Oberschmidt, J. 303, 358
Lazarus, D., see Ott, A. 315
Lazarus, D., see Pierce, C.B. 312
Lazarus, D., see Pontau, A.E. 317
Lazarus, D., see Yoon, D.N. 267, 287, 289, 291, 292, 293, 294, 305, 366, 368, 413
Le Claire, A.D. 117, 191, 272
Leadbetter, A.J. 14
Ledbetter, H.M. 228
Legnidi, D.G., see Fluss, M. 173
Lehman, C., see Dederichs, P.H. 367, 368, 371, 397
Leib, R. 131
Leibfried, G. 251, 353, 367
Lengeler, B., see Bourassa, R.R. 176, 177, 178, 181
Lenting, B.P.M., see Van Weperen, W. 136
Lesage, B. 322
Lessman, E., see Maier, K. 128
Libanati, C.M., see Walsöe de Reca, N.E. 278
Lichtenberger, P.C., see Mackenzie, I.K. 233
Lidiard, A.B. 248, 413
Lidiard, A.B., see Barr, L.W. 351
Lidiard, A.B., see Catlow, C.R.A. 44, 130, 297, 416
Lidiard, A.B., see Faux, I.D. 268, 270, 287, 297, 413
Lidiard, A.B., see Howard, R.E. 144

Lilley, E., *see* McKeever, S.W.   136
Lindberg, M., *see* Pelleg, J.   314
Lindemann, F.A.   352, 353
Lindström, R.   329, 343
Lindström, R., *see* Kantola, M.   336, 337
Link D., *see* Andeen, C.   305, 361
Linkens, A., *see* Vanderschueren, J.   136, 423
Lischer, E.J., *see* Killean, R.C.G.   174, 192, 200, 222, 308
Littleton, M.J., *see* Mott, N.F.   383
Lodding, A.   208, 209, 315
Lodding, A., *see* Mundy, J.N.   315
Lodding, A., *see* Ott, A.   315
Lohstötter, H., *see* Spalt, H.   287
Loje, K.F.   262, 296, 305, 341, 347, 370
Lord, A.E.   308
Lories-Susse, C., *see* Bastide, J.P.   18
Los, V.F., *see* Sobolev, G.A.   403
Losee, D.L.   245, 282
Losee, D.L., *see* Batchelder, D.N.   243
Loussier, R., *see* Bui Ai   137
Lowell, R.C., *see* Tomizuka, C.T.   273, 305
Löwenberg, L., *see* Mundy, J.N.   315
Löwenberg, L., *see* Ott, A.   315
Lowndess, R.P.   361, 362
Lowrie, R.   222
Lü Dajiong   410
Ludwig, W.   4, 19, 125, 353, 357, 367, 368, 369, 380, 381, 382, 397, 410
Ludwig, W., *see* Leibfried, G.   251
Ludwig, W., *see* Varotsos, P.   60, 62, 206, 287, 290, 293, 305, 368, 413, 417, 422
Luescher, E., *see* Adlhart, W.   195, 196, 197, 198, 199, 201, 204, 205
Luescher, E., *see* Fritsch, G.   200, 201, 203, 205
Luescher, E., *see* Killean, R.C.G.   200
Luescher, E., *see* Korpiun, P.   16, 41, 109, 243, 244
Luescher, E., *see* Ritter, M.   195, 196, 204
Luescher, E., *see* Stetter, G.   214
Lundy, T.S.   169, 170, 174, 175, 176, 273, 313, 314
Lundy, T.S., *see* Federer, J.I.   319
Lundy, T.S., *see* Murdock, J.F.   317
Lundy, T.S., *see* Pawell, R.E.   128
Lupis, C.H.P.   144

MacDonald, D.K.C.   214, 215

MacDonald, D.K.C., *see* Dugdale, J.S.   270, 387
MacHargue, G.J., *see* Lundy, T.S.   314
Macht, M.P., *see* Döhl, R.   407
Mackenzie, I.K.   233
Macrander, A.T.   112, 244, 282, 283, 284
Maier, K.   128, 177, 179, 183, 219, 273, 351
Maier, K., *see* Bihr, J.   312
Mainwood, A.   144
Mallejac, D., *see* Bianchi, G.   169, 170, 172
Manasreh, M.O.   302
Manghnani, M.H.   19
Manghnani, M.H., *see* Albert, H.L.   19
Manghnani, M.H., *see* Fischer, E.S.   273, 277
Manning, J.R., *see* Butrymowicz, D.B.   310, 311
March, N.H.   353
March, N.H., *see* Bhatia, A.B.   412
Martel, P., *see* Kim, S.M.   173
Martin, D.L.   199
Martin, G.   266, 287, 291, 292, 293, 366, 368, 413
Martinson, R.H.   199, 200
Mashida, H.   253, 254, 255, 257
Matthews, G.R., Jr.   136
McArdle, P.B., *see* Emrick, R.M.   366
McFall, W.D., *see* Mundy, J.N.   315
McGervey, J.D.   185
McGervey, J.D., *see* Trifthäuser, W.   182, 185
McHargue, G.J., *see* Lundy, T.S.   314
McKee, B.T.A.   173, 191, 305
McKeever, S.W.   136, 422
McKinstry, H.A., *see* Slagle, O.D.   289, 326, 327, 329, 332
McLean, K.O.   305
McLean, K.O., *see* Smith, C.S.   332, 370
McLellan, R.B.   144, 363
Meakin, J.D.   233, 236, 237
Mehrer, H.   117, 120, 122, 128, 192, 197, 198, 208
Mehrer, H., *see* Backus, J.G.E.M.   183, 184, 186, 272, 273
Mehrer, H., *see* Bihr, J.   312
Mehrer, H., *see* Göltz, G.   197, 198
Mehrer, H., *see* Lam, N.Q.   183, 184, 186, 187

Mehrer, H., see Maier, K. 128
Mehrer, H., see Seeger, A. 119, 172, 188
Mehrer, H., see Steiner, E. 118
Mellander, B.E. 301
Melville, J.G., see Decker, D.L. 313
Merchant, H.D., see Rapp, J.E. 257
Merchant, H.D., see Srivastava 251
Messer, R. 208, 209, 210, 211, 213, 315, 421
Meyer, K. 410
Migault, A., see Romain, J.P. 19
Miliotis, D., see Varotsos, P. 131, 132
Miller, G.R., see Shelley, R.D. 136
Miller, R.A. 192, 246
Miller, T.E., see Mundy, J.N. 214, 216, 217, 218, 277
Mints, R.G. 14
Mitchell, J.L., see Martin, G. 266, 287, 291, 292, 293, 366, 413
Mitra, S.S., see Ferraro, J.R. 357
Mitra, S.S., see Postmus, C. 357
Moghaddam, H.R., see Shaw, D. 135
Moleko, L.K. 352, 353
Monfort, C.E. 199, 204, 206, 207
Moores, B.M., see Titman, J.M. 208, 209
Moran, P.R. 136
Mori, S. 19
Morizur, G. 222
Morrison, J.A., see Chadwick, A.V. 243, 245
Morse, P.L.R., see Denham, P. 361, 362
Mott, N.F. 129, 383
Mourikis, S., see Kostopoulos, D. 136, 136, 258
Mourikis, S., see Varotsos, P. 136, 247
Mozgovoi, A.G., see Petukhov, V.A. 225
Mrowec, S. 309
Mukherjee, K. 353
Mulimani, H.G., see Sastry, P.V. 353
Mullen, J.P. 117, 120
Müller, P. 133, 134, 260
Mundy, J.N. 124, 127, 195, 196, 197, 200, 203, 205, 206, 207, 214, 215, 216, 217, 218, 219, 220, 221, 222, 223, 224, 277, 305, 315
Mundy, J.N., see Dworschak, F. 191
Mundy, J.N., see Einziger, R.E. 225, 307, 308, 314
Mundy, J.N., see Lodding, A. 208, 209, 315

Mundy, J.N., see Ott, A. 315
Murarka, S.P., see Agarwala, R.P. 314, 319
Murdock, J.F. 317
Murdock, J.F., see Lundy, T.S. 169, 170, 174, 175, 176, 273, 313
Murieta, H., see Hernandez, J. 136
Murnaghan, F.D. 19, 21, 369, 370, 379

Nachtrieb, N.H. 191, 194, 196, 197, 204, 272, 274, 280, 305, 349
Nachtrieb, N.H., see Coston, F. 233, 236, 237, 238
Nachtrieb, N.H., see Döhl, R. 410
Nachtrieb, N.H., see Lawson, A.M. 349
Nachtrieb, N.H., see Resing, H.A. 313
Nadler, C. 253, 254
Nanao, S. 182
Nanao, S., see Doyama, M. 182
Nasimov, R.N., see Sobolev, G.A. 403
Nasser, M.S., see Donovan-Vojtovic, B.F. 313
Naundorf, V., see Döhl, R. 406
Neb, D.K., see Singh, R.K. 412
Neidhardt, A., see Krautheim, G. 311
Neighbours, J.R., see Alers, G.A. 229
Nerenberg, M.A.H., see Haridasan, T.M. 417
Nerenberg, M.A.H., see Jacobs, P.W.M. 417
Neumann, G. 177, 178, 179, 407
Neumann, G., see Bartdorff, D. 177, 178, 179, 183, 273
Newsham, D.M.T., see Leadbetter, A.J. 14
Nieminen, R.M., see Finnis, M.W. 275
Nigara, Y., see Kamiyoshi, K. 337
Nihoul, J., see Schepper, L. 195
Nishikawa, S., see Kusunoki, K. 313
Noack, F., see Messer, R. 208, 209, 210, 211, 213, 315, 421
Nohman, M., see Fritsch, G. 200, 201, 203, 205
Nölting, J., see Bremer, F. 14, 264
Nölting, J., see Kramer 116, 229, 236
Nomicos, K. 410
Nomicos, K., see Varotsos, P. 140, 141, 404
Norberg, R.E., see Cowgill, D.G. 243, 245
Norberg, R.E., see Henry, R. 242

Norberg, R.E., *see* Ott, A. 280, 305, 315
Norberg, R.E., *see* Sirovich, B.E. 242, 243
Norberg, R.E., *see* Yen W.M. 243, 245
Norden-Ott, A., *see* Ott, A. 280, 305, 315
Norgett, M.J. 268, 270, 297, 413
Norgett, M.J., *see* Catlow, C.R.A. 129, 130, 252, 254, 259, 268, 270, 287, 292, 297, 412, 413, 416
Nowick, A.S. 125, 131, 168, 388
Nowick, A.S., *see* Feder, R. 190, 191, 192, 194
Nowicki, L.J., *see* Lam, N.Q. 177, 179, 182, 183, 184, 186, 187, 273
Nowicki, L.J., *see* Mundy, J.N. 219, 220, 221, 222, 223, 224
Nunes de Oliveira, L. 133

Oberschmidt, J. 303, 358, 364
Ockers, S.T., *see* Berger, A.S. 171, 172, 173, 177, 178, 180, 181, 182
O'Connell, R.J., *see* Spetzler, M. 253, 285, 289, 290, 291, 320, 342, 370, 415
Ong, S.H., *see* Jacobs, P.W.M. 136
Oshcherin, B.N. 353
Ott, A. 280, 305, 315
Ott, A., *see* Lodding, A. 208, 209, 315
Ott, A., *see* Mundy, J.N. 315
Owens, C.W. 313

Pantelis, P., *see* Alnatt, A.R. 253, 254, 289
Pantelis, P., *see* Jacobs, P.W.M. 248, 393
Papaioannou, G., *see* Varotsos, P. 141
Pare, V.K., *see* Peters, R.D. 19, 157
Parker, E.H.C. 243
Parkhomenko, E.I. 418
Parot, P., *see* Vanderschueren, J. 136, 423
Pathak, P.D. 226
Pavlinov, L.V., *see* Korolev, A.A. 219, 220
Pawell, R.E. 128
Pawell, R.E., *see* Lundy, T.S. 314
Peart, R.F. 277, 314
Pederson, D.O., *see* Manasreh, M.O. 302
Peisl, H., *see* Guerard, R. 169, 170, 171, 172, 173
Peisl, H., *see* Raab, R. 154, 292
Peisl, H., *see* Spalt, H. 287
Pelleg, J. 314
Pelleg, J., *see* Ben-Abraham, S.I. 353
Peo M., *see* Maier, K. 219, 351
Pershits, Y.N. 257

Peters, R.D. 19, 157
Peterson, N.L. 118, 120, 122, 127, 128, 173, 227, 228, 229, 231, 273, 313
Peterson, N.L., *see* Rothman, S.J. 177, 178, 179, 183, 184, 185, 186, 273, 312, 320
Petukhov, V.A. 225
Pfister, G. 136
Pfundstein, M., *see* Neumann, G. 407
Philibert, J., *see* Adda Y. 117
Pierce, C.B. 312
Pistorius, H. 367, 397
Pochapsky, T.E. 192
Poivier, J.P., *see* Ho, P.S. 305
Pokorny, M. 214
Ponomarev, A.V., *see* Sobolev, G.A. 403
Pontau, A.E. 317
Porte, R.J., *see* Mundy, J.N. 214, 215, 216, 217, 218, 277
Postmus, C. 357
Prasetyo, A., *see* Fritsch, G. 199, 200
Prokhvatilov, A.I. 403
Puma, M., *see* Figueroa, D. 136
Puma, M., *see* Laredo, E. 136
Puri, D.S. 19
Puri, D.S., *see* Gard, V.K. 285

Quattrochi, A., *see* Ferraro, J.R. 357

Raab, R. 154, 292
Raaen, A.M. 261
Rabinovitch, A., *see* Ben-Abraham, S.I. 353
Rabovskii, Ya.A., *see* Gorbachev, V.A. 310, 311
Radenac, A., *see* Morizur, G. 222
Radhakrishna, S. 134, 135, 422
Radhakrishna, S., *see* Corish, J. 129
Rao, C.N.R., *see* Beniere, M. 320
Rapp, J.E. 257
Rasch, K.D. 219, 220, 222
Read, M.E., *see* Butrymowicz, D.B. 310, 311
Reddy, K.V., *see* Kostopoulos, D. 257, 258, 259, 260, 320
Rehn, L.E., *see* Holder, J. 367, 399
Reif, F. 352
Reimers, P., *see* Bartdorff, D. 177, 178, 179, 183, 273
Reimers, P., *see* Neumann, G. 407

Rein, G., *see* Maier, K. 351
Reinhold, U., *see* Krautheim, G. 311
Resing, H.A. 313
Resing, H.A., *see* Nachtrieb, N.H. 272
Revel, G., *see* Tzanetakis, P. 173
Rice, S.A. 80
Rice, S.A., *see* Lawson, A.W. 349
Rice, S.A., *see* Nachtrieb, N.H. 272
Rimai, D.S. 302, 303, 363
Ritter, M. 195, 196, 204
Roberts, R.W. 289, 327, 332, 334, 336, 342, 347, 370
Roberts, R.W., *see* Smith, C.S. 332, 336, 370
Robinson, L.C., *see* Rothman, S.J. 183, 184, 185, 186, 312, 320
Robrock, K.H. 367
Rokbani, R., *see* Beniere, F. 253, 254, 255
Rolfe, J. 256
Rolfe, J., *see* Chandra, S. 252, 256, 260
Romain, J.P. 19
Ross, R.A., *see* Decker, D.L. 307, 313, 322
Rossel, J., *see* Nadler, C. 253, 254
Rothman, S.J. 177, 178, 179, 183, 184, 185, 186, 273, 312, 320
Rothman, S.J., *see* Lam, N.Q. 177, 179, 182, 183, 184, 186, 187, 273
Rothman, S.J., *see* Mundy, J.N. 219, 220, 221, 222, 223, 224
Rothman, S.J., *see* Peterson, N.L. 227, 228, 229, 231, 313
Roux, F. 314
Rubio, J., *see* Hernandez, J. 136
Ruoff, A.L., *see* Day, J.P. 210
Ruoff, A.L., *see* Ho, P.S. 200, 206, 207, 305
Ruoff, A.L., *see* Kohler, C.R. 276

Sachdev, M., *see* Finnis, M.W. 275
Saghafian, R., *see* Azimi, A. 302
Saghafian, R., *see* Fontanella, J. 123, 368, 420
Sahni, V.C. 417
Sahni, V.C., *see* Jacobs, P.W.M. 417
Saile, B., *see* Maier, K. 219, 351
Salimäki, K.E. 342, 343
Salov, B.G., *see* Sobolev, G.A. 403
Samara, G.A. 285, 286, 290, 301, 302, 303, 305, 364, 366, 408, 409

Sammis, C.G., *see* Spetzler, H. 253, 285, 289, 290, 291, 320, 342, 370, 415
Sankaram Bhima 336, 337
Sanyal, S.P., *see* Singh, R.K. 412
Sastry, P.V. 353
Sathyaray, K.V. 314
Sawatzky, A. 312
Sawatzky, A., *see* Jaumot, F.E. 312
Sawayanagi, F. 313
Scalon, R.D., *see* Frankel, J. 19
Schaefer, H.E., *see* Maier, K. 219, 351
Schilling, W., *see* Robrock, K.H. 367
Schmunk, R.E. 273
Schmutz, J.D., *see* Vanfleet, H.B. 313
Schober, H.R., *see* Dederichs, P.H. 367, 368, 371, 397
Schoknecht, W.E. 109, 112, 242, 282, 283
Scholz, R. 177
Scholz, R., *see* Dederichs, P.H. 367, 368, 371, 397
Schoonman, J. 408
Schouten, D.R. 214, 215, 216, 276
Schuele, D.E., *see* Bartels, R.A. 251
Schuele, D.E., *see* Felice, R.A. 102, 210, 211
Schuele, D.E., *see* Miller, R.A. 192
Schuele, D.E., *see* Loje, K.F. 262, 296, 305, 341, 347, 370
Schuele, D.E., *see* Wong, C. 305, 361, 362
Schuele, W., *see* Maier, K. 128, 177, 179, 183, 273
Schuele, W., *see* Scholtz, R. 177
Schuele, W., *see* Wong, C. 362
Schulte, C.W., *see* Campbell, J.L. 185
Schultz, R.J., *see* Hood, G.M. 173, 407, 410
Schultz, R.J., *see* Rasch, K.D. 219, 220, 222
Schwalbe, L.A. 244, 282
Seeger, A. 119, 172, 174, 175, 188, 208, 227, 280, 421
Seeger, A., *see* Göltz, G. 197, 198
Seeger, A., *see* Maier, K. 219, 351
Seeger, A., *see* Mehrer, H. 122, 192
Seeger, A., *see* Steiner, E. 118
Semercham, A.A., *see* Sobolev, G.A. 403
Sen, S.K. 311, 312
Sen, S.K., *see* Dut, M.B. 311, 312
Seran, J.L. 311

Serin, B. 16
Settatree, G.R., see Leadbetter, A.J. 14
Shahi, K. 410, 412
Shakhova, A.G., see Dobrovinskaya, O.R. 320
Shanker Jai 270, 387, 412
Sharma, O.P., see Shanker Jai 270, 387
Shaw, D. 135
Shaw, G.H. 296, 305
Shea, N., see Buescher, B.J. 228, 231
Shelley, R.D. 136
Shih, S. 311
Shirn, G.A. 228
Shukla, A.K., see Beniere, M. 320
Shull, H.E., see Barsch, G.R. 257, 260, 370
Siegel, R.W. 171, 172, 173, 177, 182, 305
Siegel, R.W., see Berger, A.S. 171, 172, 173, 177, 178, 180, 181, 182
Siegel, R.W., see Fluss, M. 173
Siegel, R.W., see Gupta, R.G. 219
Siegel, R.W., see Rasch, K.D. 219, 220, 222
Sigvaldason, H., see Balzer, R. 226, 227, 228, 229, 232, 233, 234, 235, 236, 237, 238
Sime, S.J., see Alnatt, A.R. 253, 254
Simmons, R.O. 88, 112, 162, 169, 170, 172, 176, 177, 178, 180, 181, 184, 185, 186, 187, 190, 242, 244, 393, 403
Simmons, R.O., see Batchelder, D.N. 243
Simmons, R.O., see Fouchaux, R.D. 266, 267
Simmons, R.O., see Heald, S.M. 414
Simmons, R.O., see Losee, D.L. 245, 282
Simmons, R.O., see Schoknecht, W.E. 109
Simon, J.P. 227
Singh, R.K. 412
Singh, S.P. 370
Sinha, A.P.B., see Bharati, S. 312
Sirovich, B.E. 242, 243
Siu Li, M. 137
Sladek, R.J., see Rimai, D.S. 302, 303, 363
Slagle, O.D. 289, 326, 327, 329, 332, 363
Slater, J.C. 270, 271, 356, 357, 387
Slater, N.B. 80, 81
Slichter, C.P. 174
Slichter, C.P., see Ailion, D.C. 208, 209
Slifkin, L.M. 129, 347, 410, 411, 416

Slifkin, L.M., see Alexander, W.B. 313
Slifkin, L.M., see Batra, A.P. 121, 122, 130, 131, 252, 260, 261, 262, 263, 264, 266, 267, 305, 320
Slifkin, L.M., see Cain, L.S. 338, 339, 412
Slifkin, L.M., see Crawford, J.H. 128, 132, 289
Slifkin, L.M., see Sonder, E. 312
Slifkin, L.M., see Tomizuka, C.T. 312
Smedzkjaer, L.C., see Fluss, M. 173
Smith, B.L., see Parker, E.H.C. 243
Smith, C.A., see Smith, P.A. 276
Smith, C.S. 251, 255, 260, 326, 332, 334, 336, 337, 354, 401, 370
Smith, C.S., see Daniels, W.B. 273, 305
Smith, C.S., see McLean, K.O. 305
Smith, C.S., see Miller, R.A. 246
Smith, C.S., see Roberts, R.W. 289, 327, 332, 334, 336, 342, 347, 370
Smith, C.S., see Schmunk, R.E. 273
Smith, F.A., see Mundy, J.N. 196, 197
Smith, P.A. 276
Sobolev, G.A. 403
Solunskii, V.I., see Dobrovinskaya, P.R. 320
Sonder, E. 312
Sonder, E., see Tomizuka, C.T. 312
Sondergeld, C.H., see Sobolev, G.A. 403
Sorojic, V., see Wenzl, H. 311
Spain, I.L., see Chiarodo, R. 273
Spalt, H. 287
Spetzler, H. 253, 285, 289, 290, 291, 320, 342, 370, 415
Spetzler, H.A., see Sobolev, G.A. 403
Srivastava, K.K. 251
Stakhovskii, I.R., see Sobolev, G.A. 403
Stal'gorova, O.V., see Voronov, F.F. 19
Stals, L., see de Schepper, L. 195
Stansbury, E.E., see Murdock, J.F. 317
Stark, J.P., see Shih, S. 311
Steichele, E., see Stetter, G. 214
Steinberg, D.J., see Guinan, M.W. 280, 305
Steinemann, S., see Talmor, Y. 225, 307
Steiner, E. 118
Stetter, G. 214
Stewart, A.T., see McKee, B.T.A. 173, 191
Stöckmann, H.-J., see Heitjans, P. 414
Stoneham, A.M., see Harding, J.H. 417
Stoneham, A.M., see Mainwood, A. 144

Stott, J.P.   136
Strelkow, P.G.   262
Strelkow, P.G., see Kraftmakher, Ya.A.   214, 215
Suarez N., see Laredo, E.   135
Sullivan, G.A.   195
Sun, Ch.Y.   174
Suryanarayana, S.V., see Singh, R.K.   412
Suzuki, K.   136
Suzuki, K., see Ihara, S.   412
Svare, I., see Raaen, A.M.   261
Swartz, K.D.   232
Swenson, C.A.   210, 354, 355
Swenson, C.A., see Anderson, M.S.   243, 244, 245, 282, 369, 406, 407, 415, 421
Swenson, C.A., see Beecroft, R.   199, 200, 204, 206
Swenson, C.A., see Fugate, R.W.   16
Swenson, C.A., see Monfort, C.E.   199, 204, 206, 207
Swenson, C.A., see Schouten, D.R.   214, 215, 216, 276
Swiderski, T., see Zielinski, M.   136

Tahara, R., see Furakawa, K.   173
Takamura, J., see Furakawa, K.   173
Talinskii, V.K., see Gorbachev, V.A.   310, 311
Tallon, J.L.   19, 169, 174, 273, 311
Talmor, Y.   225, 307
Tanigawa, S., see Doyama, N.   182
Tanigawa, S., see Nanao, S.   182
Tannhäuser, D.S.   263, 321
Taylor, R., see Jacucci, G.   275
Teisseyre, R.   410
Teltow, J.   36, 260
Teltow, J., see Ebert J.   260
Teltow, J., see Müller, P.   133, 134
Terentiev, V.A., see Sobolev, G.A.   403
Tharmalingam, K.   384
Timofeev, A.N., see Gobachev, V.A.   310, 311
Titman, J.M.   208, 209
Tomizuka, C.T.   273, 305, 312
Tomizuka, C.T., see Abey, A.E.   266, 296, 299
Tomizuka, C.T., see Sonder, E.   312
Tomlin, D.H., see Peart, R.F.   314
Trappeniers, N.J., see Van't Klooster   19, 157

Triantis, D.   136, 247, 257
Trifthäuser, W.   173, 182, 185, 191
Trifthäuser, W., see McGervey, J.D.   185
Trifthäuser, W., see McKee, B.T.A.   173, 191
Trivisonno, J., see Diederich, M.E.   200, 201
Trivisonno, J., see Felice, R.A.   102, 210, 211
Tsui, R.T.C., see Gupta, D.   128
Turetskii, I.M., see Sobolev, G.A.   403
Turnbull, D., see Baughman, R.H.   351
Turnbull, D., see Dyson, B.F.   240, 313
Turnbull, D., see Owens, C.W.   313
Turnbull, D., see Warburton, W.K.   238, 315, 316
Tzanetakis, P.   173

Ul'yanov, V.L., see Cyrbu, I.N.   257
Uyeda, S.   410, 411

Vaidya, S.N.   232, 238, 280, 305
Vajda, P., see Simon, J.P.   227
Valenta, P., see Maier, K.   351
Van Liempt, J.   349
Van 't Klooster, P.   19, 157
Van Vechten, J.A.   84
Van Weperen, W.   136
Vanderschueren, J.   136, 423
Vanfleet, H.B.   191, 192, 193, 272, 312, 313, 323
Vanfleet, H.B., see Decker, D.L.   313, 322
Varotsos, C.   137, 303, 364, 412
Varotsos, C., see Lazaridou, M.   408, 409
Varotsos, P.   5, 60, 62, 86, 95, 129, 130, 131, 132, 136, 137, 138, 140, 141, 165, 193, 198, 206, 247, 249, 252, 258, 264, 266, 267, 284, 287, 290, 292, 293, 297, 298, 305, 309, 332, 337, 345, 357, 359, 360, 361, 362, 366, 368, 370, 386, 403, 404, 405, 406, 407, 408, 409, 413, 416, 417, 419, 420, 422
Varotsos, P., see Alexopoulos, K.   167, 195, 308, 309, 408, 409, 410
Varotsos, P., see Dologlou, E.   419
Varotsos, P., see Eftaxias, K.   407
Varotsos, P., see Kostopoulos, D.   136, 258
Varotsos, P., see Lazaridou, M.   408, 409
Varotsos, P., see Varotsos, C.   137, 303, 364
Varotsou, M., see Varotsos, P.   141

Varshni, Y.P. 186, 187, 189, 273
Vegard, L. 328
Veisman, V.L., see Pershits, Y.N. 257
Vempati, C.S., see Jacobs, P.W.M. 417
Verma, M.P., see Gard, V.K. 285
Verma, M.P., see Puri, D.S. 19
Vignes, A., see Roux, F. 314
Vineyard, G.H. 74
Vold, C.L. 192, 194, 272, 312
Vold, C.L., see Kammer, E.W. 235, 236, 280
Vollmer, O. 273
von Stebut, J., see Ganne, J.P. 32
Voronov, F.F. 18
Vostry, P., see Simon, J.P. 227
Vukcevich, M.R. 19

Wagner, C. 144
Wagner, J.B., see Shahi, K. 412
Waida, E.S., see Shirn, G.A. 228
Waida, E.S., see Talmor, Y. 307
Walker, E., see Talmor, Y. 225, 307
Wallmark, G.N., see Gilder, H. 226
Walsöe de Reca, N.E. 278
Warburton, W.K. 238, 313, 315, 316
Warburton, W.K., see Cohen, B.M. 191
Warita, Y., see Ihara, S. 412
Weil, J.A., see Nachtrieb, N.H. 196, 197, 204
Weiss, Jonathan D., see Decker, D.L. 313
Wenzl, H. 311
Wert, C.A. 74
Weymouth, J.W., see Sullivan, G.A. 195
Whitmore, D.H., see Bauer, C.F. 247, 248, 250
Wilkinson, G.R., see Denham, P. 361, 362
Winslow, F.R., see Lundy, T.S. 314

Wintersgill, M.C., see Andeen, C.G. 137, 358
Wintersgill, M.C., see Figueroa, D.R. 137, 358
Wintersgill, M.C., see Fontanella, J. 123, 131, 305, 368, 422
Wolf, D., see Göltz, G. 197, 198
Wolf, W., see Brünger, G. 197, 198
Wolfenden, A., see Tallon, J.L. 169, 174, 273, 311
Wong, C. 305, 361, 362
Wubbenhorst, R., see Hehenkam, T. 312
Wylde, L.E. 257

Yeh, C.C. 115
Yeh, C.C., see Brooks, C.R. 115
Yen, W.M. 243, 245
Yli-Pietila, H., see Lindström, R. 343
Yoon, D.N. 267, 287, 289, 291, 292, 293, 294, 305, 366, 368, 413
Yoshizawa, H. 248

Zehe, A., see Krautheim, G. 311
Zeller, R. 367, 368
Zeller, R., see Dederichs, P.H. 367, 368, 371, 397
Zener, C. 5, 130, 171, 349, 354, 355, 363, 364
Zener, C., see Lawson, A.W. 364
Zener, C., see Wert, C.A. 74
Zielinski, M. 136
Zierau, W. 264
Zieten, W. 345, 346
Zitzmann, R., see Guerard, R. 169, 170, 171, 172, 173
Zschau, J. 410

SUBJECT INDEX

Activation
  enthalpy, 121–125, 163, 164, 165, 274, 304, 306, 309, 322, 349, 354, 356, 388
  entropy, 121–125, 163, 164, 165, 309, 322, 350, 354, 356
  Gibbs energy, 120, 124–126, 147, 165, 269, 271, 274, 364, 377
  volume, 125–128, 271, 274, 304, 306, 364, 365, 368, 377
Ag
  bulk modulus, 185, 186, 187, 272, 305
  defect volume, 272, 275, 304–306
  diffusion in
    AgBr, 321
    Al, 313
    β-Ti, 317
    Cu, 311
    Li, 315
    Pb, 313, 323
    Sn, 238
  expansion coefficient, 187, 188, 190
  self-diffusion, 168, 186–188, 365
    parameters, 184, 185, 189, 190, 272, 304–306
  vacancy
    concentration, 184
    formation parameters, 184, 185, 190
    migration parameters, 189
AgBr
  bulk modulus, 168, 262, 263, 265, 296, 299, 370
  compressibility, 333
  conductivity, 131, 260–267, 299
  defect entropy and enthalpy, 261, 263, 264, 267, 296, 299, 304–306, 339, 340
  defect volume, 267, 295–301, 304–306, 339, 341, 346, 377
  expansion coefficient, 262, 263, 345
  heterodiffusion of various impurities, 321
  microscopic calculations, 264, 298, 412, 413, 416, 417
  Na$^+$ diffusion, 260, 261, 267
  self-diffusion, 122, 264, 265, 267
  specific heat, 105, 264
  temperature variation of defect entropy and enthalpy, 105, 252, 261, 263, 264, 267, 347
AgBr–AgCl mixed crystal
  compressibility, 333
  conductivity, 338, 412
  defect parameters, 339, 341
AgCl
  bulk modulus, 296, 299, 370
  compressibility, 333, 338
  conductivity, 130, 260, 299
  defect entropy and enthalpy, 266, 267, 295–297, 299, 304–306, 339, 340
  defect volume, 266, 295–301, 304–306, 339, 341, 377
  expansion coefficient, 266
  microscopic calculations, 264, 267, 298, 412, 413, 416
  Na$^+$ diffusion, 261, 266, 267
  self-diffusion, 122, 261, 267
  temperature variation of defect entropy and enthalpy, 252, 264, 267, 412
AgF, diffusion study by NMR, 261
AgI
  bulk modulus, 296–298, 370
  conductivity, 296, 297
  defect entropy and enthalpy, 296
  defect volume, 5, 296, 297, 304–306, 377

expansivity of defect volume, 297, 298

Al
bulk modulus, 169, 170, 272, 275
compressibility, 367
defect volume, 272, 275, 276
expansion coefficient, 169, 174
heterodiffusion of various impurities, 311, 312
self-diffusion parameters, 169, 170, 172, 174–176, 272, 275, 276
vacancy
concentration, 170, 171, 172
formation parameters, 169–174
migration parameters, 176, 273

Alkali halides
analysis of conductivity and diffusion plots, 120, 128–130, 252, 254, 255, 259
bulk modulus, 285, 288, 291, 370
cohesion, 382, 385
compressibility of defect volume, 292–294, 366, 368
defect volume, 131, 132, 266, 267, 284–295, 305, 306, 360, 413
dielectric constant, 359, 360, 384–387
expansivity of defect volume, 105, 132, 290–292
microscopic defect calculations, 252, 267, 287, 290, 412, 414, 416, 417
temperature variation of defect parameters, 129, 130, 252, 267, 290, 412, 413, 416
thermocurrents, 132–137

Anderson–Grüneisen parameter, definition, 150

Anelastic effects, 130

Atomic or ionic
migration, dynamics, 71–84, 117–121, 125, 127, 357, 358, 387
polarizability, 130, 359, 360, 386
radius, 385

Ar
bulk modulus, 103, 282, 369
expansion coefficient, 244
defect formation volume, 282, 283, 377
self-diffusion parameters, 243, 244
vacancy
concentration, 244
formation parameters, 103, 244, 282, 377, 409

As, diffusion in
Ag, 312
Cu, 311

Attempt frequency, 71, 72, 119, 124, 135, 138, 228, 241

Au
bulk modulus, 375
diffusion in Sn, 238–241
self-diffusion, 365, 366

BaF_2
bulk modulus, 407
defect entropy and enthalpy, 361, 408
dielectric relaxation, 361
thermocurrents, 136

Ballistic model of migration, 84

bcc metals
defect volume, 275–279, 304–306, 377
self-diffusion, 165–168, 195, 350

Be, diffusion in Cu, 311

Bi, diffusion in Cu, 311

Born–Mayer, repulsive parameters, 382–384

Born–Oppenheimer approximation, 72

Bound
vacancy, 131, 132, 361
interstitial, 361, 362
motion parameters, 131, 132, 139, 361, 362

Br, diffusion in
AgBr, 264, 265, 320, 321
NaCl, 320

Bulk modulus of the solid, 15, 19, 20, 354, 415

CaF_2
bulk modulus, 305, 361
defect entropy and enthalpy, 305, 362, 409
defect volume, 305, 408
dielectric relaxation, 361, 362
doped with trivalent earth ions, 362
thermocurrents, 136

$cB\Omega$-model
expressions for defect parameters, 125, 147–161, 163–166, 269–271, 274, 303, 304
theoretical foundation, 363–401

Cd
defect volume, 122
expansion coefficient, 61, 122

Subject Index

in AgBr, expansivity, 345–347
in KBr, solubility, 256
self-diffusion parameters, 122
CdF_2, 303
Cerium
  bulk modulus, 19, 279
  defect enthalpy and entropy, 278, 279
  defect volume, 56, 271, 279
Chemical potential, 13, 34, 35, 142–144
Cl, diffusion in
  AgCl, 265
  KCl, 252,
  NaCl, 254, 255, 320
Co, diffusion in
  Al, 313
  β-Ti, 317, 318
  Cu, 311
  γ-U, 316
  Nb, 314
  Pb, 313, 323
Cohesive energy, 380–383, 385
Compressibility
  contribution of defects, 33, 34, 41, 65
  defect
    activation volume, 123, 127, 206, 292, 293, 365, 366, 368, 369, 398, 399
    formation volume, 4, 32, 33, 41, 42, 56, 65, 125, 156, 157, 367, 368, 399
    migration volume, 139, 368, 369
  perfect crystals, correlation, 47
  solid, definition, 15, 32, 40
  X-ray study, 111
Concentration
  Frenkel defects, 67
  Schottky defects, 64
  vacancies, 27, 39
Conductivity
  correlation with diffusion coefficient, 119, 120, 420
  ionic, analysis of, 128–130, 267, 413, 416
Configurational entropy
  Frenkel defects, 66, 67
  Schottky defects, 63, 64
  vacancies, 26, 27, 38,
Correlation
  effect in diffusion, 117–119, 197
  factor, 118–121, 128–129, 197
Correlation between
  activation parameters resulting from dielectric loss or ITC-experiments, 361, 362

defect enthalpy and
  Debye temperature, 353
  defect volume, 303–306, 352, 354, 358
  entropy, 150, 163, 164, 169, 309, 407
  expansion coefficient, 355, 356
  melting point, 274, 349–353
defect
  entropy and defect volume, 354, 355, 393
  entropy and Grüneisen constant, 45, 356,
  volume and defect Gibbs energy, 269–303, 363, 364, 377, 407–410
  diffusion coefficients of atoms diffusing in a given matrix, 309–323
  isobaric and isochoric defect parameters, 44–47
Cr, diffusion in
  β-Ti, 317, 318
  Cu, 311
  γ-U, 316
  Nb, 314
Cs, diffusion in NaCl, 320
CsBr
  bulk modulus, 285
  defect entropy and enthalpy, 284, 285
  defect volume, 285–287
CsCl
  bulk modulus, 285
  defect entropy and enthalpy, 148, 284, 285
  defect volume, 5, 285–287
Cu
  bulk modulus, 178, 179, 180, 272, 275
  defect volume, 272, 275, 304–306, 377
  diffusion in
    Al, 313
    β-Ti, 317
    γ-U, 316
    Li, 315
    Pb, 313, 323
  expansion coefficient, 179, 180
  heterodiffusion of various impurities, 310, 311
  self-diffusion, 182, 183
    parameters, 177, 178, 182
  vacancy formation parameters, 176–182
  vacancy migration parameters, 178, 182

Debye, Hückel theory, 259, 260, 393
Debye frequency, 82, 83, 119, 121, 189

Debye temperature, 102, 135, 158, 395
Defect formation
 isobaric, 23–37, 63–65, 148–158
 isochoric, 25, 37–47, 65, 69, 158–161, 413, 416, 417
Defect migration
 isobaric, 74, 78–80, 81
 isochoric, 74–78, 81
Defect volume
 pressure variation, 2–4, 32, 33, 41, 56, 57, 123, 125, 126–127, 154, 156, 157, 292–294, 365–369, 413, 417
 temperature variation, 2, 56, 57, 60–62, 123, 153–156, 290–292, 345–347, 413, 417
Depolarization currents
 pressure stimulated, 137–140, 403, 417
 thermally stimulated, 132–137, 361, 422
Dielectric
 constant, ionic crystals, 359, 360, 384–387
 loss, 131, 361, 421
 polarization, 131, 133–137, 384
 relaxation, 130, 362
Diffusion
 activation parameters, 120–127, 163
 analysis, 121–127, 165
 coefficient, macroscopic, 117
 coefficient, microscopic, 119,
 effect of correlation, 117–121
 effect of pressure, 126, 127
 in ionic crystals, 120
 in metals, 119, 120, 165
 isotope effect, 127
Dipole
 interstitial–impurity, 361
 vacancy–impurity, 131
Divacancies, 122, 128, 172, 197, 208
Double jumps in Na, 198
Dynamical theory
 correlation with $cB\Omega$-model, 387
 of atomic migration, 80–84

Earthquake prediction, piezostimulated currents, 141, 403–406, 410, 411, 417–420
Effective charge, see Szigeti charge
Einstein approximation of solids, 49, 71
Elastic models applied to defects, 82, 171, 363, 364
Electrical conductivity, 119, 128–130, 254, 259, 260–267

Electronic
 configuration of dopants, 131
 polarizability, 130
Enthalpy of a solid, 12, 14, 17, 28, 39, 89, 415
Enthalpy of defect formation
 correlation with
  Debye temperature, 353
  formation volume, 304–306
  melting temperature, 349–353
 $cB\Omega$-expression, 148, 151, 152, 160, 163
 definition
  isobaric, 28
  isochoric, 40
 microscopic calculations, 54, 130, 252, 264, 267, 292, 412, 413, 416
 pressure variation, 55, 57, 60–62
 temperature variation, 53–55, 89, 96–97, 99, 105, 114, 116, 123, 129, 151, 252, 264, 413, 416
Enthalpy of defect migration, 79
Entropy of a solid, 11, 14, 16, 17
Entropy of defect formation
 $cB\Omega$-expression, 149–151, 392–394
 configurational, 26, 27, 38
 isobaric, definition of, 28,
 isochoric, definition of, 38
 microscopic calculations, 129, 417
 pressure variation, 55–57, 62, 156
 statistical definition, 50–52
 temperature variation, 2–4, 53–55, 90, 114, 116, 129, 151, 264, 267
Entropy of defect migration, statistical definition, 77, 80
Expansion coefficient, see thermal expansion coefficient
Expansivity of a solid
 defect contribution to
  isobaric perfect crystal, 15, 31, 32
  isochoric perfect crystal, 42
 pressure variation, 18
 temperature variation, 15
 X-ray, 108–110, 112

fcc lattice
 correlation, 120
 diffusion, 120, 121, 168
fcc metals
 correlation of defect enthalpy and melting point, 349–351
 defect volumes, 272, 275, 276, 365

Subject Index

self-diffusion, 168, 272, 275, 365
Fe, 195, 308, 309
Fe, diffusion in
  Ag, 312
  Al, 313
  β-Ti, 317, 318
  Cu, 311
  γ-U, 316
  Nb, 314
Formation parameters, *see* vacancy formation
Frenkel defects
  definition of, 66
  in silver halides, 260–267, 295–301
  simultaneous formation with Schottky defects, 70

Ga, diffusion in
  Al, 313
  Li, 315
Ge, diffusion in
  Ag, 312
  Al, 313
Gibbs defect energy
  $cB\Omega$-expression, 147
  connection with defect volume, 153, 269–303, 363–365, 377
  isobaric formation, definition, 26
  isochoric formation, definition, 40
  migration, definition, 79
  pressure variation, 29, 30, 62, 79
  temperature variation, 28, 50, 51, 79, 80
Gibbs energy, perfect solid, 26, 40, 49, 50
Gibbs energy, real solid, 12, 15, 17, 50, 415
Green's function, 367, 382
Grüneisen constant, 3, 45, 126, 127, 131, 138, 140, 159, 271, 357, 369, 387

Harmonic lattices, properties, 48–52
Harmonic microscopic defect calculations, 287, 290, 413
^{3}He, vacancy concentration 412
Hg, diffusion in
  Al, 313
  Li, 315
  Pb, 313, 323
High-temperature
  curvature of
    diffusion plots, 121–125, 165–168, 260
    ionic conductivity plots, 128–130, 259, 262–264, 267, 292, 413, 416
  variation of defect Gibbs energy, 149, 150, 258, 262
Hydrostatic pressure, effect on
  bulk modulus, 19, 20, 369, 370
  dielectric loss, 131, 132, 137
  diffusion, 126, 127, 292–294, 365, 366
  ionic conductivity, 287–295, 366
  thermocurrents, 137, 138
  vacancy concentration, 32, 58
  vacancy formation parameters, 55–58
Hydrostatic pressure experiments, analysis, 55–58, 126, 127, 131, 132, 137–142, 287–295, 417, 421, 422

I, diffusion in
  AgBr, 321
  NaCl, 320
  NaI, 257
Impurities
  influence on expansion coefficient, 343–347
  in ionic crystals, 131
In
  bulk modulus, 280
  defect volume, 279–281, 304–306
  diffusion in
    Ag, 312
    Cu, 311
  melting temperature, 280
  self-diffusion, 168, 280
    parameters, 280, 281
  vacancy
    formation parameters, 280, 281, 304–306
    migration parameters, 281
Interactions among point defects, 248, 260
Interatomic force model, 381–383, 412
Interstitial formation, 66, 69, 70
Interstitial migration, 261, 299–300
Interstitialcy mechanism, 129, 259, 261
Ionic conductivity
  analysis, 128–130, 258, 259, 267, 292, 413
  pressure variation, 290–295
  relation with diffusion, 119, 120
Ionic crystals
  dielectric constant, 359, 360, 384–387
  Frenkel defects, 66–70
  Grüneisen constant, 357, 387
  impurities, 131, 343–347, 361

Jost defect model, 358–360, 384–387
microscopic defect
   energy calculations, 105, 130, 254, 264, 267, 292, 412, 413, 416
   volume calculations, 267, 287, 290, 413, 417
   relaxation defect volume, 44, 267, 287, 413
Schottky defects, 63–65
Ionic thermocurrents
   analysis, 132–134, 137–140
   applications to Geophysics, 141, 142, 403–406, 410, 411, 417–420
   depolarization, 132, 138
   interconnection of parameters, 361–362
   linear heating rate, 134, 135
   polarization, 136
   variable heating rate, 133, 134
Isothermal
   bulk modulus, 15, 19, 20, 369, 370
   pressure experiments, 55, 126
Isotopic effect in diffusion, 127, 128, 198, 199, 421

Jost model
   calculation of defect volume, 360
   calculation of entropy, 359
   compatibility with $cB\Omega$-model, 372, 384–387
Jump frequency, 71, 72, 75–84, 119, 120

K
   bulk modulus, 215, 216, 217, 276
   defect volume, 277
   expansion coefficient, 216
   self-diffusion, 116, 168, 214–218
      parameters, 215, 217
   vacancy
      concentration, 214
      formation parameters, 214, 215, 217
KBr
   bulk modulus, 255, 288
   conductivity, 256, 257
   defect entropy and enthalpy, 256, 288
   defect volume, 288, 294, 295, 304–306
   doped with Ca^{2+}, 256
   expansion coefficient, 255
   solubility parameters of divalent impurities, 256
KBr–KI mixed crystals, conductivity, 412

KCl
   bulk modulus, 251, 288
   conductivity, 252, 393
   defect entropy and enthalpy, 251, 252, 288, 393
   defect volume, 288, 293–295, 304–306
   diffusion, 122, 252, 253
   doped with Sr^{2+}, 252
   expansion coefficient, 251
   thermocurrents, 136
KCl–KBr mixed crystal
   compressibility, 327, 329–334
   conductivity and diffusion, 336
   dielectric constant, 337
   expansion coefficient, 342, 343
   lattice constant, 328, 329
KI
   bulk modulus, 260
   conductivity, 260
   defect entropy and enthalpy, 259, 260
   doped with Sr^{2+}, 260
   expansion coefficient, 260
   thermocurrents, 136
KI–NaI
   density, 329
   diffusion, 336
Kr
   bulk modulus, 244, 369
   defect entropy and enthalpy, 243, 245, 407
   defect volume, 41, 112, 283, 409, 410
   expansion coefficient, 244
   vacancy concentration, 244
   vacancy formation parameters, 245, 409

Lattice statics technique, 413
Lead, see Pb
Lennard–Jones potential, 369
Li
   bulk modulus, 210, 211, 406
   expansion coefficient, 210, 211
   heterodiffusion of various impurities, 315
   self-diffusion, 208, 211–213, 407, 414
      parameters, 209, 212, 421
   vacancy concentration, 102, 208, 212, 213
   vacancy formation parameters, 209, 214
LiBr
   bulk modulus, 409
   defect parameters, 409
LiD thermocurrents, 136

LiF
  bulk modulus, 305, 407, 409
  defect enthalpy and entropy, 304–306, 408, 409
  defect volume, 304–306, 409, 410
LiH, 136
Lorentz–Lorenz equation, 386

Madelung energy, 382, 383
Maxwell's equations, 37, 43
Mg, diffusion in Al, 313
MgO, thermocurrents, 136
Mg_2SiO_4
  conductivity under pressure, 409
  defect enthalpy and volume, 409, 410
  dislocation migration, 409, 410
Migration
  atomic, dynamics of, 71–84
  enthalpy, 79, 358
  entropy, 77, 79, 80, 84, 392
    orientation of a dipole, 131, 134, 361
  f-energy, 76, 77, 84
  g-energy, 79
  internal energy, 76, 79, 81
  isobaric, 78–80
  isochoric, 74–78
  rate theory, 74–80
  volume, 79
    bound vacancy, 131, 132, 137–141
    compressibility, 139, 368, 369, 421, 422
    expansivity, 132, 391, 392, 421
Mixed alkali halides
  bulk modulus, 329, 330, 332
    pressure variation, 334
  compressibility, 327, 329, 331, 332, 423
  condition for maximum conductivity or diffusivity, 336, 338
  conductivity, 335, 336, 338
  expansion coefficient, 342, 343, 423
Mixed silver halides
  bulk modulus, 333, 341
    pressure variation, 341
  compressibility, 333, 338, 347
  condition for maximum conductivity, 337, 338–341
  expansivity, 344–347
Mn, diffusion in
  Al, 313
  β-Ti, 317
  Cu, 311

γ-U, 316
Mn-Zn-ferrite disaccomodation, activation parameters, 409
Mo, diffusion in
  β-Ti, 317
  Nb, 314
Molecular crystals, 351, 352
Morse potential, 367
Motion, see migration
Mott–Littleton method, 383
Murnaghan equation of state, 21, 369, 370

Na
  bulk modulus, 199–201, 204
  defect volume, 205–207, 275, 304–306
  expansion coefficient, 198, 201, 204
  isotope effect, 198, 199
  self-diffusion, 116, 165, 166, 168, 197, 200, 202, 203
    parameters, 196, 197, 200, 204
    pressure variation, 126, 127, 205–207
  vacancy
    concentration, 195, 196, 205
    formation parameters, 196, 204, 205
    migration parameters, 205
NaBr, elastic and defect properties, 288, 294, 295
NaCl
  bulk modulus, 253, 288, 291
  conductivity, 254, 255, 292, 293
  defect entropy and enthalpy, 253–255, 288, 304–306
  defect volume, 288, 290–294, 304–306
  doped with
    Ca^{2+}, 254
    Sr^{2+}, 254
    Y^{3+}, 254
  expansion coefficient, 253
  heterodiffusion of various impurities, 320
  microscopic defect calculations, 254, 267, 287, 290, 292, 412, 413
  pressure variation of defect volume, 292–294
  self-diffusion, 122, 254, 292
  solubility parameters, 253, 254, 255
  thermocurrents, 136
NaCl–NaBr conductivity, 336
NaF
  bulk modulus, 246–249
  conductivity, 247, 248, 250

defect entropy and enthalpy, 247, 248
doped with
  Ca^{2+}, 247
  Mg^{2+}, 247
expansion coefficient, 246, 249
temperature variation of defect entropy and enthalpy, 249, 250
thermocurrents, 136
NaI
  bulk modulus, 257
  conductivity, 257–259
  defect entropy and enthalpy, 258
  doped with Ca^{2+}, 257, 258
  expansion coefficient, 257
  heterodiffusion of various impurities, 320
  self-diffusion, 257–259
  temperature variation of defect Gibbs energy, entropy and enthalpy, 258, 259
  thermocurrents, 136, 257, 258
NaI–KI diffusion, 336
Nb
  bulk modulus, 307, 308, 314
  heterodiffusion of various impurities, 314
  self-diffusion, 225
    parameters, 225
  Zr, diffusion in, 307, 308
Ne
  bulk modulus, 243, 282
  defect volume, 282, 283
  expansion coefficient, 243
  self-diffusion, 242, 243
  vacancy formation parameters, 242, 282, 283
Nernst–Einstein equation, 119, 120, 420
NH_4Cl, thermocurrents, 136
Ni, diffusion in
  Ag, 312
  β-Ti, 317, 318
  Cu, 311
  γ-U, 316
  Nb, 314
  Pb, 313, 323
Noble gas alloys, 412
Noble gas solids
  bulk modulus, 244, 245, 282, 369–370, 409
    pressure variation, 282, 369–370
  cohesive energy, 381
  expansion coefficient, 244, 245

self-diffusion, 243–246
vacancy formation
  parameters, 243, 282, 407, 408, 409
  volume, 282–284, 409, 410
Nuclear magnetic resonance studies, 129, 174, 208, 242, 261, 414

Optic mode frequency
  effective ionic charge, 359
  relation with migration entropy, 361, 362
  role in ionic migration, 357, 358, 362, 369, 387, 388
  volume dependence, 357, 387
Organic molecular crystals, 351, 352

P, diffusion in β-Ti, 317, 318
Pb
  bulk modulus, 192, 194, 275
  defect volume, 275, 276, 304–306
  diffusion of various impurities, 313, 323
  expansion coefficient, 190, 192
  self-diffusion, 168, 191–194
    parameters, 191, 192, 275, 304–306
  solubility parameters of various impurities, 191
  vacancy concentration, 191, 194
  vacancy formation parameters, 191, 192, 194
$PbBr_2$
  bulk modulus, 409
  conductivity under pressure, 409
  defect enthalpy, 409
  defect volume, 409, 410
$PbCl_2$
  bulk modulus, 409
  conductivity under pressure, 409
  defect enthalpy, 409
  defect volume, 409, 410
PbF_2
  bulk modulus, 301, 363
  conductivity under pressure, 301–303
  defect enthalpy and entropy, 302, 303, 364
  defect volume, 5, 302, 364, 417
Pd, diffusion in
  Cu, 311
  Nb, 314
  Pb, 313, 323
Perfect crystal, definition
  isobaric, 24, 87

isochoric, 25, 87
Phonon
 entropy, 48
 frequencies, 48–50
Plutonium, self-diffusion, 271, 278, 279
Polarizability, atomic and ionic, 130, 359, 360, 386
Polarization in ionic lattices, 133, 135, 384, 417, 418, 422
Positron annihilation (and spectroscopy), 173, 182, 184, 185, 191, 208, 219, 227, 233
Pressure induced telluric current, 140–142, 403–406, 410, 411, 417–420
Pt, diffusion in Pb, 313, 323

Quantum effects, 135
Quasi-elastic neutron scattering, 128, 197, 248
Quasi-harmonic approximation
 bulk modulus, pressure derivative, 19, 21
 defect formation parameters, 51, 52
 migration theory, 77, 79
 self-diffusion analysis, 124, 125, 388, 390, 417
 temperature variation of defect formation entropy, 4, 125, 389
Quench, vacancy loss, resistivity, 171, 173, 176, 177, 205, 218, 219, 233, 366

Random walk, 118, 119
Rare-earth impurities in fluorides, 136, 361, 362
Rare gas solids, *see* noble gas solids
Rate theory of migration, 74–80, 83, 84
Rb, diffusion in
 AgBr, 261, 321
 NaCl, 320
RbCl–KCl diffusion, 336
RbI–KI diffusion, 336
Relaxation processes involving migration of an atom or ion, 131, 132, 138
Reorientation of dipoles ("complexes"), 131, 361
Rh, diffusion in Cu, 311
Ru, diffusion in
 Ag, 312
 Cu, 311

S^{2-} in alkali halides, 254
Saddle-point, 74, 82, 83

Sb, diffusion in
 Ag, 312
 Cu, 311
Schottky defects
 concentration in a crystal A^+B^-, 64
 configuration entropy, 63, 64
 contribution to specific heat, 104
 formation enthalpy and entropy, 64, 65, 412, 413, 416
 formation volume, 268, 287, 288, 413
  pressure variation, 292–294
  temperature variation, 290–292
 Jost model, calculation of entropy and volume, 359, 360, 386
 microscopic calculations
  formation energy, 254, 267, 292, 412, 413, 416
  formation volume, 267, 287, 292, 412, 413
 specific heat per defect, 103, 104, 105, 291
Self-diffusion
 activation parameters, 120–125
 alkali halides, 119–121, 292–294
 analysis, 125–126, 165–168, 415, 417
 coefficient
  calculation from a single measurement, 165
  macroscopic and microscopic, 117–119, 420
 metals, 120, 121, 123, 124
 parameters, connection with melting temperature, 350–352
 pressure variation, 3, 126, 127, 206, 292–294
 rare gas solids, 243–246
Shear elastic modulus, connection with defect parameters, 5, 171, 363, 364
Silver halides
 bulk modulus, pressure variation, 296, 299, 370
 conductivity, 260–264
 correlation effect, 118
 defect concentration, 66–70, 262, 264, 266
 defect entropy and enthalpy, 261, 267, 296, 299
  temperature variation, 264, 266, 296
 defect volume, 295–301, 304–306
 diffusion, 260, 264, 265, 267
 microscopic calculations, 106, 264, 267, 412, 413, 416

Slifkin, pioneering experiments, 130, 260, 266, 267
Simmons–Balluffi technique, 88, 112, 170, 172, 176, 178, 184, 195, 226, 232, 244, 403, 414
Sn
  bulk modulus, 236, 280
  defect volume, 238, 280, 304–306
  diffusion in
    Ag, 312
    β-Ti, 317, 318
    Cu, 311
    Li, 315
    Pb, 313
  expansion coefficient, 232, 236
  fast diffusion of Au and Ag, 238–241
  self-diffusion, 226, 235–238, 280
    parameters, 233, 234, 236, 238, 304–306
  vacancy
    concentration, 232, 234, 237
    formation parameters, 233, 234, 237
Solubility
  of divalent ions in alkali halides, 251–256
  of impurities in lead, 191, 192
  parameters, 142–144, 251–256
Specific heat of a solid
  analysis of measurements, 112–117
  contribution of
    Frenkel defects, 104–105
    Schottky defects, 103–104
    vacancies, compared with isobaric perfect crystal, 88, 89, 92, 93
    vacancies, compared with isochoric perfect crystal, 88, 94, 95
  isobaric curve of the isochoric specific heat, 85, 86
  pressure variation of contribution of vacancies, 90, 91
  relation between the isobaric and isochoric specific heat
    of the isobaric perfect crystal, 92
    of the isochoric perfect crystal, 96
  pressure variation, 17, 86
  temperature variation, 13, 86
Specific heat of one vacancy, 94–102, 105, 122, 157, 161, 395, 417
$SrCl_2$ thermocurrents, 136
SrF_2
  defect entropy and enthalpy, 407, 408
  defect volume, 137, 408, 409
  doped with trivalent rare earths, 137, 361
  elastic and expansivity parameters, 408, 409
Statics, lattice, 413
Strain energy model of diffusion processes, 82, 363, 364
Szigeti charge, 359, 422

Ta diffusion in Nb, 314
Tb dopant in CaF_2, dielectric relaxation, 362
Tl, diffusion in
  Ag, 312
  Al, 313
  Pb, 313, 323
Thermal expansion coefficient
  activation volume, 122, 123, 167, 168
  bulk solid, 15, 18
  constant "c" of $cB\Omega$-model, 390
  contribution of Schottky defects, 65
  contribution of vacancies compared with
    isobaric defect crystal, 31, 32
    isochoric perfect crystal, 42
  defect volume, connection with specific heat of one vacancy, 98, 100, 102, 105, 122, 123, 292, 417
  explanation of $(1/T)$-behaviour observed by Gilder and coworkers, 60
  formation volume, 31, 60–62, 123, 287, 291, 292
  ionic solids doped with aliovalent impurities, 343–347
  migration volume, 123, 132
  mixed alkali halides, 341–343
  X-ray study, 108–110
Thermal pressure, 16, 34, 59
Thermally stimulated
  depolarization currents (TSDC), 132–137, 422
  polarization currents (TSPC), 136, 137
Thermodynamic
  equilibrium, 11–13, 23, 72
    number of defects, 27, 38
  functions, 6, 415
    pressure dependence, 16–21
    temperature dependence, 13–16
  potentials, 6, 12, 415
  properties of dilute solutions, 142–144
Ti
  bulk modulus, 277, 318
  defect enthalpy and entropy, 277, 278
  defect volume, 277, 278

heterodiffusion of various impurities, 317, 318
self-diffusion, 166, 168, 277, 319
  parameters, 277
Tin, see Sn
TlBr
  bulk modulus, 285, 286
  conductivity under pressure, 285
  defect entropy and enthalpy, 284, 285
  defect volume, 285, 286
Tracer diffusion coefficient, 117, 120, 121
Tungsten
  bulk modulus, 221, 222
  diffusion in Nb, 315
  expansion coefficient, 222
  positron spectroscopy, 219
  self-diffusion, 219, 221, 224
    parameters, 219
  vacancy
    concentration, 222, 223
    formation parameters, 219, 223
    migration parameters from quenching experiments, 218, 219

U, diffusion of various impurities, 316
Uniaxial stress, 137

V, diffusion in β-Ti, 317, 318
Vacancy
  concentration, 27, 39
    calculation from a single measurement, 164, 165
    variation with pressure, 29, 58
    variation with temperature, 29, 39
  divalent impurity dipoles, 131, 361
  self-diffusion activation
    enthalpy, 121–123
    entropy, 121
    Gibbs energy, 120, 165
Vacancy formation
  comparison of the two formation entropies, 44–45
  enthalpy
    isobaric, 28, 416
    isochoric, 40, 416
    temperature variation, 29, 53, 54, 97, 99, 413, 416
  entropy
    isobaric, 28
    isochoric, 38
    temperature variation, 3, 4, 29, 39, 53–54
  Gibbs energy, isobaric, 26, 46

Gibbs energy, isochoric, 40, 46
internal energy, 30, 38, 45, 46
parameters from $cB\Omega$-model, 147–165
specific heat, isobaric, 97, 99, 101
specific heat, isochoric, 97, 99, 101
statistical definition of parameters, 50–52
volume, 29
  pressure variation, 4, 32, 33, 55–57
  temperature variation, 31, 57, 60–62
Vacancy migration, see migration
Vacancy pairs, 121, 259
Vegard's law, 328
Vibrational
  energy, 48
  properties of defect lattice, 50
  properties of perfect lattice, 49

W, see tungsten
White tin, see tin
Wigner formula, 29

X-ray study of
  compressibility, 111
  expansion coefficient, 108–110
  formation volume, 112
  lattice parameter, 107, 108
  vacancy concentration, 110, 112, 412
Xe
  bulk modulus, 245, 282
  expansion coefficient, 245
  self-diffusion parameters, 243, 246
  vacancy formation
    entropy and enthalpy, 246, 282
    volume, 282–284

Y^{3+}, solubility in NaCl, 254
Yb
  bulk modulus, 277, 278
  defect volume, 277, 278
  Grüneisen constant, 278
  self-diffusion, 278

Zener theory of defect parameters, 5, 171, 363, 364
Zn
  bulk modulus, 229
  defect enthalpies and entropies, 227
  defect volume, 60, 122
  diffusion in
    Ag, 312
    Al, 313
    Cu, 311
    Li, 315

expansion coefficient, 229
positron technique, 227
self-diffusion, 228–232
    calculation from $cB\Omega$-model, 229, 231
    parameters, 227, 228, 231

vacancy concentration, 226
vacancy formation parameters, 226, 227
Zr, self-diffusion, 116, 166, 168, 199, 319
    diffusion in Nb, 307, 314